GREEN PLANET
The Story of Plant Life on Earth

GREEN PLANET

The Story of Plant Life on Earth

Edited by David M. Moore

Cambridge University Press

CAMBRIDGE
LONDON NEW YORK NEW ROCHELLE
MELBOURNE SYDNEY

Contributors

Project Editors: Graham Bateman
Peter Forbes
Production: Clive Sparling
Picture Research: Christine Forth
Design: John Fitzmaurice

Published by the Press Syndicate of the
University of Cambridge.
The Pitt Building, Trumpington Street,
Cambridge CB2 1RP
32 East 57th Street, New York, NY 10022,
USA
296 Beaconsfield Parade, Middle Park,
Melbourne 3206, Australia

First published 1982

AN EQUINOX BOOK
Planned and produced by Equinox
(Oxford) Ltd, Mayfield House, 256
Banbury Road, Oxford OX2 7DH
Copyright © Equinox (Oxford) Ltd 1982

Library of Congress catalogue card
number: 82—4287

British Library cataloguing in publication
data
Green Planet: the story of plant life on
earth
1. Botany—History
I. Moore, David M.
581 QK15

ISBN 0 521 24610 5

Origination by Art Color Offset, Rome,
Italy.
Filmset by Keyspools,
Golborne, Lancashire, England.
Printed and bound in Japan by
Dai Nippon, Tokyo.

Frontispiece Banff National Park in the
Rocky Mountains, Canada.

P.B.   P. D. W. Barnard PHD
Department of Botany
University of Reading

B.N.B. B. N. Bowden DPHIL
Department of Applied Biology
Chelsea College
University of London

N.J.C. N. J. Collins PHD
Kidderminster

P.R.C. P. R. Crane PHD
Department of Biology
Indiana University USA

A.C.   A. Cronquist PHD
The New York Botanical Garden
New York, USA

T.T.E. T. T. Elkington PHD FLS
Department of Botany
University of Sheffield

O.H.F. Sir O. H. Frankel DSC FAA FRS
Canberra City
Australia

P.F.G. P. F. Garthwaite OBE MA FIFOR
Godalming
Surrey

C.H.G. Professor C. H. Gimingham
Department of Botany
University of Aberdeen

L.B.H. L. B. Halstead PHD DSC FGS FLS
Department of Geology and Zoology
University of Reading

J.B.H. Professor J. B. Harborne PHD DSC
Department of Botany
University of Reading

J.L.H. Professor J. L. Harper MA DPHIL FRS
School of Plant Biology
University College of North Wales
Bangor

V.H.H. Professor V. H. Heywood
PHD DSC FLS FIBIOL
Department of Botany
University of Reading

L.H.   Leonie Highton
The Condé Nast Publications
Limited
London

C.J.H. C. J. Humphries PHD
British Museum (Natural History)
London

C.B.J. C. B. Johnson PHD
Department of Botany
University of Reading

D.M.M. Professor D. M. Moore PHD FLS
Department of Botany
University of Reading

P.D.M. P. D. Moore PHD
Department of Plant Sciences
University of London
King's College

P.H.N. P. H. Nye PHD
Department of Agricultural
Sciences
University of Oxford

B.P.   B. Pickersgill PHD
Department of Agricultural Botany
University of Reading

J.O.R. J. O. Rieley PHD
Department of Botany
The University of Nottingham

G.D.R. G. D. Rowley BSC
Department of Agricultural Botany
University of Reading

B.S.   B. Seddon PHD
Birmingham City Museum
Birmingham

D.W.S. D. W. Shimwell PHD
School of Geography
University of Manchester

N.W.S. N. W. Simmonds SCD AICTA
FRSE FIBIOL
The Edinburgh School of
Agriculture Edinburgh

R.M.W. R. W. Wadsworth PHD
Department of Botany
University of Reading

J.Wn.   G. J. Waughman
The Marine and Technical College
South Shields, Tyne and Wear

B.Wr.   B. D. Wheeler PHD
Department of Botany
University of Sheffield

T.C.W. T. C. Whitmore MA PHD SCD
Department of Forestry
University of Oxford

S.W.   S. R. J. Woodell PHD MA
Botany School
University of Oxford

Contents

Preface

In the pages of this unique book, Professor David Moore and his colleagues have made a timely and valuable contribution that will be of interest to readers throughout the world. The authors have brought together in one volume an authoritative and attractive account of the history, structure, and distribution patterns of the Earth's vegetation. There is nothing directly comparable, and the book should be read with profit and enjoyment by many.

All life on Earth, including our own, ultimately depends upon the energy-capturing activities of approximately 300,000 species of green land plants and algae. In view of this, nothing is more critical for human prosperity than a proper understanding of these organisms and the ways in which they are distributed over the surface of the Earth. The properties of plants, the ways in which they have evolved, and the kinds of communities that they comprise—all are of fundamental interest.

We live at a time when the human population of the globe has already exceeded 4,500 million and is expected to reach at least 8,000 million during the first half of the next century. By that time, some two-thirds of all the people on Earth will be living in the tropics, where their activities will have largely eliminated undisturbed tropical vegetation as we know it today.

Virtually all of the land on Earth that can be cultivated on a sustainable basis, using the technology that we now possess, is already being utilized, and unless a significant advance is made in the near future, much of the land that is being cleared in the tropics will simply be abandoned after a few years or decades of productive use. In addition, many authorities believe that the widespread clearing of tropical forests, coupled with the difficulty many plants and animals have in reproducing outside undisturbed forest, will lead to the extinction of perhaps a quarter of all the kinds of organisms on Earth during the next few decades. Only a greater knowledge of the world's green plants, to which this book makes such a valuable contribution, can provide a basis for either conserving or utilizing these species, and thus helping to alleviate the problems of hunger that already beset at least a tenth of the people on Earth and threaten many more.

We derive some 85% of our food from no more than 20 species of plants at present. About most of the remainder, we know virtually nothing. There are doubtless many plants that would be of great value to human welfare if we understood their properties; only a greater appreciation of this fact by educated people throughout the world will lead to an appropriate acceleration of our inventory of their relevant characteristics. The permanent alteration of many types of vegetation and widespread extinction that are taking place highlight the urgency of these matters. Nothing comparable has occurred in the history of life on Earth at least since the end of the Permian era, some 230 million years ago. In fact, as this book develops in more detail, the human population has now reached the level where it is permanently restructuring the global ecosystem for its own purposes, and it badly needs the tools to do this properly, so that total productivity may continue or be increased in the future. Increased knowledge of the green planet described in the following pages is essential if these tools are to be developed with due regard for Man's aspirations and the plant cover of the world.

PETER H. RAVEN
(Director, Missouri Botanical Garden, St. Louis, USA)

Opposite Various environmental factors are at work in this savanna landscape in the Serengeti Plains, Tanzania. Browsing by ungulates and a prolonged dry season with fires restrict the tree cover. Here, a thunderstorm is about to end the dry season, bringing the rains that sustain the savanna grassland.

Introduction

It is the common experience of all of us that as we move from one part of the world to another we encounter differences in climate and scenery, peoples and cultures, animals and plants, the study of which constitutes geography. Biogeography is concerned particularly with the distribution of all living organisms (usually excluding Man) over the face of our planet, while plant geography, or phytogeography, concentrates on describing and interpreting the similarities and differences shown by plants and vegetation occurring in the world.

Plant geography has two principal branches which, although tending to draw on different data and requiring somewhat different techniques in their study, are nevertheless inextricably intertwined. The geographical study of vegetation has its roots in ecology, which is concerned with the way plants live in communities, interacting as individuals or species with each other and the environment, as well as with the animals that ultimately depend upon them for all their food. Floristic plant geography, or chorology, on the other hand, deals with the distribution of taxa—species, genera and families—and is based on taxonomic and evolutionary principles and data.

During the 400 million or so years since terrestrial plant life first appeared on the Earth there have been major changes in the climate patterns, while mountains have been raised, eroded away and raised again, and the very continents we once thought so permanent, are known to represent one pattern in a continuous shuffling of land and water. Furthermore, the plants themselves have been evolving, migrating, becoming extinct and further diversifying, so that the kinds and distribution of the plants and vegetation we see today are but the latest products of an ever-changing scene. The ever-increasing impact of Man on this world has led to a proper concern with conservation and environmental planning, especially in recent years, but it is necessary to relate the loss of species and habitats we so lament to the changing panorama plants and animals have provided over long ages.

Green Planet, then, begins by outlining the phases of exploration which increasingly revealed the Earth's plant cover to Man (Chapter One: Charting the Green Planet), and then considers the techniques and concepts that have been developed to describe and understand this cover (Chapter Two: Tools of the Trade). The book continues by examining the factors determining the diversity and distribution of plants and the communities they form, both in the past (Chapter Three: Evolution of the Green Planet) and at present (Chapter Four: Environmental Factors); in each case the features of the physical environment and those derived from the properties of the plants themselves are considered separately, although they are clearly not entirely disassociated. Against this background it is then possible to describe and understand the distribution of plants in the modern world, whether as constituents of complex communities (Chapter Five: Vegetation Today) or as members of related groups, eg species, genera and families (Chapter Six: The Realms of Plants), and, finally, to assess the changes wrought by Man (Chapter Seven: Man and the Green Planet), good as well as bad.

The text is soundly based on contributions from an international group of 30 acknowledged experts. Since the subject matter is highly visual, the text is complemented by a high proportion of illustrations, mostly in color, in a manner representing a major new departure in books on ecology and geography. It is hoped that the close collaboration between editors and designers has resulted in a balance of text and illustrations that will give the reader some idea of the beauty and fascination of this green planet we are privileged to inhabit.

DAVID M. MOORE

Opposite The impact of Man is demonstrated in this false-color image of Imperial Valley, California, USA, taken by the Landsat earth-resources satellite. This area is the largest expanse of irrigation in the Western Hemisphere. The fields of crops show up as red and the source of the irrigation, the All-American canal, can be seen as a blue line winding through the red. The large blue area is an artificial salt lake.

CHAPTER ONE

Charting the Green Planet

The Historical Background to the Ecology and Geography of Plants

Some knowledge of the diversity of plants must be almost as old as mankind, and early Man must have soon become aware that useful and dangerous plants varied in their occurrence and distribution. After Neolithic men became cultivators rather than hunters, the discovery of the staple crops, the search for others and the colonization of new areas all drew attention to the conditions necessary for successful growth. As human horizons expanded, exploration continued in the search for plant materials whose properties, real or imagined, were of value to the physical or social well-being of society.

Early explorers no doubt mostly accepted unreflectingly the differences of vegetation which they observed, but in the 17th century the emerging scientific community began to take an interest in the collections of dried plants and seeds, occasionally living plants, which they brought home. The increasing amount of new information progressively led to speculation about the diversity of plants, the reasons for their distribution across the world and the ways in which they were grouped into often startlingly distinct kinds of vegetation. In the second half of the 18th century, the Swede Carl Linnaeus' Sexual System for expressing the relationships of species and genera permitted an orderly documentation of the numerous new plants discovered during the great age of European exploration of the world, and paved the way for the later "natural" systems of, for example, the Frenchman Antoine-Laurent de Jussieu, out of which have grown the modern, broadly based, taxonomic studies of plants.

In the late 18th and early 19th centuries, earlier information and speculation on plant distribution began to crystallize into the sciences now known as plant geography and plant ecology. This was linked to the great scientific voyages of discovery, which combined acute observation of the plants and animals encountered with high adventure. Thus, the travels of the German Alexander von Humboldt in South America in 1799–1804 not only gave rise to a vivid account of the problems of exploring the Andes but also led to the development of his ideas on the structure and aspect (physiognomy) of vegetation, which has been so important in its subsequent study. The English botanist Joseph Hooker, who traveled on James Clark Ross's expedition to Antarctica in 1839, laid the foundations for his contributions to understanding the geography of species and genera with his collections and observations made in these southern lands. Humboldt and Hooker were early exponents of the two main approaches—structural and floristic—to studies of the world's vegetation. Hooker's experiences led him to support strongly Charles Darwin's evolutionary ideas, themselves partly developed during his extensive travels (1831–1836) aboard the *Beagle*, a surveying ship sent out by the British Admiralty.

As the role of natural selection in the diversification, adaptation and extinction of organisms was realized, set against a backcloth of changes in the physical world revealed by geological and other information becoming available, the outlines of modern plant geography and ecology became apparent. Twentieth-century advances in biology and geology have led, inevitably, to specialization. Within the scope of this book, which the earlier naturalists would have recognized as being concerned with one subject—the plant cover of the earth—several modern disciplines are embraced. Thus, for example, taxonomy, genetics and evolution are concerned with the development and description of plant diversity, and geology and paleobotany are concerned with the world and its plants in earlier times, while physiology and biochemistry provide information on the mechanisms permitting plants to live under various conditions. These approaches, and more besides, all contribute to an understanding of the world's plant cover, and the organization of information about it, which are the concerns of plant geography and plant ecology, but first we shall outline the plant exploration of the world, not all of it primarily for scientific purposes, out of which these interests developed.

Opposite "Snowbeds at 13,000 feet in Ionok valley with rhododendrons in blossom, Kinchinjunga in the distance". Original caption to this illustration which appeared as the frontispiece of Volume II of Sir Joseph Hooker's *Himalayan Journals*.

Plant Exploration

Until the Renaissance, plant exploration seems to have been largely a haphazard affair, more often a by-product of other ventures, such as military, missionary or trading expeditions, than an activity in its own right. A famous exception was the expedition mounted by Queen Hatshepsut to find frankincense trees. The explorers traveled from Egypt to Somalia ("the land of Punt") and, as shown on a bas-relief of about 1500 BC, returned in triumph with 31 young trees packed in soil-filled wicker baskets. Edible, medicinal and ornamental plants were all sought after in the Chinese, Egyptian and Minoan civilizations. In particular, the Chinese emperors were cultivating highly sophisticated ornamental gardens some 5,000 years ago, and undoubtedly sent their horticultural emissaries to neighboring provinces to acquire plants for the embellishment of their estates. The Greeks laid the foundation stones of modern botany through a scientific interest in plants for their own sake, regardless of their usefulness: Theophrastus cataloged some of the plants encountered on Alexander the Great's campaigns; botanical descriptions are also found in the works of Aristotle, Dioscorides and Pliny.

The Ancient Romans, Greeks and Persians all sought, and traded in, plants which they found useful for sustenance as well as beauty. Greece, Persia and Afghanistan possessed fine nurseries, many specializing in roses, the roots of which were transported to Rome, where they were cultivated for food (jellies, candied petals and so on), wine and decoration. While roses and peonies were being brought from the Middle East, Rome's empire-builders were taking their indigenous plants for cultivation in the newly-conquered territories. The madonna lily, for instance, was carried across Europe by legionnaires who used it as a medicine, while favorite vegetable plants, such as radishes, were raised wherever the armies were stationed.

Plant exploration as we know it seems to

have been born in the 16th century. One of the earliest types of book made available by the invention of printing was the *herbal*, and in the 16th century there was a sudden spate of these collections of plant descriptions both from the Classical literature and from medieval traditions. As the limits of existing (often very corrupt) descriptions became obvious, so too did the need to make new drawings and descriptions. Thus the writers and illustrators had also, eventually, to be field botanists. In this way exploration began in search of the known and, as new species began to turn up in large numbers, went on to search for the unknown.

The first serious plant collectors had incalculable problems concerning communications, frequently hostile natives and, above all, the actual transportation of living plants. Before the revolutionary invention of the Wardian case (see p. 13) in the 19th century, travel had devastating effects on live collections. Most early botanists had to content themselves with collections of dried plants (herbaria).

Europe and Mediterranean Countries

Many existing herbals were examined, and translated, by Charles de l'Ecluse, also known as Clusius (1526–1609). Clusius visited Spain in 1564–1565, although his Flora of Spain was not published until 1576. During his journey, he is said to have discovered over 200 species that had hitherto been unknown. In 1573 he was appointed to the Court of the Holy Roman Empire; while there he went on numerous expeditions to list the plants of Austria and Hungary, publishing his Flora for the region in 1583. His next task was no less than a Flora of Europe, for which he received specimens from many other botanists, notably Bellus. The result of his labors was *Rariorum Plantarum Historia* (1601). Four years later his *Exoticorum Libri Decem* listed all the tropical plants then

A plant collector and his equipment, by the German illustrator Hendschel, published about 1870 in the London magazine *Graphic*.

known. The latter part of the century was a period of increased interest in flowering plants, with flower gardens becoming widespread. It is possibly owing to Clusius' encouragement that Ghislaine de Busbecq (1522–1592), the Emperor's Ambassador to Turkey, introduced to the gardens of western Europe various Turkish plants, such as the anemone, ranunculus, tulip, hyacinth, lilac and horse chestnut.

Turkey seems to have exerted a profound spell upon Western botanists. Another who journeyed there at about this time was Pierre Belon (1517–1564), a Paris-educated doctor whose particular interest in trees was fostered by his work in developing a living tree collection or arboretum for the Bishop of Mans. Belon made accurate notes concerning all his sightings (later to be published in his *Les Observations* in 1554) but he did not attempt to send home any living specimens or seeds.

By the end of the 16th century, the herbarium, or collection of dried plants, had become quite common. The German physician Leonhardt Rauwolf (c. 1537–1596), for example, who set out in the 1570s to gain a clearer knowledge of herbs described in Classical texts by studying them in their natural habitats, carried special papers for preserving his horticultural finds. Rauwolf's long journey, which extended to Aleppo, Baghdad and beyond, was beset with adventure and danger. He and his party were attacked by robbers, suffered hardship in the deserts, narrowly escaped charges of spying and were almost shipwrecked. Despite these manifold difficulties, he managed to make a considerable collection of dried specimens.

By the beginning of the 17th century, interest in plant exploration was two-fold: botanical and horticultural. Not only did

One of the earliest recorded plant-hunting expeditions was mounted by the Egyptian queen Hatshepsut around 1500 BC. In this contemporary bas-relief, porters are depicted returning from Somalia with a frankincense tree.

botany become a study in its own right, distinct from that of medicine, but as botanic gardens developed, it became important for the botanist, as well as for the horticulturalist, to bring back living specimens. The first really elaborate expedition was funded by Louis XIV of France, who was also curious to know more about the "Plants of the Ancients". In 1700 he sent Joseph Pitton de Tournefort (1656–1708), then Professor of Botany at the Jardin du Roi, to Greece, Turkey and the Levant. In two years he collected over 1,300 plant specimens and seeds, many of which, including the oriental poppy (*Papaver orientale*), were successfully raised in France. In his book, *A Voyage into the Levant* (1718), he was the first botanist to describe the common azalea and rhododendron.

In the early 18th century, a more scientific approach to plant exploration began to be adopted. It is worth pausing here to consider what such an approach entails. The explorer, whose sole concern is with dried specimens, must be present when the plants are in flower so that suitable pressings can be made, while the plant collector has to mark the plants that he is interested in and return when the seeds are ripe. However, if a full botanical record of a given area is to be made, the expedition would last not less than a year, since in midsummer, for example, the spring-flowering plants will be bearing capsules, the summer-flowering ones will be in flower, and those that flower later will be, at best, in bud (this problem is usually solved by making a number of expeditions, from which a true picture emerges).

Another difficulty facing early scientific plant explorers was that of the transportation of specimens. Despite all the efforts of living-plant collectors, until well into the 19th century only a small proportion of living plants arrived safely. It was once reckoned that for every successful importation from China, for example, a thousand were lost. Of course, ships from China to Europe had to pass twice through the tropics, a circumstance that was also likely to affect the viability of seeds. In addition to climate and

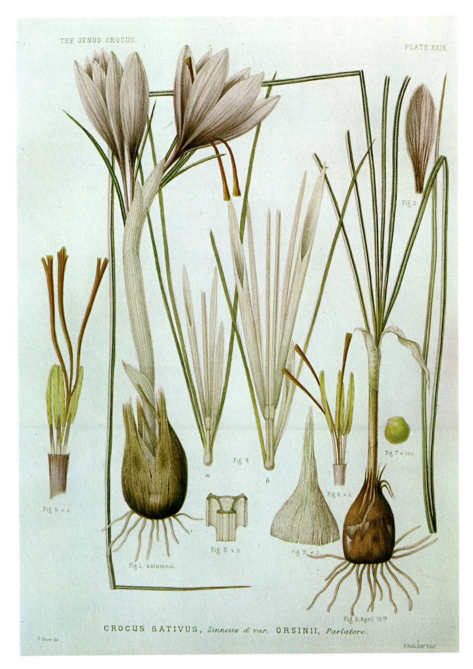

THE GENUS CROCUS. PLATE XXIX

CROCUS SATIVUS, *Linneus & var.* ORSINII, *Parlatore.*

A plate from George Maw's *A Monograph of the Genus Crocus* (1886).

The Wardian case, a sealed glass container creating a stable microclimate, made it possible for many plants to survive the extreme conditions of long sea journeys.

time, there was the problem of vermin: many of the early shipments were found to have been devoured by rats and cockroaches. This problem was solved by packing the specimens in glass jars and, eventually, in the hermetically sealed glass containers known as Wardian cases. These were invented by Dr Nathaniel Bagshaw Ward (1791–1868), a doctor of medicine who lived and worked in the East End of London. He observed that plants which could not survive in the City's polluted atmosphere flourished in a sealed (or virtually sealed) glass bottle, provided some moisture was present in the soil. His findings were used to develop a traveling-case designed to protect plants from temperature and atmospheric changes during extended botanizing expeditions.

Europe and the Near East were virtually botanical training-grounds for many of the ever-increasing army of European plant hunters who then moved on to botanize the newly discovered regions in the west and the Far East. The Frenchman André Michaux (1746–1802), traveled extensively in Persia, rediscovering many of de Tournefort's plants, before setting out for his great work in North America (see p. 15). Similarly, Francis Masson (1741–1806) made botanical explorations in Portugal and Spain between 1753 and 1785, which no doubt stood him in good stead for his later achievements in the Cape. The pioneering work of plant explorers in the Mediterranean countries has tended to be overshadowed by that of the travelers to more distant lands, but one botanist whose main work was in Europe and who achieved worldwide fame was Henry John Elwes (1846–1922). Based in Smyrna, he traveled extensively in Turkey, which, even in the 1870s, had many new species awaiting botanists. At about the same time, George Maw (1832–1912), a rich tile-manufacturer from Shropshire in England, was traveling in France, Switzerland, Italy, Iberia, Greece, Morocco and Turkey in search of new plants, particularly crocuses. He discovered *Chionodoxa lucilliae* and introduced *Saxifraga maweana*.

In Morocco, Maw worked with Sir Joseph Hooker (1817–1911), who had already done pioneering work in the Himalayas (see p. 19). They were joined by John Ball (1818–1889), who specialized in alpine plants. They found the only way they could travel unmolested was by adopting local

A 19th-century engraving of a plant-hunting expedition in the South American jungle.

costumes, and, even then, the party met with many difficulties, all recorded in Ball's *Journal of a Tour in Morocco and Great Atlas* (1878). The expedition resulted in the introduction to Europe of *Linaria moroccana*. This North African expedition was to be overshadowed in the 1930s by that of Ernest Balls (1892–   ), who also specialized in alpines.

The Americas

With the greater intellectual enlightenment and scope for travel in the 17th century, botanical discovery increased dramatically and a new breed of professional plant hunters appeared. One of the earliest of these was John Tradescant "the Younger" (1608–1662), a considerable traveler, whose interest took him to North America on three

occasions between 1637 and 1654. Little is known of his exploits but, as many North American plants are listed in the *Musaeum Tradescantianum* (1656), it is likely that he was responsible for these, which makes him among the first of the steadily rising numbers of plant hunters who now began to close in on the New World. Another Englishman, the Reverend John Banister (1654–1692), was sent out by the garden-loving Henry Compton (1632–1713), Bishop of London and head of the church for the American Colonies. Banister's successful tabulation of indigenous plants in Virginia was published in England by John Ray (1627–1705) in his *Historia Generalis Plantarum* (3 vols, 1886–1704), generally considered to have been the first printed record of the plant life of North America.

In South America, José Celestino Mutis (1732–1808), the Spanish botanist and plant collector, did much work in Colombia. He first arrived in Bogotá in 1760 and was employed as physician to the Viceroy. He was the first Professor of Natural History at the University of Bogotá and in 1783 organized a large "Expedición Botánica". He never traveled far, remaining in Bogotá to supervise the botanic garden and a team of artists executing a large number of plant drawings, but his students and associates collected from all regions of the country. In 1778 the Spanish botanists Hipolito Ruiz López (1754–1816) and José Antonio Pavón y Jiménez (1754–1840) collected extensively in Peru, Chile and Ecuador. They suffered many privations and misfortunes, although possibly their greatest obstacle was the Spanish authorities. Although they intended to be in the field for no more than about four years, so many incidents delayed their plans that Ruiz and Pavón did not return to Spain until 1788. Their specimens increased the known flora of the countries they visited by over 50% and their collections were so well preserved that A. B. Lambert, who purchased duplicates between 1816 and 1824, was able successfully to germinate some Peruvian plants brought back by Ruiz and Pavón nearly 30 years before.

Both the Spanish and the Portuguese governments were loath to allow foreign botanists to explore their South American territories, but they did make exceptions. Thus, for example, Alexander von Humboldt (1769–1859) and Aimé Bonpland (1773–1858) were allowed into Mexico and Venezuela, while in 1801 the two explored the Andes from Columbia to Peru. The end of the Napoleonic wars brought some relaxation, however, and Brazil in particular saw a veritable army of commercial hunters of exotic plants in the 19th century.

One of the great names in botany is linked with the West Indies—that of the Irish-born doctor, Sir Hans Sloane (1660–1753), whose collection of books and specimens was to form the basis of the British Museum. As physician to the Duke of Albermarle, *en route* to Jamaica in 1687, he sedulously recorded the local flora at each of his ports of call:

John Tradescant the Younger (1608–62): a portrait in the Ashmolean Museum, Oxford, England. The genus *Tradescantia* was named in his honor by Linnaeus. Tradescant discovered and listed many new North American plants.

Madeira, Barbados, Nevis and St. Kitts, and he brought back to England some 800 plants and seeds. One of Sloane's greatest attributes was his talent for inspiring younger botanists and travelers. His *Voyage to Jamaica* became a classic of its kind and was used as a copybook by many later botanical travelers.

Important for botany in the eastern states of North America was Mark Catesby (1682–1749), who arrived in America in 1712 and sent seeds back to nurseries in England throughout his seven-year stay. His work was so highly esteemed that, when making a further visit to America in 1722, patrons including Sloane commissioned him to collect North American plants on their behalf. Catesby's plates of plant and bird life in his *Natural History of Carolina, Florida and the Bahama Islands* (1730–47) are of unusual beauty.

Catesby was followed by the Quaker John Bartram (1699–1777), from Pennsylvania, described by Linnaeus as "the greatest natural Botanist in the world". Bartram had his own botanic garden near Philadelphia, but was also responsible for the introduction to Europe of an enormous number of American plants, including rhododendrons, lilies, ferns and trees.

A contemporary who tried to outdo the generous spirited Bartram was William Young (fl. 1740); although something of an upstart in botanical circles, he did introduce numerous plants to Europe, including the extraordinary Venus' fly trap (*Dionaea muscipula*) and the water lily, *Nuphar advena*.

Meanwhile, further south, the Caribbean islands were being horticulturally explored by Dr William Houston (c. 1695–1733) and, later, Robert Millar (fl. 1730–1740), two Scottish surgeons who sent plants and seeds

back to the Chelsea Physic Garden in London. Still further south, in the Andes, the French doctor Joseph de Jussieu (1704–1779) discovered the beautiful *Heliotropium peruvianum*.

Six years after the death of de Jussieu, the French botanist André Michaux arrived in New York. He and the Scottish-born John Fraser (1750–1811) were to bring about significant changes in the history of plant hunting. Michaux was a highly practical man who soon established two gardens, one in New York and one in Charleston for the more exotic species, from which to dispatch his plants to Rambouillet and the Jardin des Plantes, Paris. The two men met in Charleston and, notwithstanding an inevitable rivalry, decided to combine forces on a tour to the Indian country as far as the Mississippi. They later made extensive separate journeys in America. Michaux's botanizing took him to Florida in the south and Hudson's Bay in the north. Apart from collecting plants for export, he also imported plants to America from the Far East and Europe. These include the maidenhair tree (*Ginkgo biloba*).

Fraser made expeditions to Newfoundland and Cuba, during one of which he found the spectular *Rhododendron catawbiense*. As gardener and botanist, he had a sound instinct for what would transplant well in Europe. His introductions include *Rhododendron speciosum*, *Zenobia pulverulenta* and *Euphorbia marginata*.

The year after Michaux's death, two American army captains, Merriwether Lewis (1774–1809) and William Clark (1770–1838), started out from St. Louis on their historic expedition to find an overland passage to the West. Both men were amateur naturalists and collected many seeds and dried specimens during their 18-month return journey, including *Philadelphus lewisii* and *Gaillardia aristata*.

Thomas Nuttall (1786–1859), however, has claim to be the first serious botanist to work in the far West. He was born in Liverpool, England, and traveled to America as a journeyman printer in 1808. He became so fascinated by the local flora that he even financed two early plant hunting expeditions from his own pocket. Traveling by ship, canoe, horse and on foot, he made extended journeys throughout North America, publishing his findings in *Genera of North American Plants* (1818), which became the most comprehensive book on the subject. He was curator of Harvard University Botanic Garden between 1822 and 1834.

Next on the scene was the great David Douglas (1799–1834), who was sent in 1823 by the Horticultural Society in London to collect plants in North America, an expedition followed by several others to Columbia River (Oregon), California and the Fraser River. His travels resulted in over 200 new species being sent to Europe, the many conifers among them including the Douglas fir. Douglas met with near starvation, hostility from the Indians, a period of virtual blindness and a tragic early death after

falling into a pit dug to trap animals and already occupied by an incensed bull.

On one of his forays Douglas encountered another Scottish plant hunter, Thomas Drummond (c. 1790–1835), who, from 1831, on behalf of the Glasgow Botanic Garden, botanized from the Arctic to the Southern States, discovering *Orobus niger* (*Lathyrus niger*), *Herbertia drummondiana* and *Callirhoe papaver*.

Scotsmen were particularly prominent in these exciting days of plant hunting. Two others, John Jeffrey (1826–1854) and William Lobb (1809–1864), worked in the far West in the second half of the century. Jeffrey, sent out by the Oregon Association, founded by a group of Scottish landowners who wanted conifers for their estates, brought back *Abies grandis*, *Tsuga heterophylla*, *Pinus balfouriana* and *Penstemon jeffreyanus*. Lobb, who was employed by Veitch's Exeter nursery, also sought new conifers, and introduced *Abies venusta*, *Thuja plicata* and *Ceanothus lobbianus* to Britain.

All nations participated in this great quest for new plants. The Belgian horticulturist Jean Linden (1817–1898) not only collected in the Americas, but set up a nursery in Luxembourg for the introduction of new plants, particularly orchids, while a Bohemian, Benedict Roezl (1824–1885), who had worked in some of the foremost gardens in Europe, founded a nursery in Mexico in 1854, exporting great numbers of lilies and orchids to Europe. More recently, the American landscape designer and garden author Carl Purdy (1861–1945) also specialized in collecting lilies, which he shipped to Europe and the western states.

Africa

The Americas offered rich rewards to the plant hunters, but as Africa and the Far East began to be opened up by the mercantile adventurers, so the missionaries and botanists followed, frequently under one cassock, as it were. Michel Adanson (1727–1806), who had studied under the French botanist Bernard de Jussieu, a brother of the renowned Joseph, worked as a missionary in the wild and dangerous Senegal hinterland and found an astonishing range of hitherto unknown tropical plants, taking back to France a large collection of seeds and plants in the mid 1750s.

Some 20 years later, two plant hunters from northern Europe were working independently of each other at the Cape: Carl Peter Thunberg (1743–1828), a Swede, and an Englishman, Francis Masson (1741–1806). Meeting, they decided on a joint expedition. In spite of appalling difficulties, they made an impressive range of discoveries, most notably, perhaps, Masson's discovery of *Nerine saariensis* on Table Mountain and Thunberg's discovery of *Disa caerulea*.

The most intrepid of all the South African plant hunters was probably William John

Burchell (1782–1863), who, between 1810 and 1815, collected and brought back to London over 40,000 specimens.

Plant explorers (obviously Victorian gentlemen) examining the remarkable *Welwitschia bainesii*, a desert plant of southwestern Africa, which is thought to attain 2,000 years. An engraving from *The World of Wonders* (London, c. 1890).

China and Japan

The Far East, thanks to the comparative sophistication of its peoples and their profound resentment of Western materialism, proved more difficult for botanical penetration than Africa, with all that continent's perils of jungles, natives and disease.

Carl Thunberg and a Dutchman, Engelbert Kaemfer (1651–1715), had been among the few Europeans to make early contact with the Japanese, although Franz von Siebold (1796–1866) was probably the first to enjoy any kind of status in Japan, possibly due to his skill as an eye-surgeon. A colorful, unscrupulous character, Siebold arrived in Japan in 1826 and set up in practice in Deshima where he also cultivated a considerable garden. Eventually he was expelled, but on his delayed return to Holland he set up a "Jardin d'Acclimatation" at Leyden with some of the plant species he had managed to retain.

Gradually, China also relaxed her adamantine rules against foreigners, and plant hunters were permitted to explore, albeit on extremely circumscribed itineraries. The Scottish sea surgeon, James Cunninghame (d. 1709) and the French missionary Pierre Nicholas le Cheron d'Incarville (1706–1757), collected assiduously despite these limitations. Another Scot, James Main (c. 1770–1846), went out to China in 1792 in search of plants for his employer, Gilbert Slater of Essex, in England, and met the British Government's Embassy botanists, David Stronach and John Haxton. Between them, they made a number of discoveries; Main introduced to Europe *Clerodendron fragrans* and *Spiraea crenata*.

In 1803, it was decided that the botanic gardens at Kew should sponsor a collector in the Far East and William Kerr (d. 1814) was chosen. Kerr's attempts to introduce new species to Europe were frequently unsuccessful, due to the rigors of long sea-journeys, but he was successful with *Pieris japonica*, *Rosa banksiae* and *Begonia evansiana*.

Not until the middle of the 19th century were Japan and China explored horticulturally to any substantial degree. Robert Fortune (1813–1880) arrived in China in 1843, less than a year after the signing of the Treaty of Nankin. He was employed as a collector by the Horticultural Society in London, but in spite of some improvements in East–West relations he ran into numerous local difficulties and obstructive regulations. Nevertheless, due to the recently invented Wardian case, he was able to send his discoveries back to Europe. Fortune arrived in Hong Kong, then proceeded to Amoy and the Island of Chusan, an island paradise for a plant collector. There he found *Wiegela rosea* (*Diervilla florida*). Dressed in local costume and with shaven head, Fortune traveled to Soochow in 1844, whence he introduced a white wisteria. He also brought back to England winter jasmine (*Jasmimum nudiflorum*) and two chrysanthemums which have led to the "pompom" strains in Europe, the first forsythia (*Forsythia viridissima*) and the winter-flowering honeysuckle, *Lonicera fragrantissima*. On subsequent visits to China and India he found *Mahonia bealei* (*Berberis bealei*), one of his most celebrated finds, as well as *Skimmia reevesiana* and *Clematis lanuginosa*. In 1858, Fortune was employed by the American government to explore China for tea plants which would grow in the Southern States.

A few years later, the French missionary Armand David (1826–1900) arrived in Peking. Despite ill health, David made several journeys into Mongolia and as far as Tibet, seeking new species. Another French missionary, Jean Marie Delavay (1838–1895), made three visits to China, discovering and sending back numerous seeds and specimens to France. Although many of these were lost, his name is remembered in *Incarvillea delavayi* and *Iris delavayi*.

By the end of the 19th century, the great world upsurge in plant hunting was beginning to subside, although the rarities of China and Japan still challenged the botanist until well into this century. Charles Maries (1851–1902), an Englishman collecting for

Below Ernest Wilson (1876–1930) on a plant-hunting trip in Formosa, now Taiwan. Wilson introduced very many Chinese species into Western cultivation, including *Davidia involucrata*, commonly known as the dove, ghost or handkerchief tree.

the firm of Veitch, arrived in Japan from China in 1877. Some of his first finds were *Aronia asiatica* and *Chimonobambusa quadrangularis*, a square-sectioned bamboo, and his name is remembered in the fir tree, *Abies mariesii*.

In 1892 the great American collector Professor Charles Sprague Sargent (1841–1927) arrived in Japan, at the same time as James Harry Veitch (1868–1907), a member of the famous family of nurserymen and son of the English plant explorer John Gould Veitch. Their main contribution to plant hunting was their popularization in the West of the Japanese flowering cherries. Sargent, Director of the Arnold Arboretum, discovered many new trees, two of which— *Malus sargentii* and *Prunus sargentii*—bear his name. After Sargent's return to the Arboretum, he sent out an English collector, Ernest Wilson (1876–1930), specifically to collect cherry trees. Wilson's first object during this expedition was to find the davidia tree, which had been seen by an amateur botanist, Augustine Henry (1857–1930), some 12 years earlier. After searching over

Above left One of Robert Fortune's most famous discoveries made in China, *Mahonia bealii*, here depicted in the *Botanical Magazine* (1855).

Above right Part of the title-page of Robert Fortune's *Visit to the Tea Districts of China and India* (1852), showing tea plantations in China.

Right George Forrest and his dog Lassie sitting outside a tent, photographed in Yunnan Province, China in the early 1920s. In all he made seven expeditions from Yunnan, collecting vast amounts of plants.

many square miles he succeeded, even finding a specimen bearing seeds.

George Forrest (1873–1932), a tough and resilient Scot, survived ambush, injury and malaria to become one of the most successful of all plant hunters, making primulas and rhododendrons his speciality.

One of the greatest collectors in the Far East was Frank Kingdon-Ward (1885–1958), whose accounts of his journeys are modern classics. Following Forrest, Kingdon-Ward travelled in Yunnan and Szechwan, and introduced to Europe rhododendrons, primulas, gentians and lilies, although his best-known introduction is *Meconopsis betonicifolia*.

Four other noteworthy 20th-century botanists who worked in China were Reginald Farrer (1880–1920), specializing in alpines, William Purdom (1880–1920), Frank Meyer (1875–1918), engaged by the American Department of Agriculture to collect economic plants, and an Austrian, Joseph Rock (1884–1962), who was also sent out by the Department of Agriculture, to seek seeds of the chaulmoogra tree (*Hydnocarpus anthelmintica*), the oil of which, although to some extent replaced by sulfone drugs, is used in the East to treat leprosy.

India and the Himalayas

In general, the early 19th-century rulers in India proved far more accommodating to plant explorers than the Japanese and Chinese, although in remote areas they could be as hostile as those of other lands. Exploration increased under the British Raj.

Nathaniel Wallich (1786–1854) went from his native Denmark to India to work as a doctor, but in his early 30s became the director of the Calcutta Botanic Garden, which he transformed into one of the foremost institutions of its kind in the world. Far-sighted and practical, he often sent

Above An illustration from the *Botanical Magazine* (1849) of *Amherstia nobilis*. This was the plant the Duke of Devonshire sent his gardener, John Gibson, to India to search for. Gibson duly returned with the first living specimen to reach England.

Left An illustration from the *Botanical Register* (1840) of *Clematis montana*. This plant was introduced to Europe by Countess Sarah Amherst, wife of the Governor General of India, who brought it back from an official visit to the Northern Provinces in 1826. The genus *Amherstia* which comprises the single species, *A. nobilis*, was named in her honor.

native collectors to areas inaccessible to Europeans, while making many expeditions himself. He shipped regular supplies of seeds and plants to England, as well as successfully introducing plants to the Botanic Garden in Calcutta. His more notable discoveries include a number of cotoneasters (*C. microphylla* and *C. rotundifolia*), the geranium named in his honor (*G. wallichianum*), *Lilium giganteum* and *Meconopsis napaulensis*.

Wallich's work exercised considerable influence on the life of a young English gardener, John Gibson (1815–1875). The Duke of Devonshire, having read Wallich's *Plantae Asiaticae Rariores* (3 vols, 1830–1832), decided that he would like a specimen of the amherstia for his own hothouses at Chatsworth in Derbyshire, and sent out the young Gibson to collect specimens. Thanks to Wallich and the invaluable Wardian cases, Gibson brought back over 80 species of orchid unknown in Britain, as well as the prized amherstias, *Rhododendron formosum* and *Dendrobium gibsonii*, named in his honor.

Joseph Hooker also did remarkable work in India, and after an Antarctic expedition, went to the East in 1847 as a collector for Kew Gardens. Hooker's accounts of his travels into the Himalayas make exciting reading. Despite innumerable adventures and hazards, Hooker reached Chakung, a meeting-point for the floras of the temperate and tropical zones, with oak trees and orchids growing in the same regions. His return journey was delayed and complicated by intrigue and threats, but he continued with his quest and successfully brought back to the West a vast number of plants, including primulas, species of *Meconopsis* and, most important of all, rhododendrons. His drawings and descriptions of the latter, called *Rhododendrons of the Sikkim Himalaya*, were published in England in the 1850s.

Hooker's *Himalayan Journals* inspired another Englishman to travel to those remote regions. Henry John Elwes met with similar problems to those encountered by Hooker but still managed to collect a number of arisaemas (*A. utile* and *A. griffithii*, for example) and a new species of *Hedychium*. Elwes is better known for his later work in collecting in Europe and for his twelve-volume *Trees of Great Britain and Ireland*, compiled with Augustine Henry, which remains a standard reference book.

Australia and New Zealand

Until the 19th century, Australia and New Zealand were virtually unexplored botanically, although a number of antipodean plants had been introduced to Europe, presumably by sailors and adventurers. The buccaneer William Dampier (1651–1715) had made modest expeditions at the beginning of the 18th century and Sir Joseph Banks (1743–1820) had also made researches while sailing with Captain James Cook (1728–1779), but the first person to work on any scale in the east of Australia was

the Scotsman George Caley (1770–1829), who sent back to England a number of seeds and introductions, including *Platycerium bifurcatum*. Two other Scots, Allan Cunningham (1791–1839) and Charles Fraser (c. 1788–1831) made extensive and adventurous journeys into the interior. Fraser continued in Australia; Cunningham moved onto New Zealand.

Later, Fraser also worked in the western part of the country and in 1829, possibly because of his report some three years earlier, a group of settlers arrived from England to establish the towns of Perth and Fremantle. A botanic garden was set up within a year or so by James Drummond (1784–1863), who had been supplied with seeds, mostly of food-producing plants, by the Horticultural Society. Thus, a two-way plant-transporting scheme was established between Britain and Australia. By the 1830s amateur plant hunters were regularly sending much-coveted seeds to England, many of which were successfully raised. These included *Helipterum manglesii* and *Trichinum manglesii*, both named after Captain James Mangles, an amateur plant hunter of some note.

The greatest botanist to work in Australia was the German Ferdinand von Müller (later Baron von Müller, 1825–1896), who settled in Australia in the 1840s and devoted his considerable energies to botany. By 1852, he had been appointed official botanist to the State of Victoria. On one expedition, in 1855, he collected some 2,000 plants, and by 1863 it was estimated that he had traveled more than 20,000 miles across nearly 20 degrees of unexplored country in search of plants. In 1857, Müller was made director of the Melbourne Botanic Garden. Although respected as a center for botanical studies, the gardens proved unpopular as a local amenity, and Müller left in 1873.

One of Müller's visitors to the Botanical Garden was John Veitch. He arrived in Sydney at the end of 1864 and made various trips, including a cruise in the Polynesian Islands, where he found many foliage plants ideal for the hothouses which had become such a vogue in Victorian England—

cordylines, dracaenas, plus many others, two of which commemorate his name: *Dizygotheca veitchii* and a palm, *Veitchia joannis*.

Throughout the 19th century, New Zealand was also being seriously explored, following the earlier forays of Banks. Allan Cunningham journeyed to New Zealand from Australia in 1826. Greatly helped by local missionaries, he made important discoveries, including *Viola cunninghamii*. A second trip, in April 1838, is commemorated in the epiphytic orchid *Dendrobium cunninghamii*, which he sent back to the English nursery of Loddiges.

Cunningham's good work in New Zealand was continued by a young Cornish missionary-lexicographer, William Colenso (1811–1899), who arrived in 1834 and remained for 65 years. His ostensible task was to translate the New Testament into Maori, but he spent all his leisure time in botanical enquiry. He was particularly fascinated by ferns and made arduous expeditions

to regions never before visited by white men. In the Ruahine Range, which he was the first European to cross, he found many species and unknown alpine flora. He also discovered *Ourisia macrophylla* and *Olearia macrodonta*.

Russia and the Far North

The problems of plant hunting in extreme northerly latitudes, and the limited scope for indigenous species being successfully grown elsewhere, undoubtedly curtailed the extent to which Scandinavia and Russia were explored by botanists. Yet one celebrated Swede cannot be omitted from any history of the subject. In 1732 Carl Linnaeus (1707–1778), the founder of modern botanical nomenclature, made a journey to Lapland which was to have far-reaching consequences, less for its discoveries than for the exploratory enthusiasm it kindled in others.

In the same year, the Russians commissioned a German, Johann Georg Gmelin (1709–1755), to travel with a party of scientists to Siberia, the Bering Straits and northern Japan. It was an ill-fated expedition, for Gmelin lost a large part of his collections and notes, although he salvaged sufficient to be able to publish his *Flora Sibirica*. In 1757 another Russian expedition was planned, under the aegis of Catherine the Great. Gmelin's nephew, Samuel Gottlieb Gmelin (1744–1774), and another German, Peter Simon Pallas (1741–1811), were the co-opted botanists and the latter's *Travels* give a full account of the six-year journey, or rather series of journeys. He also started, but never completed, a definitive *Flora Rossica*.

Another German botanist who figures prominently in Russia's floral history is Carl Maximowicz (1827–1891) who, at 25, was appointed Director of the Herbarium of the St. Petersburg Botanic Garden. In 1853, he started traveling widely within Russia in search of new plants, returning with some 100 new species to St. Petersburg, including *Lonicera maximowiczii*.

Botanists are among the least well-known of explorers. Of course, people like Darwin and Humboldt are famous for their exploration and descriptions, but their fame rests primarily on their work in the animal kingdom. This imbalance is in many ways an injustice. The economies of whole nations, as well as personal fortunes, have been made and broken by the work of men who discovered and, at the behest of Western scientists and governments, often introduced to other areas such fortune-spinning crops as tobacco, coffee, tea, rubber, breadfruit, opium and myriad vegetables and fruits from potatoes and tomatoes to bananas and mangoes. Not only this, but taking into account the origins of very many of our trees and flowers, botanists have probably done more than any other explorers to change our everyday surroundings.

L.H.

Plant Geography

As we have seen, much of the plant exploration of the world was related to the search for new species having industrial, medicinal, agricultural or horticultural value, a search which continues today, often at great expense. Nevertheless, increasingly over the past two centuries private and government funds have become available for more general collections of plants as scientific studies rose to prominence. Explanations of the different distributions of species, genera and families that became apparent through this exploration called for accurate information on the areas that they occupied, and even today such data are lacking for many regions and groups of plants. Filling this gap by means of sound taxonomic studies remains a prime concern of plant geography. While creationist views of plant diversity held sway, with all species considered to have spread over a stable world from one area of origin, little real progress could be made in understanding how ob-

served plant distributions could have arisen. Help in providing a rational basis for floristic plant geography, concerned as it is with these matters, followed the acceptance of Darwin's evolutionary ideas, which themselves arose in part from the facts of plant and animal distributions. Darwin was also influenced by the views of his friend Sir Charles Lyell on geological change, which further assisted the task of the student of plant distributions. Nevertheless, it is only in the past 20 years or so that the geophysicists' acceptance of continental drift (see p. 72), long demanded by many plant geographers, has shown just how much the world in-

Baron Alexander von Humboldt (1769–1859), a central figure in the development of the disciplines of plant geography and plant ecology. Based on his extensive travels, notably those to South America in 1799–1804, he wrote prolifically on the scientific challenge posed by the facts of plant distribution over the world, while simultaneously developing his ideas on the social groupings of plants in communities, probably the earliest aspect of what has now become ecology.

habited by plants has altered over the ages.

The interest, just mentioned, in the distributions of taxa (see Chapter Six: The Realms of Plants), and the underlying causes, is probably one of the oldest tenets of plant geography. However, there gradually developed ideas about the distribution of different kinds of plant cover, vegetation such as grasslands, forests and deserts, for example (see Chapter Five: Vegetation Today). This concern with plant species as communal beings seems to have begun to be formalized in the late 18th and early 19th centuries when the German botanist Carl Willdenow recognized that there are natural communities of plants which can often be characterized by one or two particularly prominent species—we still commonly talk about oak or beech forests, for example. This was expanded by Humboldt who, very shortly after his expedition to South America, wrote his *Essai sur la Géographie des Plantes* (1805), in which he lent the authority of his wide travels to provide a firm basis for plant geography.

Humboldt observed that different combinations of species occur in different regions. In this he provided a link between the preoccupation with the distributions of particular taxa already mentioned and the sociological view of vegetation, which has subsequently been formalized as phytosociology (see p. 59) and nowadays is such an important tool for describing the world's plant cover. However, Humboldt seems to have been more impressed by the fact that vegetation with similar overall appearance or aspect (see Physiognomy, p. 49), although of very different species-composition, occurs in widely separated regions of the world with comparable physical conditions, the most important being those factors collectively grouped under climate (see p. 117). This approach to vegetation has led to considerations of the habit, form and other adaptations of plants that permit them to survive under various environmental conditions, an approach which has been so important in describing the world's plant cover on a structural basis. This approach was developed by the Frenchman Alphonse de Candolle who, in the middle of the 19th century, indicated that the world's vegetation cover had been affected by such major climatic changes as those brought on by the Ice Ages, then recently described by Edward Forbes. The use of fossil plant-remains, in relation to various measures of the prevailing environmental conditions as a means of examining the development of modern vegetation was thus born (see Fossil Plants and their Study, p. 62).

Many other workers than those mentioned have contributed to the development of modern plant geography. However, in the last decade of the 19th century, the great German master of this subject, Carl Drude, clearly distinguished the factors which shape the occurrence of plants, as species and communities, over the world. Effectively, he

asked why do taxa have the distributions that we see, how have the floristic or structural communities of plants that we note come to occupy their present locations in the world and what effect has Man had on these patterns and phenomena since he evolved on this green planet? These are still the problems which students of plant geography are trying to solve, both by collecting data and by developing new concepts. As we shall see, from this approach to the world's plant cover there developed the inextricably entwined discipline now known as plant ecology.

D.M.M.

Plant Ecology

Ecology is concerned with the relationships of organisms to their environment and to each other. Although some Greek and Roman philosophers and scientists made "ecological" observations, the subject really began with the upsurge of interest in animal and plant geography in the 18th and 19th centuries. The term "ecology" was coined by the German biologist Ernst Haeckel in 1869.

As geographers considered the structure of the world's vegetation there gradually arose an interest in plant communities and their environments on a more local basis. Thus, the Frenchman Henri Lecoq surveyed his country's heathlands in 1844, estimating the relative importance of species in the vegetation, while at about the same time Hampus von Post continued this trend towards close analysis of vegetation with his studies in central Sweden. He emphasized the importance of studying the plant cover, rather than merely deducing it from supposedly important environmental factors, and his principle aims—to determine the species that live in an area and their relative abundance, coverage of the area and habit (see Life-form, p. 50)—remained fundamental to subsequent students of ecology.

The descriptive phase of plant ecology developed through the work of such people as Eugene Warming in Denmark, to reach its zenith in the 1920s and 1930s, with such workers as Frederic Clements in North America, Arthur Tansley in Britain and Josias Braun-Blanquet in Europe, each acting as synthesizers and pioneers in various approaches to the subject. The ecology of animals developed out of the studies on plants and took a somewhat different line, being more concerned with populations. However, ecologists soon began to realize that plants and animals are indissolubly linked in natural communities. In Britain, Charles Elton gave a boost to the joint approach with his ideas on production ecology, expounded in his book *Animal Ecology* (1927), and this line of interest was developed for some years. The classic work of R. L. Lindemann (1942) on the trophic-dynamic nature of natural systems was the signal for a great burst of work on the ways in

which the sun's energy is utilized by plants and animals, and nutrients are utilized by the organisms and cycled through the system. By this time also the concept of the *ecosystem*—the whole complex of living and nonliving components in any area, which interact with each other and through which energy and nutrients flow—was fully absorbed into ecological thought. It is one of ecology's few unifying concepts.

The study of whole systems and communities is termed *synecology* and that of individual species or populations *autecology*. Each has its own methods of work and techniques, but both draw upon a wide range of skills and knowledge. Herein lies a paradox for ecologists. They can bring to bear on ecological problems parts of many disciplines, including physics, chemistry, mathematics, electronics, genetics, physiology, taxonomy, geology and sociology, but at the same time they can rarely hope to attain results with the degree of precision attainable by specialists in most of these subjects, and this lack of precision is to many a severe limitation to ecology. While this is frustrating to some, to others it is a unique challenge.

Despite their difficulties, with the development of experimental approaches since the 1960s, ecologists have gone a long way toward producing testable hypotheses and generalizations about such processes as herbivore-plant interactions, competition between organisms for food, light, water and space, successional processes in natural systems, species diversity, community structure and stability, and population regulation. Yet an ecologist asked to make predictions about the potential effects of a proposed action on a natural ecosystem will be reluctant to do so without being given time to conduct studies. Why is this?

The basic unit of ecological study is the *species*, which is normally studied as a series of individuals or in populations. Until it is studied, a species has very few predictable properties. Whilst we know that a green plant will manufacture complex organic substances from carbon dioxide, water and sunlight, the great variability in most characters that exists within a single species means that conclusions drawn from a study of a population in one place have to be applied with caution to another from a different place. The ecologist is consequently at a disadvantage when compared with the chemist, who has a precise knowledge of molecular structure and can predict with some certainty the outcome of a reaction between two compounds.

Extending this to the ecosystem, with its immense complexity, the ecologist is even more at a disadvantage. If a great deal is known about a particular ecosystem, some predictions can be made as to the probable results of a proposed modification, from the knowledge built up over many years. Otherwise, predictions, even for related systems, are likely to be based on empirical considerations. What may be true for an oakwood may not apply to a beechwood, for example, since not only is the dominant and most conspicuous species different, but the soil and many of the associated plants and animals will also be different. So without much research on an area, even experienced ecologists must be cautious in their predictions, and sometimes they will be wrong. In the present time of rapidly changing pressures on so many ecosystems, ecologists must be prepared to give answers on insufficient evidence. What the ecologist *can* do is to state a range of possibilities: what *might* happen in a given situation, rather than what *will* happen.

The difficulties arising from the complexity of natural ecosystems cannot be too strongly emphasized. Few people realize, for instance, how many species of green plants and

animals, not to mention fungi and bacteria, exist in a temperate woodland. In a few hundred hectares of such a woodland there may be between 200 and 400 flowering-plant species; another hundred or so bryophytes and lichens and a comparable number of large fungi; uncountable species of microfungi (many as yet unnamed), algae and bacteria; a few dozen species of birds; a sprinkling of mammals; innumerable insects, spiders and other invertebrates. The total can be measured in thousands. All these are interacting with each other in time and space and although a general picture of their interrelationships and functioning can be obtained, it must inevitably be superficial.

All this does not mean that ecological predictions are not worth attempting. Man's

role in changing the environment is now clear to all, yet Man too is part of "nature" and of "natural" ecosystems and, while primitive people had a relatively small effect, the development of agriculture and the industrial revolution have given us a unique capacity to modify, alter or destroy the world around us (see Chapter Seven: Man and the Green Planet).

Synecology

Synecology is concerned with groups of organisms associated together; that is with communities, or better, with ecosystems, in which living plants and animals, together with their nonliving environmental complex, make up an integrated whole, interrelated and interdependent. Synecology provides a synthesis of all the information acquired from detailed study of parts of the ecosystem together with an integrated study of the whole. While the autecologist might be likened to an architect designing individual buildings, the synecologist is like the town planner, bringing all the buildings and services together into a working whole.

Synecology can be studied at several levels. Much work in the past has concerned the description and classification of vegetation, which most ecologists agree is a necessary prerequisite for subsequent functional and experimental work.

The functional approach has included studies on succession (replacement) of plants and animals in many environments and the factors, both physical and biological, that control this replacement. Apart from the changes in the numbers and types of organisms during succession, recent work has been concerned with all the other effects—on soils, nutrients, turnover, reproductive strategies and productivity—that result from the progressive changes of systems with time.

The other main line of synecological research is that of whole ecosystem energetics, which attempts to delineate the ways in which incoming energy from the sun is used by the organisms, from the primary pro-

In attempting to understand the structure and relationships of plants and animals in the ecosystem, synecology utilizes studies of contemporary communities in adjacent areas. In addition, however, synecology is also able to utilize observations on the changes in the plant cover of one area through time. The above two pictures, taken at Kingley Vale, Sussex, England, show the dramatic changes which can occur in only 16 years. In 1954 (**above**) the short grass-turf and trim wood-margins had been maintained for many, many years by the cropping of rabbits. By 1970 (**below**), with the rabbit population virtually wiped out by myxomatosis, the ecological balance has changed as the relation between plants and animals altered. Most low-growing species of grasses have disappeared, to be replaced by tall, coarse grasses (mainly *Arrhenatherum* and *Dactylis*); bramble (*Rubus*) and gorse (*Ulex*) have spread to form a scrub, while there are signs that the woodland is regenerating at its margins.

Of importance in autecological studies, which are partly concerned with the reactions of plants to their environment, is the ability to provide conditions in which the factors essential for plant growth—light, temperature, moisture, nutrients—can be controlled and monitored. This is achieved in a phytotron, such as that at Duke University, North Carolina, United States (**above**), within which (**above right**) different "walk-in" chambers provide a range of environmental conditions, and experimental plants, such as the cotton shown here, are grown in pots on movable trolleys which allow their transport between chambers.

ducers (plants) which fix the energy, through the successive groups of organisms—herbivores, carnivores and decomposers—that use it. Such studies can give a balanced picture of the functioning of a whole system. Against this background, the importance of individual species can be assessed.

A very recent development in synecology is the use of computer models. Utilizing existing data, attempts can be made to construct predictive models of whole systems and these in turn can be tested against real systems.

Autecology

Autecology is the study of individual organisms or species. In practice it is usually populations that form the basis of study. Autecological work is initially descriptive: the place of the species in its community has to be defined. Description and observation have as their main function the formation of hypotheses which can be subjected to experiment. Experiments on plants can be done in their natural environment, in experimental gardens or under closely controlled conditions in greenhouses or growth cabinets.

Field experimentation usually involves altering the environment in some way that observation has suggested may help to elucidate the plants' natural behavior. Populations may be manipulated by adding more individuals or by removing some, or by adding or removing competing species. The environment can be changed by adding, for example, nutrients or water, by increasing shade and in many other ways. Unfortunately, the changing of one factor will change many others in ways which are not immediately obvious, and change them not only for the species under observation but also for all the others in the area. Unraveling the complexity of the results may prove very difficult.

Experiments under strictly controlled conditions can eliminate most of the unwanted variation in environment and are useful in elucidating the behavior of a species in response to particular factors, but interactions of factors are less easily studied in this way. The experimental garden allows some control and, in particular, eliminates interactions with other species. Ideally, a combination of methods should be used, and field observations should be reinterpreted in the light of what has been learned from the plants subjected to experimental manipulation.

One crucial point to be remembered when investigating the autecology of a species is that the evolutionary unit is not the species as a whole but the local population. Since each local population is likely to differ from others in its inherited (genetical) constitution (see Genecology, p. 36), conclusions drawn from the study of one population must be applied with caution to others and the examination of such local variation has motivated much autecological work.

The study of a few species of similar ecology—often closely related species—in an attempt to define the factors which enable them to occupy similar habitats, has proved very fruitful. In a variety of species, it has been demonstrated that, in morphology, photosynthesis and respiration, responses to nutrients, tolerance of toxic ions etc, local adaptations are a crucial part of the plants' strategy for survival. Conversely, the few studies on species of very restricted ranges have shown that their ranges of tolerance are very narrow.

Darwin was the first to draw attention to the idea that related species would be likely to compete with each other more intensively than non-related ones, and the idea was explored by the American G. F. Gause (1934) who experimented on various organisms, including *Paramecium*, in aquaria, and showed that they occupied different habitats. Most work on these lines has been done with animals, but several flowering-plant genera have been studied, notably buttercups (*Ranunculus* spp.), poppies (*Papaver* spp.), plantains (*Plantago* spp.) and duckweeds (*Lemna* spp.). These studies indicated that, for closely related species to live together in the same habitat, they must have some controlling factors not in common, and reproductive behavior that prevents them from merging. For instance, if the seeds of two species germinate at different times, or if germination is controlled by slight environmental discontinuities, or if in some other respect the seeds exploit the environment in slightly different ways, then species with otherwise very similar ecologies can live in the same habitat.

Autecology thus has available a variety of approaches for understanding the ways in which individuals interact with the physical environment and with other members of the same and different species. It clearly has a great deal to contribute in the future and should eventually provide a complete picture of species' behavior. It is hoped that in time it will be possible to produce a synthesis of plant community behavior from a knowledge of individual species, which will complement the knowledge gained from synecology about the whole ecosystem.

s.w.

CHAPTER TWO

Tools of the Trade

The Methods and Concepts of Plant Ecology and Geography

As in most other areas of science, increasing knowledge of the world's plant cover has depended upon a continuing interplay between the expansion of concepts concerning the distribution of species and the communities to which they belong and the development of methods and procedures for documenting the information and testing the resultant hypotheses. There has been a continuing link between plant geography and ecology—the main concerns of this book—and plant taxonomy or classification. Students of vegetation and plant distribution depend upon the taxonomist to provide an ordered view of plant diversity, not only to give a set of reference points for determining the taxa which are their raw material, but also to indicate taxonomic relationships and variability. Taxonomy, conversely, is greatly aided by knowledge of the distribution and habitat preferences of species and the populations of which they are composed.

Taxonomy has periodically given rise to disputes, ranging from the view, which has its roots in doctrines of special creation, that all variation can be described and classified, to the other extreme which considers that, since evolutionary change is a constant feature of life, it is not possible or worthwhile to describe species and other taxa which are but transient phenomena. Similarly, widely differing concepts have emerged from vegetation studies. These range from the idea that vegetation comprises a series of units in which the constituent species are as closely integrated as the cells and organs of an individual organism, to considering the vegetation in any place to be composed of those species which happen to be present and are able to survive. In the latter case, no two patches of vegetation are ever exactly alike, thus making classification of doubtful value.

Taxonomy and systematics are concerned with describing plant species and interpreting their interrelationships and evolution, and provide a basis for storing information about the plants as it becomes available. After Linnaeus had provided a stable method of naming plants in a way which reflected their apparent affinities to one another,

data on the morphology, anatomy and geographical distribution of plants were incorporated into a rather formalized system of descriptions. The overall basis was little changed as the facts of evolution became known, indeed it has persisted until today. Nevertheless, awareness of evolution influenced interpretation of the diversity being charted. The rise of cytology and genetics provided new tools for investigating the basis of variation, relating it to natural selection, and thereby clarifying the role of, for example, competition and environmental factors in shaping the structure and composition of plant communities. This "experimental" approach to taxonomy, nowadays called biosystematics or genecology, has provided a further link between students of the diversity of plant life and those concerned with the structure and distribution of plant communities.

The distribution of species was initially given by briefly indicating those territories or countries where they occurred. As exploration and knowledge expanded, and the information became more detailed, the introduction of distribution maps, before the middle of the 19th century, facilitated presentation of distributional data and their comparison for different taxa. It soon became apparent that unrelated species frequently show similar distribution patterns; this led to investigations of the past migrations of plants, one of the concerns of chorology. These floristic considerations have also been developed by techniques for assessing the relative importance of species in plant communities to give the basis of phytosociology, important in describing the world's plant cover on both major regional and more local levels.

The correlation, observed early, between the structure of vegetation and the major climatic zones led to a preoccupation with the adaptive importance of plant form (physiognomy). This has resulted in increasingly refined descriptions of the life-forms shown by different plant species and their contribution to the structure of the various communities. Techniques for sampling small, representative areas of vegetation have been very important in developing ecological studies. This has been particularly so with the increasing emphasis during this century on sound statistical analysis as a basis for mathematical and biological models of vegetation. The concern with evolution and vegetation dynamics has gone hand-in-hand with the development of the techniques and concepts outlined in this chapter.

Opposite Areas such as the tropical rain forests of the Guiana Highlands shown here have the largest standing mass (biomass) of living vegetation per unit area and also represent a natural climax vegetation (or biome) that will only alter naturally if the climate changes.

The Identity and Relationships of Plant Species

All studies of plants, but particularly ecology and plant geography, depend upon accurate identification of the species. Identification, providing a name for a taxon, effectively makes accessible all the information that has been obtained for plants to which that name applies. Consequently, it is important to have as stable a system as possible for giving names to plants (and animals), and nomenclature is concerned with this. Taxonomy and systematics aim to describe plants accurately and to arrange them, by species, genus etc, in such a way as to reflect their affinities. These data are made available to the ecologist and plant geographer in floras covering the particular region(s) in which they are interested. Although there is considerable debate about the desirability, or even possibility, of producing taxonomic treatments which reflect the evolutionary histories and relationships of taxa, the study of evolutionary processes at and below the species level has led to a discipline, gen-ecology, which further stresses the close links between taxonomy and the study of vegetation.   D.M.M.

Taxonomy and Systematics

Organic evolution has produced an enormous array of diversity in the natural world. Man's instinctive reaction when faced with this diversity is to classify it, to divide it up into smaller, more manageable units. The study of classification, including its principles and procedures, is called *taxonomy*. *Systematics* refers to the scientific study of the diversity and differentiation of organisms and the various relationships which exist between them. The process of ordering plants or animals into groups of increasing size and comprehensiveness, arranged in a hierarchical fashion, is called *classification*. We also use the term classification to refer to the arrangement of groups which results from this process.

Folk Taxonomy

Early Man had to classify plants (and animals) to enable him to know which were safe to eat, which were poisonous, which could be used for shelter, and so on. As a result, each tribe or local community developed what is now called a folk taxonomy, consisting of a small number of plants and animals whose characteristics and properties were well understood by all members of the tribal group. Such restricted folk taxonomies are still found today in primitive tribes and in a way are employed by all of us today. For example, we all, wherever we live, soon learn a limited repertoire of fruits, vegetables, spices, ornamental flowers, trees etc, the actual examples depending on the area or country concerned. These plants are part of common knowledge and an important aspect of such folk taxonomies is that the name of the plant conveys its properties, so that when we talk about an apple or a lettuce we know immediately, if we belong to the appropriate community, what is meant without having to describe the object. Folk taxonomies are essentially practical and are limited to plants that have an importance for the culture concerned. Early classifications were based on the appearance, taste, smell and nutritive value of the plants.

Scientific Taxonomy

Unlike folk taxonomy, scientific taxonomy and classification are comprehensive, in that they attempt to cover all known organisms. Since the number of different plants and animals runs into many thousands—about 300,000 flowering-plant species, 10,000 ferns and fern allies, 40,000 fungi, 850,000 insects, 20,000 fishes, and so on, totalling about one and a half million described species of organisms alive today—it is clear that

To the taxonomic botanist the herbarium is like a reference library—it contains a large store of classified information, and enables plants from all over the world to be examined and compared with relative ease.

there is a basic need for taxonomy, to allow biologists and other people to identify, describe and communicate knowledge about these organisms to other people. Biological taxonomy is sometimes described as an information storage and retrieval system for biology. In order that experiments can be repeated and the results of the work communicated in a meaningful way, biologists and others have to work with named organisms, ie organisms belonging to a species, genus and family; this presupposes a system of classification.

The naming of plants serves not only as a reference system but implies that the plant belongs to an already described group which possesses certain characteristics and that all samples of the group can be expected to possess these characteristics. Thus classification in a sense constitutes a system of inbuilt guarantees for other users. The main difference between folk taxonomies and normal scientific taxonomies or classification is that the number of entities in the latter is so great that the names no longer convey anything to the majority of the users about the properties or characteristics of the plants or animals referred to. The name of a plant is in effect a key to the literature concerning the plant and it is the role of the taxonomist to know this literature and to know how to handle it, much in the same way that a lawyer knows how to find his way through the complex legal literature to find the information he needs in a particular case. Taxonomy is thus inextricably bound up with nomenclature (see p. 29).

The way in which plants and animals are classified is basically an extension of the methods applied to nonliving objects, that is

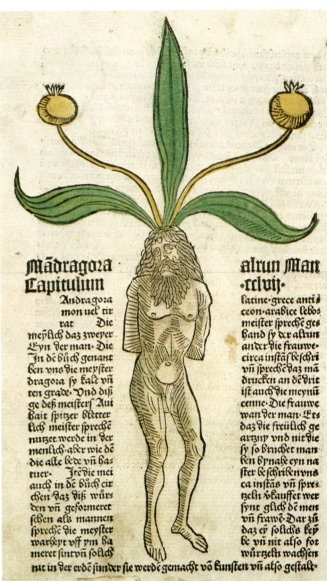

Above Illustrations have long been important in describing plants, though early examples owe more to art than science. These illustrations of the mandrake (*Mandragora officinarum*), taken from an old German herbal (P. Schoffer, Mainz, 1485), show the female (**left**) and male (**right**) forms which the roots were said to resemble. They exemplify the fancies, simplifications and inaccuracies derived from successive copying of older sources, and from attention to the demands of design (eg symmetry) and superstition (eg the medieval "doctrine of signatures").

on the basis of features they possess in common. Classifications are not unique. We can, for example, take a group of 10 children and using the simplest form of classification—one which splits them into two classes (dichotomous)—we can divide them into, say, male or female, blonde or brunette, whether at school or not, susceptible to tuberculosis or not, and so on, according to our purpose. Likewise, books can be classified according to color, type of binding, subject-matter, author, and so on. Which characters we use for our classification depends on the use that is to be made of our classification, and if we have no reason for adopting any of the available alternatives there can be no rational grounds for choosing between them.

Classifications based on one or a few features are termed *artificial*. They have a limited use in biology, such as special classifications based on, say, seedling-type for agronomists, flower color for amateurs etc. Most biological classifications are termed "general purpose" since they are used in a wide variety of contexts but without the exact use being specified. They are constructed by using large amounts of data and therefore have a high information content. An important characteristic of such classifications is that they have a high *predictive value* or *predictivity*, which means quite simply that if a group of plants has many features in common, and some characteristic, such as a particular chemical compound, *not* used in constructing the group is found in one member, then the other members are also likely to possess it. High predictivity is achieved by basing classifications on multiple character correlations but raises the question of why such correlations, which make biological classification feasible, exist. The answer is that plants and animals have an evolutionary history: the similarity of the members of groups is due to their common genetic makeup derived from their common ancestry, although there are exceptions to this rule. Evolution is the factor that causes characters to hang together and makes biological classification possible; which is not the same, however, as basing classification on evolutionary history or phylogeny.

The Taxonomic Structure

The raw material of taxonomy comprises populations and species which are grouped together and built up into a classification following a fairly simple set of principles. *Species* are groups of more or less identical individuals forming populations in nature, and those species that show most features in common are put together in larger, more inclusive groups called *genera* (singular, genus); genera, in turn, which have most features in common are placed together in yet more inclusive groups called *families*, and so the process continues. This method is sometimes referred to as the box-within-box system—building up a series of increasingly inclusive groups or boxes on the basis of overall similarities. A taxonomic group of any rank (species, genus, family etc) is called a *taxon*. Parallel with this series of more inclusive taxa, a series of different ranks, level or categories is recognized, forming a hierarchy.

When a taxon is formed, it has to be assigned to a particular rank and it is the rank in the hierarchy that determines the group name—species, genus, family etc (see under Nomenclature, p. 29, where a table is given showing the main ranks in the flowering plant hierarchy). One of the most difficult tasks for the taxonomist is to decide on the rank to be afforded to a particular group: should it be a section or a subgenus or a genus? The number of categories recognized today has been built up over the past two centuries to accommodate the different degrees of variation found in the plant kingdom.

Types of Classification

Two major types of biological classification are recognized today. The first is termed *phenetic* and expresses the relationships between organisms in terms of their similarities in properties or characters without taking into account how they came to possess them. In preparing a phenetic classification, any kind of data or evidence (other than evolutionary) may be taken into account, such as morphology, anatomy, chemistry, embryology etc, but there is no necessary evolutionary significance attached to such classifications, although they may later be used for phylogenetic or evolutionary interpretation. Such classifications are empirical in the sense that they are based on the observation and assessment of a greater or lesser number of features possessed by the organisms to be classified. This is the commonest type of classification used by biologists, and the methods employed have not materially altered in the last two centuries. A refinement of this kind of classification, numerical phenetics, is discussed below.

Contrasting with phenetic classification is *phylogenetic* or evolutionary classification, in which some aspects of the evolution of the organisms involved are taken into account in making the classification. Since all plants have an evolutionary history or phylogeny, many taxonomists believe that the ideal and correct classification must be one that reflects this. The difficulty is that phylogenetic classification requires evidence of several different kinds. The main components of phylogenetic relationship which a phylogenetic classification should reflect are cladistics, patristics and chronistics. *Cladistic* relationship refers to the pathways of ances-

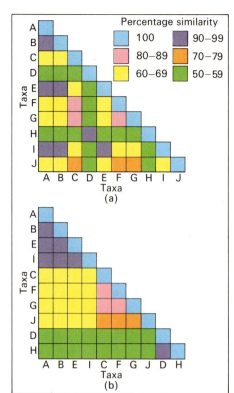

Of the two major types of biological classification, that termed phenetic depends upon the degree of similarity shared by organisms or taxa (species, genera etc). One way of achieving this is by constructing similarity tables such as those illustrated. (a) A schematic similarity table showing a matrix of hypothetical coefficients between pairs of taxa. The percentage similarity is shown by color. (b) Cluster analysis of the same coefficients achieved by rearranging the similarity matrix so as to bring the most similar taxa next to each other. It is now possible to see immediately the most similar (or dissimilar) taxa with regard to the characters tested.

try, in other words the branching patterns by which the evolutionary lineages arose and diverged from one another. This relationship is usually expressed in a tree-diagram or *cladogram*. If phenetic and cladistic relationships in a group are shown to be concordant, the similarities of the members of a group must be the result of common ancestry. *Patristic* relationship or similarity is that due to common ancestry. Unfortunately, resemblance between the members of a phene-

Below Dendrograms are a means of expressing the relationships between organisms and/or taxa, their exact construction depending upon whether phenetic or phylogenetic classificatory principles are involved. The two main types of dendrogram are shown here. (a) Phenogram indicating the overall resemblance between 11 imaginary taxa. B, C, D and E show 95% similarity for the features measured. (b) Cladogram for 11 imaginary taxa which have a common ancestry through the stem at X. Each taxon has a different degree of evolutionary (cladistic) relationship with the others.

tic group may not be due to common ancestry, as when unrelated taxa develop similar attributes (parallelism and convergence) in response to strong environmental selection; this is often very difficult to detect. A further complication is the time element involved in evolutionary relationships. The study of the time-scale of evolutionary events is called *chronistics* and is usually indicated in phylogenetic diagrams by the vertical dimension. Divergence of evolutionary lineages may take place at any time and the rate of evolution may be fast, moderate or slow.

A fully phylogenetic classification should therefore express not only the present-day phenetic relationships of the organisms being classified but should include and indicate the cladistic, patristic and chronistic dimensions. In practice this is never fully possible and most evolutionary taxonomists must content themselves, at least as regards plant groups, with a phenetic classification in which it is hoped that all members of a group have had a common ancestry (monophylectic), within an arrangement which indicates as far as possible the known or assumed evolutionary relationship of the groups relative to each other. Very seldom are the cladistic, patristic and chronistic elements recognized, simply because the information is not available.

The only sure evidence for common ancestry is an adequate fossil record, which for most groups of plants is not available (see Fossil Plants and their Study, p. 62; the Geological Record, p. 71). As a result, phenetic evidence has to be interpreted as evidence of evolution and this, as we have seen, is a difficult and fallible process.

A recent development has been the application to taxonomy of computers and

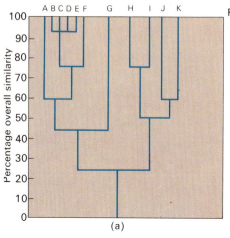

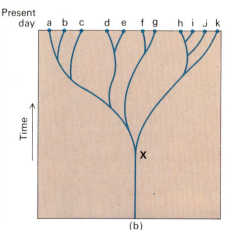

methods in which the similarity between organisms is evaluated numerically and the classification worked out in a hierarchical system based on these similarities. This allows statements about relationships to be quantified, thus avoiding imprecise statements about taxonomic groups such as "very similar" or "somewhat different". Numerical taxonomy allows the various operations involved to be formulated and repeated by other people. Numerical methods can also

Below Phylogenetic classification tries to take into consideration some aspects of the evolution of organisms, which can be expressed in the sorts of diagrams shown here. (a) Evolution by divergence through two "generations" of species. Here, four contemporary species D, E, F and G are derived from two immediate progenitors B and C, all of which are derived from an overall common ancestor A. Notice that the ancestral generations are extinct in the diagram. (b) Evolution by fusion through hybridization. In this diagram, the species A_1 and B_1 and C_1 and D_1 have given rise to a generation of hybrid daughter species E_1 and F_1 which in turn have hybridized to give the modern species G_1. (c) Cladogram illustrating the evolutionary pathway of three members of the cabbage family (Cruciferae) in relation to other flowering plant families.

be applied to working out cladistic relationships.

Taxonomy and systematics are still the basis and background for most biological work today. The task of surveying and classifying much of the world's plant life is very far from complete and the study of relationships is still in its infancy in most cases.

v.h.h.

Nomenclature

Nomenclature is the giving of names and it would be superfluous to stress the importance of names in the communication of knowledge. The feeling of many primitive peoples that the ability to "name" an organism somehow gives them power over it is not altogether far-fetched; as was noted earlier (see p. 26), the name of a plant species provides a key to everything known about it. Although the names given to plants have no ecological significance, a stable and universally agreed system of nomenclature is a necessary prerequisite for studies of vegetation.

Types of Plant Names

Plant names fall into three types: vernacular, scientific and cultivar names. *Vernacular* or common names originate early in any language, although they pay little respect to the niceties of species distinction. These vernacular names are often closer in application to generic than specific level, and a great many plants of recent discovery have no common name. They tend to be loosely applied: the bluebell of England (*Endymion non-scripta*) is not the bluebell of Scotland (*Campanula rotundifolia*). One plant species may have many vernacular names. Worse still, vernacular names often suggest false relationships. The white bryony and black bryony are both common hedgerow twiners with tuberous roots and unisexual flowers, but they are totally unrelated: the former (*Bryonia dioica*) is a dicotyledon of the cucumber family, the latter (*Tamus communis*) a monocotyledon related to the yams. Vernacular names are hence of greater historical and linguistic interest than scientific. Attempts have been made from time to time to standardize them, however,

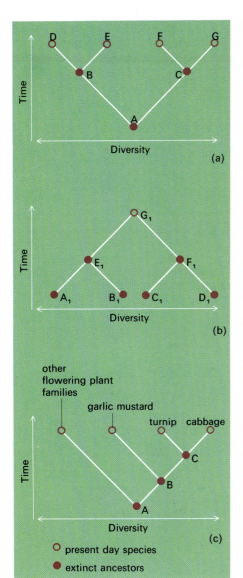

○ present day species
● extinct ancestors

THE TYPES OF PLANT NAMES

CATEGORY	COMPOSITION & TYPOGRAPHY	EXAMPLES	EXPLANATION
1. Vernacular	One or more words; as for normal nouns; no typographical distinctions	Orchid family; primrose; honesty	The common name in any language; national only; often loosely applied to more than one species, and one species may have several names
2. Scientific	Two words for a species, one for other ranks; capital initial for generic names and above, small initial for specific epithets and below; italic type	Orchidaceae; *Primula vulgaris*; *Lunaria annua*	The botanical name as applied under the International Code of Botanical Nomenclature, and standard in all countries; one latinized binomial per species, and latinized uninomials for other ranks
3. Cultivar	One, two or three words; capital initial, Roman type and single quotes or "cv" to precede; non-latinized fancy names if new	'Hybrid Musks' 'Christmas Cactus' 'Royal Sovereign'	The fancy name for plants of agriculture and horticulture, usually not known in the wild and maintained only by Man's efforts; governed by the International Code of Nomenclature of Cultivated Plants and more or less international in usage

CLASSIFICATION OF THE BEAN, SHOWING THE HIERARCHY OF CATEGORIES AND THEIR NAMES

CATEGORY	SCIENTIFIC NAME OF TAXONOMIC GROUPS (TAXA)	VERNACULAR NAME
Class	Angiosperm**ae**	Angiosperms; Flowering Plants
Subclass	Dicotyled**oneae**	Dicotyledons, Dicots
Superorder	Ros**idae**	Rose Superorder
Order	Fab**ales**	Legume Order
Family	Fab**aceae** (Leguminosae)	Legumes; Legume Family
Subfamily	Papilion**oideae**	Pea Subfamily
Tribe	Phaseol**eae**	Bean Tribe
Subtribe	Phaseol**inae**	Bean Subtribe
Genus	*Phaseolus*	Bean
Species	*Phaseolus vulgaris*	French or Kidney Bean
Variety	*Phaseolus vulgaris* var *humilis*	Bush Bean
Cultivar	*Phaseolus vulgaris* 'Canadian Wonder'	

Endings in **heavy type** are standardized indications of rank

and to provide translations of the scientific names. But yellowleaf goldmoss stonecrop has never gained favor over *Sedum acre aureum* and is nearly twice the length.

Scientific and *cultivar* names, for wild and cultivated plants respectively, are international in usage and governed by their own special Codes drawn up at international congresses and reviewed and reprinted at intervals (five years commonly). These Codes have no legal standing; one cannot be prosecuted for disobeying them, but they are voluntarily accepted by nations throughout the world and transgressors stand to have their work ignored.

These Codes ultimately go back to Linnaeus himself, who in 1736 and 1737 drew up rules for the naming of plants and later gave us the binomial (or binominal) system which has stood the test of time for over two centuries. Thus the species name of a wild plant is a binomial composed of a generic name followed by a specific epithet (trivial name), for example *Quercus robur* L. (the common English oak). The name of the author (the person who first described the species) can be added in full or abbreviated after the binomial to aid precision. Here the "L." means Linnaeus, and distinguishes his *Quercus robur* from that of any other author. Scientific names are in Latin or made to look like Latin, whatever their origin, for example Commelinaceae (after Commelin), *Pitcairnia* (after Pitcairn Island), *cordiformis* (heart-shaped).

Binomials are unique and a mine of compressed information. Consider the name *Aloë arborescens* Mill. What does it convey to the botanist?

Rank. He sees that it is a taxon (taxonomic unit) of the rank of species.

Genus. It is a member of the genus *Aloë*. This immediately gives him the mental image of fleshy-leaved rosette plants of the family Liliaceae.

Species. It is one species among 400 that its author distinguished by the epithet "arborescens", meaning tree-like.

Source. "Mill." means Philip Miller and the most likely source of his name would be his *Gardener's Dictionary* of 1768—the first book to go to for further data. What other tripartite name could convey so much?

But for all its proven utility, the binomial has one disadvantage. A plant has to be classified before it can receive a scientific name. This may take time. A newly discovered cactus or orchid may not flower to reveal its genus until it has been in cultivation for some years, and meanwhile the introducer, nurserymen and public will be clamoring for a name. Worse, if further study reclassifies the plant, the name must change also and it may thus end up with a string of synonyms (see Name Changes, p. 32). Yet despite these drawbacks, no one has yet come up with a better alternative to binomials.

Cultivar ("cultivated variety") names are the outcome of a desire to segregate man-made plants from those growing sponta-

neously. They have a Code to themselves (the Horticultural Code), and the names are preferably non-latinized and in living languages, for example *Fuchsia* 'Mission Bells'.

The Birth and History of the Hierarchy

The origin of plant names and plant classification goes back to the origin of language itself. Major food plants are among the first to receive names. At first a single name suffices, until another similar species is found and then an adjective is added to distinguish one

from the other. Thus "buttercup" becomes "field buttercup" and "alpine buttercup" when a second species is found in the

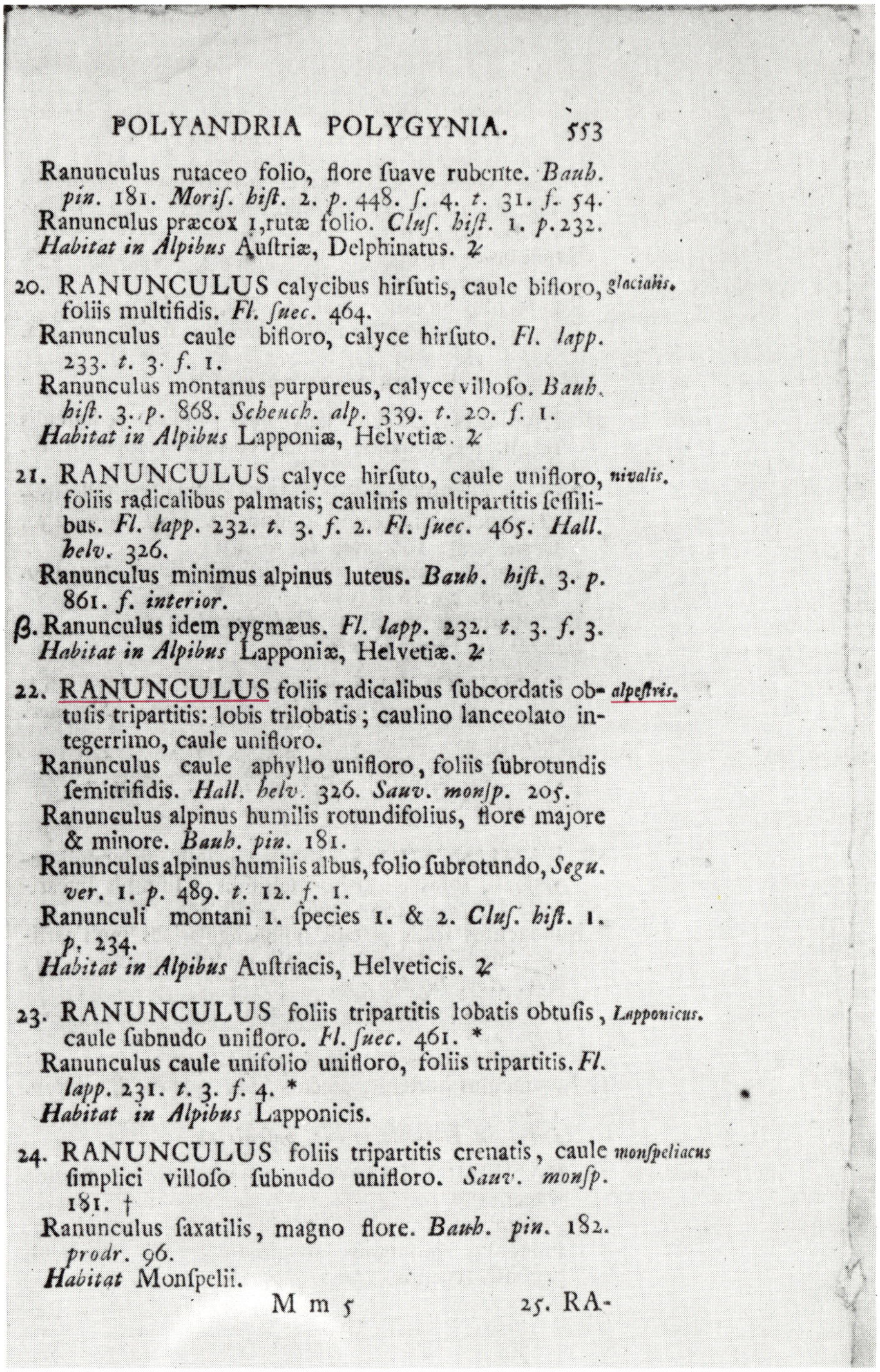

Early formal descriptions of plants were written in Latin, a language nowadays usually used only to describe new species. This page from Linnaeus' *Species Plantarum*, published in 1753, illustrates the brevity of such descriptions, which concentrate on a few distinctive characters. Linnaeus reduced the long Latin names of his predecessors to simple binomials—for example the generic name *Ranunculus*, followed by the species' name *alpestris*, as seen here in red. *Ranunculus alpestris* L. is today the accepted name for this alpine white crowfoot of southern Europe, and 1753 is taken as the starting date for the valid publication of the scientific names of plants.

POLYANDRIA POLYGYNIA.   553

Ranunculus rutaceo folio, flore fuave rubente. *Bauh. pin.* 181. *Morif. hift.* 2. *p.* 448. *f.* 4. *t.* 31. *f.* 54.
Ranunculus præcox 1, rutæ folio. *Cluf. hift.* 1. *p.* 232.
Habitat in Alpibus Auftriæ, Delphinatus. ♃

20. RANUNCULUS calycibus hirfutis, caule bifloro, *glacialis.* foliis multifidis. *Fl. fuec.* 464.
Ranunculus caule bifloro, calyce hirfuto. *Fl. lapp.* 233. *t.* 3. *f.* 1.
Ranunculus montanus purpureus, calyce villofo. *Bauh. hift.* 3. *p.* 868. *Scheuch. alp.* 339. *t.* 20. *f.* 1.
Habitat in Alpibus Lapponiæ, Helvetiæ. ♃

21. RANUNCULUS calyce hirfuto, caule unifloro, *nivalis.* foliis radicalibus palmatis; caulinis multipartitis feffilibus. *Fl. lapp.* 232. *t.* 3. *f.* 2. *Fl. fuec.* 465. *Hall. helv.* 326.
Ranunculus minimus alpinus luteus. *Bauh. hift.* 3. *p.* 861. *f. interior.*
β. Ranunculus idem pygmæus. *Fl. lapp.* 232. *t.* 3. *f.* 3.
Habitat in Alpibus Lapponiæ, Helvetiæ. ♃

22. RANUNCULUS foliis radicalibus fubcordatis ob- *alpeftris.* tufis tripartitis: lobis trilobatis; caulino lanceolato integerrimo, caule unifloro.
Ranunculus caule aphyllo unifloro, foliis fubrotundis femitrifidis. *Hall. helv.* 326. *Sauv. monfp.* 205.
Ranunculus alpinus humilis rotundifolius, flore majore & minore. *Bauh. pin.* 181.
Ranunculus alpinus humilis albus, folio fubrotundo, *Segu. ver.* 1. *p.* 489. *t.* 12. *f.* 1.
Ranunculi montani 1. fpecies 1. & 2. *Cluf. hift.* 1. *p.* 234.
Habitat in Alpibus Auftriacis, Helveticis. ♃

23. RANUNCULUS foliis tripartitis lobatis obtufis, *Lapponicus.* caule fubnudo unifloro. *Fl. fuec.* 461. *
Ranunculus alpinus unifolio unifloro, foliis tripartitis. *Fl. lapp.* 231. *t.* 3. *f.* 4. *
Habitat in Alpibus Lapponicis.

24. RANUNCULUS foliis tripartitis crenatis, caule *monfpeliacus* fimplici villofo fubnudo unifloro. *Sauv. monfp.* 181. †
Ranunculus faxatilis, magno flore. *Bauh. pin.* 182. *prodr.* 96.
Habitat Monfpelii.

M m 5   25. RA-

mountains. The process goes on until we have telegram-style descriptive phrase-names like "low-growing round-leaved alpine buttercup with a smaller flower" or "*Ranunculus alpinus humilis rotundifolius flore minore*" as Caspar Bauhin had it in 1623.

Such lengthy names are unwieldy and hard to remember. Linnaeus had developed to a high level the art of making them before he introduced his binomial system, as an aid to memory and indexing. Linnaeus did not invent the binomial—there were many earlier if haphazard examples—but he was the first to apply the idea uniformly throughout the plant and animal kingdoms.

A parallel development along with the evolution of names was the "box-within-a-box" concept of classification, culminating in the present-day hierarchy of categories. By the 18th century, Linnaeus could write: "The whole System of Botany is comprised under five appropriate divisions, namely Classes, Orders, Genera, Species, Varieties". Nowadays we need more; their order is fixed, and standardized endings are applied to several as indications of rank. It matters little whether or not these categories have any real existence. Much has been argued for and against the "naturalness" of genera, species and so on, but it need not be repeated here. No other way of classifying the animate world was possible, at least until the computer era. The future may see plant descriptions stored on tape under code numbers instead of names, with the searching and retrieval done electronically. But that day is a long way off.

The Botanical Code

The International Code of Botanical Nomenclature, the outcome of many revisions of previous editions, currently consists of six Principles setting out the basis of the text, 75 Articles and various Recommendations. This occupies only a small part of an initially daunting volume, for it is followed by full translations in French and German and, for more than half of the 426 pages, by Appendixes of conserved names (see Name Changes, p. 32).

The *Articles* are the rules themselves; they are retroactive (in the main) and mistakes of the past must be altered to accord with them. The *Recommendations*, as their name implies, are not obligatory. It cannot be stressed too strongly that a working knowledge of the Code is essential before any attempt at naming or revising the names of plants is made.

The *Principles* begin by telling us that botanical and zoological nomenclature are independent, so that the same name can be used for a plant and for an animal (eg *Corydalis* and *Ricinus*). We are next introduced to the concept of types.

The Type Concept
This is a device that permanently anchors a name to a plant or concept and hence has a

great stabilizing influence in nomenclature. Whenever a new species is described for the first time, the author must designate a particular preserved (not living) specimen as the type specimen (holotype). This is then stored with more than usual care and can be used as a standard of reference in any future controversy over the name.

Suppose a new species has been described as *Mephistophelia diabolica* Nick. A later botanist may decide that it was a mixture of two or more distinct species. Under the Code he must retain the original name for one of them—but which? With the description alone to go by, it may be inadequate to decide just what the original author had in mind, but with the type specimen before him the reviser can retain the name *M. diabolica* Nick for that and give new names—*M. nefanda*, *M. satanica*, and so on—to the segregates.

The type of a genus is one species and the type of a family is one genus. Thus, if *M. diabolica* is the type of the genus *Mephistophelia* and a subsequent author wishes to break it down into three genera, the name *Mephistophelia* remains permanently associated with its type species and the "splits" become, say, *Lucifera* and *Beelzebubia*, each with its own distinctive type species.

If a type specimen has been lost, or none was ever designated, it is permissible for a later botanist to choose a new one, which is called a *lectotype*. Locating type specimens in the various herbaria is made easier by consultation of Stafleu and Cowan's *Taxonomic Literature* (1967). Most of Linnaeus' original specimens, the foundation of our present body of plant names, are housed in the headquarters of the Linnean Society in London.

Publication and Priority
The application of plant names depends upon priority of publication: when two or more compete for the same taxon, it is the oldest that must be adopted. But a name must be validly published to qualify; mere mention at a meeting or in correspondence is not enough. Firstly, the good news must be made public (effective publication). A description must appear in a printed book or journal of botanical standing and available to the public; since 1935 nonscientific newspapers or tradesman's catalogues are ruled out, although mimeographed periodicals are permissible. Secondly, the description must comply with the Code (valid publication). This means that after 1935 there must be a diagnosis in Latin for a new species (sp. nov.), and after 1958 a citation of a type specimen; illustrations are optional, but an indication of how the novelty differs from its allies is recommended.

Alternatively, where a new name is coined, as when a species is transferred from one genus to another (comb. nov.), there must, after 1959, be a full reference to the old name and its source and date of publication. Here is an example, as published, of a name change that resulted from the author's belief that the genus *Matucana* (family Cactaceae)

is distinct from the genus *Borzicactus*:
Matucana calvescens (Kimnach et Hutchison) F. Buxbaum comb. nov. = *Borzicactus calvescens* Kimnach et Hutchison in Univ. Calif. Bot. Garden Contributions No. 147 and Cact. Succ. Journ. America XXXIX, 1957.
This leads us to the Journal in which the earlier name was published, its authors and date, but unfortunately omits the page number, which is required under Article 33. Hence the name *Matucana calvescens* was not validly published, and only became so in 1975 when the combination was republished with the missing page number; prior to 1975 it was referred to as an illegitimate name (a legitimate name being one that is valid and in full accordance with the Code). A correct name is the one that must be adopted for a taxon in any given system of classification. Thus either of the above names can be correct, according to whether or not you follow Buxbaum's or Kimnach and Hutchison's classification; but they cannot both be correct at the same time.

In matters of priority we take the starting point as 1753, the year of Linnaeus' *Species Plantarum*. Certain special groups like fungi and fossils have their own different dates. A great aid to the botanist in searching for original publications of names of species and genera is the *Index Kewensis*, which lists them all and is kept up-to-date with five-yearly supplements. A companion work, the *Index Londinensis*, lists illustrations up to 1935.

Priority has been a mixed blessing in nomenclature. It gives a clear-cut ruling for deciding which of several competing names must be adopted, but often this proves to be the least known and most hastily described.

Choice of Names
Plant names can come from any source. Specific epithets are usually descriptive (*albiflorus* for white flowers) or geographical (*hispanicus* for a native of Spain) or commemorative of persons (*bolusii* after Harry Bolus, or *bolusiae* after Mrs H. M. L. Bolus). Those taken from nouns generally add -*ii*, although different adjectival formations are used where appropriate. As in the example above, the ending -*iae* is added to the names of ladies. Species epithets, where adjectival, agree in gender with the generic name (the same is true of cultivar names of latinized form). Thus, *Azalea indica*, transferred to *Rhododendron*, becomes *Rhododendron indicum*, since that genus name is neuter. It has never been possible to reject names because they are long, difficult to pronounce or badly chosen. Thus *Austroboreoheterocephalocereinae*, a subtribe of the Cactaceae, cannot be overthrown just because it is too long to write on any normal plant label or because of the north/south contradiction at the start. It can, however, for other reasons, since it was invalidly published at the start. Misspellings of personal names must also be retained if the author used them consistently and specified no correction. But errors of typography and orthography, like "*smithi*" for "*smithii*" and

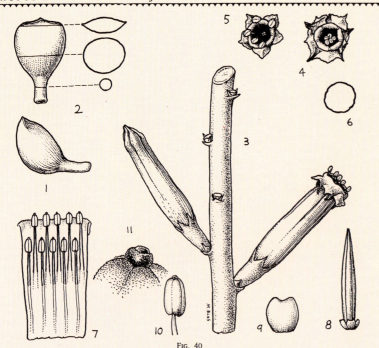

FIG. 40

Adromischus liebenbergii P. C. Hutchison, U.C.B.G. 56.076-2. 1. Leaf, side view. 2. Leaf, top view and cross-sections. 3. Flower and bud. 4. Perianth-limb, expanded. 5. Perianth-limb, reflexed. 6. Cross-section of perianth-tube at midpoint. 7. Stamen insertion. 8. Carpels. 9. Nectary-scale. 10. Anther. 11. Anther apex. 1-2, natural size. 3-8, x 3. 9-10, x 9. 11, greatly enlarged. Drawing by Mrs. M. Blos, 1958.

Jcones Plantarum Succulentarum

(a) 14. *Adromischus liebenbergii* P. C. Hutchison[1]

By P. C. HUTCHISON

(b) In 1956 I received from L. C. C. Liebenberg three plants of the new species described below, a distinctive and attractive dwarf succulent of easy culture. Two of the plants flowered for the first time in 1958. The species promises to be a worthwhile addition to collections of succulent plants.

(c) *Adromischus liebenbergii* P. C. Hutchison, sp. nov. *Planta usque ad 6 cm. alta; folia cuneata petiolata, petiolo 1 cm. longo, pallide brunnescenti-viridia ceracea, lamina 2 cm. longa lataque, 1.4 cm. crassa, obtuso apice ipso acuto, margine brunneo-flavescente; lobi corollae cuspidati connati incarnati; antherae globulis duplicibus terminalibus praeditae.*

Roots fibrous. *Stem* erect, up to 6 cm. tall. *Leaves* alternate, erect to spreading, cuneate with a petiole up to 1 cm. long and 4 mm. in diameter, unspotted, pale tannish green with a coating of wax, the blade up to 2 cm. long and wide and 1.4 cm. thick, the apex blunt with a dull point and a tan margin, the cross-sections at all points symmetrical. Inflorescence simple; peduncle tannish green, glaucous below, up to 30 cm. long, 2 to 3 mm. in diameter for most of its length; rachis 10 to 15 cm. long; flowers (d) 15 to 25, solitary, the lower about 1 cm. apart, suberect. *Pedicels* 2 to 3 mm. long, greenish. *Perianth*-tube 13 to 14 mm. long, about 3 mm. in diameter, barely constricted toward the apex, the cross-section at mid-point rounded, the sinuses indented to mid-tube and green, the tube otherwise tan, the limb-lobes united, cuspidate, *ca.* 3 mm. wide, 2 mm. long, pale whitish pink, the margins pleated and whitish, the central portion darker pink and reflexed, epapillose as is the greenish tube interior. *Stamens* biseriate, the two series inserted about 2 mm.

apart in mid-tube, the filaments 4 to 6 mm. long, the upper series the longer and broader below; 5 anthers barely exserted, whitish, the (d) terminal globule translucent, double, unstalked or shortly-stalked, the lower globule roughened, the upper smaller, smooth. *Carpels* 11 to 13 mm. long. *Nectary-scales* 1+ mm. long and wide, emarginate, white, the sides rounded.

(e) South Africa, Laingsburg Div., from the farm Varsbokkraal, near Laingsburg, growing with *Crassula rupestris* on the rocky, stony side of Witteberg Mountain, *leg.* L. C. C. Liebenberg 6186, *University of California Botanical* (f) *Garden 56.076-2* (BOL-Holotype).

At the Bolus Herbarium there is an excellent watercolor drawing of this species, without locality data, but labeled "Cook Whitehill 1737/22, 21 Feb. 1930." There is no specimen of this collection in the herbarium material I have examined from that institution.

Bryan Makin, of Alperton, England, informs me that the species is cultivated in England (g) under the epithet "turgidus." No such name has appeared in the literature on this genus.

(h) *Adromischus liebenbergii* is another new species[2] of the assemblage related to *A. hemisphaericus* (L.) Lem. and the first species of this affinity with a distinctly petioled, cuneate leaf. Similar, but not identical leaf-shapes are found in some specimens of *A. cooperi* (Bak. f.) Berg., and in undescribed material related to *A. maculatus* (Salm-Dyck) Lem.; however, all of these have larger leaves which are spotted and entirely different flowers with the perianth 'segments free. The double globule on the apex of the anthers in *A. liebenbergii* seems to be unique in the genus. On living plants of the clonotype this globule was unstalked; on a second plant it was shortly stalked.

[1]University of California Botanical Garden (Berkeley) Contribution Number 154.

[2]Cf. *Adromischus bicolor* Hutchis., Cact. & Succ. Jour. Amer. 29: 26-28, 1957.

"*alboflorus*" for "*albiflorus*", can be corrected.

Name Changes

Plantsmen are justifiably aggrieved when one of their favorites suffers a change of name, and use this as a rod to beat the "armchair botanists" who are accused of having nothing better to do. Unfortunately, the rule of "one species, one name" is bound to bring about some changes, and chaos like that of the pre-Linnaean days would result if our Codes were not strictly observed. The same plant may receive two or more names from botanists in different parts of the world who are ignorant of each other's work. Obviously only one name can stand (the earliest) and we refer to the others as synonyms. Conversely, two different species may accidentally receive the same name; these we call homonyms. The later must be renamed, even if the earlier is much less well-known. Cases like this are rare, but one recently led to one of the most widespread of all flowering plants, the common reed, long known as *Phragmites communis*, having to be rechristened *P. australis*.

Efforts have long been made to find ways of circumventing the rules to avoid name changes. Familiar names of genera and families can be protected by inclusion in lists of Conserved Names (Nomina Conservanda), whereby long usage can be preserved. Without these lists, appended to the Botanical Code, hundreds of well-known names would fall.

The above changes result from observance of the Codes; they are purely nomenclatural. It is wisest to accept these gracefully since, in time, the new names will take their place in standard reference works. But another type of name change results from quite different circumstances and remains a matter of opinion. This is where taxonomists disagree over the boundaries of their units. Thus, what one botanist calls the large genus *Anemone*, another splits up into *Anemone*, *Hepatica* and *Pulsatilla*. Both will cite an array of facts to support his case, and since nobody can define a genus unequivocally and plants are notoriously reluctant to read our textbooks, we must respect both views. Thus *Anemone alpina* L. in the book of one author turns out to be the same species as *Pulsatilla alpina* (L.) Delarbre in that of another. Such double author citations are an indication of a transfer of this sort or a change of status.

Must the change be accepted? There is a real quandary here for the gardener, because it is apparent from the above that unless the reason for a name change is known, there

First description of a new species, reproduced from the *Cactus and Succulent Journal of America*, vol. 31, no. 3 (1959). The following elements are frequently found in such descriptions, but only (a), (c) and (f) are obligatory under The International Code of Botanical Nomenclature. (a) Name; (b) source; (c) Latin diagnosis; (d) full description in a modern language; (e) distribution; (f) citation of type; (g) synonymy; (h) affinities and distinctions from other species. The illustration is also a voluntary element.

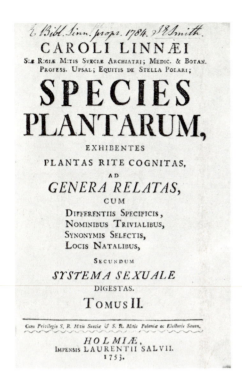

The title-page of Linnaeus' *Species Plantarum* (1753), a flora of all the species then known, which is recognized as the starting point for the modern system of binomial nomenclature.

is no means of telling whether it is nomenclatural or taxonomic. If it is for sound nomenclatural reasons it is far more likely to be permanent than a change arising from taxonomic study. The latter poses the question: do I favor Professor Lump's classification, or that of Dr Splitz? To answer that might require much delving in libraries, unless there is an authoritative monograph or live specialist on hand to consult.

But take heart! Name changes are not nearly as common as the few much-publicized ones suggest, and in any case plenty of name changes occur in everyday life without anyone turning a hair. The day that Miss Brown becomes Mrs Green will surely be celebrated with cork-popping jubilation, and who would dare suggest that she remain "Miss Brown"?

G.D.R.

Floras

The principles of taxonomy and nomenclature most commonly find expression in the production of printed floras. The Goddess of Flowers has lent her name to botany in two different senses: a flora can be defined as (a) the plants that grow spontaneously in a given region or time period (eg the Cretaceous flora) or (b) a book in which the plants of a given area or period are described in systematic order. In this section, however, we are concerned only with floras in the latter sense.

An exact definition is scarcely possible, although at its best a flora has a recognizable systematic layout, terseness of style and a functional format that makes it an important component of botanical literature. The classic definition was given by the 19th-century English botanist George Bentham when he wrote: "The principle object of a Flora of a country is to afford the means of determining [ie ascertaining] the name of any plant growing in it, whether for the purpose of ulterior study or of intellectual exercise". It therefore consists of descriptions of all the wild or native plants found in the country in question, so drawn up and arranged that the user may identify any specimen by checking with the corresponding descriptions in the text.

A more recent definition is one supplied by Stanwyn Shetler, writing in connection with a proposed flora of North America: he regards it as "a time-honored information retrieval system ... a physical respository of data about plants, which are organized and formatted so as to answer a time-tested set of prescribed questions. It is as old as taxonomy itself and equally indispensable".

It is generally agreed, therefore, that the prime purpose of a flora is to document the plants of a chosen area and to enable them to be identified as quickly and easily as possible. But beyond this, a great deal of accessory information may be included, according to the choice of the author and the extent to which the plants have been studied in the chosen area. Thus some floras include data on ecology, soil preference, local and wider distribution (with or without maps), flowering season, folk uses and names, chromosome numbers and so on. Variation and natural hybrids may also be recorded.

G. H. M. Lawrence draws a distinction between a flora, which catalogs the plants of a given area with location (and sometimes also citations of actual herbarium specimens) and a "manual", which is primarily concerned with the practical problem of identification, for which some sort of key is essential. In practice, however, this distinction would be difficult to maintain. Some floras contain keys but no descriptions and some contain descriptions but no keys; some have neither and some combine the two in one. There are many intermediates between a simple list of species names and the elaborate and sophisticated floras referred to above. Some floras rely partly or wholly on pictures for identification. Typical are the tourist floras aimed at the non-botanical traveler anxious to pin names on the wild flowers he encounters on his rambles. Examples of this type are Polunin's *Flowers of Europe* and Huxley's *Mountain Flowers*. Although often scorned by the botanical purist these serve their purpose well and are referred to by professionals more often than they would admit.

A good modern flora has something in common with a monograph—the rigid formalized layout and telegram-style descriptions with standardized abbreviations, the keys or synopses, the documentation by references to books and specimens, and the paring down of information so as to avoid overlap and repetition. A flora differs from a monograph, however, in being concerned only with the plants of a specified area, omitting species and characters not present within that area. Thus descriptions of families and genera taken from a flora may not cover all the known species. Further, while a monograph is presumed to be based upon original research using actual plants, with a checking of type-specimens and a sifting of all past literature, a flora is more or less derivative, making full use of existing monographs and rarely venturing on changes in classification. Indeed, the bulky paraphernalia of Latin diagnoses and new names is out of place in a book designed for practical purposes and styled to fit the pocket where possible. A flora is thus synthetic rather than analytic; as a storehouse of information it is less consistently helpful than a monograph because of the abridgements demanded by compactness and the selectiveness forced on it by the boundaries of the area being covered. These boundaries are usually politically defined—a necessary but unhappy situation, since plants are notoriously unconcerned with human politics and define their own distributional ranges. A flora is the better if it is devoted to a natural botanical region, such as Malaysia.

Types of Flora

Floras are written for a wide range of different people: other botanists, foresters, agronomists, horticulturists, ecologists, gardeners, amateurs, students etc. They may be designed for use in the field or in herbaria. Looking at an array of floras in a library, there are few generalizations that can be made about them apart from their common aim to list and describe the plants of an area. They differ in size, shape, weight, illustrations, language and so on. Some are produced in limited edition folio size, others in editions of one or two thousand; popular floras with colored plates may sell by the tens of thousands.

Floras may be local, national, regional or continental. Local floras cover the plants of a county, province or district of a country and may be derived from a national flora or the national flora may use it and others as a basis. County floras have long been a tradition of the British Isles and are often compiled by amateurs. Local floras do not usually contain descriptions or keys but are largely listings of the species concerned and their geography and ecology. National floras may be abbreviated or extensive, sometimes running to several volumes.

While a local flora can be condensed into a relatively inexpensive volume to fit the pocket, those for whole countries are usually too bulky. Ideally there should be two versions of the same work: one for the fully-fledged herbarium-orientated specialist, the other for use in the field. For Great Britain, for example, there is Clapham, Tutin and Warburg's 1,269-page single-volume *Flora of the British Isles*, which has been conveniently condensed to give an *Excursion Flora* by the omission of the rarest and alien species and the shortening of descriptions.

Floras of less well-known areas such as the tropics are research or general floras and contain full descriptions, citation of specimens and critical notes. They are often published in family parts. Regional floras cover several countries or parts of them, eg van Steenis' *Flora Malesiana* and Hegi's *Illustrierte Flora von Mitteleuropa*. The latter is a gigantic compendium, now in its second edition, comprising detailed descriptions, keys, information on chemistry, cell structure (cytology), pollen analysis (palynology), variation and cultivated plants. Floras of cultivated plants are not unknown, and popular floras, aimed mainly at the tourist, contain a selection of the most common or conspicuous plants of a country or region.

History and Development of the Flora

A tool so refined as the modern flora did not develop overnight. Its history is one of a slow and gradual development rather than dramatic advances, although a few landmarks, like the introduction of binomial nomenclature, mark the course. The story can be said to begin at the very start of botanical inquiry, for the earliest observers were of necessity studying the plants of a fairly small area. Thus the monumental *Flora Europaea*, started in 1964, completed in 1980 and covering an estimated 13,000 species, can trace its descent from Theophrastus (c. 372–c. 287 BC), who described the plants of his native Greece and adjoining countries in southern Europe, North Africa and part of the Near East. He dealt with some 500 species from an area roughly equal to that covered by *Flora Europaea*. The well-known and justly popular picture flora of Britain by W. Keble Martin likewise has its prototype on the walls of the Festal Temple of Karnak in Egypt, where the flora of Syria is immortalized in a series of bas-reliefs dating from the 14th century BC.

Floristic studies leading to the production of floras began in Europe in the 16th century, when modern botany grew out of herbalism into an independent field of study. Previously, botanical studies had been closely associated with medicine through *materia medica* and herbals; the number of species recorded or used was very limited to begin with and handed down from generation to generation. The need, therefore, at the beginning of the 16th century, was for botanists to describe the plants of their own native country. Progress was slow in those European countries, such as Great Britain, which have poor floras: only about 500 species of British plants were known and described by the year 1600; the figure had doubled by 1700 after a great deal of effort by botanists, medical men and apothecaries.

A double-page spread from Sowerby's *English Botany* (1790–1814), the most celebrated of British floras, which attempted to illustrate every British plant. The work contains over 2,500 hand-colored engraved plates.

The total reached no more than 1,145 during the course of the following 100 years.

The first flora of the British Isles was *Phytologia Britannica*, produced in 1650 by William How, a physician of London. The first British flora using Linnean binomials was by Linnaeus himself, *Flora Anglica*, which appeared in 1754; this was merely a list of names with a brief preamble. Hudson's *Flora Anglica*, a more substantial book, followed in 1762. The first serious flora to be written in North America was T. Walter's *Flora Caroliniana* of 1788.

In Europe, many countries remained largely unknown floristically until the beginning of the 19th century, notable examples being Spain and Greece, although the latter can be regarded as the cradle of scientific botany. Even today the flora of these and other Mediterranean countries remains imperfectly studied.

The production of floras has continued at an increasing if uneven rate. The 19th century was a period of intense activity, with mighty works like Martius' *Flora Brasiliensis*, Boissier's *Flora Orientalis* and the British colonial floras, based on Kew, in preparation. In the early part of the century, the floristic exploration of Europe intensified and the middle of the century heralded the first great age of floras. This was not restricted to Europe but was worldwide, although largely as a result of European botanists exploring their then dependent colonial or imperial territories (see Plant Exploration, p. 12). A large proportion of the effort of European taxonomic botanists has been devoted during the past century and a half to writing floras of extra-European territories. For

example, British botanists, largely working from Kew, concentrated their efforts on flora of the British Empire. Examples were W. J. Hooker's *Flora Boreali-Americana* (1829–40), George Bentham's *Flora Hongkongensis* (1861) and *Flora Australiensis* (1863–78).

Floras in the 20th Century

There are few parts of the world where European taxonomic botanists are not currently engaged in writing floras. As a consequence of the intensive floristic activity over the past 100 years or more, Europe now boasts about a third of the world's herbaria and about half of the collections in terms of numbers of specimens (about 80,000,000). Europe is still the most intensively studied continent and other areas are more or less neglected. Tropical countries are, on average, twice as rich in species as temperate countries, area for area, yet are much less studied botanically. This is especially true of tree species, which in tropical forests pose problems for the collector. Sidney Blake and Alice Atwood, in the first attempt to catalog all the floras of any consequence issued up to 1939, listed 3,025 titles for America, Africa and Australasia, and in 1961 added 6,841 titles for western Europe. This works out as a ratio of about 60:1 in favor of western Europe, based on land areas involved; since then the ratio has increased.

To look through a list of titles of floras reveals one surprising and disconcerting feature: the large number that remain incomplete. This is not only a failing of older compilations; many floras started in the

Opposite A plate from H. R. Descole, *Genera et Species Argentinarum* (1944), a 20th-century flora in the sumptuous style of the earlier period.

present century have similarly fallen by the wayside and will never be finished. While a flora of an English county, say, or of a small island can reasonably be undertaken by a single botanist and completed in a few years, those of larger areas require teamwork and even then the task is often under-estimated and far too broad a basis adopted on which to work. The scale on which a floristic survey is undertaken should be related to practical considerations; this means a preliminary estimate of the number of species involved and the time span required to see the work through to completion. This will determine the amount of detail that can be included. For example, of the admirable flora of South Africa, commenced in 1965, so far only volumes 1, 13 and 26 have been published. G.P. De Wolf calculates that the compiling of

674. Gemeine Haselnuß – *Córylus avellána* L. (♄ – 2/5, ähnlich 675. Lamberts-H., Große H. – *C. máxima* MILL.) — 676. Moor-Birke – *Bétula pubéscens* EHRH. (♄ – 1/3) — 677. Hänge-B. – *B. péndula* ROTH (♄ – 1/3)

Above Many modern illustrations of plants, such as these of hazel (*Corylus*) and birch (*Betula*) from *Atlas der Gefasspflanzen* (1959), are simple line drawings which draw attention to the distinguishing features of the species.

Schubertia grandiflora

a flora in a fairly economical format, like that of Clapham, Tutin and Warburg referred to earlier, can proceed at the rate of about 250 species per year per taxonomist. For a more expansive treatment, as in van Steenis' over-ambitious *Flora Malesiana*, the production rate falls to about 100 to 50 species per year per taxonomist.

An outstanding example of teamwork is seen in *Flora Europaea*, a five-volume concise account of all the flowering plants and ferns growing wild or naturalized in Europe. Volume 1 was published in 1964 and the fifth and last volume in 1980. The background to *Flora Europaea* was the enormous amount of specimens and published material available. Herbarium specimens of European plants ran into millions, although the total number of species is little more than 13,000. In terms of literature, there are thousands of floras of European plants—local, national and regional—published over the past 200 years. And the number of papers on the floristics and taxonomy of European plants published in journals runs into many thousands, written in more than a score of languages. Naturally enough, the same species are often treated differently in these different works owing to the adoption of different national or local viewpoints. As a result there was a need for an overall synthesis of this vast corpus of information. Relying on a network of advisers representing each European country or state, and tight editorial control, authors' manuscripts were extensively reviewed and revised to achieve completeness and accuracy, and effectiveness of keys, synonymy, geographical distribution and style. The role of the author was in many ways subservient to the system, but it is difficult to know, with the data-base available, whether any other method would have been possible. No individual or group of two or three could have written such a flora of Europe or had access to a substantial

216

lateral honey-leaves *c*. 1¼ times as long as wide, ovate or sub-quadrate; follicles glabrous. *Greece*. Gr.

Sect. *Staphisagria* DC. Annual or biennial. Honey-leaves glabrous, the same colour as the perianth-segments, the upper paler or yellowish, shortly clawed, unwinged. Seeds few.

24. D. staphisagria L., *Sp. Pl.* 531 (1753). Stems 30–100 cm, stout, simple, patent-pilose. Leaves palmately 5- to 7-fid or -lobed, pubescent on both surfaces with mixed very short and longer hairs; segments entire or 3-lobed, with ovate-lanceolate, or oblong, acute lobes. Flowers deep blue. Perianth-segments 13–20 mm. Limb of lateral honey-leaves gradually narrowed into claw. Follicles 8–11 mm wide, inflated. *Mediterranean region*. Al Bl Co Cr Ga Gr Hs It Ju Lu Sa Si.

25. D. pictum Willd., *Enum. Pl. Hort. Berol.* 574 (1809). Stems pubescent, sometimes with mixed long and short hairs. Rhachis, pedicels and outside of perianth-segments shortly pubescent. Inflorescence usually dense; pedicels shorter than flowers (except sometimes the lower). Perianth-segments 9–14 mm, pale blue; spur 6–8 mm. Limb of lateral honey-leaves suborbicular, abruptly contracted into claw. Follicles up to 6 mm wide, inflated. *W. Mediterranean islands, Italy*. Bl Co It Sa.

26. D. requienii DC. in Lam. & DC., *Fl. Fr.* ed. 3, 5: 642 (1815). Like 24 but rhachis, pedicels and outside of perianth-segments villous-hirsute, inflorescence usually lax; lower and middle pedicels distinctly longer than flowers; bracteoles inserted distinctly above base of pedicels; limb of lateral honey-leaves obovate, more or less gradually narrowed into claw. *S. France (Isles d'Hyères), Corse, ?Sardegna*. Co Ga ?Sa.

13. Consolida (DC.) S. F. Gray[1]

Like *Delphinium* but always annual; the 2 upper honey-leaves coalescent into a single structure (nectary), with a single spur; lateral honey-leaves absent; stamens in 5 spirally arranged series; carpel 1.

Literature: E. Huth, *Bot. Jahrb.* 20: 365–91 (1895). R. de Soó, *Österr. Bot. Zeitschr.* 71: 233–46 (1922).

1 Nectary ovate, entire; follicle appressed-silky   **3. hellespontica**
1 Nectary lobed; follicle glabrous or pubescent but not silky
　2 Spur circinate-involute at apex
　3 Intermediate lobes of nectary acute; follicle 15–25 mm
　　　　　　　　　　　　　　　　　　　　　　　1. aconiti
　3 Intermediate lobes of nectary obtuse; follicle *c*. 7 mm
　　　　　　　　　　　　　　　　　　　　　　　2. thirkeana
　2 Spur straight or scarcely curved
　4 Follicle glabrous
　　5 Perianth-segments 9 mm or more   **9. regalis**
　　5 Perianth-segments less than 9 mm   **10. tenuissima**
　4 Follicle pubescent
　　6 Spur less than 7 mm
　　7 Lower pedicels not more than 5 mm   **8. tuntasiana**
　　7 Lower pedicels more than 7 mm   **6. brevicornis**
　　6 Spur 7 mm or more
　　8 Lower bracts entire
　　9 Inflorescence racemose, not or little-branched
　　　　　　　　　　　　　　　　　　　　　　7. uechtritziana
　　9 Inflorescence paniculate, much-branched
　　10 Upper lobe of nectary more than 2 mm wide; plant
　　　　sparsely pubescent; flowers dark blue; seeds black
　　　　　　　　　　　　　　　　　　　　　　9. regalis

---

1 By A. O. Chater.

Above While many modern descriptions of plants continue the Linnaean tradition of brevity, they generally aim to deal with all major features of the species, and to do so equally for all related species; furthermore, a key for identifying the species by effectively asking short questions with mutually exclusive answers is an essential part of any taxonomic account. This excerpt from *Flora Europaea* (1964–1978) shows half a page with descriptions of *Delphinium* species and part of the key to the related genus *Consolida*.

amount of the literature and material available.

A very different approach is needed in writing what is often termed a research or critical general flora, such as *Flora Malesiana* and other tropical floras. There, the documentation and materials available are limited and have to be verified personally by the author concerned and listed in the text, with the specimens examined cited so that the data-base is known. In tropical countries, one of the limiting factors is the amount of herbarium material available, and a major effort is usually required to mount field expeditions to explore poorly known or totally unexplored territory to collect specimens for study and comparison. The number of botanists engaged in this inventory work

in what are, after all, natural resources, is regrettably small and many of the tropical floras in preparation are progressing so slowly that completion cannot be envisaged this century. In many parts of the tropical world, there is still no national flora in preparation and very few local floras, as in several South American countries. The flora of Venezuela is probably twice as large as that of the whole of Europe and that of Brazil three or four times as large. Many thousands of plant species are at risk in such areas and may be destroyed through fire, clear-felling of forests, urban and industrial development and other pressures of civilization before they are discovered and described.

Writing a flora is only the first stage of resources studies, yet on a world basis it is still desperately incomplete and time is rapidly running out. The rational exploitation of resources, conservation programs and land usage depend largely on a knowledge of the species comprising the vegetations concerned. The need for floras is therefore self-evident yet at the same time greatly underestimated.

V.H.H.

Genecology

We have seen that the basic unit of classification is the species, but it is also apparent that species are not immutable and are continually evolving in a response to the pressure of natural selection. It has long been known that the vigor and form of a plant can be profoundly modified by the environment—seeds falling on poor, stony soil give rise to much smaller plants than those germinated in rich, well-fertilized garden loam, while plants have larger, thinner leaves when grown in shade than when exposed to full sunlight. Experiments carried out by the British Ecological Society in southern England during the 1930s showed that species differ in their ability to respond to different environments. Some, such as *Centaurea nigra*, grow reasonably well on a variety of soil types, while others, such as

Plantago major, are very "plastic" and exhibit marked morphological differences in response to the physical and chemical properties of the soil in which they grow.

For a long time it has also been known that certain plants are inherently adapted to certain habitats owing, as Darwin pointed out, to the environment favoring certain traits; in other words, natural selection. With the development of genetics early in this century, there was increased awareness that the *phenotype*, the plant we see, results from

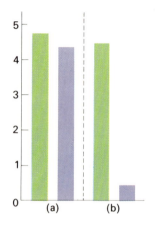

Above Comparison of growth between serpentine (a) and non-serpentine (b) races of *Gilia capitata*. When grown on non-serpentine soil (green) the yield of both serpentine and non-serpentine races is high; however, when grown on serpentine soil (purple), the yield of the non-serpentine races is much reduced. Thus, whereas the serpentine race is adapted to growth on serpentine soils, but still thrives on non-serpentine soils, the non-serpentine race has no such adaptation to serpentine soils. (Serpentine soils are mineral-rich, containing particularly large amounts of magnesium.)

Below Effects of environment on the growth of a single clone of *Potentilla glandulosa* subspecies *typica*; ie all plants were vegetatively derived from a single plant. The plants were grown under artificial conditions equivalent to: (a) dry and sunny; (b) moist and sunny; (c) dry and shady; (d) moist and shady. Although genetically identical (ie having an identical genotype) each plant develops different morphological forms (phenotypes) in response to the environmental conditions.

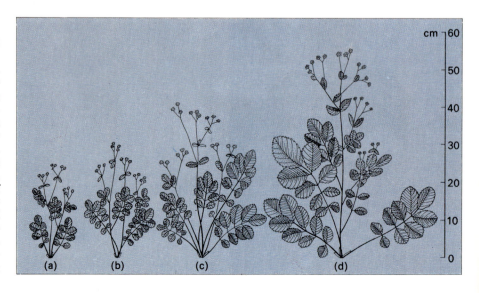

Transplant gardens, such as those run by the Carnegie Institution in California at Stanford, near sea-level, Mather (**right**), at 1,400m, and Timberline (**below right**), at 3,000m, in the Sierra Nevada, have been important tools in testing, in a range of environmental conditions, the genetic differentiation of local races in such species as *Achillea borealis*.

the interaction of the *genotype* (the plant's genetic make-up) and the environment. The word genecology was introduced in 1923 by the Swedish botanist Göte Turesson to denote the study of variation within plant species in relation to the environment. The effect of the environment, already noted, can be studied by growing plants of the same genetic constitution under a variety of conditions; such plants are usually obtained by vegetative reproduction, to give a clone of genetically identical individuals. The genotypic component of the variation between the plants is assessed by growing them under uniform environmental conditions, so that any phenotypic differences must be the result of genetical differences between them.

In his pioneer genecological studies, Turesson collected living plants and seeds of many common species from their natural habitats and grew them together in experimental gardens in southern Sweden. In many instances he found that the distinctive phenotypes occurring in the different habitats were maintained under the more or less uniform garden environment. Typical of his results are those on the hawkweed, *Hieracium umbellatum*, which, in southern Sweden, has three principal habitats: woodlands, sandy fields and dunes. Woodland plants are glabrous (lacking hairs), erect, with broad leaves; those from sandy fields are hairy and prostrate, with leaves of medium width, while plants on the dunes are glabrous, erect, with narrow leaves and, unlike the others, regenerate by basal shoots produced in the autumn. All these features were maintained in cultivation and, furthermore, different populations from the same sort of habitat invariably exhibited the characteristic morphology. Turesson called these local races *ecotypes*, and subsequent work on many species has shown that such habitat-correlated variation is widespread, supporting the view that it results from the selection of genotypes adapted to each

habitat. Frequently, a certain habitat favors selection of particular features in unrelated species and we find, for example, that alpine ecotypes of many species have fewer leaves, larger flowers and a greater propensity for vegetative propagation than their lowland ecotypes.

The features selected are not necessarily obvious from the gross morphology. Thus maritime ecotypes of bent grass (*Agrostis stolonifera*) are less susceptible to damage by salt spray than inland ecotypes because waxes on their leaf-surfaces give them a lower wettability. Not surprisingly, ecotypes have been formed in response to Man's manipulation of plants. For example, cocksfoot grass (*Dactylis glomerata*) has different ecotypes suited to hay meadows and permanent pastures.

Some of the most extensive genecological studies have been carried out in California, United States, by the Carnegie Institute of Washington. A series of experimental gardens, along a transect from the mild oceanic conditions on the coast at Stanford to the

Left Effects of altitude on the form of *Achillea borealis* from Kiska Island (Aleutian Islands). These altitudinal ecotypes were grown under uniform garden conditions and the morphological form of each did not vary from that in its original habitat, thus indicating that the variation was due to the genetic makeup (genotype) not just an adaptation to the environment.

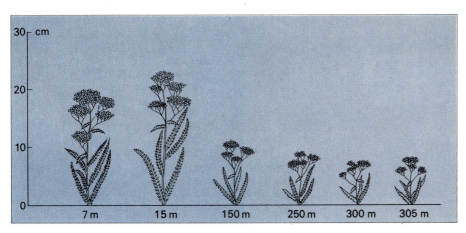

extreme alpine climate of the Timberline station, at over 3,000 m (10,000 ft) in the Sierra Nevada, was used to test ecotypic differentiation in a large number of species. Virtually all species studied showed morphological and physiological adaptations to the range of conditions along this climatic transect. This work demonstrated that apparently identical populations in a particular habitat could result from either genetically determined characters (*genecotypes*) or the environmentally induced response of "plastic" genotypes (*phenecotypes*). This parallelism between genecotypes and phenocotypes (also called *ecads*) was also noted within the dune ecotypes of *Hieracium umbellatum*. Furthermore, by cultivating for several generations hybrids between ecotypes in the gardens along the Californian transect, the genetical basis of the differences between the ecotypes was clearly established and shown in several cases to result from the interaction of many genes. More refined studies of these ecotypes and their progeny became possible with the increasingly controlled environments made available by using growth cabinets and their more sophisticated elder brothers the phytotrons, whose development was pioneered in California.

Detailed studies of the sea plantain (*Plantago maritima*) on the coasts of Scotland revealed genecological differentiation even within restricted areas. In salt marshes, for example, *P. maritima* grows under various conditions, from waterlogged mud near the sea to stable grassland on the landward side. These effectively constitute an environmental gradient and the plantains in many salt marshes studied show an associated gradient in several characters with, for example, prostrate, short-leaved, short-seeded plants near the sea and progressively more erect plants with longer leaves and seeds as the coastal meadow is approached. The term *cline* was introduced in 1938 for any gradient in measurable characters, and variation such as that shown by sea plantains, in response to ecological factors, is said to be ecoclinal. *Topoclines*, correlated with geographical gradients, are also widely known; there is, for example, a gradual decrease in the width of leaf lobes in the bloody cranesbill (*Geranium sanguineum*), from west to east in Europe, a gradual increase from northeast to southwest Europe in the frequency of plants of white clover (*Trifolium repens*) which contain cyanogenic glucosides, and a gradual increase from Florida to New York in the amount of chilling required to break dormancy in the red maple (*Acer rubrum*).

Clines and ecotypes both clearly result from the selective effects of the habitat, and nowadays it is usual to refer to them under the general term of ecotypic differentiation. Whether this results in discontinuous or continuous variation in some or all characters can depend upon a number of factors, including the intensity of sampling, the uniformity of the habitat, the breeding system and the degree of gene exchange between plants in a given area. Genecology, in summary, has yielded and is yielding information on how plants are adapted to their habitats, by the selection of genotypes which either are suited to a particular set of conditions or, where "plasticity" is involved, are able to operate over a range of environments. It is thus important in understanding one of the sources of intraspecific variation which has to be handled and interpreted by taxonomists. Probably most importantly, however, genecology is fundamental to any appreciation of how particular plants are able to flourish in particular environments and communities—the concern of ecology—and what attributes are of value in adapting to new conditions as they extend their distribution (see Colonization and Establishment, p. 145), an aspect of equal interest to phytogeographers.

D.M.M.

Diagram showing how maps and diagrams are used to demonstrate different aspects of plant distribution. Shown here are maps and diagrams of the distribution of the sea buckthorn (*Hippophae rhamnoides*). (a) Distribution in Europe and Asia showing the sea buckthorn to inhabit coastal regions of western Europe, but also to be a steppe plant in the east. (b) In the British Isles, it occurs mainly on the coasts, where it is native to the east coast (open circles) elsewhere. (c) and (d) The typical habitat of western European sea buckthorn is on the landward side of young, unstable sand dunes where competition from other species is not too severe and trampling is not intense. (e) Fruiting shoots of the sea buckthorn.

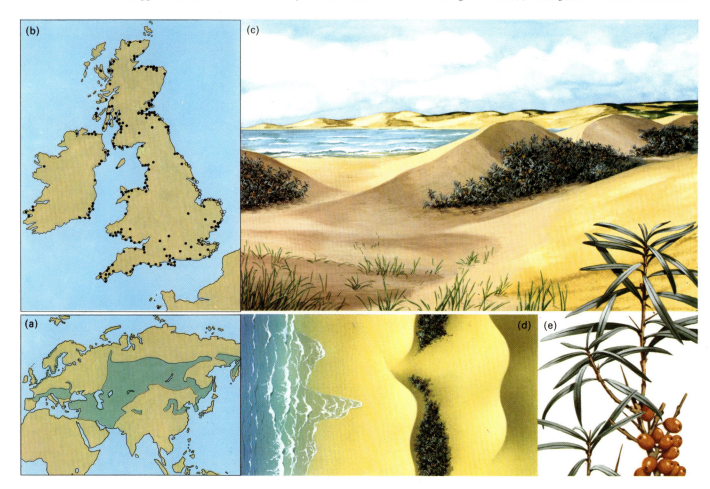

(b)

(c)

(a)

(d) (e)

Distribution of Plant Species

Chorology

The study of the geographical distributions of animals and plants is termed chorology. It is a study that has occupied biologists since the 18th century, when taxonomy had developed to such an extent that organisms derived from different areas could be identified accurately. The extensive travels of 19th-century naturalists provided the basis for the scientific study of distributions and it was soon realized that such a study could provide information not only about the physical requirements of different species, but also about the possible evolutionary relationships of species.

Both casual and systematic collecting of plant material from various parts of the world has provided information for the construction of world-scale distribution maps of plant species, but lack of information from remote and inaccessible areas means that the picture produced is still often incomplete. Surveying under such conditions is extremely difficult, both from a geographical and a taxonomic point of view. The use of aerial photography has proved extremely valuable as a tool for the mapping of habitats and, when coupled with selective ground surveys, it can provide a means of mapping plant communities. It may be possible from such habitat maps to infer the distribution of individual species which are typical components of such habitats, but the inference is not always justified even for large and abundant components of the vegetation. There is, therefore, no adequate alternative to a ground survey.

World-scale distribution maps are of considerable interest, but intensive studies of distribution patterns may require a higher degree of resolution than can be provided by these. Subdivision of an area and intensive survey within the subunits are the obvious means of satisfying this need.

The growth of amateur botany has subsequently permitted surveys to be conducted at a more detailed level, using 10km² (nearly 4mi²) samples laid in a contiguous grid over an entire country. Obviously, this more intensive type of survey provides not only a more detailed knowledge of a plant's distribution, it also permits more effective correlation with the physical and biological factors of the environment. Thus, maps to the same scale marked with isotherms, rainfall or geology, can be compared with plant distribution maps and suggestions can be made regarding the factors which may be critical in determining the distribution of a species. The taxonomic problem may still be a serious one and many of the more difficult groups of plants, including many bryophytes and lichens, are still inadequately recorded, even on this fairly gross scale, within a

country having a relatively high density of well-informed naturalists.

The finer points of a plant's environmental requirements, however, cannot even be resolved at this level of sampling. Habitat diversity may still be high even within a 10km square and more precise information requires a smaller sample area, which in turn usually necessitates a smaller gridded study area. This reduction can continue until one is eventually observing individuals of the species within a community.

Some local floras give detailed information on distribution and habitat, especially in the British Isles, where many amateurs are involved in collecting the basic data. This computer map of Warwickshire in the Midlands of England is taken from *Computer Mapped Flora: A Study of the County of Warwickshire* (Academic Press, 1971).

In this type of study the sheer bulk of data obtained can be very considerable and the use of computerized retrieval systems becomes essential.

P.D.M.

Mapping

The methods used for presenting information on the distribution of taxa depend upon the data available and on the requirements of the user. In early publications on plants we often read that a particular species occurs in, for example, North Africa, reflecting the small amount of information then known, while today we usually only require a plant label in a botanic garden to inform us that the species comes from, say, Burma and Tibet. However,

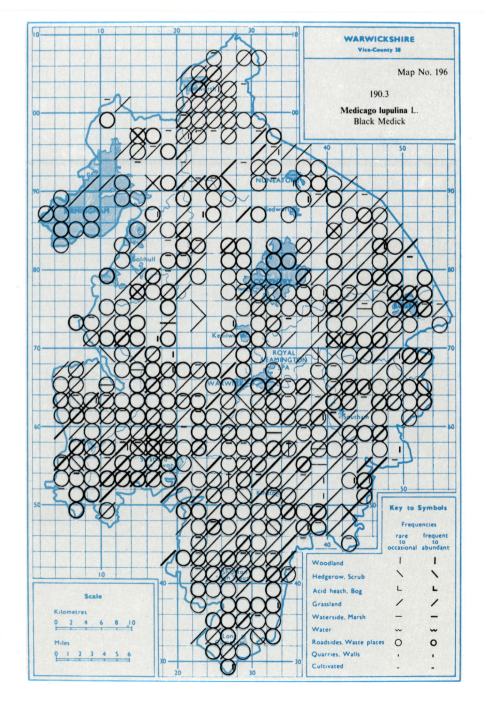

as more data are available and required, it becomes increasingly difficult to convey a clear impression of distribution by listing localities; this can be achieved more rapidly and economically by means of distribution maps.

The first distribution maps, prepared by the Danish botanist Joachim Schouw in 1823, indicated the general limits of species, and such outlines are still used, as we shall see later. The general range of a taxon may also be indicated by shading the areas of the world where it occurs, and this is still employed when the data are imprecise. However, much the most common kind of distribution map is that in which a spot is placed on the map for every locality or group of localities where the taxon is known to occur, the degree of precision depending upon the size of the spots in relation to the scale of the map. By using a number of different symbols instead of the conventional black circles it is possible to change a static representation of a taxon's distribution into a much more dynamic tool. Thus, for example, by simply indicating with an open circle localities in which a species has not been recorded in, say, the last 50 years, with black circles for more recent records, the changing distribution of the species becomes immediately apparent; in such cases it is usually the result of extinction by Man. By using different symbols for modern and fossil records of various ages the changing distribution over much longer time-scales can be readily demonstrated. In a similar manner, chromosomal or morphological features can be incorporated into a map so that their relationship to a species' distribution can be seen, thus providing a valuable tool for examining such important phytogeographical processes as migration and colonization.

When using a map to compare the distributions of several taxa, such as species within a genus, different symbols can be used as just described. However, this can become confusing, especially with numerous sympatric taxa, and it is usually clearer and equally informative to retreat from the precision of "dot maps" by outlining the ranges of the various taxa with distinctive lines—continuous, dotted, dashed etc.

For maximum reliability, a distribution map should be as accurate for the localities where a taxon is absent as for those where it is present. This is best achieved by using a grid over the map, with each grid square being systematically studied by field workers. The size of the squares is determined by the size of the area under consideration and the availability of workers. Some measure of the reliability of the map can be shown by presenting the number of times each grid square has been surveyed. Mapping by means of such grids has the added advantage that it can easily accommodate the use of computers for data storage and mechanical plotting of the maps, which is one reason for their importance in modern studies. Clearly, the finer the grid, the more detailed the map, and there is no reason why such maps should

not bridge the gap between the geographer's approach to a species' distribution and the metre quadrat (see p. 46) of the ecologist—except personnel and cost! Indeed, some modern, computer-produced maps are three-dimensional, with the vertical scale indicating a species' abundance in various parts of its range.   D.M.M.

Vegetation and its Description

Plants only occasionally grow as isolated individuals. More usually they are found in some degree of spatial association with other plants, either of the same or of different species. This corporate growth of plant material—which mantles most of the Earth's surface—is referred to as *vegetation*. The basic unit of vegetation is the plant individual and the nature of the vegetation at any one site is both defined and characterized by the component individuals.

It is necessary to distinguish clearly between vegetation and *flora*. The flora of an area comprises the species of plants that are found growing there. The vegetation, however, also includes the quantitative contribution of the species to the total plant cover, their spatial organization and their arrangement into plant communities. Vegetation is normally composed of individuals belonging to many different species and it may be usefully regarded as an interacting system of populations of species. The species present in a particular type of vegetation constitute its *floristic composition*. The prime determinant of this is the available flora and consequently the study of vegetation requires an understanding of plant geography and consideration of the floristic regions within which the different types are developed (see Chapter Six: The Realms of Plants).

One of the principal aims of vegetation science has been the description of the different types of vegetation that cover the Earth's surface. Work along these lines has been in progress since the beginning of the 19th century, and currently continues, though there still remain large areas of the world where remarkably little is known about the vegetation.

There is a whole range of features that can be used as a basis for describing and classifying vegetation: physiognomy, structure, dominant species and floristic composition. In addition, classifications based upon criteria outside the vegetation—such as the habitats or the presumed climax (see p. 43) of the community—are also possible. All of these approaches have been used and the progress of vegetation science has been marked by the proliferation of different descriptive techniques. This has led to the development of distinct "traditions" of vegetation description, based upon par-

ticular centers, each with their particular approaches and emphases.

The earliest descriptions of vegetation were made by naturalists, explorers and geographers. Many of these were based upon the structure and general appearance of the vegetation, using broad categories such as scrub, forest etc. Thus Humboldt, at the beginning of the 19th century, attempted to classify vegetation on the basis of its major growth-forms and by 1872 August Grisebach had produced his classic work describing some of the major physiognomic types of world vegetation. Physiognomy is a useful character to employ in large-scale vegetation surveys, especially in regions where the flora is little known, as it can be readily recorded without a detailed knowledge of the species involved. Physiognomic descriptions are still required for parts of the world and at present there are several sophisticated schemes of structural classification in use, in which the basic unit of vegetation is the *formation* (see Physiognomy, p. 49).

The other main approach to vegetation description, which in fact developed out of the structural and physiognomic tradition, has been to use the floristic composition as the basic descriptive tool (see Phytosociology, p. 59). The simplest floristic approach—which again is particularly suited to large-scale descriptive work—has been to use the dominant species to characterize vegetation. More complex is the use of the full floristic complement—mosses and herbs as well as trees and shrubs. This demands a detailed knowledge of the plant species and is also more time-consuming, both in the recording and in the processing of the data. It is thus especially suitable for regional studies made within the broad physiognomic framework of world vegetation. The basic floristic unit of vegetation is the *association*. On a more detailed level still is the quantitative descriptive approach involving the precise measurement of vegetation features such as abundance, height, biomass, age, structure etc. This is particularly appropriate for intensive studies on small areas of vegetation, where it may be especially useful for answering specific ecological questions.

B.Wr.

Plants in the Ecosystem

As we have just noted, the study of vegetation tends to crystallize around the structural and floristic approaches. However, before considering the terms, concepts and procedures associated with these it is necessary to deal with a number of topics which are common to the two, either because they are of such wide currency that they apply to both or because they are basically irrelevant to such distinctions; perhaps indicating the partial artificiality of trying to divorce the structure of vegetation from its floristic composition.

Within any area being studied, the role of plants in relation to all the living organisms has to be recognized. Thus, any study of vegetation is set against that of the *ecosystem*, which is an entire community of animals and plants interacting with its particular environment. It consists of "producers" (mostly green plants), "consumers" (animals), and

This diagram gives an impression of the energy flow through a typical plant-herbivore-carnivore food chain. At each step in the chain considerable quantities of energy are either used up by respiration or are wasted because it is not assimilated. For these reasons, the farther down a food chain an organism exists the greater the area of primary producer required to sustain it.

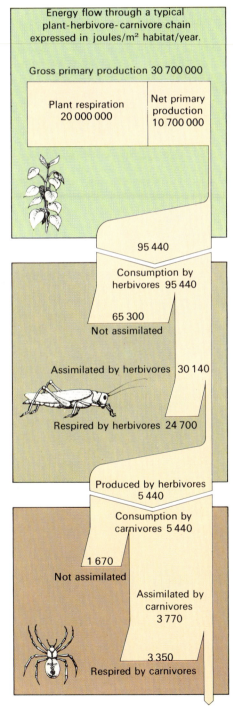

Energy flow through a typical plant-herbivore-carnivore chain expressed in joules/m² habitat/year.

Gross primary production 30 700 000

Plant respiration 20 000 000

Net primary production 10 700 000

95 440

Consumption by herbivores 95 440

65 300

Not assimilated

Assimilated by herbivores 30 140

Respired by herbivores 24 700

Produced by herbivores 5 440

Consumption by carnivores 5 440

1 670

Not assimilated

Assimilated by carnivores 3 770

3 350

Respired by carnivores

Produced by carnivores 420

"decomposers" (bacteria and fungi). Within an ecosystem, or indeed any given part of the Earth's surface, the total amount of living matter, measured by dry weight, is termed the *biomass*. Because the contribution of animals is small, many figures stated for biomass include only the *phytomass* of the vegetation (see p. 47). Biomass is of limited value as a measure of productivity, which is a function of both the standing crop and the rate of metabolic activity. Detailed measurements of the biomass of different components of an ecosystem can, however, provide a valuable insight into its structure and function. For example, in any stable ecosystem, the biomass of any one trophic level must be much greater than that of the level above; there must be more plants than herbivores, and more herbivores than carnivores. Biomasses of different trophic levels are also useful when comparing different ecosystems.

Within an ecosystem, a group of plants occupying a specific habitat is termed a *synusia*. The species in a synusia resemble each other in life-form (see p. 50) and have a similar role in the ecosystem, but they may be quite unrelated taxonomically. The term can be used with reference to any ecosystem, but it is especially useful in discussions of the tropical rain forest, where many synusiae have evolved. Examples of synusiae in the rain forest are the large epiphytes, epiphytic microorganisms, climbing plants, strangling plants, parasites, saprophytes, herbs, shrubs and up to three layers of tree canopy. The large epiphytes are an illustration of the fact that a synusia may be composed of many different taxonomic groups: in the rain forests of South America, for example, orchids, ferns, bromeliads and cacti are all to be found growing as epiphytes.

D.M.M.

Community

One of the terms most commonly applied to plants (or animals) living together is the *community*. Within a given landscape it is often apparent that the plant species which comprise the vegetation are not distributed independently of one another but that approximately similar combinations are repeatedly found growing together. Such recurrent groups of species tend to be associated with particular habitats characterized by a certain combination of environmental conditions. They serve, often, to differentiate the vegetation cover into a patchwork of *stands* (see p. 42), each with a particular floristic content and, frequently, a distinct physiognomy as well. This has led to the concept of the plant community.

Two important usages of the term "plant community" may be distinguished. One—the "real" community—refers to the plant assemblage that occupies a particular portion of vegetation. The other—the "abstract" community (or community-type)—refers to some phytosociological unit of vegetation

A tropical plant community: a palm tree festooned with creeper and supporting a bromeliad higher up. The relationship is plainly one of dependence.

that has been abstracted by the comparison of samples of "real" communities (see Phytosociology, p. 59). To clarify the distinction the term *phytocoenose* has been proposed for the first and *phytocoenon* for the second.

The designation "community" can be applied to any assemblage of plants found growing together. However, the term is suggestive of some degree of interaction and integration between the components and it may be desirable to restrict its application to situations displaying these features. The community may sometimes be clearly demarcated spatially in that particular communities may appear as comparatively uniform areas of vegetation clearly separated from one another by sharp boundaries. In other cases, however, the essential continuity of the stands may be evident with the gradation of communities into one another.

The degree of integration within the community has provoked considerable discussion. Several models have been proposed. One of the first was provided by the American ecologist, Frederic Clements, in 1916. He proposed the *organismal model*. This stressed the unity of the community by regarding it as a superorganism, with the component parts (species) interacting to produce a functional whole. Just as an organism grows and matures, so does the community—following a reproducible pathway—by the process of *succession* (see p. 149). This viewpoint has been widely criticized as being too hypothetical, since there are clearly many features in which a community is not analogous to an organism. One of the severest critics was another American, Harold Gleason, who put forward a completely different viewpoint: an *individualistic model* which, by contrast, emphasized the coincidental nature of the community. In this, the phenomena of vegetation were considered to depend entirely upon the phenomena of component individuals. Communities were thus chance aggregations of individuals, fortuitously generated by the coming together of such species as were available to survive in the particular environmental regime and in the presence of one another.

Although the nonliving (abiotic) environment exerts a prime influence in determining the species that it will support, it is also evident that within an area of vegetation there are interactions between the species that are, or have been, important in determining its floristic composition. The species with most influence upon the others is called the dominant (see Dominance, p. 43); this is usually the species with greatest biomass. Three types of interactions may be recognized.

First is *competition*, in which the plant individuals are in some way competing for a scarce environmental resource. Competition for light is particularly important and, in the absence of any limiting factors, the most successful species are often those that can grow the fastest or the tallest, or which produce dense shade or large accumulations of litter. Competitive exclusion may often serve to prevent species from naturally occupying environments which would otherwise be quite suitable for them. Or it may create a dynamic equilibrium in which the performance of the species is less than their potential. As competitive species usually have high growth rates, they are only likely to be successful in suitably productive environments which are able to sustain this. If their competitive ability is in some way reduced (by an unfavorable or less productive environment or through regular disturbance, as with grazing) this may permit the growth of associates which would otherwise be absent. Species diversity may thus be enhanced.

Second, species may be *complementary* to one another, co-existing by occupying different niches, not demanding the same environmental resources. This is well shown in the stratification of vegetation, where the plants occupying the lower strata are capable of surviving in the shade cast by the taller species or where rooting zones occupy different levels and permit the exploitation of different areas of soil. It is also shown in *aspectation*, in which complementary species have their main growth periods at different times of the year, thus reducing interference. Thus, some herbs of deciduous summer forest have their main growth and flowering in spring before the main expansion of the leaf canopy. Such interactions give rise to a subunit of the plant community known as a *society*. This consists of a group of plants holding similar positions within the community as, for example, when they are components of the same stratum, develop in a particular season or occupy the same microhabitat.

The third category of interaction is *de-pendence*. Here a species can only occur through the presence of others. For example, certain hygrophilic plants can only survive in the humid microclimate of a forest floor. In this case, the occurrence of the species is not dependent upon a particular species creating the appropriate microclimate, but more specific dependencies are usually found in parasitic, symbiotic and, to a lesser degree, epiphytic relationships.

Community composition is thus regulated by environmental factors in association with sociological interactions. The range of communities a species occupies is its *sociological amplitude*; the extent to which it is confined to a particular community is its degree of fidelity (see p. 61).

Communities are not static. The species populations possess an age structure and there is a continual flux of individuals. In a stable community, the proportion of each species is usually subject only to minor fluctuations, although several examples are known of stable communities which display a cyclic change in the abundance of some of the components. In contrast, progressive compositional changes, associated with a changing environment, lead to a new community as part of the process of succession.

Interrelated with the idea of the community is that of the *stand*, and the two are sometimes confused. A stand is an area of same-aged trees in a forest or of plants in an agricultural crop; it is also a term applied to any more or less uniform area of vegetation. The use of the term in ecology emphasizes that a definite area of vegetation is meant, not an abstraction, and as such it is important in the classification of vegetation, which

The section of a landscape where one type of vegetation gives way to another is called the ecotone. In the Upper Orinoco region of southern Venezuela there is a marked change from savanna (open grassland) to forest.

is partly concerned with the similarity between stands in different places.

In agriculture and forestry, the limits of the stand are usually clearly defined by the edges of the planting. In natural vegetation the limits are frequently not so clear. Where sudden change does occur, it is usually produced by Man's activities, the effects of animals or by abrupt discontinuities in such environmental factors as the depth, moisture and composition of the soil. This problem of determining the limits of stands clearly relates to the organismic and individualistic views about the community referred to earlier, the latter suggesting that sharp demarcation would not be expected. Whatever the merits of these two rather extreme views, however, students of vegetation have been able to recognize situations in which it seems to comprise a continuum and those in which units exist with intergrading zones (ecotones) of various extent between them.

An *ecotone* is a zone of change between one vegetation type and another. In many places, transitions between communities are so gradual that one cannot say clearly where one changes to another, while in other areas transitions are relatively abrupt. Ecotonal situations are ecologically very interesting: they contain organisms from both of the adjoining communities and these organisms are under stress as conditions change from favorable to unfavorable for them. In such areas, much can be learned about the factors which limit organisms. Further, there are organisms which are confined to ecotones and can exploit the special conditions there. The result is that ecotones often contain a greater variety of species than the stands or communities on either side. An example of a plant exploiting an ecotone is the Venus' flytrap (*Dionaea muscipula*) which, together with several other carnivorous plants, occupies a narrow zone between wet

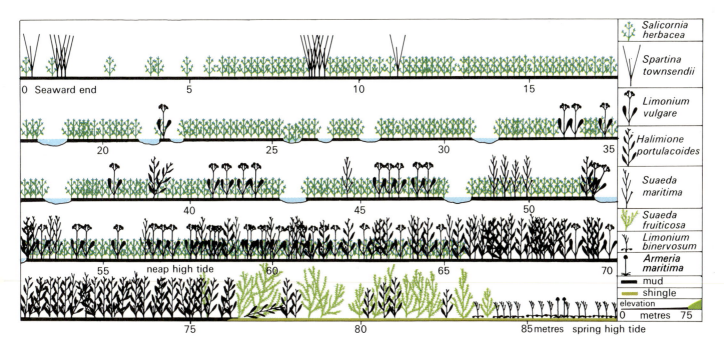

Salicornia
herbacea

Spartina
townsendii

Limonium
vulgare

Halimione
portulacoides

Suaeda
maritima

Suaeda
fruiticosa

Limonium
binervosum

Armeria
maritima

mud

shingle

elevation

0 metres 75

Transect across part of a small salt marsh at Scolt Head Island, Norfolk, England. The transect starts at the seaward limit of plants (om, **top left**) on a nearly flat muddy shore, and continues up to the level of the spring tides on a fringing shingle bank (88m. **bottom right**). Each plant that touched the transect line is represented by a stylized representation of its form. Many meandering drainage channels cross the marsh. The vegetation on the mud of this marsh appeared at first sight to be composed of three zones: Salicornietum at the lowest end, a Limonietum in the middle and a Halimionetum higher up. However, the transect shows that there is a gradual transition from one "apparent" zone to another. ie a continuum, in which glasswort (*Salicornia*) becomes less frequent with increasing numbers of sea lavender (*Limonium*), itself gradually replaced by sea purslane (*Halimione*).

evergreen-shrub bogs and wire-grass savannas in the southeastern United States. Here the special conditions provided by a fluctuating water table, low soil nutrients and frequent fires, which are unfavorable to most other plants, enable the insectivorous plants to survive.

The abundance of species along ecotonal boundaries is often referred to as the "edge effect" and although many organisms (trees for example) show a fall-off in species number at the forest edge, the importance of the habitat for other organisms, including animals such as birds, makes for much greater variety. Hedges can be thought of as "linear edges" and hence much attention has been given to them as potential sources of diversity in man-made landscapes.

Vegetation in which the species-composition changes gradually from one point to another is called a *continuum*. It thus has some relation to the broad (or shallow) ecotones just mentioned except, of course, that the vegetational gradient never ceases. While detailed studies on a relatively small scale suggest that plant species comprise part of a continuum, it is not so obvious that this is the case on a larger scale. It is probably true

that few species have exactly the same tolerance of the various environmental factors (see Chapter Four: Environmental Factors) governing their occurrence in a particular habitat or area, and such similarities as may exist will be further complicated by the various interactions between species and between individuals, already referred to. Such factors as these will tend to produce an endless changing combination of species which will be seen as a continuum. However, when some species are particularly prominent in the vegetation (See Dominance, below), they can so modify the local environmental conditions as to determine to some extent what other species may or may not survive in their company (see under Climate, p. 117; Pattern, p. 48), while major environmental changes will cause sufficiently great discontinuities in a continuum to permit the recognition of distinct vegetation-types.

D.M.M.

Dominance

In a plant community, one or several species occupy more space, absorb more nutrients and contribute more biomass than the rest; such species are called the *dominants*. There may be one, two, or, as in the tropical rain forest, many species making a large contribution to the community; when more than one species dominates, the term *co-dominant* is used. Dominant species are almost always the tallest plants in the community, and as such exert considerable influence on the environment within the community; dominant species do this in several ways: by affecting the light regime beneath their canopy, by creating areas of protection from exposure, and by utilizing a larger proportion of the available nutrients. This effect on the character of a plant community is due to the dominance of the dominant.

The way in which a dominant species can

influence the nature of a community is well illustrated in north temperate regions by the differences between oakwoods and beech-woods. The leaves of beech trees (*Fagus* spp.) are orientated so that the blades lie predominantly in the same plane, which is roughly horizontal; this greatly reduces the amount of light penetrating to the lower layers of the community, so that a shrub layer is completely absent, and the ground layer is very impoverished with regard to species and cover. By contrast, the leaf blades of oak trees (*Quercus* spp.) do not naturally lie in the same plane, so that the amount of light reaching the lower layers of an oakwood is considerable, and in consequence oakwoods have richly developed herb and shrub layers.

Dominance is usually associated with stable climax vegetation; in the case of grassland, the smaller stature of the dominant will obviously reduce its influence on the environment of the community, compared with a woodland climax. However, dominant species are also found in many successional communities (see p. 149). In these cases, the dominant species have some effect on the environment of the community, but they do not impart stability; their dominance results from the ability to exploit the prevailing conditions, and is therefore transient, because conditions do not stabilize prior to climax.

J.WN.

Climax

The term climax is applied to the state of equilibrium which a plant community reaches with the prevailing environment. That is, as a succession (see p. 149) develops from the initial colonization of open ground, and one type of vegetation succeeds another, the changing combination of plants will ultimately give rise to a type of vegetation which does not develop further. It appears to be permanent, and becomes characteristic of

Above A mangrove swamp at low tide. Mangrove is the natural climax vegetation in sheltered muddy inlets and river estuaries in the tropics. The dominant plants usually belong to the family Rhizophoraceae.

Opposite The apparently unchanging pattern of fields and hedgerows in western Europe and other parts of the world is a dis-climax vegetation imposed by human activity, which began with the large-scale removal of the forests that constitute the climatic climax. This scene in southern England also reminds us that many of man's crop fields provide examples of uniform, well-delimited stands.

a particular region. This is the climax vegetation.

It is our common experience that certain regions of the world support apparently stable types of vegetation which are related to the prevailing climatic conditions, principally temperature and rainfall with their seasonal distribution. In equatorial regions, for example, we expect to find tropical rain forest, dominated by a considerable diversity of broad-leaved evergreen trees. In many cool temperate parts of North America and Eurasia we expect to find deciduous forests adapted to cool winters and warmish summers, while in regions with long, cold winters and short, cool summers we expect to find tundra vegetation dominated by low, often cushion-forming herbs and with no trees. These vegetation types, which are considered to represent the most advanced state which succession can reach under the prevailing climatic conditions, are called climatic climaxes and are the basis for the generalized maps of the vegetation of the world that are widely published.

This general pattern of climatic climaxes led some people to consider that within an area of uniform climate all vegetation must eventually develop by successional stages into a single uniform stable community, the climax, the characters of which will depend on the type of climate. Indeed, it was even maintained by some that the plant association was actually a superorganism which developed through successive phases from youth to maturity, just like an individual plant or animal. On the other hand, it has been suggested that vegetation is in a constant state of flux and that it changes continually, never reaching a stable climax. Most modern views would lie somewhere between these two extremes.

A *biome* is a natural climax community of plants and animals covering a large geographical area. Biomes have some relationship to the climatic life-zones which range from the Arctic to the tropics, but differ in that within any one climatic life-zone several climax formations can develop, and therefore several different biomes. The boreal region of eastern Canada, with its ubiquitous moose populations, is a good example of a biome. Other examples are the tropical rain forest in Southeast Asia, the savanna of southern Africa, the steppes of the USSR and the South American pampas.

Useful though the climatic climax is for giving a broad outline of the major vegetation types of the world, it is obviously not an inevitable sole end product in a particular climatic regime. Within the broad climatic zones, for example, local topographic factors can cause major departures from the expected climate and most areas comprise a mosaic of microclimates which affect the plants profoundly. Thus, the north and south sides of a hill may have climatic differences which allow the development of forest on one side and grassland or scrub on the other, the south-facing side being much warmer and drier than the other. Change in altitude has a profound influence on the local climate, and

Below Rain forest, the climax vegetation of the humid tropics, seen here in eastern Panama with the Gulf of San Miguel in the distance.

this is made obvious on many mountains by the clear timberline (see p. 186) below which forest develops but above which trees seem unable to survive. These *subclimaxes* are all apparently stable, as is the particular variant of them known as the *edaphic climax*, which is dependent upon changes in the soil. Soils developed over serpentine rock, for example, which are deficient in certain nutrients and rich in some heavy metals that are toxic to many plants, often bear an open herbaceous vegetation contrasting markedly with adjacent forests developed over non-serpentine soils in the same general climatic region. Other local vegetation variants have been given the term *postclimax*, where it is considered that special factors favor the development of a type of vegetation more advanced than the climatic climax. For example, woodland fringing watercourses in an area which is climatically grassland or prairie is considered a postclimax, woodland being a more "advanced" vegetation type than one dominated by herbs. Because of the occurrence of these apparently stable types of vegetation within a broad climatic zone, the idea of the *polyclimax* finds a lot of support, since it is easier to reconcile with the observed facts. When the vegetation of a climatic region has become stable and has reached its full development, there will be a number of different types of vegetation produced by the interaction of the climate with the other environmental factors. The region is then said to have polyclimax vegetation. It is contrasted with the idea of a *monoclimax*, where, within one climatic region, only a single vegetation type is expected. While admitting that the climatic climax can still be considered as the highest type of prevailing vegetation within a climatic area, providing there are no contrary conditions, the polyclimax concept permits recognition of the diverse climaxes without the forced interpretation often necessary to link up all stable communities with the climatic climax.

To the terms already introduced must be added the *dis-climax*. This is applied to communities produced as a result of the distortion by animal or human interference of the succession that is normal for a particular region. Much of the vegetation of the world has been affected by Man and consequently the dis-climax is familiar to all of us. The apparently unchanging pattern of fields and hedgerows in western Europe is a dis-climax vegetation imposed by Man, who first cut down and now largely prevents the renewal of the woodlands that constitute the climatic climax. Again, open chalk grassland is a dis-climax resulting from early clearance of the climax woodland and is maintained by continued grazing by animals such as sheep and rabbits.

As has been noted above, climax vegetation has been recognized by its stability. However, stability is a relative concept, dependent on our human experience of time. While there is evidence of long-term stability of some areas supporting tropical rain forest,

and some associations of plant species in California seem not dissimilar from those in fossil deposits of Pliocene age, the vegetation of any area of the world seems to change, slowly or more rapidly, with time. Throughout the long history of plants on this Earth there have been changes in the positions of the continents (see Mesozoic Environments and Continental Drift, p. 72) and the climatic zones, mountains have risen and been eroded away, and glaciers have advanced and retreated over enormous tracts of land, all profoundly affecting the vegetation time and time again. Such changes, obvious when viewed in terms of geological time, may appear minute within the human life span, but they do still seem to be continuing. In consequence, absolute stability of vegetation seems impossible and there have been efforts to find other means of recognizing or defining the climax.

In the early stages of a succession, the change in the structure of the vegetation and the species composing it is rapid. During this phase it has been shown that there is marked interaction between the component species, which are not distributed at random. However, as the succession proceeds, the rate of change in the vegetation and the species-composition decreases, though never reaching zero change, while the interaction between the species becomes reduced to the point where they are distributed at random in the community. This last phase of the succession can be covered by the term climax, which can be recognized by the relative stability of the vegetation and, more objectively, by the random distribution of the species. Furthermore, if the climax community is destroyed, it regenerates a community of the same kind, whereas this is not necessarily the case for the earlier, seral stages of the succession, which may develop in a new direction.

All these criteria are helpful in permitting the recognition of the terminal phase of a succession, which, covered by the term climax, provides a useful unit for describing and understanding the potential vegetation cover of the various parts of the world.

D.M.M.

Sampling Vegetation

Since it is seldom possible to examine a whole community or larger area of vegetation in its entirety, it is customary to study a part or parts and then extrapolate from these to obtain a picture of the whole—this is sampling. Samples may be taken at random over a particular area or they may be related to features of, and differences in, the vegetation that become apparent from preliminary observations. The number and size of the samples necessary to clearly define the structure or floristic composition of the vegetation of a particular region is, of course, necessarily a balance between what is ideally desirable and that which is practicable. It is beyond the scope of this book to delve into the

various ways for determining the level or pattern of sampling, but two tools for taking samples of vegetation—quadrats and transects—are so widely used that they will now be considered.

D.M.M.

Quadrat

The quadrat is a most important ecological tool. It consists of any four-sided structure which can be used to delimit a plot of ground within which observations are to be made. Frequently it is a metal frame, but it can also simply be four corner posts connected by string. Although, strictly speaking, the term refers to a four-sided structure, sampling plots of other shapes (eg circles) are sometimes used and these are also often called quadrats. The size, shape and siting of a quadrat depends entirely upon the purpose for which it is to be used.

The size of the quadrat used depends both upon the nature of the vegetation to be sampled and the measurements that are to be made. Thus for examining areas of forest, large quadrats (several hundred square metres) may be appropriate, whereas for herbaceous vegetation smaller quadrats are employed. A fifty-centimetre (20-inch) or metre-square (40-inch) is often a convenient size.

Several approaches are possible when using quadrats to sample vegetation. One is just to use a single plot to represent a particular type of vegetation (see Relevé, p. 59). In this case, the sample site is subjectively selected to give such representation of the vegetation as is required. Another possibility is to make a number of quadrat samples, scattered throughout the vegetation. These may be arranged randomly or regularly (ie with a fixed distance between the samples). An advantage of this approach is that it helps to avoid much of the bias that may be introduced by the subjective selection of sample plots, though it tends to be more time-consuming.

In the *point quadrat* the plot area is reduced to a single point and the plant material intercepted by this is recorded. The points are usually represented by long pins suspended in a quadrat frame or by the intersections of a grid strung across it. *Permanent quadrats* are plots demarcated over substantial periods of time to permit the long-term recording of changes within them. As an alternative to the quadrat, plotless sampling techniques, where there is no rigidly defined sample plot, are sometimes used.

Quadrats may also be arranged at intervals along a *transect* to provide an examination of variation in the plant cover along an environmental gradient.

B.Wr.

Transect

A transect is a series of samples taken in one direction across vegetation. The most common use of a transect is to study zonation. A real or imaginary line is laid at right-angles to the boundaries between the

Structure of Vegetation

The plant material that comprises vegetation may, of course, be derived from any of the major taxonomic groups—angiosperms, gymnosperms, ferns, mosses and liverworts. Lower groups, such as the fungi and algae, are additional components, although they usually receive very little attention. The different species contributing to the vegetation cover have a characteristic size and shape and, although this may be a rather variable property, nonetheless it is possible to classify plant material into broad morphological or growth-form categories (see Life-form, p. 50). When, as is usual, plants of different growth-forms grow together they generate *structure* within the vegetation. The concept of structure in vegetation has been expressed in several different ways and consequently there has been some confusion as to the precise compass of the term. Here, vegetation structure is defined as the spatial organization of the component plant material, and it encompasses the abundance of the individuals, their distribution and pattern (horizontal structure), and their vertical organization into layers or strata (vertical structure).

Abundance

The total amount of plant material comprising vegetation constitutes its *phytomass*, and this is usually partitioned into above-ground and below-ground components. It shows enormous variation between different types of vegetations: estimates range from below 1 tonne/hectare (0.4 ton/acre) for some unproductive and structurally simple vegetation of the arctic tundra to values in excess of 400 tonnes/hectare (160 tons/acre) for the productive and structurally complex rain forests of the wet tropics.

The contribution each species makes to the phytomass may be expressed by some measure of its abundance. Estimates, based upon some subjective rating of abundance, have often been employed. One system adopts five classes of abundance—dominant, abundant, frequent, occasional and rare—and incorporates the additional qualifications of "very" and "locally" when these are required. This scale has been widely used although it has many limitations, not least the lack of precise definition of the compass of the classes.

Cover
Cover refers to the proportion of ground that is covered by the aerial parts of the plant material growing upon it. When these are in lateral contact and form a continuous cover the vegetation is said to be closed. When the vegetation cover has gaps, which could be colonized by other individuals, it is said to be open, and if the amount of space is much greater than that occupied by plants the term sparse is used.

Students recording the plants from metre-square quadrats of vegetation taken along a transect (line) from the edge of a fluctuating lake or *turlough* in County Clare, Eire.

zones of vegetation and contiguous or spaced samples are taken along this line. Each sample may include measurements made on a single plant or on all the plants in a quadrat of suitable size, together with data relating to the immediate environment.

The environment in which plants grow varies from place to place and normally the presence and abundance of plant species in vegetation varies accordingly. These variations are often greatest in one direction due to the influence of such things as the presence of a spring, a stream, a fall in the ground, the direction of the prevailing wind, or changes in the underlying rock.

To study these areas where vegetation is changing, it is clearly useful to sample vegetation and habitat in known spots. These samples can then be analyzed statistically and maps and diagrams made of the distribution of the plants in relation to the habitat. If the samples are placed along a line at right-angles to the direction of change, it is easier to represent visually the way vegetation changes with the habitat. Additional transects parallel to the first may be needed to

confirm any relationships found. The data from the transect may be less suitable for statistical analysis than data from samples distributed in other ways but the pictorial representation possible has an assured place in ecology and often cannot be replaced by a statistical summary.

The size of the samples and their spacing along the transect should be related to the way that the vegetation changes along the transect and to the variability in the vegetation. In a salt marsh, the representation of all the plants that touch the transect line may show the zonation adequately but more plants must usually be sampled. In grassland, if a slow visible change takes place over 100m (330ft) or so then quadrats 1m² (10ft²) placed every 5 or 10m (every 15 or 30ft) may be sufficient to show the changes in the vegetation clearly. If, however, the same change takes place over 5m, these quadrats must be modified without increasing the sampling errors too much. In this case, the vegetation may be sampled sufficiently by contiguous quadrats each 0.5 or 0.25m (20 or 10in) by 1 or 2m (3 or 6.5ft) placed with their short axes along the transect so that 10 or 20 are needed to fill the 5m. Habitat measurements would normally be closely related to these samples of the vegetation.   B.Wr.

The amount of cover contributed to the vegetation by an individual species may be estimated in several ways. The simplest is the use of a crude, subjective estimate derived from a visual impression of its aerial extent. Various scales have been devised for this purpose.

One of the best-known is the scale introduced by K. Domin. This eleven-point scale is as follows: + = isolated, cover small; 1 = scarce, cover small; 2 = very scattered, cover small; 3 = scattered, cover small; 4 = abundant, cover 5–10%; 5 = abundant, cover 10–25%; 6 = cover 25–33%; 7 = cover 34–50%; 8 = cover 51–75%; 9 = cover more than 75%; 10 = complete cover.

This scale can be applied easily and, through practice, with a high degree of consistency. Consequently, it has received wide usage in estimating plant cover and has been particularly useful in making rapid surveys of extensive areas of vegetation. It should be noted that it is not, strictly speaking, purely a scale of cover but also, in the lower ratings at least, it combines cover with abundance (the number of individuals). It is thus referred to as a cover-abundance scale. Also, the classes are not of equivalent range: at the lower end of the scale, the divisions are finer and more precise.

Another scale that is widely used is the Braun-Blanquet scale. This six-point scale is as follows: + = less than 1% cover; 1 = 1–5%; 2 = 6–25%; 3 = 26–50%; 4 = 51–75%; 5 = 76–100%. It permits a rapid estimation of cover to be made and, although subject to the obvious limitations of all subjective scales, usually the six cover classes can be applied consistently.

A more accurate method for cover estimates is by the use of point quadrats (see Sampling, p. 46). Here, a number of long pins suspended in a frame are lowered into the vegetation and the species "hit" by each pin are recorded. The cover is given by the number of "hits" for each species expressed as a percentage of the number of pins lowered.

In closed vegetation the sum of the cover values of all the species will often exceed 100%. This is because the aerial parts of the plants frequently overlap vertically, often to a very considerable extent, thus giving rise to stratification. To indicate the area exclusively occupied by a species at ground level, measurements of *basal area* can be made. This is particularly appropriate for trees, where it is normally measured as the cross-sectional area at 1.5m (5ft), about mean breast height, and is an important measure used by foresters for assessing the amount of a species in a forest stand.

Density

Density is the number of individual plants of a particular species within a given area. It is usually estimated visually by ecologists using a frame quadrat (see p. 46). It allows accurate and direct comparisons of the abundance of species between different areas. One disadvantage is that it is often difficult to decide what is meant by an individual plant in the case of rhizomatous or stoloniferous species.

Frequency

Frequency is a measure referring to the chance of finding a species within any one sample of a piece of vegetation. Two forms of frequency are used: *shoot frequency* refers to the chance of finding any part of the species within the sample; *rooted frequency* is the chance of it being rooted there.   B.WR.

Pattern

Pattern, as used in plant ecology, generally refers to some aspect of vegetation which is repeated in a predictable manner, either in space or in time. Firstly, it may refer to the distribution of different plants with respect to each other. These are patterns in real space, which in many cases can be discerned without the aid of measurement and statistics. Secondly, there are changes in the relative abundance of different populations with time. Such time patterns are an important feature of population dynamics, competition, and succession. Finally, there are the relationships which exist between the spatial distribution of a population, or community, and the various environmental factors which influence the distribution. This is essentially an abstract concept, and the relationships may become apparent only if sophisticated statistical procedures are applied; where, however, the environmental factor is some geomorphological feature (such as a mountain slope or lake-side) these patterns may be easily detected without such methods. Topics in the second and third of the above categories are dealt with in chapters Four and Six.

Clumping

Examination of a plant community will generally show that individual members of the various populations are neither equally spaced nor spaced at random: individual plants are usually arranged in clusters or clumps. Sometimes such clustering of individuals can be discerned with the unaided eye, on other occasions methods designed specifically to detect pattern must be utilized. In general, the most satisfactory methods are those that do not depend upon the size of each area used in the sampling procedures, because the scale of the clustering pattern will vary from species to species. Clustering commonly results from the process of reproduction, either by vegetative propagation, or by heavy seeds which are not widely dispersed; in either case the progeny will be clustered around the parent. Clustering is

The periodic freezing and thawing of the tundra creates polygonal patterns on the ground which can be clearly seen by aerial photography, the hollows between the mounds being colonized by different plants.

occasionally caused by some environmental factor such as the varying soil structure, or depth of the water table.

Vegetation Cycles

Changes in vegetation may be categorized according to whether they are directional with respect to time, that is, part of a succession from pioneer to climax, or cyclic, in which case the same pattern of vegetation-change is repeated again and again at any one location. The cyclic changes consist of several different plant assemblages following each other in a regular manner. Because all the phases of a cycle can usually be found in an area at one time, the vegetation has the appearance of a mosaic. Possibly the best-known of these cyclic mosaics is the hummock and hollow complex which occurs on bogs. Initially, bog pools are colonized by bog mosses, which require more or less continual immersion in water (eg *Sphagnum cuspidatum*). This initial species provides a substratum for species which can tolerate only periodic inundation, such as *Sphagnum recurvum* or *S. papillosum*. When the bog surface has been raised to a sufficient height above the water table, the hummock is colonized by other sphagnum mosses which prefer even drier conditions, as well as by various heath plants. With the gradual erosion of the hummock surface, and the general raising of the surrounding water table caused by the growth of other hummocks, the eroded hummock is replaced by a pool, and so the cycle starts again. A mosaic of *Sphagnum* hummocks may also be found on fens, in which case the hollows are occupied by fen plants. These are sometimes called mixed mires because the mosaic is composed of bog and fen communities (mixed mires are not strictly a cyclic phenomenon, but they do resemble the hollow and

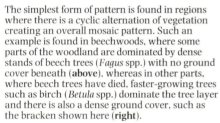

The simplest form of pattern is found in regions where there is a cyclic alternation of vegetation creating an overall mosaic pattern. Such an example is found in beechwoods, where some parts of the woodland are dominated by dense stands of beech trees (*Fagus* spp.) with no ground cover beneath (**above**), whereas in other parts, where beech trees have died, faster-growing trees such as birch (*Betula* spp.) dominate the tree layer and there is also a dense ground cover, such as the bracken shown here (**right**).

hummock of bogs in being a mosaic, as well as being a process of producing dry land which involves sphagnum mosses).

Another example of a cyclic mosaic can be found in beechwoods. The shade created by beech leaves (*Fagus* spp.) prevents the regeneration of any tree species. When a beech tree dies, however, the space becomes occupied by a community dominated by faster-growing species such as birch (*Betula* spp.). Birchwood is an environment in which seeds can germinate and grow, so the birch is slowly eliminated and the beechwood community is re-established. Thus a mosaic of different phases is an integral part of beechwood regeneration.

Cyclic mosaics can also be found on heathland, grassland and moorland. In all cases these patterns appear to be associated with community regeneration.

Vertical Structure

In all but the most simple vegetation (such as monospecific stands) the juxtaposition of species of different sizes and growth-forms imparts a vertical structure. This is readily seen in forests where several strata of vegetation can usually be recognized, the structural complexity often being further enhanced by lianas and epiphytes. Vertical structure also occurs in low-growing shrub and herbaceous vegetation. The subter-

ranean layering of the below-ground parts of plants is an aspect of vertical structure that, for obvious reasons, has not been widely studied.   J.WN.

Physiognomy

Physiognomy is a concept which has frequently been confused with that of structure. The term is applied to the general appearance of vegetation and, as such, it incorporates a number of different concepts relating to structure. For example, such broad vegetation classes as forest, grassland and savanna are essentially physiognomic terms. They refer to the general appearance given to the vegetation by the dominant species, hence they are essentially structural rather than taxonomic terms, and do not take account of floristic composition.

When analyzed, physiognomy can be seen to contain a number of more closely defined biological concepts. It includes an element of vertical-structural description of the living plant material in space. For example, the term forest immediately produces an impression of a vegetation type dominated by trees (ie the phanerophyte life-form) and which therefore possesses a particular layering of biomass above the ground into a characteristic and rather complex system. The term grassland, on the other hand, suggests a simpler layering system in the vegetation type, with the dominant species being hemicryptophytes. Desert implies open, sparse vegetation cover. One should not confuse this concept with that of Christen Raunkiaer's "biological life-form spectrum" (see p. 57), however, where the entire flora is analyzed into its life-form composition. Here the life-form of the dominants is the most important contribution to the physiognomy of the vegetation.

Physiognomic accounts of vegetation may

also contain an element of lateral, spatial description as well as a vertical, structural one. For example, the term mixed deciduous evergreen forest conveys information relating to life-form, vertical layering and also to the fact that two life-form types among the dominants are intermixed in a lateral, spatial sense. Thus the dispersion and density of individuals within a population can be included in a physiognomic description, for example in considering the density of trees in savanna.

The physiognomy of a particular vegetation type is a product not only of its structure—as reflected by its density, height, stratification, dominant growth-forms and species diversity—but it also results from further features such as the color, luxuriance and periodicity. Periodicity in vegetation is usually displayed by a marked seasonality in its overall appearance. Such a seasonal change usually reflects a response to periods unfavorable to growth, an obvious example being provided by the deciduous nature of the broad-leaved forests of temperate climates. Vegetation subject to seasonal change often shows a characteristic sequence of germination, growth, flowering and fruiting for each of its component species, and this can produce a markedly different overall appearance at different times of the year. The study of such changes is called *phenology*.

P.D.M.

Formation

One of the basic physiognomic units of vegetation is the formation, a term introduced by the German August Grisebach in 1838, although it has been used in a variety of ways. In Europe, by and large, it has mainly been used to refer to a physiognomic category of vegetation, the formation com-

Physiognomy, the overall appearance of vegetation, depends upon a number of features such as its structure and the characteristics of its most prominent species. This is well-illustrated by these two vegetation-types; the semi-desert scrub, dominated by *Haloxylon* and *Zygophyllum* shrubs, in northeastern Iran (**above**) is clearly distinguishable from the clover and ryegrass pastures in the foothills of the Tararua Range, New Zealand (**top**).

prising the plant communities of a particular geographical region which have a similar physiognomy and occur in similar habitats. A rather different usage was adopted by the American ecologist Frederic Clements, who regarded the formation as comprising all the vegetation of a given climatic region regardless of its physiognomy. He considered that, within a particular climate, all the different physiognomic types of vegetation were part of a series in a succession leading to a single, climatically determined climax vegetation and that consequently they could be linked together into a single formation.

However, it seems most appropriate to restrict the use of formation to a regional category of vegetation consisting of plant communities dominated by particular life-forms. In different regions, similar formations may occur with related physiognomies but often a very different species composition. These may be joined together into a higher unit, the *formation-type*. Where this corresponds to a major zonal type of world vegetation it may be designated *climatic formation-type*. The formation may also be subdivided, with *subformations* being recognized on the basis of minor, though consistent, physiognomic distinctions. Furthermore, a formation may also be divided into physiognomically related classes of the Zürich-Montpellier school of phytosociology (see p. 59), thus serving to link physiognomic and floristic classifications of vegetation.   B.WR.

Life-form

Of great importance in the physiognomic and structural description of vegetation are the growth- and life-forms of species. The *growth-form* of a plant individual or species is its general habit of growth. The architecture of the shoot system of plants gives rise to numerous growth-forms, distinguishable by the direction of main axes (erect, lateral,

creeping), and the frequency and spacing of branching. The growth-form of shaded plants differs from those grown in full light, and stunted growth-forms due to severe wind exposure are well known in coastal and timberline trees. The *life-form* of a plant also includes its typical habit of growth, its stature, shape and conspicuous features of form and structure but, in addition, its characteristic duration and periodicity. The description of life-form is a part of the full botanical description of species and genera but differs from morphological accounts in selecting for emphasis certain features of structure and habit which provide a ready means of characterization and comparison.

Although there is no universally agreed code of life-form description, there are strong resemblances between the different schemes. This is due to common recognition of the features of principal importance, such as height, form and structure of stem, branching habit, foliage characteristics, underground organs and the duration or relative permanence of these. Some schemes include reference to external features of the plant surface, for example bark, cuticle, hairs, thorns, and to special organs such as tendrils, leaf-pitchers and aerial roots. The situation typically occupied by the plant is also recognized in the case of water plants and epiphytes.

Opinion on the relative importance of these attributes has varied and this is particularly evident in the features chosen to classify life-forms into serial or hierarchical order for convenience in summary, reference and comparison. There are also two quite different principles of classification current in modern practice, namely the conceptual and the objective. Examples which follow, treated in historical order, illustrate these alternative approaches.

The significance of plant physiognomy in its relation to the landscape was first recognized by Humboldt in 1805. He described 19 distinctive plant forms, each being typified and named after the genus or family in which that form is most clearly represented. These were the forms exemplified by the palms, banana, tropical trees of Malvaceae, *Mimosa*, Ericaceae (heaths), cactus, orchid, *Casuarina*, aroids (Araceae), lianas, *Aloë*, lily, willow, myrtle, laurel, Melastomaceae, conifer, grass and fern. His identification of these vegetative forms laid a foundation upon which others were to build and many of the types he named retain their integrity in later schemes.

In 1872 August Grisebach extended and refined ideas on life-form by introducing greater precision in discriminating types, although still based upon external characters; for example he distinguished the "laurel-form" with stiff, evergreen undivided broad leaves, from the otherwise comparable small-leaved "olive-form", and separated the "liana-form", with reticulate-veined leaves, form the "rattan-form", with parallel-veined leaves. Altogether he described 60 vegetative forms and attempted to demonstrate a

This bottle tree (*Brachychiton rupestre*), growing in Sydney Botanical Garden, Australia, is an example of a bottle-stem crown tree (life-form class 1c).

connection between these and the climatic conditions in which they occur.

Awakening interest in plant physiology prompted Carl Drude in 1876 to apply its principles to the understanding of plant form and in the next decade the idea of adaptation to environmental factors became established. This had a great influence on the formulation of life-form descriptions as it introduced behavioral criteria and consideration of internal plant anatomy to supplement the purely morphological grounds of earlier accounts. Thus Drude, in a later work, regarded as of major importance "the duration of organs and protective measures against injury during the resting period" (ie the season of vegetative dormancy) as well as "the position of the renewal-shoot and the main axis in relation to hibernation". He therefore anticipated the work of Raunkiaer, who has been credited with the discovery of adaptations to climatically controlled conditions (see p. 57).

In the conceptual view of life-form, each form includes, according to H. Reiter, "organisms which with reference to their life functions and the structures relevant to these correspond in all essential respects". Thus each life-form is conceived as the collective attributes of plants that are judged to bear a strong resemblance to one another. When such groups have been assembled and described it remains to arrange them into classes of a higher rank, again aggregating those life-forms that share the same principal features and segregating those that differ in

Two typically coniferous life-forms are illustrated by these two close-ups of shoots: (**right**) the needle-leaf of the Scots pine (*Pinus sylvestris*) (life-form class 1g); and (**far right**) the scale leaf of the Lawson cypress (*Chamaecyparis lawsoniana*) (life-form class 1h).

fundamental respects. This process of classification results in different arrangements, even of predominantly the same life-forms, according to the criteria chosen; it has always been regarded as a difficult task. Humboldt declared that life-forms are "by their nature not capable of strict classification". Again, a century later it was still, according to Johannes Warming, "considered an intricate task to arrange the life-forms of plants in a genetic system, because they exhibit an overwhelming diversity of forms, ... also because it is difficult to discover guiding principles that are really natural".

With ecological relevance in mind, Warming, Andreas Schimper, Josef Schmithüsen and others classify life-forms partly by an assessment of the plant's response to availability of water supply and severity of water loss (by transpiration). The importance of moisture conditions was first recognized in 1849 by Jules Thurmann, who used the terms hygrophilous and xerophilous for plants inhabiting moist and dry soils respectively. By 1898 Schimper and Warming used mesophyte and xerophyte to denote plants adapted to atmospheric conditions of moisture and aridity, respectively. The term tropophytic indicated plants with deciduous foliage which experience the conditions of hydrophytes and xerophytes in alternate seasons, with consequent modification of form. Various adaptations of structure and function can be accommodated in classification by grouping those life-forms which qualify plants for survival under like conditions of abundance, shortage or fluctuation in water supply.

Twenty life-form classes are now recognized, comprising up to about 100 life-forms. Many of them were named by Drude, who also defined as principal life-form types: woody plants (classes 1, 3, 5, 9), semiwoody (classes 2, 6, 7), perennial herbs (classes 11, 12, 13, 14), monocarpic herbs (classes 16, 17), water plants (classes 18, 19) and thalloid plants (classes 17, 20).

Class 1. Crown trees. Distinct trunk supporting a crown of branches, stem, branch and root woody due to secondary thickening: many dicotyledonous angiosperms and conifers, few (if any) monocotyledons.

1 a) Evergreen crown-trees; hygromorphic to mesomorphic, very tall smooth trunk, some with plank-buttresses, thin bark, cauliflory (formation of flowers on old wood) common, buds without bud-scales, leaves of large to middle size, entire, long-oval, often with apex extending to a long point ("drip-tip"): Tapang type, principal form of tropical rain forest.

1 (b) Tropophytic crown-trees (ie trees adapted to a moist summer and a dry winter); seasonally deciduous, leaves of varied size and form, mesomorphic to hygromorphic, falling in dry season or in winter, thick protective bark and bud-scales:
  (i) rain-green trees of the tropics and subtropics, eg teak.
  (ii) summer-green trees of the temperate zone, eg beech.

1 c) "Bottle-stem trees"; tropophytic trees with grossly thick flask-shaped trunk acting as a water reservoir, predominantly deciduous with firm rather large leaves: trees of the seasonally dry tropics, especially members of the families Bombacaceae and Sterculiaceae, including the baobab tree (*Adansonia digitata*).

1 d) Thorn-trees; extremely xeromorphic trees typically with wide-spreading flat-topped crown, gnarled branches, thick bark and commonly thorny twigs, some evergreen with small stiff leaves, others deciduous, pinnate with small leaflets: semi-arid regions of tropics and subtropics, eg the umbrella acacia (*Acacia oswaldii*) of East Africa.

1 e) Sclerophyll trees; xeromorphic, evergreen crown-trees of low to medium height with rather small, hard or stiff leaves, often prickly on margins, bark thick, buds protected: characteristic of Mediterranean, eg *Quercus ilex*, and similar climatic regions.

The Australian cycad *Macrozamia riedlei* is a drought-resistant palm fern illustrating life-form class 2b: xeromorphic tuft-trees.

1 f) Evergreen broad-leaf trees; mesomorphic with firm, more or less leathery leaves of medium size, bark thin and buds unprotected: laurel type of humid warm temperate regions, eg Canary Islands (cloud forest) and of tropical montane forest.

1 g) Needle-leaf trees; xeromorphic, predominantly coniferous, with tough, usually persistent needle leaves (evergreen), some species deciduous, all with scaly or thick protective bark and bud scales: pines and firs of boreal and semi-arid montane regions, eg Afghanistan.

1 h) Scale-leaf trees; extremely xeromorphic, with leaves reduced to closely overlapping scales, represented by the cupressus type among conifers and by the tamarisk type among angiosperms: semidesert.

1 i) Succulent-leaf trees; leaves usually thick, glossy with tough cuticle, trees with low-branching stems and typically either stilt-roots or pneumatophores ("knee-roots"), seedlings germinating from fruits while still on the tree: mangrove swamps of coasts in the tropics and subtropics.

1 j) Xylopod trees; trees with underground water storage in distended woody stem base of rootstock (xylopodium or lignotuber): known mainly from dry northeast Brazil, eg members of family Sterculiaceae.

1 k) Strangler-trees; unique characteristics give them affinities with lianas and epiphytes (see classes 3 and 4) in the early stages of their growth but at maturity they are crown-trees; they germinate as epiphytes then extend aerial roots down to the soil, twisting round the host tree and thickening until they support the strangler-tree on multiple or latticed trunks: banyan form.

Class 2. Tuft-trees. Distinct trunk typically unbranched and terminating in a dense rosette of very large leaves; in other cases sparingly forked or even openly branched, each stem ending in a single tuft of leaves; leaf-forms varied: found only among ferns (Cyatheaceae), gymnosperms (cycads), monocotyledons (palms, screw-pines, Liliaceae) and a few genera of dicotyledons showing convergent evolution.

2 a) Hygromorphic tuft-trees; up to 10–20m (33–66ft) high, unbranched, leaves soft, large and repeatedly pinnate: tree-ferns, especially in frost-free shaded habitats, mainly tropical, but note *Dicksonia antarctica*, 3m (10ft) high, in Tasmania.

2 b) Xeromorphic tuft-trees; with leaves

Kingia australis represents the "grass-tree" life-form, class 2d. In the picture, the tip of the stem is shown bearing numerous stout-stalked flower heads.

divided either pinnate or fan-like, firm, leathery to stiff: palms and palm-ferns (cycads).

2 c) Candle-trees; single trunk, unbranched, bearing at the top a dense tuft-rosette of broad strap-like leaves from which arises a candle-like flowering axis: "frailejon" type of high altitudes on tropical mountains, eg *Espeletia* (Andes), giant lobelia (East Africa).

2 d) Grass-tree type; relatively short trunk with forked branching and terminal tufts of long, narrow overhanging leaves (genera of Liliaceae: *Xanthorrhea*, *Kingia*, *Dasylirion*) or more richly branched and up to 18m (60ft) high: *Dracaena*, *Yucca*, *Aloë*, and some euphorbias (candelabra form).

2 e) Screw-pine type; hygromorphic tuft-trees with stilt-roots supporting a stout forked trunk clothed at the top by long leaves spirally arranged and drooping at the tip: *Pandanus*, tropical lowland and montane rain forest.

2 f) Succulent-leaved tuft-trees; short trunk bearing erect rosettes of thick massive spade- or sword-like leaves: some species of *Aloë* (South Africa), *Yucca* (Central and North America), *Puya* (South America).

2 g) False-trunk trees; unbranched, non-woody plants up to 10m (33ft) high with false stems of densely imbricate fibrous leaf sheaths supporting a spreading crown of very large mesomorphic leaves: banana type, in moist shaded tropical conditions.

2 h) Cane-trees; stems up to 25m (82ft) high and 25cm (10in) thick, woody but without secondary thickening, "jointed" by ring-like leaf scars, with tufts of leafy twigs on the upper nodes: bamboo type.

Class 3. Lianas. Climbing woody plants with long branching stems not capable of supporting themselves, rooted in the soil, their stems ascending by attachment to

Lianas in semi-evergreen forest, Antigua, West Indies, illustrating life-form class 3: climbing woody plants with long branching stems not capable of supporting themselves.

A staghorn fern (*Platycerium*) growing epiphytically on a rain-forest tree. This is an example of life-form class 4b: funnel epiphytes with leaves modified to catch water.

Genista lydia, a switch-shrub of the broom type with grey-green photosynthetic stems—an example of life-form class 5h. The leaves (not visible in the picture) are minute and deciduous.

trees etc by various special means.

3 a) Twining lianas.

3 b) Root-climbers, eg ivy.

3 c) Tendril-climbers.

3 d) Scrambling lianas, with wide-angled branching and hooks or spines that prevent downward slipping: pre-eminently in the wet lowland tropics.

Class 4. Epiphytes. Germinating from seed or spores on the bark, branches and forks of trees and shrubs, possessing means of attachment and special means of water absorption or tolerance of temporary dessication.

4 a) Semiparasitic (eg mistletoe).

4 b) Funnel epiphytes with modified leaf forms to collect rainwater.

4 c) Nest epiphytes with richly branching roots that collect humus.

4 d) Epiphytes with pendant aerial roots anatomically modified to absorb water from their surfaces.

4 e) Bryoid epiphytes including mosses, leafy liverworts and filmy ferns.

4 f) Thalloid epiphytes including nonleafy liverworts and lichens.

Class 5. Shrubs. Woody stems branching from the ground, no distinct trunk, generally 50cm to 5m (20in–16ft) high.

5 a) Hygromorphic, evergreen shrubs with large broad leaves, buds unprotected: shade plants of wet tropical forest.

5 b) Tropophytic, seasonally deciduous shrubs with buds protected:

(i) summer-green shrubs of temperate regions, including species important at subalpine boundaries of forest, eg *Betula tortuosa* (Finland), *Nothofagus antarctica* (Chile).

(ii) rain-green shrubs of the seasonally dry tropics and subtropics.

5 c) Mesomorphic, evergreen shrubs with medium firm leaves; laurel-leaf forms of

moist warm-temperate (montane) regions, eg *Polylepis* (Andes), *Hagenia* (East Africa), *Hebe* (New Zealand).

5 d) Scleromorphic, evergreen shrubs with small rigid hard leaves or leaf substitutes (phyllodes), eg *Phillyrea* (Mediterranean), *Acacia aneura* (Australia), common in semi-arid regions.

5 e) Xeromorphic ericoid shrubs; leaves as in the heath family (Ericaceae), short linear with inrolled margins protecting stomata: also occurring in other families, eg Frankeniaceae, Epacridaceae, Myrtaceae.

5 f) Xeromorphic coniferous shrubs; leaves small, closely overlapping, scale-like or short, needle-like, including plants of semi-arid habitats, eg junipers, and some timberline genera of tropical mountains, eg *Libocedrus*, *Dacrydium*.

5 g) Thorn-shrubs; xeromorphic, both evergreen and deciduous: regions with extreme heat and aridity, eg *Prosopis* (North and Central America), *Acacia* spp. (South America).

5 h) Switch-shrubs; xeromorphic, either leafless or with scale-like minute leaves: eg Spanish broom (*Spartium junceum*), tamarisk.

5 i) Tola-shrubs; xeromorphic, with hard small, scale- or needle-leaves, resinous, aromatic and inflammable: puna regions (South America).

5 j) Succulent shrubs; a woody branch system bears succulent leaves of various forms: salt marshes and salt-deserts.

5 k) Xylopod shrubs; with water-storing underground tuberous stems from which woody branch systems arise, eg shrub species of *Eucalyptus*.

Class 6. Semi-shrubs. Weakly lignified stems, much of the shoot system of one- or two-year duration.

6 a) Mesomorphic type; leaves soft, not

rolled or hairy, persisting into second season, stems biennial, arising from a woody rootstock (lignotuber): bramble type (*Rubus* spp.) of humid temperature regions.

6 b) Sage-leaved type; xeromorphic, with soft leaves often gray-green, partly rolled, more or less hairy or densely woolly, persisting except in severe drought, eg *Artemisia*, *Lavandula*, *Salvia*: semi-arid regions.

Class 7. Succulents. Included in this class are plants more than 50cm (20in) high which are either stem-succulents or rosette leaf-succulents, the former leafless, with photosynthesizing green tissue just beneath the stem surface, the latter with grossly thick massive leaves; both weakly lignified and with water-storing tissue.

7 a) Candelabra form; erect trunk branching into upturned thick stems.

7 b) Columnar form; erect trunk unbranched or bearing only pear-shaped shoots, height up to 10m (33ft), eg *Cereus*.

7 c) Shrubby form; all shoot branches are succulent stems, eg *Opuntia*. All three types represented pre-eminently in Cactaceae (New World) and Euphorbiaceae (Old World).

7 d) Agave form; low stocky shoots (no distinct trunk) bearing erect, succulent sword-like leaves up to 2m (6.5ft) high.

Class 8. Dwarf succulents. Generally less than 0.5m (1.6ft) high (rarely up to 1m or 3.3ft), though often wider than this.

8 a) Globose-columnar form; unbranched leafless, globe- or club-shaped.

8 b) Shrubby form; branched, shoots leafless cylindrical, flattened, ribbed, eg Asclepiadaceae.

8 c) Branched trailing form; shoots leafless cylindrical, spherical, oval etc.

8 d) Leafy stem-succulents; tubular stems

The banana "tree" (*Musa acuminata*) is a giant herb (life-form class 11) with enormous leaves. It originated as a weed of tropical forest clearings. The flowers develop at the tip of the inflorescence with the males in the axils of the red bracts and the females behind.

Dwarf willow (*Salix herbacea*), a creeping dwarf shrub no more than 5cm (2in) tall that forms dense carpets. This arctic and temperate mountain species forms a close mat of short shoots from an underground stem, and is an example of life-form class 9: dwarf shrubs. The dwarf habit is an adaptation to the harsh conditions of low temperature and little available water typical of the habitats in which it grows.

Gypsophila imbricata is a cushion plant (life-form class 10a), which forms extremely hard cushions up to 1m (3.3ft) across. It grows on cliffs or flat rocks where it is often exposed to drying winds and high temperatures.

bearing mesomorphic leaves after rains, eg *Pelargonium* (South Africa), *Oxalis sepalosa* (Central Andes), *Senecio* (Mexico, Canary Islands).

8 e) Dwarf leaf-succulents; either with trailing branched shoots, eg *Sedum* (Crassulaceae), *Carpobrotus* (Aizoaceae), or with leaf-rosettes on the ground surface, eg *Sempervivum* and *Mesembryanthemum*.

8 f) Pebble-plants; obovoid succulent leaves embedded in the ground: extreme hot desert conditions (Southwest Africa).

Landscape in the Organ Pipe Cactus National Monument, Arizona, United States, with organ-pipe cactus (*Lemairocereus thurberi*), barrel cactus (*Echinocactus acanthodes*), saguaro (*Cereus giganteus*), cholla (*Opuntia bigelovii*) and ocotillo (*Fouquiera splendens*), examples of succulents from life-form classes 7a, 7b and 7c.

Class 9. Dwarf shrubs. Generally less than 50cm (20in) high but often spreading extensively. In addition to dwarf shrubs corresponding to the shrub categories 5 b), d), e), f), h), j), k) and distinguished only by their lesser height, we have:

9 a) Creeping dwarf-shrubs; richly branched and spreading in a more or less dense carpet often only a few centimetres high, stems usually on the ground surface, eg *Loiseleuria* (evergreen), *Arctostaphylos* (deciduous) but more rarely below the surface, eg *Salix herbacea*: important in subpolar and subalpine environments.

Class 10. Cushion plants. The shoot system, either woody or not, is richly branched but the branches are short and densely packed so that the entire plant assumes a mound-like form situated directly on the ground surface. This compact habit withstands strong winds; it also increases the temperature within the cushion in cold climates, and within the cushion the water loss by transpiration is much reduced—an adaptation to dry air conditions.

10 a) Hard-cushion plants; extremely xeromorphic, the woody stems densely compact, small-leaved, resinous, forming semiglobose hummocks up to 1m (3.3ft) high; some even resemble boulders, eg *Bolax gummifera* (Umbelliferae): regions of severe wind exposure and frost, deserts, subantarctic zone of the Andes and New Zealand, eg *Azorella*, *Raoulia*, *Haastia*, *Oreobolus*, *Fredolia* (Sahara).

10 b) Thorn-cushion plants; growth mainly prostrate with much-branched thorny shoots spreading close-pressed to the ground to form flat cushions of irregular outline, small-leaved foliage: chiefly in high mountain regions with summer drought, eg Afghanistan, western Himalayas, northern Chile. Examples in Caryophyllaceae, Cruciferae, Leguminosae.

10 c) Herb-cushion plants; stems scarcely woody or nonwoody but shortly branched to form dense mounds or flat cushions: many rock plants of high mountains in temperate regions, eg *Androsace helvetica* (Alps), *Silene acaulis*, *Armeria maritima* (northwest Europe).

10 d) Moss-cushion plants; hemispherical mounds and hummocks formed by species of mosses having various habits of growth, eg erect (*Polytrichum commune*), densely compact (*Leucobryum glaucum*), hygromorphic (*Sphagnum* spp.): may be subdivided accordingly.

Class 11. Giant herbs. Hygromorphic, soft-stemmed plants often up to 4–5m (13–16ft) high, with large or even huge leaves supported on massive stalks: shade plants of tropical wet forest, eg *Begonia* spp. and various Araceae.

Common vetch (*Vicia angustifolia*) is a grassland herb belonging to life-form class 13e: trailing-herbs with weak ascending stems bearing tendrils or minutely toothed stems which enable them to gain support from other plants.

Tulipa michaeliana, a bulbous perennial (life-form class 14a), is the commonest tulip in northeast Iran, growing generally in wheat fields and flowering while the wheat is only a few inches tall.

Class 12. Tussock herbs. Perennial, non-woody plants with renewal-buds above the ground surface within tufts of dense shoots or on the stump-like base from which the shoots emerge.

12 a) Hygromorphic tussock herbs; broad-leaved herbs with seasonal foliage (tropophytic); ground ferns with leaves arising from buds protected by dense scales on the stump-like upright stem.

12 b) Tall tuft grasses; tropophytic grasses with persistent tufts of narrow inrolled leaves (xeromorphic) from which tall culms bearing mesomorphic leaves sprout in the rainy season: savanna grasses of the seasonally humid tropics, eg *Panicum, Paspalum, Sporobolus, Aristida.*

12 c) Bunch grasses; persistent in low grass tufts with hard (scleromorphic) narrow inrolled leaves and extensive root systems; important in the steppe of seasonally hot semi-arid regions, eg esparto grass (*Lygeum spartum*) in North Africa, ichu-grass (*Stipa ichu*), in the Andes, "Spinifex" (*Triodia basedowii*) in Australia.

12 d) Tussock grasses (and sedges); forms whose erect and tight-packed culms build tussocks varying from low stumps to pillars 1.5m (5ft) high. Leaves seasonally deciduous (*Molinia caerulea*) or tough and persistent (*Carex paniculata*). Characteristic in humid, oceanic, cool-temperate regions and in the corresponding high altitude conditions (paramo), eg *Poa flabellata* (subantarctic), *Stipa brachychaeta* (Argentina), *Mesomelaena sphaerocephala* (Cyperaceae, Tasmania).

Class 13. Turf-dormant herbs. Herbaceous plants, generally tropophytic, with renewal-buds situated at the ground surface, protected within the tuft or (on the forest floor) by leaf litter: pre-eminently in regions where the annual period of growth is interrupted by a season of low temperatures when the aerial shoots die down.

13 a) Rhubarb-herbs; hygromorphic, large-leaved herbs with the leaves carried on stout petioles springing directly from the thick rhizome at soil level, intermediate between giant herbs and forms with buried rhizomes: eg *Rheum* (Asia), *Gunnera* (South America) in moist habitats.

13 b) Stem-herbs; mesomorphic, with erect leafy stems; many tall meadow and forest herbs with annually renewed stems, leaves stalked or sessile, chiefly on the erect stem though some may be basal (ie including "semi-rosette" herbs): eg willowherb, goldenrod, hemp agrimony.

13 c) Rosette-herbs; all leaves in a basal rosette, and attached at virtually the same level. The flowering stems have few or no leaves. These herbs resemble in their growing season the conditions of biennial herbs in the wintering season, eg many *Saxifraga* species, dandelion, *Primula, Plantago.*

13 d) Creeping herbs; lacking erect leafy stems, possessing stolons or other means of spreading vegetatively either along the soil surface or closely beneath it in the turf, eg *Rumex acetosella, Agrostis stolonifera, Potentilla anserina*; or with creeping leafy stems, eg *Lysimachia nummularia, Veronica repens.*

13 e) Trailing-herbs; with weak ascending stems, some with tendrils or minutely toothed stems aiding their support among other plants, eg many species of bedstraw (*Galium* spp.) and vetch (*Vicia* spp.).

Class 14. Underground perennials. Herbs possessing underground organs from which the aerial shoots are produced annually during a limited growing season, then either withering or persisting functionally into the next growing season (ie tropophytic or evergreen): plants occupying extremely diverse situations and showing corresponding variety of shoot forms but in temporal behavior all capable of avoiding or tolerating periodic adverse conditions.

14 a) Bulb-tuber plants; herbs with shoots arising annually from bulbs, stem- or root-tubers and dying above ground at the end of their growing season: varied forms include ground-rosettes (Liliaceae, Amaryllidaceae) and leafy stem herbs (Orchidaceae). Habitats include the forest floor, especially seasonal forests (eg *Arum maculatum, Ranunculus ficaria*), open ground in winter-rain climates (*Urginea maritima*, some *Iris* spp.) and even deserts where growth is ephemeral (*Pancratium* spp.).

14 b) Rhizome herbs; with rhizomes buried in the soil below the turf or litter layer; rhizomes may be wiry, cord-like or thick and more or less fleshy: this group also includes forest herbs (eg *Polygonatum, Convallaria*), plants of seasonally arid regions (*Iris* spp.), reed and swamp plants

Tussock grassland (*Poa* spp.) in New South Wales, Australia, with eucalyptus woodland beyond. The tussocks are an example of life-form class 12d.

The handsome spear thistle (*Cirsium vulgare*) is a biennial herb (life-form class 15b). It produces a rosette of leaves in the first year, and a tall flowering stem in the second, after which it dies.

Chickweed (*Stellaria media*) is a creeping annual (life-form class 16d) that can produce several generations in a single growing season. Such species are ideally adapted to the rapid colonization of disturbed ground.

Lichens are classified as mat-forming cryptogams (life-form class 17c). The yellow species in this picture is *Rhizocarpon geographicum*, so named because of the resemblance of its uneven outline to a map.

(*Phragmites australis, Sparganium erectum*). In some, the aerial shoots die and disappear seasonally but in others they persist and remain functional year-round (eg *Ammophila arenaria, Cladium mariscus*).

14 c) Root parasites and saprophytes; nongreen plants nutritionally dependent either on living plants by attachment to their root system or on decomposing dead plant matter in the soil: eg toothwort (*Lathraea squamaria*), *Cytinus hypocistis*, *Rafflesia arnoldii*.

Class 15. Winter annuals and biennials. Herbs with a one- or two-year life span: transitional between perennials and those

A papyrus (*Cyperus papyrus*) swamp fringing Lake Victoria in Africa. Portions of the papyrus mat are broken off by rises in the water level and form floating islands. This is an example of life-form class 18: sudd-plants.

enduring for only the length of a single growing season (see Class 16).

15 a) Winter annuals; from autumn germination these produce a leaf rosette at ground level which survives the winter (as in turf-dormant perennials), then gives rise to a flowering stem and ripens seed early in the next growing season, passing the summer period of drought as seeds.

15 b) Biennials; similar to the above but requiring two full growing seasons with an intervening winter to complete their life-cycle, first producing large-leaved rosettes and in the second year tall leafy stems (as in turf-dormant stem-herbs): eg foxglove (*Digitalis*), thistles (*Cirsium*).

Class 16. Ephemeral herbs and summer annuals. Completing their life-cycle from germination to seed production within a short growing season (as brief as 8–15 days), or several generations within a growing season of longer duration, summer annuals requiring the full grow-

ing season to reach flowering and to set seed. A wide range of shoot forms includes:

16 a) Stem-herb annuals, eg sunflower (*Helianthus annuus*).

16 b) Rosette annuals, including desert plants with succulent leaves which germinate after rainfall and then complete their growth using stored water.

16 c) Trailing annuals, eg *Fumaria vicia*.

16 d) Creeping annuals, eg *Stellaria media*.

16 e) Climbing annuals, eg *Polygonum convolvulus* and *Echinocistus lobata*.

16 f) Parasitic annuals, eg *Cuscuta*.

16 g) Annual grasses.

Thus most of the shoot forms found in herbaceous perennials are represented.

Class 17. Mat cryptogams. Clothing the surface of soil or rock, stones etc in a close mantle or mat.

17 a) Leafy cryptogams: mosses, leafy liverworts and filmy ferns (Hymenophyllaceae).

17 b) Green thalloid types: thallose liverworts (Hepaticae).

17 c) Lichen mat, including the so-called foliose, fruticose and thallose lichens.

Class 18. Sudd-plants. Plants with floating stems forming rafts from which tall erect leafy shoots arise, capable of withstanding large fluctuations of water level including summer drought: usually grasses and sedges, occurring in the tropics, eg *Cyperus papyrus, Vossia cuspidata*.

Class 19. Water-herbs. Restricted here to plants whose entire form is adapted to life in water and which therefore differs uniquely from the forms of terrestrial and semi-aquatic plants.

19 a) Floating-leaf water-herbs; these are rooted in the sediment below water and long ascending stems or petioles carry the leaves to the water surface: eg *Victoria regia, Nymphaea, Potamogeton natans*. Some also possess submersed leaves and these belong to the following subclass

Salvinia cuculata, a floating aquatic cryptogam (life-form class 20a). It is an attractive water fern of tropical regions.

Ranunculus penicillatus, pictured here in the River Wear at Durham, northern England, is anchored to the sediment in the river bottom, and is classified under life-form class 19a: floating-leaf water-herbs.

because they are capable of living in conditions where formation of floating leaves is prevented.

19b) Submersed water-herbs: usually rooted in the underwater sediment, the vegetative shoots not projecting above the water surface, leaves of various forms and arrangements: eg rosette type, *Lobelia dortmanna*, *Luronium natans*; long-stemmed type, *Myriophyllum*, *Potamogeton*, *Batrachium*, buttercups.

19c) Natant water-herbs; the whole plant body floats at the water surface with leaf surfaces partly exposed to the air, roots pendant in the water: eg *Stratiotes aloides*, *Trapa natans*, *Lemna*.

Class 20. Aquatic cryptogams. Comprises nonflowering plants whose form often differs remarkably from their terrestrial allies.

20a) Natant floating cryptogams: eg *Salvinia natans* (water fern), *Azolla*, *Riccia fluitans*.

20b) Benthic (bottom-living) cryptogams: eg *Isoetes*, *Pilularia* (water fern), *Fontinalis* (moss):

The foregoing synopsis relies upon many of its predecessors, especially the authors previously mentioned; indeed a large measure of agreement as to what constitute recognized life-forms is essential if the concept is to have any valuable application. However, the arrangement of these forms into a structured classification is still subject to reasoned opinion.

By contrast, objective classification commences with a predetermined series of major categories to which any plant may be allocated by observing usually just one or two major characteristics which offer several clear-cut alternatives, eg woody or non-woody, with a single stem, multiple stems or no visible stem. The primary categories are subdivided by reference to another characteristic, which ideally should be measurable,

eg plant height, leaf size. Much depends upon the choice of criteria at each stage for the naturalness or artificiality of the resulting classification.

Raunkiaer (1903) devised a scheme based on the position of the perennating buds or organs in relation to the ground surface, ie whether underground, at soil level, or in various height intervals above it. Using a single primary criterion enables all plants to be compared in this one respect independently of form. The position of renewal buds and the measure of protection afforded by, for example, snow cover, leaf litter and the soil itself at progressively lower levels were believed to reflect the climatic severity of the season unfavorable to growth. Plants with buds exposed on twigs high above the ground he named phanerophytes, subdividing them into measured intervals by height: megaphanerophytes taller than 30m (98ft), mesophanerophytes 8–30m (26–98ft), microphanerophytes 2–8m (6.5–26ft) and nanophanerophytes 25cm–2m (10in–6.5ft). Plants with buds exposed on stems near to the ground up to 25cm (10in) he named chamaephytes. Cryptophytes have their winter buds and other surviving organs situated either beneath the soil (geophytes) or below water in semi-aquatic (helophytes) or aquatic (hydrophytes) situations. Plants

which survive the unfavorable season as seeds Raunkiaer termed therophytes.

Raunkiaer further proposed a series of leaf-size classes. These are leptophyll, up to 25mm²; nanophyll, 25–225mm²; microphyll, 225–2,025mm² (0.35–3in²); mesophyll, 2,025–18,225mm² (3–28in²) and including the subdivisions nothophyll, 2,025–4,500mm² (3–7in²); macrophyll, 18,225–164,025mm² (28in²–1.7ft²); and megaphyll, over 164,025mm².

Most widely used of the objective classifications are those based on the position of the perennating buds. They may be broadly related to the majority of life-forms detailed above. Thus phanerophytes include crown-

A representation of the life-form system of Raunkiaer based on the position of perennating buds. 1) Phanerophytes (buds on twigs high above the ground level); 2a) and 2b) chamaephytes (buds near to the ground); 3a), 3b) and 3c) hemicryptophytes (b is at soil level); 4a) and 4b) geophytes (buds below the soil surface); 5) therophytes (surviving the unfavorable season as seeds).

trees, tuft-trees, shrubs, succulents, lianas; chamaephytes include tuft-herbs, cushion plants, dwarf shrubs; hemicryptophytes correspond largely with turf-dormant herbs, geophytes with underground plants, and therophytes with summer and winter annuals and ephemerals. However, defined height-classes cannot be exactly reconciled with life-forms founded on physiognomy.

Life-forms of all designs show some correspondence with major climatic regions of the world. Thus, excepting stem-succulents, phanerophytes are most abundant where warmth and moisture are prevailing features of climate; in the humid tropics without a marked dry season, phanerophytes comprise 60–90% of the flora. Chamaephytes are characteristic of arctic and alpine regions where snow gives winter protection to buds close to the ground, while in cool temperate regions, such as Denmark, 50% of species are hemicryptophytes. Therophytes include summer annuals of cold-winter climates, winter annuals of dry-summer climates and ephemerals of deserts. Although few life-forms, however delimited, are exclusively found in a single region, they are valuable since they recur in regions showing some resemblance or equivalence in climatic conditions, and they are essential aids in describing the world's extraordinary range of vegetation.　　　　　B.S.

Phytosociology—the Floristic Composition of Vegetation

Phytosociology (sometimes called phyto-coenology) is difficult to define exactly as the word has been used in several different ways. Taken at face value it simply means plant sociology, that is the study of plants as gregarious organisms. It is thus concerned with the sociological interactions between plants (both of the same and of different species) and the way in which they grow together to form plant communities. Hence it can be considered to be the science of vegetation, concerned with the social organization of plant material. In this sense phytosociology has an enormous scope, ranging from primary descriptions of stands in terms of their structure, physiognomy and floristic composition, to the detection of which species are recurrently found growing together and the detailed determination of the factors and processes that are instru-

mental in causing this (see Pattern, p. 48). Such sociological studies can involve not only a consideration of the response of the individual species to their environment and the social interactions between them (competition, dependence etc) but also a whole range of community phenomena such as the change in species' performance with seasonal change, population age structures, natality, mortality etc, together with longer-term studies on compositional changes.

However, the term phytosociology is often used in a narrower sense to refer to the floristic description of vegetation (community-types), their classification, naming and distribution. Often the aim has been to describe and classify the vegetation of particular regions. Such broadly-based phytosociological surveys have been of considerable importance in documenting the vegetation of various parts of the world and when they have been conducted in sufficient detail they have led to the production of vegetation maps for the areas concerned. Such surveys also provide the descriptive framework into which more detailed studies on particular plant communities can be located.

Approaches to Phytosociological Surveys

The recognition of units of vegetation requires several stages: firstly, the selection of the characters of vegetation that are to provide the basis of the unit, then the sampling of an appropriate range of vegetation, followed by the comparison of samples, leading to the extraction of vegetation units and then their characterization.

Various properties of plant material have been used for defining vegetation units. Plant growth-form and appearance have frequently been employed as the basis of physiognomic vegetation types. Many of the earliest descriptions of vegetation were along physiognomic lines and, indeed, this has proved to be particularly successful in categorizing global types of vegetation. But it is the component plant species that have been most generally employed in various ways to generate vegetation units based upon their floristic composition.

The Zürich–Montpellier Approach
Perhaps the most fully developed school of phytosociological research in Europe is the Zürich–Montpellier (ZM) school (or Braun-Blanquet school). It developed around 1913 out of the physiognomic tradition of early European ecology by a shift of emphasis to the use of floristic composition as the basis for vegetation units. It has subsequently been refined into a comprehensive system of plant sociology which has found widespread application not only throughout most of Europe but also in various other parts of the world.

In the ZM approach, vegetation is described subjectively by means of the sampling

unit called a relevé (or Aufnahme). This is usually related to the concept of the minimal area, which is the smallest area needed for a plant community to attain its full expression. It is usually estimated as the size of the quadrat of half the area. For a particular type of vegetation, such as oak woodland, the minimal area is found from the examination of a number of examples of that type. The minimal area of oak woodland is about 200m² (2,150ft²), for heather heath about 16m² (170ft²), and for acid grassland about 9m² (100ft²). In the classification of plant communities, only those plant stands which cover at least the minimal area can be usefully considered for classification because only these will contain the full complement of species in their characteristic abundance. The relevé, which must be larger than the minimal area of the community, is located in a uniform stand. It consists of a list of all the species present, with estimates of their cover and sociability by means of two code numbers, annotated with observations on the characteristics of the vegetation and habitat. Sociability is used to indicate how the species are distributed within a stand of vegetation. It is estimated for each species on a five-point scale: (1) growing as solitary plants; (2) forming clumps, tufts, tussocks or dense groups; (3) forming small patches or cushions; (4) forming small colonies or large patches; (5) forming very large colonies, almost pure populations. Recently, however, there has been a tendency to discard sociability estimates, largely because sociability tends to be a characteristic feature of each species.

The lists of flora obtained by this means are arranged into a matrix (or raw table) in which columns represent the samples and rows represent the species. The intersections contain information on cover (see p. 47) and sociability for each species in the samples. The positions of the rows and columns are then subjectively re-ordered to reveal correlations between the records of different species which can be used to partition the samples into distinct groups. This is often achieved by the recognition of mutually exclusive sets of differential species (species of restricted occurrence in the raw table which tend always to occur together) followed by tabular re-ordering based upon these. The process thus achieves a simultaneous classification of both the samples and the species, and uses the floristic information itself to display that classification in the final differentiated table.

The samples, or relevés, are grouped into classes of unspecified rank, called phytocoena. The percentage occurrence of a species in the samples of such a class constitutes its constancy or presence. Each phytocoenon may be condensed to form a single synoptic list in which the constancy of each species comprising the unit is indicated. A number of such lists can be incorporated into a synoptic table. This may be manipulated in the same way as the raw table. It permits the comparison of a large number of phytocoena, and hence the assessment of the sociological amplitude of

Opposite Raunkiaer's classification of the life-forms of plants depends upon the position of their perennating buds. In this winter scene in southern England it is clear that the trees and shrubs (oak and hazel) are phanerophytes, since their next year's growth will start from buds held more than 1m (3.3ft) above the ground surface, while the grasses in the foreground are obviously hemicryptophytes, with their growing buds at soil-level. Geophytes such as the bluebell, with their buds below the soil surface, are hidden by the snow.

RELEVÉ NUMBER

20 31 23 34 33 22 21 1 3 8 2 6 7 11 32 35 36 37 38 40 39 41 24 25 26 27 28 29 30 9 10 5 16 17 18 19 13 12 4 15 14

Association
Festuco-Nothofagetum antarcticae
    Festuca purpurascens
    Luzula chilensis

Association
Triseto-Nothofagetum antarcticae
    Trisetum caudulatum
    Antennaria chilensis
    Ranunculus peduncularis
    Geranium patagonicum

ALLIANCE
Agropyro-Nothofagion antarcticae
    Agropyron fuegianum
    Hieracium antarcticum

ALLIANCE
Escallonio-Nothofagion antarcticae
    Nothofagus antarctica
    Escallonia virgata
    Hypochoeris arenaria
    Stellaria media
    Baccharis patagonica

ALLIANCE
Viola-Nothofagion pumilionis
    Nothofagus pumilio
    Viola magellanica

Two of the alliances determined from the differential table are shown here.

Top *Escallonio-Nothofagion antarcticae*. Lago Porteño, Ultima Esperanza, Magallanes, southern Chile. In moist areas by the lake there is woodland dominated by *Nothofagus antarctica* (behind) and *Escallonia virgata* (flowering, foreground). March, 1964.

Center *Viola-Nothofagetum pumilionis*. Dos Lagunas, Ultima Esperanza, Magallanes, southern Chile. Tall forest dominated by *Nothofagus pumilio*, with the well-drained ground covered with *Viola magellanica* (non-flowering in autumn, March, 1964).

Order
Nothofagetalia pumilionis antarcticae
    Poa nemoralis
    Poa patagonica
    Berberis ilicifolia
    Poa oligeria
    Agoseris coronopifolia
    Cystopteris fragilis
    Bromus coloratus

other Orders

CLASS
Nothofagetea pumilionis antarcticae
    Osmorrhiza chilensis
    Calceolaria palenae
    Galium aparine
    Cardamine glacialis
    Ribes magellanicum
    Codonorchis lessonii
    Geum magellanicum
    Vicia magellanica
    Blechnum penna marina
    Chloraea magellanica
    Bromus setifolius
    Acaena ovalifolia
    Cotula scariosa
    Galium fuegianum
    Poa scaberula

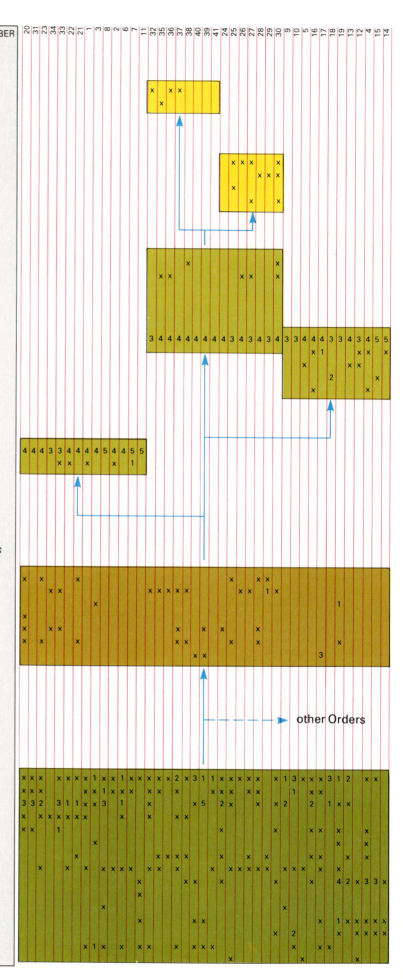

each species together with an estimation of its *fidelity* to each phytocoenon.

Five grades of fidelity are recognized. In the first (and lowest) grade are species normally occurring optimally outside the community; these are called *accidentals*. In the second grade are species showing no strong preference for any one community: these are called *companions*. The third grade comprises species occurring in several communities but slightly better developed in one; they are called *preferential species*. In the fourth grade are species occurring in several communities but with a clear preference for one. These are called *selective species*. The fifth grade comprises species which are restricted to one community, called *exclusive species*. Species of fidelity grades three to five are of particular value for characterizing units of vegetation and are known as *character-species*.

Having been extracted and compared, phytocoena are then arranged within a hierarchical classification of vegetation. When thus formalized, they are called *syntaxa*. The syntaxa of the ZM system, in ascending order of rank, are associations, alliances (which usually possess a characteristic physiognomy and structure), orders and classes. The basic unit is the *association*, a recurrent grouping of certain species of plants, which may be subdivided into subassociations, variants and facies (in descending order of rank). An important feature of the classification is that all syntaxa are given names and that the rank of a particular

syntaxon is specified by a suffix. These are: for the class: *-etea*; order: *-etalia*; alliance: *-ion*; association: *-etum*; subassociation: *-etosum*.

The distinctve features of the ZM approach are thus the floristic basis for the phytocoena, the use of selected diagnostic species for characterizing these, and the arrangement of syntaxa into a formal hierarchical classification.

The Uppsala Approach
At the same time as the ZM tradition was expanding under the influence of Josias Braun-Blanquet, there was a parallel growth of phytosociology in Scandinavia. The Uppsala school was pioneered by Rutger Sernander and subsequently developed by Gustav Du Rietz. Particular emphasis was placed upon the stratification of vegetation and the dominant species of the strata. The basic vegetation unit was the *sociation*. This was composed of a particular combination of layers, each characterized by a specific dominant. Sociations having a common dominant in the uppermost layer were sometimes united into *consociations*.

From about 1935 onwards, Uppsala has increasingly moved towards some of the concepts of the ZM school. More attention has been paid to the whole composition of plant communities, and floristic units (comparable with associations) have been recognized. Earlier sociations have, in some cases, been grouped into associations and occasionally the ZM hierarchy of syntaxa has been adopted. However, several workers have preferred to arrange phytocoena along gradients of changing floristic composition.

The British Approach
Early work in the present century, associated particularly with Arthur Tansley, led to a broad description of British vegetation in which the basic units ("associations") were defined by their dominant species. However, as one species may dominate several floristic communities, some of the units thus defined were rather broad in scope. Recently, several workers have successfully applied the techniques of the ZM school to British vegetation.

The Numerical Approach
In recent years, encouraged by the availability of digital computers and the growth of the parallel science of numerical taxonomy, ecologists have applied numerical procedures to the recognition of phytocoena. These are based upon some quantitative assessment of similarity between vegetation samples, the similarity measure usually being based on the total floristic content of each sample. Two procedures have been particularly important. The first is association analysis, a *divisive* procedure which aims at successively subdividing a set of samples into a number of smaller classes of increasing homogeneity. The second is information analysis, an *agglomerative* procedure, which starts with a number of individual samples and progressively fuses those that are most

similar to one another into groups. Such numerical methods, which require a computer for their successful application, are normally used with samples that have been collected with a minimum of user bias (ie randomly), but they can also be used with relevé data and in conjunction with some of the concepts and procedures of the ZM tradition.

The Synusial Approach
As well as the classification of whole stands of vegetation, attempts have been made to recognize subunits within plant communities and to classify on the basis of these. For example, the Estonian school of phytosociology, influenced by Theodor Lippmaa, classified the different strata of communities independently of one another. Such subunits of the stand are often referred to as *societies*, in this case *layer societies*, and the abstract vegetation units derived from them are *synusiae* (see p. 41). Other types of societies include microcommunities which occupy specific microhabitats within the stands, such as epiphytic mosses growing on trees. Also, plants that are important in a community for only part of the year such as spring herbs in woodlands, can be grouped into *aspect societies*. This approach thus emphasizes the relative independence of certain elements within the community. It suffers, however, from disagreement upon what synusial categories can be recognized. As the basic unit, the *union* (corresponding to association) has been used. Unions have been grouped into *federations*.

B.WR.

Ordination

The various attempts to classify plant communities referred to above have stemmed from a desire to simplify and order the complexity of composition displayed by vegetation. However, another approach, seeing vegetation more as a phenomenon showing continuous variation in composition, has been that of ordination, much favored by North American ecologists. *Ordination* is a phytosociological term meaning "ordering". It may be defined as an arrangement of units whereby their positions are determined by their properties. The units involved may be samples of stands or species. The properties on which their ordination is based can be various and depend upon the precise purpose of the ordination. In principle, at least, any relevant property may be used. Thus, stand samples can be ordered on the basis of such features as a measured environmental factor, the amount of a particular species, or on their total floristic composition.

The development of ordination as a phytosociological technique has been closely associated with the recognition that plant communities do not necessarily have a discrete identity but frequently display a continuum of gradation into one another.

Opposite Simplified differentiated table showing the floristic groupings revealed by a phytosociological study of the deciduous beech forests of southern Patagonia. Within such forests species present in 41 sample areas (relevés) were estimated for their cover (× : < 1%; 1: 1–5%; 2: 6–25%; 3: 26–50%; 4: 51–75%; 5: 76–100%). The sequence in which the relevés were studied is given by the numbers 1–41 in the top line. This information permitted the production of an unsorted matrix or "raw table". The relevés were then subjectively re-ordered so as to bring together those with the most similar species-composition, while the sequence of taxa was changed to show more clearly mutually exclusive groups of species. This then permitted the ordering of the relevés into groups of unspecified rank called phytocoena. Comparison between these phytocoena and others from surveys elsewhere lead to the hierarchical classification of vegetation shown here. These are formally termed syntaxa, and are listed to the left of this diagram. In all the samples the deciduous southern beeches *Nothofagus pumilio* or *N. antarctica* figure prominently, hence the name of the class, which is also characterized by the last 15 species in the list. Within the class, several mutually exclusive sets of species are recognized as orders, one of which is shown in this table. Within this order, three alliances of different floristic composition can be delimited; two of them have *Nothofagus antarctica* as a prominent exclusive species, the other has *N. pumilio*, and this is reflected in the names applied to the alliances. Two associations are shown within one of the alliances (*Agropyro-Nothofagion antarcticae*); these are distinguished by exclusive species of lesser importance and the exact delimitation of them is determined largely by preferential species not listed here.

This means that it may sometimes be difficult to apply the other, and more familiar, technique of ordering phytosociological data, namely classification. In classification, individuals with certain shared features are grouped together into a class and are thus distinguished from other groups of individuals with different shared features which occupy other classes. If, however, a set of individuals shows continuous intergradation in their features it can be difficult to recognize discrete classes with distinct boundaries between them. This does not mean that classification is impossible (the intergrading spectrum of color can be very conveniently classified, for example) but it does mean that the boundaries between the classes may be somewhat arbitrary.

In ordination, however, no attempt is made to recognize classes of similar individuals. Instead, emphasis is placed upon the individual and its relationships to other individuals. This arrangement is achieved geometrically. Simplest is the unidimensional ordination, in which an axis is constructed with a linear ranking of values from one end to the other, the values being those of the particular feature upon which the ordination is based. Plotting the samples along this will show their relations; those most dissimilar will be the furthest apart. Then, if desired, a second axis can be constructed, based upon another attribute, at right-angles to the first and the samples can be ordinated with respect to both axes. This process can be continued and additional axes constructed, although more than three axes (dimensions) are difficult to express graphically.

Stands may be ordinated along axes of particular environmental values, thus dis-

Sphenopteris elegans, a primitive seed-bearing plant found in Lower Carboniferous sandstone from a Scottish quarry.

playing the relationships of the stands to their gradients. The amounts of each species present in the samples can then be plotted onto the ordination to show the associated changes in composition. This is sometimes called direct gradient analysis.

Another approach is to use ordination to display, geometrically, the relations between stand-samples based upon their complete floristic composition. It is possible to arrange samples geometrically in such a way that the distance between them reflects their floristic similarity, although this can normally only be achieved by using more than three dimensions. Techniques are available, however, which are designed to extract one or a few axes along which the samples can be arranged to reflect, with as little distortion as possible, their compositional relationships. The simplest method is *Wisconsin comparative ordination*. Here the two most dissimilar samples are selected to form the two end points of one axis. The other samples are then plotted along this in positions defined by their similarity to both of the end samples. Then, if required, the two most dissimilar samples in the middle of this axis may be selected to form the end points of a second axis, and the samples are replotted to give a two-dimensional ordination. Again, the number of particular species at each site can be plotted on to the ordination as also can be associated environmental measures.

A more sophisticated approach is *principal components analysis*. This aims at extracting axes along which samples can be ordinated to express the principal trends of their floristic variation. The first axis accommodates the main direction of variation, the second axis the next most important and so on. It is then possible to try to correlate these main axes with environmental gradients.

Ordination may thus be considered, like classification, as a procedure aimed at simplifying a complex set of phytosociological data. But whereas classification attempts to extract classes from a set of samples, ordination is concerned with identifying the major axes of variation within them. The two techniques should be regarded as complementary rather than as alternatives.

B.WR.

Fossil Plants and their Study

The plant cover of the Earth today cannot be understood without reference to its evolution against a background of geological change (see Chapter Three: Evolution of the Green Planet). The study of fossil plants (paleobotany) is the principal means of obtaining information about that early plant cover.

Fossils may be defined as the remains of any organism which has been interred in and recovered from a sediment. The word is derived from the Latin *fodere*, meaning "to

A coal-measure fern of the genus *Mariopteris*, found at Monkton main colliery near Barnsley, Yorkshire, England.

dig", and originally anything extracted from the earth was termed a fossil. Its usage is now generally restricted to those remains of past forms of life that are found in rocks, with the implication that fossil remains must have considerable antiquity. Fossils may be direct remains of organisms, the marks they left in the rocks or even chemical precipitates. Thus fossil plants include remains such as roots and tree stumps, trunks and logs, leaves, twigs, flowers, cones and fruits as well as casts and impressions left in the rocks of such parts before they themselves became removed by chemical processes. Coal and petroleum, although formed from organic remains, are often so altered that they are not usually regarded as fossils in a direct sense. They are, however, often referred to as "fossil fuels".

As the process of fossilization is slow and prolonged, a dividing line cannot easily be drawn between a fossil and something on the way to becoming one. Usually the process of fossilization involves changes of a chemical nature in both the fossil and in the surrounding sediment. These changes include the processes of natural decay which start as soon as death occurs. Normally these post-mortem changes are helped by a variety of animals and fungi so that the organic compounds forming the plant body are broken down and the chemical elements recycled. However, the organisms that are responsible for decay may be prevented from acting if the plant remains fall into water; they become waterlogged and then buried in a rapidly forming sediment at the bottom of a sea or lake, where there is a low supply of oxygen. As they break down and decay, plant remains produce humic acids; under certain soil and climatic conditions these may accumulate and, by raising the soil acidity, prevent the actions of the decay organisms. In this way, plant debris may accumulate as peat, which may later become buried beneath a layer of sediment, due to seasonal flooding of a river or by sinking of the land surface. Plant remains thus buried undergo further slow chemical changes as part of the process known as *diagenesis*.

History of Paleobotany

Although fossils have been known from early historical times, it was only during the early

A maple-like angiosperm leaf preserved in chalk, showing the vein structure in fine detail.

17th century that it began to be appreciated that they were the remains of real organisms. At that time their presence in western Europe was explained by reference to the Biblical Flood (the Diluvial Theory) and they were regarded as representing unfortunate creatures annihilated by the deluge. This idea had been put forward by Martin Luther as early as 1539 but was not widely accepted at that time. In the course of the 18th century, supporters of the Diluvial Theory became divided in their opinions. Some believed in the Indigenous Theory, which maintained that the fossil remains represen-

ted plants and animals typical of the region in which they were discovered. Others like the German philosopher Gottfried Leibniz (1646–1716) held that there was an exotic or foreign element amongst the fossil remains. The Exotic Theory explained these by assuming that the foreign plants and animals had been transported from their place of death to the place of burial. Thus the presence of tropical plant remains in European Tertiary floras was explained. Towards the end of the 18th century, the Extermination Theory replaced these ideas, and the presence of extinct plants, as found in the coal measures of Europe, was explained by the assumption that they were exterminated by the Flood. It was only in the 19th century that scientists discarded the view that fossils represented a single prehistoric episode.

During the early 19th century, as a result of the observations of Lamarck, Cuvier, William Smith and Brongniart, it was appreciated that the sedimentary rocks were formed in a sequence and that they contained a succession of plants and animals (see the Geological Record, p. 71). To explain this series of fossil forms, the Frenchman Baron Georges Cuvier (1769–1832) developed the Catastrophic Theory, in which it was proposed that a series of floods had extinguished life at different times in the past. Cuvier's contemporary and compatriot, Alexandre Brongniart (1770–1847), on the basis of his studies of the strata of the Paris Basin also believed in the Catastrophic Theory but with less conviction. The British geologist Sir Charles Lyell (1797–1875) proposed in his *Principles of Geology* (1830–1833) a steady-state, Uniformitarian Theory as an alternative to the Catastrophic Theory.

These theories were familiar to Charles Darwin when he sailed in the *Beagle* in 1831 to South America. Here he observed, among other things, a series of Tertiary fossil faunas. These observations eventually led him to the theory of evolution published in the *Origin of Species* (1859). Paleobotanists subsequently came to explain the exotic element in fossil floras as the result of the evolutionary process.

Types of Fossil
Impressions and Casts

Fossil plant remains may occur as impressions, such as external molds in the rock, or as casts of an internal cavity, such as a pith cast. This form of preservation is often the least informative, as such fossils contain nothing of the original plant. Exceptionally well-preserved casts may retain microscopic details of the original plant surfaces and these can be observed by making a rubber cast and examining this replica. When these casts are examined in a scanning electron microscope such details as surfaces, hairs and stomata may be revealed, but it is more usual for such fine details to have been lost. Leaf impressions, for example, usually only reveal the outline and more major details of the venation. These can be copied using photographic techniques and compared to the leaves of better-known fossil or living forms. In this way it may be possible to match the fossil to a known form and so tentatively identify and name it.

Pith casts of *Calamites* (giant horsetails) of the Carboniferous show a characteristic pattern of longitudinal ridges and nodes, which has enabled the recognition of a large number of species. Some interesting casts were formed within the empty shells of Carboniferous seeds of pteridosperms. These are known as *Trigonocarpus* and they have revealed that the shell of the seed inside which they formed was pierced by a dragonfly-like insect and its contents sucked out; later the cavity so formed became filled by sediment to form the cast.

Preservation and recovery of fossil leaves. (a) A leaf shed from a tree is carried by wind to a river, where it sinks and becomes buried in the sediment. (b) Three forms of fossil can now form: **top**, the leaf becomes a coaly compression—it is flattened and loses water to form a carbon-rich coal; **middle**, the leaf substance is completely removed by oxidation leaving a hollow mold of the leaf surface printed in the rock; **bottom**, the mold in the rock is filled in with a replacement mineral which forms a cast of the original leaf surface. (c) When the rocks are split the type of fossil will therefore vary: **top**, a coal compression will give a leaf compression consisting of the coaly remains of the original leaf substance and a counterpart which is the impression of the surface along which the rock has separated from the leaf; **middle**, a hollow mold will only give two impressions; **bottom**, a cast will give a replacement cast and an impression counterpart.

Compressions

These consist of the nearly flattened remains of a plant in which the organic material has been altered by decay and diagenesis but not removed from the rock. In the case of bulky objects, the compression may not reveal much more than the original outline shape at first inspection. Much more, however, can be learned by employing techniques for transferring the fossil. In transferring an impression, the fossil is stuck to a glass slide with a resin; the slide is coated in paraffin wax, then scraped off the rock and the whole immersed in hydrofluoric acid. This dissolves the rock from the slide, thus leaving the fossil attached by the resin and in this way both sides of the original can be observed. It can then be photographed with ordinary light, or by reflected or transmitted infrared, which may reveal differences in the internal tissues. Portions of the fossil can be studied through a light microscope and a scanning electron microscope, and microscopic surface features of the original are often revealed. Sporangia, hairs, papillae, stomata and even wax deposits are often visible.

These features of the epidermal surface are more commonly studied after maceration, which is the chemical treatment of a portion of the fossil to separate the original cuticle (a noncellular waxy layer secreted by the epidermis) for microscopic examination. To do this, the fossil is normally oxidized in a solution of concentrated nitric acid to which a little potassium chlorate has been added. After this oxidation has swollen and broken down the altered remains of the internal tissues of the fossil, it is washed and treated with a dilute alkaline solution, usually ammonia. This renders soluble the oxidation

A light micrograph of the cuticle from the underside of a Jurassic leaf (*Otozanites pulcher*), showing stomata (dark red), with associated subsidiary cells and papillate epidermal cells (orange).

products, which are washed away to reveal the cuticle. The cuticles are then mounted for microscopic examination. Examination of the fossil during the process of maceration often reveals details of the internal tissues, as these oxidize differentially. By varying the techniques it may be possible to recover portions of the internal tissues for separate study. Maceration also reveals the presence of resinous materials which may be contained within the fossil. The study of compressed plants remains, notably leaves and reproductive structures, has proved most informative in recent decades.

The leaf cuticles of Mesozoic cycadophytes were first recognized and figured by J. Bornemann in 1856. Their study by macer-

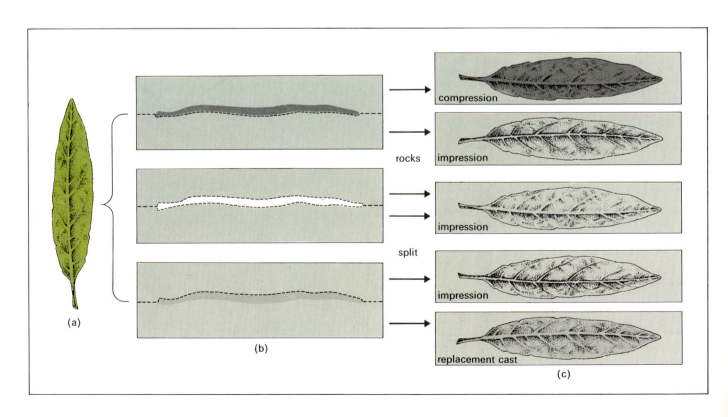

(a)

(b)

(c)

compression

rocks   impression

impression

split   impression

replacement cast

ation, pioneered by Alfred Nathorst in 1907–1912, led H. H. Thomas and H. Bancroft in 1913 to demonstrate the major characters of the epidermis by which the leaves of the two fossil orders the Bennettitales and Nilssoniales could be distinguished. This step proved to be a major advance in the understanding of Mesozoic floras. Similarly with Tertiary floras, H. Bandulska in 1923 began a series of studies on angiosperm cuticles which has been widely followed in Europe and more recently in North America. As a result, ideas about the identity of many Tertiary fossil leaves have had to be revised and it is now recognized that the fossil record of angiosperm leaves can reveal more about their evolution than was formerly appreciated.

Mummified Remains

These are normally only found in relatively young sediments that have been subjected to a minimum of compaction and compression. They consist of lumps of wood, logs, twigs, roots, stumps, fruits, cones and leaves. These remains may have become partially altered and humified and may show some compaction due to compression, but usually they retain their original three-dimensional form. They may be either brown or black in color. Mummified plant remains are quite common in parts of the Rhine Valley, being found in considerable quantities in some of the opencast Brown Coal mines which are in Miocene strata. It is often possible to treat such remains in the same way as material from living plants and to prepare them for microscopic examination. Thus it has proved relatively easy to identify sections of wood prepared from such remains with living sections, for example Taxodiaceae and Hamamelidaceae. The same applies to other plant organs such as cones, fruits and seeds.

Petrified Remains

These are possibly the rarest of all fossil plants. One of the oldest examples of a fossil plant is a petrified Bennettitalean trunk discovered in an Etruscan tomb which was opened a century ago. Similar trunks were regarded by the farmers of South Dakota, United States, as fossilized bees' nests. It was only after the 17th century that petrified remains came to be systematically collected and only since the early 19th century that their scientific study has made significant progress.

Petrifaction occurs when a mineral becomes precipitated inside the cell of the plant. Thus the petrified plant retains its three-dimensional integrity and often to a remarkable degree the fine structural details of its cell wall. The minerals commonly responsible for this process are silica, carbonates and iron sulfide. Silicified plant remains are commonly found in association with volcanic sediments or cherts formed in

areas of geothermal activity or more rarely in arid regions where pure siliceous sands are the predominant sediment. Carbonates are precipitated under special circumstances in peat deposits where they form coal balls such as those found in certain Carboniferous (Pennsylvanian) coal seams. Iron sulfide in the form of pyrites is often found in decaying vegetation and may be a major petrifying agent in anaerobic muddy sediments formed along the seaward margin of some deltas.

With petrified remains, the external surface may be studied where this is preserved and readily separated from the surrounding matrix, as for example the pyritized fruits and seeds washed out of the London clay and found on the shores of the Isle of Sheppey in

England. More usually the petrified plant remains are sectioned, usually by using a diamond-impregnated circular saw. The sections may be cut very thinly and then stuck to a glass slide and ground down until they transmit light. They are then studied in a light microscope as anatomical materials.

Another commonly employed technique is to dissolve away a thin layer of the petrified matrix and to replace this with cellulose acetate, thus transferring the remains of the fossil to a transparent acetate film. This is peeled off the surface and mounted on a slide for microscopic examination. With pyritized material these techniques seldom give satisfactory results, so the material is usually embedded in a synthetic resin and successively ground and polished surfaces are photographed using a reflecting light microscope.

It is from petrifactions that paleobotanists

Stages in the petrification (1, 2) and subsequent study (a–d) of a piece of fossilized woody tissue.

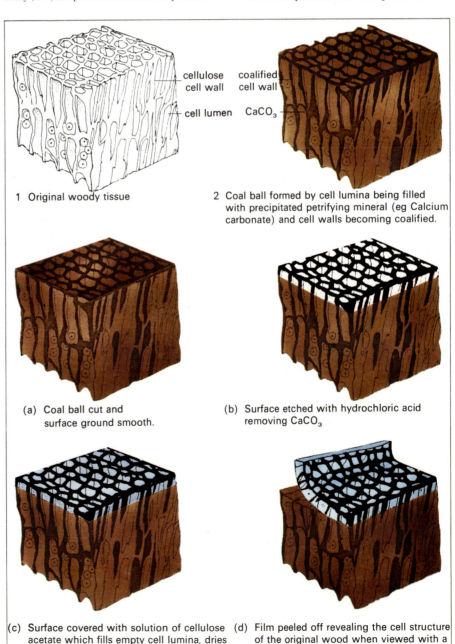

1 Original woody tissue

cellulose cell wall

coalified cell wall

cell lumen

$CaCO_3$

2 Coal ball formed by cell lumina being filled with precipitated petrifying mineral (eg Calcium carbonate) and cell walls becoming coalified.

(a) Coal ball cut and surface ground smooth.

(b) Surface etched with hydrochloric acid removing $CaCO_3$

(c) Surface covered with solution of cellulose acetate which fills empty cell lumina, dries and hardens as a clear plastic film.

(d) Film peeled off revealing the cell structure of the original wood when viewed with a microscope.

have learned most about the anatomical structure of extinct plant groups. It has proved possible to trace the evolutionary development of the vascular (water-conducting) tissues from the earliest Devonian land plants up to those found in club mosses, ferns, horsetails and conifers. Among the earliest petrified vascular plants, those preserved in the chert at Rhynie in Aberdeenshire, Scotland, are the best known. Detailed knowledge of Upper Carboniferous (Pennsylvanian) plants comes from the prolonged study of coal balls.

In the Mesozoic, knowledge of the finer details of the Bennettitalean flowers has come from a study of their petrified remains. This has suggested the possibility of beetle pollination and has shown the presence of dicotyledonous embryos. The Cretaceous record of the Palmae is readily substantiated by petrified wood showing their characteristic arrangement of vascular bundles. In the Tertiary, some of the best evidence for the antiquity of other families of flowering plants has come from petrified wood and fruit or seed remains.

Palynology

A specialized branch of paleobotany, palynology is concerned with the study of microscopic plant remains. This deals with the remains of microscopic unicellular algae such as diatoms, desmids, dinophyta, fungi and bryophyte spores, together with spores and pollen grains produced by higher land plants. The outermost walls (exines) of spores and pollen grains are remarkably resistant to decay and diagenesis in many sediments, so they are widely preserved. This, coupled with their wide distribution by air and in water, has made them unique as a record of past

Transverse section through the stem of the fossil plant *Asteroxylon mackei* which existed in the Lower Devonian period (about 400 million years ago) (×25). The fossil was obtained from Rhynie, Aberdeenshire, Scotland. The first part of the name refers to the star-like appearance of the water-conducting "wood" cells, as seen here in black. The polar cells between the arms of the "star" consist of food-conducting cells.

An ancient bristlecone pine (*Pinus aristata*), appropriately called the Patriarch, in the White Mountains, California, United States. Such specimens, up to 8,000 years old, have been invaluable for establishing tree-ring chronologies.

plant life. They are relatively easily extracted from small quantities of rock where, due to their microscopic proportions, they may be present in considerable abundance. They are removed by dissolving the rocks to release them, as they are unaffected by most acids. They can then be concentrated by centrifugation and filtration before they are mounted, either for direct examination or, usually after further treatment to clean them, for study by light or scanning electron microscopy. They are most abundant in organic deposits such as peat and coal. Their extraction usually requires a special treatment to separate them and remove organic plant debris.

With peat, brief boiling in 10% potassium hydroxide, followed by sieving, removes most of the humic acid material; this is then followed by acetolysis, which removes the cellulosic remains and leaves clean pollen grains, which are concentrated by centrifugation. Coals usually require a strong oxidizing treatment to dissolve them and so free the spores and pollen they contain.

The widespread use of palynology in stratigraphic geology has refined the methods and procedure of both geologists and paleobotanists. It has led to a greater awareness of the clear successions of plant assemblages and to an appreciation of sedimentary and ecological control of the fossil remains in any one locality; it has also shown up regional differences in past vegetation. Palynological results have had a

major influence on theories of the evolution of the flowering plants. It is now appreciated that the first angiosperm fossils of the Middle Cretaceous were unique extinct forms. During the course of the Upper Cretaceous and early Tertiary these fossil forms gave rise to the ancestors of the families of flowering plants that grow in the countryside and gardens of today.

P.B.

Dating

Accurate dating by either relative or "absolute" methods is fundamental to understanding changes in past environments and vegetation. Relative dating techniques depend on establishing a basic stratigraphic sequence from the simple principle that under normal circumstances younger rocks will lie above older ones. Other deposits may then be correlated with this sequence by features which they both share. Fossil animals and plants are themselves used as evidence for such correlation, but many other features, including lava flows, volcanic ash bands, paleomagnetic measurements, and the types of rock-forming minerals, are also widely used. An "absolute" chronology can be established either by detailed analysis of fine variations in features such as tree-rings, which are known to reflect annual periodicity, or by calculating ages from the known rates of decay of certain radioactive elements.

Although it is possible to use any kind of fossil for crude correlation, the most useful are those which are abundant and widely dispersed, but which occur only over a short stratigraphic range. As many fossils have long ranges in relation to the precision of dating required, whole fossil assemblages are commonly used for stratigraphic correlation. Pollen analysis is a good example of this technique. The major drawback of using fossils for correlation is that they are at least partially linked, through the tolerance of the living organisms, to sediments deposited under specific environmental conditions. It will therefore be difficult, for example, to correlate between deep ocean and freshwater lake deposits. For sediments deposited on or near land, fossil pollen and spores are extremely useful in correlation, but for marine deposits there is a wider range of suitable organisms including ammonites, foraminifera, graptolites and trilobites.

"Absolute" dating, based on laminated lake sediments (varves), or tree-rings (dendrochronology) relies on the correlation of the observed cycles with annual temperature fluctuations. These cycles may then be counted, and a careful analysis of their minute variations may be used to produce a standard sequence with which other varves or tree-rings can be correlated. Tree-rings are formed from the different sizes of water-conducting cells produced during the growing season; cells with a large cross-section in the spring wood, cells with a small cross-section in the later wood. Varves are formed in lakes from the different types of sediment deposited at different times of the year. In high latitude lakes, only fine particles are deposited during the winter when the infill streams are frozen, in contrast to the much coarser material deposited by the spring meltwater. Both of these techniques have been most widely used for dating events since the last glaciation. In western North America, tree-ring sequences of over 8,000 years have been obtained from the long lived bristle-cone pine (*Pinus aristata*), and in Scandinavia varves have provided a detailed chronology of the ice retreat following the last glaciation.

"Absolute" dating by radiometric (isotopic or radioactive) methods has only become possible in the last few decades and depends on the precise measurement of naturally occurring radioactive isotopes of various elements. Each of these isotopes decays at an exponential rate, and for each the half-life—the time taken for the number of radioactive atoms in a given sample to decrease by half—is characteristic and constant. From the proportions of radioactive isotopes to their decay products in a given sample, and with a knowledge of the half-life, the time since the original incorporation of the isotope in a rock or fossil can be calculated. Five of the major radioactive isotopes used are uranium[238] which decays to lead[206], uranium[235] to lead[207], thorium[232] to lead[208], rubidium[87] to strontium[87], and potassium[40] to argon[40]. All have half-lives of millions of years and are useful for dating over long time spans. Over shorter periods, the decay of carbon[14] to nitrogen[14], with a half-life of 5,568 years, is more useful and has been widely applied to dating Quaternary plant material. The technique depends on the small amounts of radioactive carbon dioxide produced in the upper atmosphere being incorporated into plant tissue by photosynthesis. In living plants the ratio of radioactive carbon[14] to nonradioactive carbon[12] is equal to that in the air, but following death the amount of radioactive isotope decreases such that the age can be calculated from the ratio of carbon[14] to carbon[12] in the sample today. The carbon[14] method has been calibrated against material of known age and yields accurate results back to 40,000 years; other radiometric techniques are less precise. All of these methods require sophisticated equipment, and rest on certain crucial assumptions, but the results they produce are close to those predicted from other evidence.

P.R.C.

Building a chronology from wood samples by cross-dating the sequences of narrow and wide rings of a living tree with the rings from wood cut in the past.

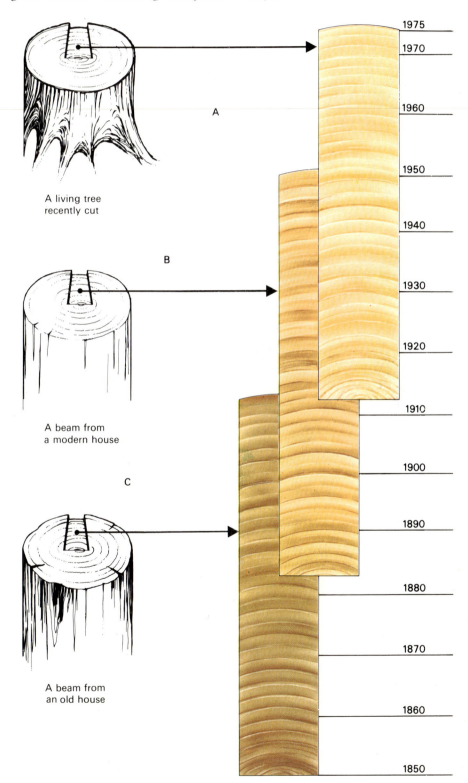

A

A living tree
recently cut

B

A beam from
a modern house

C

A beam from
an old house

1975
1970
1960
1950
1940
1930
1920
1910
1900
1890
1880
1870
1860
1850

CHAPTER THREE

Evolution of the Green Planet

The Origin of Plant Life and its Subsequent Development

The plants that inhabit the world today are only the survivors of the myriad forms that have arisen and become extinct during the 630 million years since plant life first appeared on this planet. Not only have the plants evolved during this long history but also the physical world inhabited by them has been subject to tremendous changes throughout geological time. To understand our present green planet fully would need a complete record of all these physical and organic changes, and this is not available. For example, Darwin's "abominable mystery" as to the origin of flowering plants is scarcely less of a mystery today, almost a century and a half after his despairing comment. Nevertheless, despite the many lacunae, knowledge of the history of the Earth and its inhabitants continues to grow. Although they are inevitably intertwined, it is most convenient in this chapter to deal separately with the physical and evolutionary backgrounds to the modern plant cover of the world.

From the viewpoint of plant distribution, the geological history of the Earth may generally be divided into two main parts. The most recent of these reaches back some 2 million years when the onset of the Pleistocene glaciations heralded the Quaternary era, in which we still live. Ice sheets extended southwards to about 40° latitude in the Northern Hemisphere and comparable conditions occurred around Antarctica, the glaciers advancing and retreating four times. Not only did these events profoundly affect the environments of the present temperate regions but they also had a much wider effect on the world's climate, with lowered temperatures and changed rainfall patterns extending into equatorial regions. Furthermore, sea-levels fell as water was locked up in the glaciers, thus exposing wide areas of land, which were again covered as the ice melted.

The earlier part of the geological record having importance for the subject of this book is the period reaching back 190 million years into the Jurassic period, when many modern plant groups were apparently beginning to emerge. Although there is still much uncertainty, the major features of the changing topography of the world are clear. Without doubt, of major importance to studies of plant distribution has been the confirmation during the last 20 years or so of plate tectonics, whereby our modern continents arose by fragmentation of preexisting supercontinents (continental drift), with profound influences on their climate and plant cover.

For knowledge of the changes in plants and vegetation over these time-scales we are dependent upon two major sources of information. The first is the direct evidence of the fossil record. This is inevitably very incomplete, and the immediate geological past, particularly the period encompassed by the Quaternary era (2 million years ago to the present), provides the fullest record of the changing plant cover, at least in areas such as Europe and North America which have received something like detailed study. Furthermore, the plants are externally sufficiently similar to their modern counterparts to give some confidence that deductions as to their physiological and other attributes may be valid. Further back in time, the fossil record becomes increasingly fragmentary and the plants more different from those present today, so that deductions about the relationships of the plants and the properties of the communities in which they lived must be increasingly imprecise.

The second source of data on the changes undergone by plants is by reference to known or inferred evolutionary processes. There is now abundant evidence, both experimental and observational, on the means by which speciation can take place and communities change. What is still much disputed is whether these processes, and the concepts developed from them, can be considered relevant to plant diversification and vegetation changes throughout evolutionary time. The consensus seems to be that they can, though with many reservations, and it is to be hoped that increased knowledge of the fossil record may help clarification of some of these points in the future.

Opposite This photograph, taken from a satellite (Meteosat) stationed above the equator, vividly illustrates the varied physical environments with which plants have to cope. The clear, sunny deserts of North Africa and Arabia contrast markedly with the clouds over the hot, humid equatorial forests and the cloudy polar regions. The remarkable similarity in shape between northeastern Brazil (**left**) and the Gulf of Guinea on the west coast of Africa (**centre**) is a reminder that the barrier of the Atlantic Ocean developed by the rafting apart of Africa, Europe and the Americas by continental drift.

The geological time-scale. In the left-hand column of this diagram are the eras of Earth history with a discursive account of their main features; in the next column the periods into which the eras are divided. The Archaean and Proterozoic are sometimes considered as eras. The Carboniferous period is often considered to comprise two periods, the Mississippian and Pennsylvanian, while the Tertiary and Quaternary are subdivided into seven epochs. In the third column are shown the notable events in the Earth's history; ice ages are shown by the white ice symbols. On the far right are shown the evolution and interrelationships of the major plant groups. (All dates are in millions of years before the present.)

Cenozoic

Life on the land in the Cenozoic has been characterized by the radiations of the angiosperms (flowering plants) in the plant kingdom, and of the mammals and insects in the animal kingdom. Forests of broad-leaved trees dominate the vegetation. There has been a continuation of sea-floor spreading, perhaps the major results of which have been the separation of Australasia from Antarctica and the convergence of Africa-Arabia on Eurasia throughout the era, the climax of the deterioration being the ice age, still continuing, of Quaternary times.

Mesozoic

On land, life in the Mesozoic was dominated by the dinosaurs and the gymnosperms: dominant in the sea were the marine reptiles among the vertebrates and the ammonites among the invertebrates. The Mesozoic saw the progressive breakup of Pangaea, and late in the era, the creation of most of our modern continents. The climate was warm and equable throughout.

Paleozoic

During the lower Paleozoic no terrestrial life forms were known. First land plants appeared in the lower Devonian, with the vegetation dominated by ferns, progymnosperms and club mosses by late Devonian. Giant horsetails, lepidodendroid trees, zygopterid ferns and early seed plants dominated the Mississippian, while in the Pennsylvanian much of the land was covered with swamp forests of giant horsetails and lepidodendroid trees, which produced the coal of today. During the Permian, early conifers dominated the flora. Most invertebrate groups evolved early in the Paleozoic era, the faunas being characterized by, for example, the trilobites, graptolites, primitive mollusks, arthropods and brachiopods. The first vertebrates (the jawless fish) evolved in the Ordovician, and primitive amphibians took to the land during the late Devonian/early Carboniferous. Towards the end of the era, the amphibians were replaced by the reptiles as the dominant land animals. Climate alternated between warm equable periods and rather shorter ice ages. The seas were comparatively wide-spread, especially in the early part of the era. Successive closures of ancient oceans resulted in the welding together of continental masses to form Pangaea.

Precambrian

This is a long, complex and comparatively poorly understood phase of Earth history. There were a number of important mountain-building episodes and at least one major ice age, late in the Proterozoic. In the early Archaean occurred the differentiation of the lithosphere, hydrosphere and atmosphere, and that of the crust, mantle and core. Later, free oxygen became available in the atmosphere for the first time, almost certainly as a result of photosynthesis on the part of primitive plants (algae). The first cells with nuclei appeared in the mid Proterozoic and the first multicellular organisms appeared close to the end of the Proterozoic.

Period	Epoch		Events
Quaternary	Holocene	0.01	— Early civilizations
	Pleistocene	2	
Tertiary	Pliocene	7	— Emergence of Man
	Miocene	26	
	Oligocene	38	— Start of main Himalayan folding
	Eocene	54	
	Paleocene	65	
Cretaceous		135	— Main fragmentation of Pangaea; transgression of sea over land
Jurassic		195	— Start of fragmentation of Pangaea
Triassic		225	— Worldwide regression of sea from land
Permian		280	— Formation of Pangaea
Carboniferous	Pennsylvanian	325	— Coal measures formed
	Mississippian	350	
Devonian		400	— Animal life takes to the land
			— First land plants
Silurian		440	
Ordovician		500	— Major transgression of sea over continents
Cambrian		600	
			— First multicellular organisms
Proterozoic		1000	— First cells with nuclei (eukaryotic)
		2000	— Free oxygen in atmosphere
Early Proterozoic		2600	
		3000	— First unicellular organisms (prokaryotic)
Archaean		3760	— Age of oldest known terrestrial rocks
		4000	— Origin of protolife
		4600	Formation of Earth
	million years 10 000		?"Big Bang"

Blue-green algae | Algae | Liverworts | Mosses | Club Mosses | Horsetails | Ferns | Conifers | Ginkgoales | Gnetales | Cycadales | Angiosperms

Bryophytes | Pteridophytes | Gymnosperms

Pteridosperms | Bennettitales | Nilssoniales | Lepidodendroid trees | Calamites | Zygopterid ferns | Cordaitales | Psilophytes | Progymnosperms

Present
Numerous
Conjectural

PLANTS | ANIMALS
EUKARYOTES

PROKARYOTES

PROTOLIFE

The Physical Background

Although our planet is almost 5,000 million years old, it has probably supported plants for only the past 600 million years, the so-called Phanerozoic period, and its greenness does not extend back more than the 400 million years since the first land plants are believed to have evolved. During this time—one-twelfth of the Earth's history—there have been tremendous changes in the topography of the world, which have profoundly modified the environments available to plants and animals.

We begin this first part of the chapter, therefore, by outlining the *geological record*, which charts the major changes that have taken place in the world during the Phanerozoic period. Since most of Earth's modern plant cover, dominated by flowering plants, appears to date from the middle part of the Phanerozoic period—the Mesozoic era—the chapter continues with a consideration of perhaps the most dramatic topographical event of that era, *continental drift*. By this process our modern continents arose by the fragmentation of earlier "supercontinents" at a time when flowering plants were becoming important. Because continental drift has continued until today our consideration of it brings us into the most recent era—the Cenozoic. It is only for this era (beginning 65 million years ago) that information begins to become available on the whole environment, topography and climate, and its effect on the vegetation. As will be seen, the first part of this era, the Tertiary system, is less well known than the later Quaternary system, which covers the last two million years of the Earth's history, but a fairly clear picture of the environments and vegetation is really only available for the last million years or so.

Climate, of course, is a major component of plant environments and attempts to understand its changes through time are among the oldest geological problems. A number of factors, such as annual and seasonal variation in temperature, rainfall, wind, etc, constitute climate but only one or a very few of these variables can be measured in any particular geological deposit. A further problem is that although past water temperatures, for example, can be deduced rather accurately from the proportions of oxygen isotopes in deep-sea sediments, there must be some uncertainty about how adequately conditions on land are reflected. Fossil plants and animals have been widely used to deduce paleoclimates, but clearly great care has to be taken to avoid circular reasoning when using such data to assess past environmental requirements of plants and vegetation.

D.M.M.

The Geological Record

The geological record is a chronological record of life on this planet as interpreted from the rocks. It is divided into periods, eras, systems, stages and finer units. Here only the Phanerozoic period will be considered, comprising the Paleozoic, Mesozoic and Cenozoic eras. Rocks formed prior to this—called Precambrian—may contain microscopic evidence of past life up to about 3,000 million years ago (see the Origin and Early History of Life, p. 84), but fossil life first became abundantly preserved in sedimentary rocks in the Lower Paleozoic. This era began about 600 million years ago with the Cambrian system, which was succeeded by the Ordovician and Silurian systems. During this time no terrestrial life-forms are known and probably all organisms were marine.

The Upper Paleozoic era began about 400 million years ago with the Devonian system. In the Lower Devonian, primitive land plants, the psilophytes *Rhynia* and *Psilophyton*, appeared. The Upper Devonian had a rich pteridophyte flora dominated by primitive gymnosperms (progymnosperms) and early heterosporous club mosses.

In the Lower Carboniferous or Mississippian stage (350 million years ago) the flora became progressively dominated by giant horsetails (*Calamites*), lepidodendroid trees, tree-ferns and early seed plants (gymnosperms).

In the Upper Carboniferous or Pennsylvanian stage (325 million years ago) much of Europe and western North America was part

A reconstruction of *Rhynia*, a fossil psilophyte that appeared some 375 million years ago.

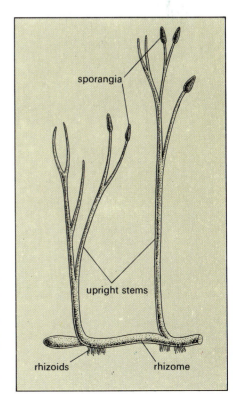

sporangia

upright stems

rhizoids

rhizome

Left An artist's reconstruction of a typical Cathaysian lakeside scene from the Upper Carboniferous period, with *Calamites*, a giant extinct relative of the horsetail (*Equisetum*), growing in the water. On the banks are some tree-like seed ferns (pteridosperms) that were common in the subtropical floral province of Cathaysia. In the background are *Tingia* trees with fan-like clusters of leaves.

Below An artist's reconstruction of a typical Euramerican swampy tropical forest of the Upper Carboniferous period, from the remains of which were formed the coal seams of Europe and North America. The larger columnar trunks belong to the giant lycopod trees *Lepidodendron* and *Sigillaria*, and among these are *Calamites* trees with whorls of branches. In the foreground are tree-like ferns, smaller seed ferns (pteridosperms) and true ferns.

of a vast tropical lowland covered by the swamp forests that produced the coal of today. These forests were dominated by *Calamites* and lepidodendroid trees together with pteridosperms, many exotic ferns, the earliest conifers, and trees of the Cordaitales.

Pangaea, the supercontinent into which earlier, separate continental masses were united, formed in the Permian system (280 million years ago), when the coal-forming lowlands turned into arid mountainous uplands. Pteridophyte forests were replaced by gymnosperms in which the early conifers, Lebachiaceae, and new pteridosperm groups became dominant.

The Mesozoic era (225 million years ago) began with the Triassic system, when much of Europe and North America had become part of a vast desert region in which the flora consisted of Voltziaceae (conifers) and strange horsetails. Towards the end of the Triassic, the North Atlantic ocean began to open, the climate moderated and new plant groups appeared, especially the cycadophytes.

In the Jurassic system (195 million years ago) rich and varied forests were developed from eastern Siberia across northern Eurasia to Yorkshire in England. Ferns were represented by primitive members of living genera like *Osmunda*, *Cyathea* and *Dicksonia*, *Equisetum* formed "reed" beds, pteridosperms, including groups like *Caytonia*, and cycadophytes of the order Bennettitales and Nilssoniales were abundant, as were ginkgophytes and conifers.

Many of the gymnosperm groups of the Jurassic were replaced in the Cretaceous system (135 million years ago) by broad-leaved trees of the flowering plants (angiosperms), which began to appear in the fossil record in the Early Cretaceous and were dominant by the Middle Cretaceous. The breakup of Pangaea was well under way, with the Atlantic Ocean separating Europe from North America and Africa from South America.

The Cenozoic era (70 million years ago) is conventionally divided into the Tertiary and

Quaternary systems. The early part of the Tertiary (Paleocene, Eocene and Oligocene) is often called the Paleogene, which was followed by the Neogene (Miocene, Pliocene and sometimes, the Pleistocene). During the Paleogene, much of Europe and North America was warmer than today and broad-leaved evergreen forests with palms flourished to latitudes above 50°N. During the Neogene, mountain ranges such as the Alps formed, the climate in the polar regions cooled progressively and rich deciduous forests covered the northern temperate zone.   P.B.

Mesozoic Environments and Continental Drift

The present-day physical separation of landmasses obviously has an important effect on the distribution of plant life, but this distribution may have evolved when the landmasses were very different. The possibility that the Americas had once been joined to

Europe and Africa was first discussed by Francis Bacon in the early 17th century. Since then, various people have been struck by the close fit between some of the present continental margins on opposite sides of the world's oceans, and have suggested that the continents we know today resulted from the breakup of ancient landmasses. By the early 19th century the attention of biologists was attracted to this concept of "continental drift" by the similarities between American and European fossil plants of the Carboniferous period (about 300 million years ago), and the strikingly disjunct distributions of various modern plant groups in, for example, Australia, New Zealand and South America. Geologists came seriously into the discussion at the end of the last century when the Austrian geologist Eduard Suess noted such a close correspondence of geological formations in the lands of the Southern Hemisphere that he proposed their derivation from a single supercontinent which he called Gondwanaland. The German geologist Alfred Wegener, who suggested mechanisms that could account for continental drift, demonstrated a remarkable number of detailed geological and paleontological correlations which indicated a common historical record on the two sides of the Atlantic Ocean. He visualized all the continents joined in a single vast landmass, Pangaea, before the onset of the Mesozoic era (about 225 million years ago). Subsequently the South African geologist Alexander du Toit favored two large landmasses: Gondwanaland in the Southern Hemisphere and Laurasia in the Northern Hemisphere.

By the 1960s a compelling amount of evidence had been produced to support the occurrence of these ancient landmasses. By using computers to obtain the best fit, the continents were shown to match well along the continental slope at the 500 fathom line (914m or 3,000ft). Ancient mountain chains, fossil beds and geological strata also matched when the continents were brought together. Previously, geophysicists had been unable to agree on a mechanism by which continental drift might take place, but, at this time, the theory of plate tectonics became almost universally accepted as providing such a mechanism.

Plate Tectonics

It is now generally accepted that the Earth's crust is composed of a few rigid plates that move relative to each other at a rate of up to 10cm (4in) per year. The plates are 80–100km (50–62mi) thick and include both continents and oceanic basins. Earthquakes and active volcanoes are largely restricted to the margins of the plates where they jostle one another as they move. Plates can slide past one another, as they do along the San Andreas fault system in California where the Pacific and West Atlantic plates adjoin. Plates can also move apart and, as they do so, lavas well up, solidify and are

A map showing the coal areas of today and the floral provinces of the Upper Carboniferous period when the coal measures were formed.

added to the plate on each side of the rift. As the lavas emerge and solidify they are magnetized by the Earth's magnetic field, which periodically reverses its polarity. Consequently, by examining the ancient magnetization, "paleomagnetism", of the lavas it is possible to determine their age and hence the rate at which the plates have moved apart. The mid-oceanic rises are important sites of lava extrusion and, as the plates grow, the older sea-floor, oceanic islands and continents are rafted to new positions. Since about half of the present ocean floor is of Tertiary age, less than 65 million years old, and is nowhere more than 200 million years old, this drift of the present continents is relatively recent in the history of the Earth, the oldest rocks of which are

A reconstruction of the coal-measure world of the Late Carboniferous period, based on the plate-tectonic outline map of Smith, Briden and Drewry (1973) showing the five major floristic provinces together with the major coal basins of the Euramerican province. Blue indicates the area occupied by the sea and white land areas outside the main provinces.

about 4,500 million years old.

The continued accretion of new material to the crustal plates would be expected to lead to an overall increase in size of the Earth, but this is not so. The reason is that moving plates can be thrust deep into the Earth's mantle along what are called subduction zones. These lie in the deep oceanic trenches of the world, usually 8–10km (5–6mi) deep, where the oceanic crust is carried down into the Earth by a form of descending convection current. The bending of the crust as it descends and its subjection to increased pressures and temperatures results in earthquakes and volcanic activity, which are largely concentrated along the subduction zones.

Timing of Continental Drift

With the general acceptance of plate tectonics and the use of paleomagnetism and other techniques for dating the different parts of the Earth's crust, the past disposition of the world's landmasses has become increasingly well known. The general sequence of events is fairly clear from the Permian period onwards. During Permian and Triassic times, about 200–260 million years ago, most of the world's landmasses were joined together into one supercontinent, Pangaea, shaped like a U on its side. The northern and southern landmasses (Laurasia and Gondwanaland respectively) were widely

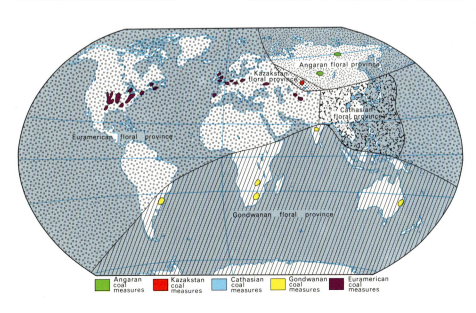

separated by the Tethys Ocean, except for present-day southwestern Europe and southern North America, which were joined to present-day northwest Africa and northern South America in the region of the equator. Laurasia comprised what is now Eurasia and North America. It was flanked on its southern, tropical shore by the Tethys Ocean, stretching from southwest Europe to Southeast Asia. Gondwanaland included what are now South America, Africa, India, Australia, New Zealand and Antarctica. It was named

A series of maps illustrating continental drift from the Permian period to the very recent past.

after the ancient kingdom in east central India where a distinctive fossil flora, characterized by the early gymnosperms *Glossopteris* and *Gangamopteris*, present in all these continents, was first found. These fossils were underlain by glacial deposits (tillites) left behind 250–350 million years ago, and the direction of the flow of the glaciers could only be explained by bringing together the continents into one mass. Southern Gondwanaland seems to have been in cold polar regions, the *Glossopteris* flora occupying the area left free by the retreating ancient glaciers. Northern Gondwanaland was in tropical areas.

There is evidence that the supercontinents were themselves composed of an agglomeration of earlier plates. In the late Cambrian period (about 510 million years ago), for example, Asia and Europe were widely separated, while the Caledonian Sea lay between Europe and North America, although what is now Northern Ireland and the Scottish Highlands were included with the latter.

The breakup of Pangaea seems to have begun in the Early Jurassic period (about 190 million years ago), when the Atlantic Ocean evidently began to open, with the rotation of Africa and South America away from North

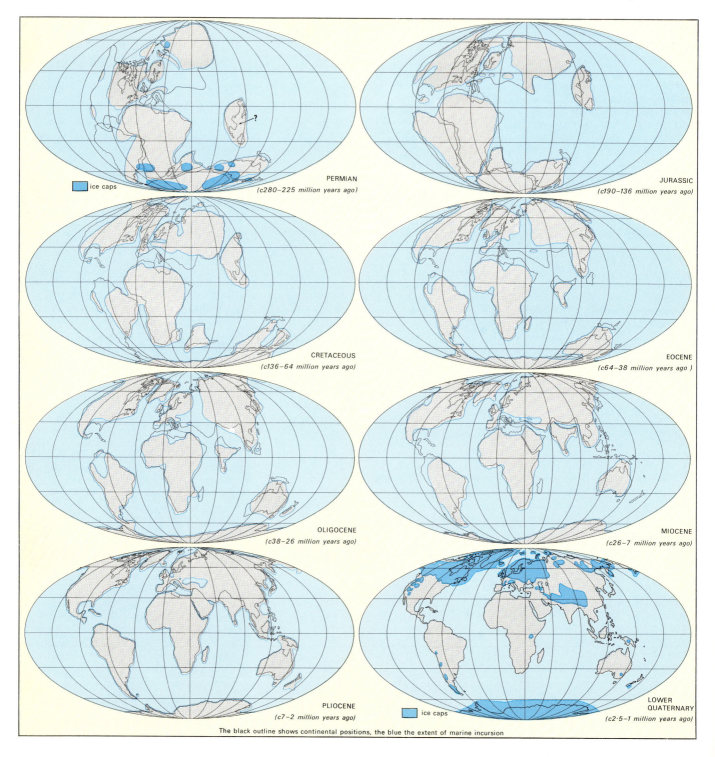

PERMIAN
(c280–225 million years ago)

ice caps

JURASSIC
(c190–136 million years ago)

CRETACEOUS
(c136–64 million years ago)

EOCENE
(c64–38 million years ago)

OLIGOCENE
(c38–26 million years ago)

MIOCENE
(c26–7 million years ago)

PLIOCENE
(c7–2 million years ago)

ice caps

LOWER QUATERNARY
(c2·5–1 million years ago)

The black outline shows continental positions, the blue the extent of marine incursion

America. The connection between Europe and Africa was also reduced until the last link between North Africa and Spain was severed in the early Paleocene epoch (63 million years ago) and did not rejoin until about 17 million years ago. Europe and North America began to separate in the late Cretaceous (81 million years ago) but a connection across the North Atlantic remained for a further 30 million years.

Africa, with India, began to separate from Antarctica (with Australia), as the latter moved towards the pole about 150 million years ago and then rotated away from South America to initiate the opening of the South Atlantic about 125–130 million years ago. India remained connected to Africa until at least 100 million years ago, when it separated and moved northwards to collide with Asia in about the middle Eocene (about 45 million years ago). New Zealand and New Caledonia separated from Australia about 80 million years ago but more or less direct connections were maintained between Australia and South America via Antarctica until about 40 million years ago. Australia then moved northwards to abut onto Southeast Asia and Malaysia in the middle Miocene epoch (about 15 million years ago), while North and South America finally became united about 5.5 million years ago after the uplift of southern Central America.

Continental Drift and Plant Distribution

Throughout the years when geophysicists declared that continental drift was not possible, many biologists maintained that it explained how similar living and fossil plant groups could be separated by the world's oceans. With the general acceptance of plate tectonics some but not all of these problems are resolved. Just as the movement of plates in the late Cambrian period probably could not affect the distribution of the earliest vascular plants, which did not appear until the Silurian period about 70 million years later, so the breakup of Gondwanaland and Laurasia can only account for disjunction in those plants which had already evolved. The angiosperms seem to have evolved about the time that the modern continents were separating, so that continuing uncertainty about their time and place of origin places obvious restrictions on interpreting the influence of continental drift on the distribution of modern flowering plants. Its effect on the distribution of ancient floras is also obscure, since there is incomplete information on the occurrence of fossil plants and the location of landmasses more than 250 million years ago. Despite these difficulties, it is possible to outline the general sequence of events over the past 300–400 million years.

During the Devonian period (about 340–380 million years ago) there is evidence of a rather widespread flora, occurring from paleolatitudes of 30°N–45°N to 50°S–80°S,

An artist's reconstruction of a riverside scene in the southern temperate region of Gondwana. In the middle distance is a forest of glossopterid trees and in the right foreground a *Glossopteris* twig bearing fruit. The banks of the rivers are lined by the extinct *Phyllotheca* and *Schizoneura*.

with, towards the end of the period, a preponderance of ferns and club mosses such as *Adiantites*, *Lepidodendropsis*, *Rhacopteris* and *Stigmaria*. This suggests the union of at least some modern continents. By Early Carboniferous times (about 330 million years ago) a distinctive assemblage of plants had evolved, apparently in response to the cold conditions of high northern latitudes and continental interiors, to give the so-called Angara flora, which occurred between paleolatitudes of 35°N–70°N. With Pangaea essentially formed, the late Carboniferous-early Permian floras showed a degree of regional differentiation unrivaled in any other geological time except perhaps during

A reconstruction of a shoot of *Nilssoniocladus*, a member of the Nilssoniales found in early Lower Cretaceous (about 130 million years old) rocks in Japan. It has spirally arranged leafy dwarf shoots on a long slender shoot.

the last 65 million years. The Gondwanaland flora (see p. 74) was quite distinct from the floras of Laurasia. It seems likely that this period marked the beginning of the long-standing differences between the Northern and Southern Hemisphere conifers: Taxodiaceae and, subsequently, the Pinaceae in the north; Araucariaceae and Podocarpaceae in the south. Within Laurasia, four principal floral provinces are generally recognized: the Euramerican in eastern North America and Europe, the Angaran in northern Eurasia, the Cathaysian in eastern Asia and the Kazakstan in central Asia. Two of these five late-Paleozoic floras have been particularly important in early studies of continental drift. Indeed, as we have seen earlier, ideas about drift were given impetus by the striking separation of the fragments of the Gondwana (*Glossopteris*) flora and the disjunction of the Euramerican flora by the opening of the North Atlantic, evidenced by the closely similar coalfield floras of Europe and the eastern United States, with their characteristic giant club mosses and horsetails, *Calamites* and *Sigillaria*.

Throughout much of the Mesozoic era (160–225 million years ago) floras typified by ferns of the families Dipteridaceae and Matoniaceae, seed-ferns (pteridosperms), Ginkgoales and small-leaved conifers were widespread. The climate over much of the globe seems to have been warm and equable between 60°N and 60°S paleolatitude, giving a broad equatorial floristic belt with two polar floristic regions in southern Gondwanaland and northern Eurasia.

It seems clear that there was differentiation of floras within each of these regions but they require a great deal more paleobotanical work. During the Late Mesozoic, interest increasingly centers on the evolution and distribution of the flowering plants which dominate the world's flora today. This is because, although the angiosperms might be older still, undoubted angiosperm pollen is only known from the early Cretaceous period (126 million years ago). It is evident that angiosperms gradually increased in diversity until by the start of the late Cretaceous period (110 million years ago) they were more abundant than gymnosperms and ferns, and probably included primitive members of several orders and even some families which exist today. There is also some evidence from pollen studies that the angiosperms migrated northwards from West Gondwanaland (Africa and South America) to temperate and then arctic Laurasia during the mid-Cretaceous period (125–100 million years ago).

Although still obscure, it certainly appears from paleobotanical studies and plate tectonics that the initial radiation of the angiosperms took place when direct migration was still possible between South America, Africa, India, Antarctica, Australia and, via Africa, Laurasia; the diversity of climatic conditions resulting from the earlier stages of the sundering of Pangaea may be of evolutionary importance. The late Cretaceous similarities between the pollen floras of West Africa and eastern Brazil suggest that at this time such families as Annonaceae and Lauraceae could have migrated directly in tropical or subtropical conditions before the South Atlantic opened to any extent; the occurrence of both in the Eocene of southern England also suggests migration to Laurasia. The Proteaceae,

Some of the earliest mid-Cretaceous flowering plants. **Below** One of the earliest leaves of flowering plants from the Dakota sandstone of central Kansas. It belongs to an extinct flowering plant, but probably one rather like the modern *Platanus*—the planes and buttonball trees. **Bottom center** One of the earliest flowers from the mid-Cretaceous of central Kansas. This extinct plant resembles existing species of *Magnolia* and is shown here at the fruiting stage. In comparison (**bottom right**) is the mature fruiting head of the living species *Magnolia wilsonii*.

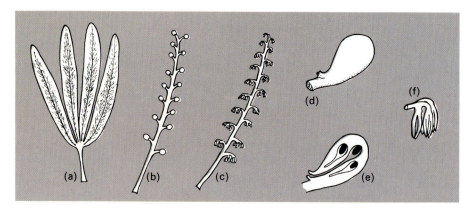

A reconstruction of *Caytonia*. The Caytoniales were a rather angiosperm-like group of fossil gymnosperms of the Mesozoic age. The leaves (a) were palmate and net-veined. The "fruits" (d) were carried on pinnately compound megasporophylls (b) and so too were the pollen-capsules (c) and (f). The "fruit" contained several seeds deeply embedded and communicating to the outside at the lip by narrow tubes; (e) shows the "fruit" in vertical section.

which initially diversified in Australia, also seems to have migrated by a tropical route via Africa to South America at this stage during the Upper Cretaceous.

Other lines of Proteaceae apparently spread to South America along a temperate route via Antarctica, pollen similar to South American *Embothrium* being known from very early Tertiary deposits (65 million years ago) in Patagonia. *Nothofagus*, the southern beech, is a key genus in considering this southern temperate route, since fossils about 80 million years old are known from Antarctica, New Zealand and South America at a time before they had separated. Furthermore, the fruits of southern beeches are not suited to dispersal over long distances.

With the initial breakup of Gondwanaland, Africa moved northwards into more tropical latitudes, losing many of the groups now shared by the southern continents, although some persist in the rich endemic flora of the more equable climate of Madagascar. The dramatic shift of India across the

torrid tropics to the Northern Hemisphere led to the extinction of virtually all plant groups of ancient southern affinities.

About 80 million years ago, then, Australia, New Zealand and the adjacent areas were still joined to Antarctica and South America and enjoyed a cool climate which favored temperate forests of such genera as *Nothofagus*, *Podocarpus*, *Weinmannia* and *Drimys*. About that time New Zealand separated and moved slowly northeastwards but, since it stayed in a generally temperate climate, it has retained a good representation of the ancient forest, as, to a lesser extent, has temperate South America. During the past 45 million years, after its separation from Antarctica, Australia has moved northwards through about 15° of latitude, thus moving from a basically temperate belt to subtropical regions; the increased aridity fostered the development of the distinctive flora we know today. At the same time, Antarctica moved southwards into a more polar position, so that forests were gradually eliminated as temperatures progressively lowered during the Tertiary and, finally, the expanding ice-fields eliminated even the evergreen scrub and rosette-cushion herb communities. It has recently begun to be recolonized from South America.

Until about 50 million years ago, South America was still closer to Africa than to North America. As the two Americas gradu-

ally approached, particularly from about 26 million years ago, dispersal of plants between them increased. The establishment of direct contact about 5 million years ago and the subsequent elevation of mountains allowed the southward spread of, for example, *Juglans*, *Alnus* and *Quercus* into northern South America and the northward migration of austral genera such as *Gunnera* and *Drimys*.

The collision of middle India with Asia during the Eocene, about 45 million years ago, led to the progressive enrichment of the Indian flora by basically tropical groups of angiosperms such as the Dipterocarpaceae, which had largely evolved in Laurasia. Similarly, Australia moved northwards until, about 15 million years ago, the Australian and Asiatic plates collided and permitted the mixing of plants (and animals) that had evolved separately in the south and north. This area of mixing has long been recognized as Wallace's line, after the great 19th-century naturalist Alfred Russell Wallace, who first recognized it. The mountains of Malaysia and Australia-New Zealand, elevated during the past 2–3 million years, formed a series of stepping stones for the migration of temperate Asiatic plants into Australasia.

As noted above, early angiosperms were present in northern Laurasia by the mid-Cretaceous, before the opening of the North Atlantic. Migration between North America and Eurasia was progressively pushed northwards as the Atlantic opened from the south, which may account for the fact that there are more affinities between temperate plants than between tropical ones on either side of the North Atlantic. However, as the Bering bridge moved southwards with the breakup of Laurasia, it became increasingly important as a migration route. During the early Tertiary, at least, there seems to have been a tropical flora extending from Southeast Asia along the shores of the Tethys Ocean to Britain. For example, during the Eocene epoch, palms of the genus *Nipa* were known in southern England. They are now restricted to Southeast Asia where, like other members of this early Eocene flora, they were able to survive the aridity of central Asia which followed the closing of the Tethys Ocean as India collided with Asia in the middle Eocene. Such tropical groups became extinct in Europe with the climatic cooling of the later Tertiary period.

Continental drift can thus be seen to provide a firm basis for understanding the discontinuous distributions of various groups of modern plants. In all cases, the plants on either side of the gap must have a taxonomic relationship which goes back as far as the rifting of the continents involved. Because of their relative youth, species or closely related groups of species which are discontinuous must have achieved some sort of transoceanic long-distance dispersal. Only continued studies of modern and fossil floras, however, will provide further information on the relative importance of moving continents

A flower of the living species *Drimys winteri* (family Winteraceae), which is essentially primitive in structure, ie the stamens (yellow) are broad and short and the carpels (green) are free and stalked. Present-day distribution of members of the ancient family Winteraceae (in South America and Australasia) is often cited as evidence for the joining of the southern continents through Antarctica some 80 million years ago.

and intercontinental transport in giving the discontinuous distributions shown by many families and genera in the present flora of the world.   D.M.M.

Cenozoic Environments and Vegetation

When the Cenozoic era began, some 65 million years ago, the arrangement of the continents as we know them today was still taking place by continental drift. Early in the era, for example, the Indian subcontinent was pushed against Asia with the result that the land crumpled to give rise to the Himalayas. Later, in a similar manner, the Alps arose when Italy, then part of Africa, was forced against Europe, while the Andes continued to rise higher as the Pacific and South American plates pushed closer together.

Most of these events took place during Tertiary times, which ended about two million years ago when a period of worldwide cooling led to the periodic advance of enormous ice sheets from the north and south polar regions. The advent of the glaciations signaled the beginning of Quaternary times, in which we still live. The tremendous effect of the glaciers on the environment and plant cover of our planet is the best understood part of the world's early history. Consequently, it must be realized that the consideration of the Tertiary world and its plant cover, which follows, must

inevitably be less confident than the description of Pleistocene conditions which completes this first part of the chapter.

The Tertiary Period

The flora of the Tertiary period (70 to 2.5 million years ago), like that of the Upper Cretaceous, is dominated by flowering plants. The earliest Tertiary marks an approximate mid-point in the evolution of this group and although some now extinct Cretaceous relicts persisted into the Paleocene (70 million years ago) many of the species are sufficiently like modern plants to be placed in Recent families and genera. The number of modern plant groups which can be recognized gradually increases throughout the Tertiary, and this change in species composition is clearly reflected in the plant communities which also become more like those with which we are familiar today. Although even some of the earliest Tertiary vegetation is reminiscent of modern types it is doubtful whether the history of recent vegetational units can be traced in anything but the most general terms.

Although there has been much work on Tertiary floras, the major barrier to our understanding of Tertiary environments and vegetation is simply lack of knowledge of Tertiary plants themselves. Research in the last two decades has shown much of the earlier work to be misleading, particularly in the identification of Recent genera in fossil floras. Most problems arise from the difficulties of accurately determining the modern relatives of fossil plants known only from dispersed and often uninformative fragments. Where differences between modern and fossil plants are detected, or where the comparisons are inaccurate or seriously incomplete, it is difficult to extrapolate from the nearest living relatives of the fossils in reconstructing the vegetation and climate. A

Some typical plants from the early Tertiary. **Top left** A common type of leaf from high latitude early Tertiary floras in the Northern Hemisphere. The leaves, although rather like those of *Cercidiphyllum*, are of an extinct plant. This specimen is from Spitzbergen, although similar material occurs in North America, the British Isles and continental Europe. **Above** A transverse section of a silicified palm wood from Herne Bay, Kent, England. Palms were a common component of early Tertiary floras throughout the world, and they have been particularly well investigated in Europe and North America. The section shows the numerous scattered fibrovascular bundles of typical monocotyledons. **Top right** A leaflet from a member of the walnut family (Juglandaceae), known to be a part of the Paleocene flora of southern England.

different, but equally problematic approach, compares the physical characteristics of the leaves in fossil assemblages with those in different Recent vegetation types. The technique relies on the broad correlation in modern vegetation between climate and leaf characters such as size, texture and margin: areas of high precipitation and temperature typically have a greater proportion of large, leathery leaves with entire margins than areas of lower temperature and precipitation, in which smaller, thinner leaves with lobed or toothed margins are more common. Using leaves in modern floras for comparison, the proportion of these different leaves in fossil assemblages can be used to give a general indication of the vegetation and climate. Despite the considerable inherent difficulties, both these approaches give similar in-

dications of Tertiary climatic and vegetational change which are well supported by evidence from animal fossils and physical techniques such as oxygen isotope analysis.

The most complete information on Paleocene vegetation comes from fossil localities in arctic Canada, Greenland, northern Siberia and northern Europe, all of which yield a fundamentally similar mixed gymnosperm and angiosperm flora. Gymnosperms, including *Ginkgo* (maidenhair tree) and a variety of taxodiaceous conifers such as *Glyptostrobus*, *Metasequoia*, *Sequoia* and *Taxodium*, were particularly abundant at the higher paleolatitudes. One of the most distinctive is a species of *Metasequoia* which differs only in minor characters from the single Recent species, *M. glyptostroboides* (dawn redwood), native to central China. In addition to several Cretaceous relicts, the angiosperms include a mixture of presumed deciduous species such as plants similar to living *Cercidiphyllum* (katsura tree), Betulaceae (birches and hazels), and *Platanus* (plane trees), as well as a few apparently evergreen types resembling modern Lauraceae (laurels). The vegetation was a rich mixed evergreen and deciduous broadleaved forest most similar to that occurring today in southern Japan and the central Chinese provinces of Hupeh and Szechuan. The climate was probably cooler than that of the Upper Cretaceous and seems to have been seasonal, warm temperate or subtropical, with only minor frosts, and perhaps a mean

annual temperature of about 15 °C. Several of these floras at paleolatitudes of over 60°N contain plants with large, apparently evergreen leaves. Under present conditions, even with favorable temperatures, such plants would experience considerable physiological difficulties during the low light conditions of the Arctic winter and changes in the angle of the Earth's axis of rotation have been suggested as a possible mechanism by which winter light levels may have been significantly increased at high latitudes during the early Tertiary.

The best studied sequence of Eocene plant assemblages is that of southern England, where fossil plants, known mainly from fruits and seeds, occur in sediments ranging in age from uppermost Paleocene (54 million years ago) to lowermost Oligocene (38 million years ago). Traditionally, they have been interpreted as varied samples of fundamentally similar vegetation, but there is increasing evidence that the Lower and Middle Eocene floras in particular are relatively distinct from those of the Paleocene and Oligocene which precede and follow them. The London Clay flora (Lower Eocene) has been particularly thoroughly examined and over 400 species from more than 40 families have been described. The plants were deposited in marine muds probably at the mouth of large rivers draining land surfaces to the north and west of London. The most characteristic and abundant plant fossils are the fruits of a stemless palm, *Nipa*, which are known over a wide area from southern England to Borneo, and Poland to Egypt, during the Eocene. Today, the closest living species, *Nipa fruticans*, forms extensive stands

Collecting fossils on the beach at Sheppy, southern England, the source of the best known and understood of Eocene floras. The fossils occur as isolated pyritized fruits and seeds among the shingle on the Sheppey beach. The fruits of *Nipa* are particularly abundant, but over 300 species have been reported from this locality.

Nipa fruticans is a short-stemmed palm now commonly found in coastal habitats in Southeast Asia. Members of this genus were an important and dominant constituent of many early Tertiary, principally Eocene, floras in southern England and continental Europe, where the fruits are extremely well known. Such distributions are a major argument for a higher mean annual temperature during the Eocene.

around the coasts of Southeast Asia and several of its present-day associates in "mangrove" habitats (see p. 206) are also known to occur in the London Clay. The flora contains many other typically tropical elements, including plants like *Dracontomelon*, *Spondias* (cashew family), *Tinospora*, *Tinomiscium* (moonseed family), *Phytocrene* (Icacinaceae), *Caryota* and *Oncosperma* (palms), but several montane or more temperate plants resembling *Magnolia* (Magnoliaceae), *Platycarya* (walnut family) and *Trochodendron* (Trochodendraceae) are also present, and raise problems for interpreting the source vegetation. Original ideas envisaged a nonseasonal, evergreen tropical rain forest in which the more temperate elements were minor and perhaps derived from upland sources. Others have interpreted the vegetation as quite unlike any still in existence today, and probably growing under seasonal subtropical or paratropical conditions. There is agreement, however, that Eocene temperatures were substantially higher than those in the Paleocene throughout the Northern Hemisphere, and the vegetation, at least in present temperate regions, appears to have been more luxuriant than at any other time in the Tertiary. A flora very similar to that of the London Clay is known from North America, and the leaf floras of Alaska suggest a mean annual temperature of about 25°C (77°F).

At the close of the Eocene, temperatures began to decline and although this has often been regarded as a gradual process there is increasing evidence to suggest a much more rapid, dramatic deterioration. This so-called "Terminal Eocene Event" is thought to have resulted in a much lower mean annual temperature (about 10°C; 50°F) but a much higher mean annual temperature range of over 30°C (54°F). Whatever the rate of change, it is clear that there was a widespread decline of the Eocene plant communities, eventually allowing the development by the Miocene (26 to 7 million years ago) of temperate broad-leaved, deciduous forest of more modern aspect, in which plants securely assignable to modern genera played an increasingly important part.

The Oligocene flora and vegetation (38–26 million years ago) was more or less intermediate between the extremes of the Eocene which preceded it, and the Miocene which followed it. Two-thirds of the genera in the Lower Oligocene Bembridge flora, from the Isle of Wight, are still living, as compared to one-third in the Lower Eocene London

During the Miocene the flora and vegetation took on an increasingly modern aspect. Communities with modern counterparts arose, including species of *Nyssa* in swamp habitats in Europe. Shown here is the living species *Nyssa sylvatica*, popularly known as the black-gum or tupelo.

Clay flora. Conifers were again important in the vegetation, including some like modern *Pinus* (pines) and *Cupressus* (true cypresses), but there were also deciduous and evergreen angiosperms, including palms, *Celtis* (hackberry) and some Lauraceae (laurels). Wetter areas were dominated by *Sparganium* (burreed), *Typha* (bulrush) and *Acrostichum* (a fern of brackish and freshwater swamps); and the water plants included *Sabrenia* (an extinct water-lily) as well as the small waterfern *Azolla*. A similar flora of Upper Oligocene age near Paris includes *Sequoiadendron* (Wellingtonia) among the conifers. The only living species, *S. giganteum*, is restricted to western North America. The beautifully preserved Lower Oligocene amber flora of the Baltic coast shares many genera with these more southerly floras but conifers and Fagaceae (oaks and sweet chestnuts) seem to have been a more important component of the vegetation. On the whole, these floras reflect a warm temperate climate for the European Oligocene, and over two-thirds of the species have their nearest living relatives in the forests of North America and eastern Asia.

In the Miocene (26 to 7 million years ago) the flora and vegetation took on an increasingly modern aspect. There was a much richer and more conspicuous herbaceous flora in which grasses were abundant for the first time, and many modern arborescent genera such as *Acer* (maples), *Carpinus* (hornbeams), *Fagus* (beeches) and *Quercus* (oaks) became much more diverse on a worldwide scale. In the Miocene brown coal floras of the Rhine rift valley, various communities with clear modern counterparts have been recognized, including *Sequoia* (redwood) forests on higher ground, *Nyssa* (black-gum)—*Taxodium* (swamp cypress) swamps, as well as *Myrica* (sweet

gale), and Cyrillaceae thickets with *Osmunda* (royal fern) and *Sabal* (fan-palm). All have modern analogues in the Florida Everglades.

In southwestern North American Miocene floras the development of semi-arid oak-chapparal reflects increasingly dry conditions, and this trend is also seen in the Pliocene floras of southern Europe. More northerly Pliocene floras such as those of Poland show rich temperate mixed forest dominated by eastern North American, eastern Asian and present-day European components. In Europe, the North American and Asian elements such as *Actinidia* (kiwi fruit family), *Liriodendron* (tulip tree), *Liquidambar* (sweet gum), *Thuja* and *Tsuga* (hemlock) were gradually eliminated by the onset of cold conditions in the early Quaternary. The predominant east-west mountain chains of Europe are thought to have impeded southerly plant migration while in eastern Asia and eastern North America open migration routes to the south allowed many species to survive.

The general pattern of decreasing mean annual temperature, increasing herbaceous vegetation and increasing floristic regionality throughout the Tertiary is generally accepted but there are widely divergent opinions on the details. Our present understanding is almost entirely based on middle

A pollen diagram showing the relative proportions of tree, shrub and herb pollen obtained from varying depths of soil, and the distribution of pollen from those plants making up the tree pattern.

and high latitude Northern Hemisphere floras and although the small amount of data from the tropics and the Southern Hemisphere is in broad agreement, much further research is required before we have a general understanding of Tertiary environmental and vegetational change on a global scale.

P.R.C.

The Quaternary Period

The climatic deterioration of the later Tertiary led eventually to the "Ice Ages" of the Quaternary period (2.5 million years ago to the present), with dramatic fluctuations between warm and cold phases. Glaciers and ice sheets advanced and retreated on mountains and at high latitudes in both hemispheres, resulting in worldwide migrations of both animals and plants. In Europe, six major ice advances are known; in North America slightly fewer have been recognized, but evidence from other parts of the world indicates that further cold phases may still be undetected on both continents. The cause of the Quaternary "Ice Ages" is not fully understood, but the movement of land masses into high latitudes and changes in the Earth's orbit were probably important factors.

The environments and vegetation of the Quaternary are much better understood than those of the Tertiary, but detailed knowledge, particularly of the earlier phases,

remains incomplete and confined to a few areas of the world. The techniques for studying Quaternary vegetational change rely on the accurate identification of fossil specimens with living plants, which may then be taken as reliable indicators of ancient environments and plant communities. Pollen analysis is the most important of these techniques (see Palynology, p. 66). Fossil pollen is usually well preserved, and the proportion of different pollen types in a fossil deposit can reflect the abundance of different species in the vegetation at the time the deposit was laid down. Although subject to many limitations, pollen analysis has been the major tool in unraveling the complicated vegetational history of the Quaternary, and the results it has provided are well corroborated by other evidence from oxygen isotopes, fossil insects and sea-level changes, as well as many other biological and physical techniques.

Evidence of the early Quaternary vegetation is sparse, but fossil deposits from the Netherlands and eastern England give an indication of the environments of northwest Europe. The molluscs, foraminifera and pollen deposited in early Quaternary coastal muds off East Anglia indicate at least three temperate phases (Ludhamian, Antian and Pastonian), separated by two cold phases (Thurnian and Baventian). During the temperate periods, mixed coniferous deciduous woodland prevailed, including *Pinus* (pine), *Picea* (spruce), *Tsuga* (hemlock), *Alnus* (alder) and *Quercus* (oak), while cooler conditions

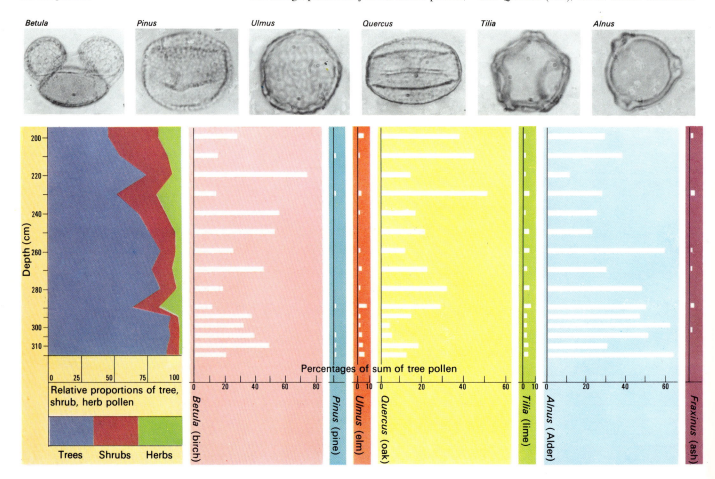

are indicated by open heathland, dominated by Ericaceae (heather and crowberry). There is no direct evidence of glaciation, but permanent soil freezing (permafrost) occurred during the Baventian. The Pastonian temperate phase shows the sequence of vegetational changes, common to all succeeding interglacials in northwest Europe, which results from a complex interaction of biotic, edaphic and climatic factors. The tundra conditions of the preceding cold phase are followed by open communities with pioneer trees, first *Betula* (birch) and then pine. Light-demanding herbaceous species flourished on the mineral-rich but organically poor soils; with continued climatic amelioration, the soils developed better structure and more humus, and tree cover increased. As the climate reached its most equable phase, the climatic optimum, the climax vegetation comprised mixed oak forest with oak, *Ulmus* (elm), *Fraxinus* (ash) and *Corylus* (hazel), often with further tree species, such as spruce, becoming important at a slightly later stage. As the climatic conditions began to deteriorate, soil leaching led gradually to the development of nutrient-deficient podsols (see p. 135) supporting increasingly open scrub and heathland vegetation. Birch and pine once again predominated, and eventually tundra conditions immediately preceded the onset of the next glacial phase.

Correlation of the British early Quaternary deposits with those from the Netherlands is difficult, but the Tegelen fossil flora is thought to be comparable with the Ludhamian. Although still including exotic elements, the flora contained a much higher proportion of present-day northwest European plants than are present in the Pliocene floras which preceded it.

The open and sparse vegetation of the Beestonian, which followed the Pastonian in eastern England, was similar to that of the Arctic today, with low shrubs such as *Salix polaris*, *Salix herbacea* (dwarf willows), *Betula nana* (dwarf birch) and herbs such as *Saxifraga* spp. (saxifrages) and *Oxyria digyna* (mountain sorrel). The Beestonian is thought to correlate with the Nebraskan glaciation in North America, during which there is evidence of similar tundra vegetation. In Europe, this cold phase eliminated further species from the Pliocene flora, and by the succeeding Cromerian interglacial the proportion of species which no longer live in northwest Europe was extremely small. The flora suggests that the climatic optimum may have been significantly warmer than at present, and in the Great Plains of North America evidence from the comparable Aftonian interglacial similarly suggests milder summers and warmer winters than today.

The Cromerian interglacial was followed by the first, and most severe of the three major Quaternary glaciations: the Anglian (Lowestoftian) of the British Isles, and the Elster and Mindel ice advances in northern Europe and the Alps. In Europe, the ice sheets

Flower of the tulip tree (*Liriodendron tulipifera*). Members of this genus were gradually eliminated from Europe by the onset of cold conditions in the early Quaternary, but they survived in southeastern North America. As the climate warmed in the Sangaman interglacial of eastern North America *Liriodendron*, along with other trees, "migrated" north again.

extended approximately to a line running from just south of Kiev to the Severn and Rhine estuaries, with sparse tundra vegetation to the south. The vegetation of the corresponding Kansas glaciation of North America is not well known, but temperate mixed forest existed in southern Indiana, and the open woodland and grassland which prevailed in northern Texas indicates cooler moister summers than at present.

The Yarmouth interglacial in North America is generally correlated with the Hoxnian of Britain and the Holstenian of Europe, during which the typical interglacial vegetational succession occurred but with *Abies* (silver fir), *Carpinus* (hornbeam) and spruce attaining particular importance in the later phases of the mixed oak woodland. In southern England. Lower Paleolithic flint artefacts occur contemporaneously with an abrupt local decline of this forest, correlated with a rise in the proportion of herbaceous plants, perhaps indicating temporary forest clearance by Man. The distribution of plants such as *Buxus* (box), *Hedera* (ivy), *Ilex* (holly) and *Taxus* (yew) in northwest Europe at this time indicates warm, moist climatic conditions.

The Wolstonian (Gippingian) glaciation, which follows the Hoxnian in the British Isles, correlates with the Illinoian, Riss and Saale ice advances in North America, the Alps and northern Europe respectively.

The vegetation was probably similar to that during the Anglian-Elster-Kansan cold phase. Warming of the climate led to the penultimate interglacial, the Ipswichian-Eemian of Britain and northwest Europe, and the Sangamon of North America. Ipswichian plants are known from various river terraces in the British Isles, and those of the River Thames in Trafalgar Square, London, include a few remaining exotic plants such as *Trapa natans* (water chestnut) and *Acer monspessulanum* (Montpelier maple), alongside a fauna containing elephants, hippopotamuses and rhinoceros. At the climatic optimum, conditions are thought to have been warm and moist. In eastern North America the Sangamon interglacial succession commenced with pine and spruce woodland. As the climate warmed, spruce decreased in importance, until, at the climatic optimum, mixed oak, *Carya* (hickory) forest had developed. Other trees included *Acer* (maple), *Fraxinus* (ash), *Juglans* (walnut) and *Liriodendron* (tulip tree), as well as *Larix* (larch) and hemlock in the later stages. Many of the taxa, such as *Fagus* (beech), holly and *Liquidambar* (sweet gum) occurred close to, or beyond, their present tolerance limits and, as in Europe, may indicate a more equable climate than at present.

The final glaciation, the Weichselian (Devensian) of northwest Europe, the Würm of the Alps and the Wisconsin of North America commenced about 75,000 years ago and was much less severe than the two major ice advances which preceded it. In Europe, the ice sheet did not extend into the Netherlands, and scarcely penetrated into lowland Britain. In North America, it extended from approximately Vancouver in the west to New York in the east. In the west, the ice sheet followed the Canada/United

States border, with a major extension down the Cascade Range and with smaller local developments in Montana and the Sierra Nevada. In the east, it extended from Wisconsin southward into central Indiana and Ohio.

On a world scale, this cold period comprised several glacial phases separated by warmer interstadials either too cold, or too short, to permit the development of mixed deciduous forest in present temperate regions. However, during at least one of these interstadials, birch, pine and spruce became re-established in the English Midlands, and there is good evidence for similar temporary afforestation in southeast Quebec. Overall, the Earth's average temperature was probably 4–5°C lower than that of the present, although the reduction may have been as much as 15–16°C close to the ice margin, or as little as 3°C in parts of central North America, eastern and central Asia. The major effect was to displace the Earth's vegetation zones in both hemispheres to lower latitudes, and to lower mountain tree- and snow-lines.

During the maximum (Middle Weichselian) extension of the ice in Europe the vegetation to the south comprised open grass and sedge tundra, with *Salix herbacea*, *Salix polaris* (dwarf and arctic willows) and *Juniperus* (juniper). The herbs included many plants of present-day arctic-alpine distribution, for example *Papaver alpinum* (alpine poppy) and *Saxifraga oppositifolia* (purple saxifrage). In places, permafrost resulted in saline conditions, and salt-tolerant plants (halophytes) such as *Armeria maritima* (sea pink) and *Suaeda maritima* (seablite) formed a conspicuous component of the flora.

Further south, the northern Mediterranean was covered by dry treeless steppe, although the southern Mediterranean was much wetter than at present, and the Sahara was displaced southward. In western North America the predominant vegetational changes seem to have been altitudinal adjustments, but in the eastern half of the continent southerly displacement of the vegetation zones occurred, although less markedly than in Europe. A narrow belt of tundra or spruce park tundra fringed the ice sheet, and in the midwest spruce and pine dominated forest existed within 300–500km (200–300mi) of the ice front. Mixed deciduous hardwood forest survived in the eastern and southeastern United States.

In northwest Europe, three phases are recognized during the latest Weichselian, which commenced about 15,000 years ago with the "Lower Dryas". The vegetation comprised open tundra, often with the characteristic *Dryas octopetala* (mountain avens), but during the succeeding Allerød interstadial (12,000–10,500 years ago) birch and *Populus* (poplar) formed open park woodland, only to be subsequently eliminated from all but the most favorable sites by a brief return to periglacial conditions during the "Upper Dryas".

The earliest phase of the present Flandrian

Sea pink or thrift (*Armeria maritima*) growing in a cliff community. This species was once a widespread tundra and pre-Boreal herb, but it is now almost entirely restricted in the British Isles to coastal or high mountain habitats, where the open or hostile conditions provide the reduced competition from other plants that it requires.

(Holocene) interglacial is the Pre-Boreal (10,000–9,500 years ago), which began in northwest Europe with subarctic conditions. As the ice retreated there was an increase in juniper, rapidly followed by an influx of birch, pine and many herbs indicative of warmer conditions, leading eventually to the development of a lightly wooded landscape on mineral rich soils. During the Boreal (9,500–7,500 years ago), *Corylus avellana* (hazel) and, later, other trees such as oak and elm became increasingly abundant. Hazel formed large areas of scrub, and in the earlier stages may have formed an understory beneath a pine or birch woodland canopy.

As the ice melted throughout the Pre-Boreal and Boreal, the sea-level gradually rose, and, towards the close of the Boreal, Ireland became separated from the rest of the British Isles, and eventually the North Sea was fully flooded. The separation of Britain from the continent severely restricted re-migration of species from the east and may account for the impoverished composition of the British flora as compared to similar areas of continental Europe.

As the climate warmed to the climatic optimum in the Atlantic period (7,500–5,000 years ago), oak and elm became more abundant and there is a marked increase in alder, probably due to increased rainfall, which is also reflected in the formation of massive peat deposits and the growth of raised bogs. *Tilia* (lime), one of the most warmth-demanding of all north European trees, appeared for the first time and in Scandinavia cold-sensitive species such as *Hedera helix* (ivy) and *Cladium mariscus* (great sedge) reached their maxi-

mum abundance. Winters were probably milder than at any other time in the interglacial. In the British Isles, dense mixed oak forest extended to altitudes of at least 750m (2,450ft) except in the Scottish Highlands and Islands where birch and/or pine remained dominant. Open habitats were almost totally confined to the highest altitudes, coastal or waterside areas, and unstable or particularly thin soils. Many of the once widespread tundra and Pre-Boreal herbs, such as *Armeria maritima* (sea pink), *Hippophae rhamnoides* (sea buckthorn) and *Silene maritima* (sea campion), still survive today in these "refuge" habitats. One particularly well known "refuge" occurs in Upper Teesdale, England, where the combination of climatic and geological conditions has preserved a whole assemblage of late Glacial and Pre-Boreal species, including *Betula nana* (dwarf birch), *Dryas octopetala* (mountain avens), *Gentiana verna* (spring gentian), *Plantago maritima* (sea plantain) and *Thalictrum alpinum* (alpine meadow rue).

In North America, the retreat of the ice was much slower than in Europe due to the enormous mass of the Laurentide ice sheet, and at the beginning of the European Boreal (9,500 years ago) the North American ice still extended to the Canada/United States border in the eastern part of the continent. As the ice retreated, the tundra belt remained narrow as spruce forest moved northward, followed by mixed deciduous oak forest from the south. Southwest North America gradually became drier, and even in the east the advance of hardwood communities was severely retarded by dry grassland spreading from the west as increasing westerly winds extended the Rocky Mountain rain shadow. At their maximum, around 7,000 years ago, such conditions probably extended as far east as Michigan, but by the climatic optimum this prairie peninsula had diminished. In some areas, spruce forest extended over 300km (200mi) further north than at present and the mixed deciduous forest probably showed a similar extension.

The pattern of postglacial climatic change elsewhere in the world seems similar to that in Europe and North America, although at high latitudes the retreat of the ice was probably more rapid. In the Sahara, southernmost South America, Australia and East Asia the wettest phase occurred about 6,000–5,000 years ago, and in most cases this corresponds with the probable postglacial temperature maximum.

The beginning of the Sub-Boreal in Britain (5,000–2,500 years ago) is marked by a dramatic decline in the abundance of elm and the re-expansion of many plants of open habitats, such as *Stellaria media* (chickweed), *Artemisia* (mugwort), *Rumex* spp. (docks) and *Polygonum* (knotgrasses). Although probably related to increasing continentality and some deterioration of the climate, this change also corresponds closely to the arrival of Neolithic agriculture in northwest Europe from its origin in the Middle East about 9,000 years ago. The earlier hunter-gatherer economies of Paleolithic and Mesolithic man had little effect on the natural vegetation, but the cereal cultivation and animal husbandry practiced first by Neolithic, and later by Bronze Age Man led to widespread forest clearance, particularly on the lighter soils and, with the deteriorating climate, hastened leaching and podsolization. A variety of crop plants and their associated weeds were introduced, and there was a marked expansion of many spiny shrubs such as *Crataegus monogyna* (hawthorn) and *Ulex europaeus* (gorse) which are resistant to grazing and seem to have exploited newly opened habitats to form scrub communities. The start of the final zone of the Flandrian, the Sub-Atlantic (2,500 years ago to the present), coincides with the origin of Celtic Iron Age culture in northwest Europe. Earlier agriculture was expanded and large permanent fields were established. The climate became cooler, but increasingly moist, so that the growth of bogs, static and degrading since the Atlantic period, was able to recommence.

In North America, the start of the Sub-Boreal was marked by a decline in elm similar to that seen in Europe. The climate seems to have been rather variable but generally becoming drier, until a rapid climatic deterioration took place at the beginning of the Sub-Atlantic. This change occurred throughout the world between about 3,500–2,500 years ago, and the lack of synchroneity suggests that a series of climatic oscillations may have been involved. The effect was of a worldwide displacement of vegetational zones to lower latitudes, approximately to their present positions. Small-scale climatic fluctuations have certainly occurred since, but from this time onward it becomes increasingly difficult to separate climatic and other effects from those due to Man. In northwest Europe, subsequent forest clearance and plant introductions over a long period have left very little natural vegetation, but only in the last 200 years has Man's technological sophistication and changing requirements established him as the dominant factor on a worldwide scale, with the result that current vegetational and environmental changes are more rapid and more extensive than those of any other period in the Earth's history.

P.R.C.

The Evolutionary Background

The origin of the enormous variety of animal and plant species has been a subject of speculation by mankind from time immemorial. Essentially there are two mutually exclusive theories: all living things were either the product of an act of creation or the result of modification by descent of preexisting forms (ie evolution). The prime example of the former view is the account of Creation in Genesis, an account which held sway for many centuries, largely on account of its adoption by the Christian faith. The beginnings of evolutionary ideas, however, can be traced back at least as far as some of the philosophers of ancient Greece, but it was not until the 19th century that they seriously challenged the creationist view.

Common gorse (*Ulex europaeus*) is one example of many species of spiny shrubs, resistant to grazing, which showed a marked increase in range during the sub-Boreal in the British Isles. Such species seem to have exploited the open habitats newly created by Man during his early farming activities.

If it is accepted that evolution has occurred, it follows that living, organic material must have been derived from inorganic matter, and it is clear that if one goes back far enough in time the physical and evolutionary backgrounds dealt with in this chapter become indistinguishable. Indeed, some of the evidence to be presented next suggests that the beginnings of life may not have been too far removed from the origin of our planet. Nevertheless, plants and animals as we know them are a more recent phenomenon, dating back no more than about 600 million years, one-eighth of the Earth's history. The general outline of their evolution has been charted, at least partially, in the geological record (see p. 71). There is still much to learn, but the picture of how our planet obtained its green cover is broadly painted in the following pages.

Following the account of the origin and early history of life, the development of modern ideas of evolution and natural selection is described. Although not the only way that evolution can occur, adaptation is certainly one of the most conspicuous evolutionary processes; indeed, the observation of adaptive radiation among the finches of the Galápagos Islands was of major help to Charles Darwin in formulating his theory of evolution and, since the much-disputed concept of genetic drift can be confused with adaptive radiation, they are linked in this account.

Since the mechanisms by which evolution takes place are best understood in relation to the formation of species, this process of speciation is considered in a separate section, while the various types of isolation are well known to be so important in speciation that they merit detailed consideration.

D.M.M.

The Origin and Early History of Life

Despite its diversity, living matter is characterized by an astonishing biochemical uniformity. All known forms of life have two features in common: they have cellular structure, in which the cells are bounded by membranes, and they are composed of two kinds of macromolecule: nucleic acids and proteins. The combination of all these features is a certain indication of life, quite apart from the consideration of function. It is thus possible to define "life" without having to define the difference between "living" and "dead".

Rival Theories of the Origin of Life

All modern theories of the origin of life stem from two papers published independently about 50 years ago, one by a young Russian biochemist, Aleksandr Oparin, in 1924, the other by the British geneticist, J. B. S. Haldane, in 1929. Although both papers have much in common, they also have points of contrast, and it is interesting that during the years that have elapsed since their publication, continuing research has emphasized not only the soundness of their common principles, but also the viability of

their contrasting points of view. These viewpoints are the basis of two modern rival theories, both championed in text-books written by leading researchers in the field: *Origin of Life and Evolutionary Biochemistry* (1972) by Sidney Fox and Klaus Dose, and *Origins of Life on Earth* (1973) by Stanley Miller and Leslie Orgel. Both theories are supported by experimental and observational evidence. Fox and Dose emphasize Oparin's view: "The fundamental proposition that 'life arose from fire' remains unshaken. Only in fire, only in incandescent heat could the substances which later gave rise to life have been formed". Miller and Orgel prefer the suggestion of Haldane that protolife was powered by solar energy, that "When ultraviolet light acts on a mixture of water, carbon dioxide and ammonia, a vast variety of organic substances are made, including sugars and apparently some of the materials from which proteins are built up ... In this present world, such substances, if left about, decay—that is to say, they are destroyed by microorganisms. But before the origin of life they must have accumulated till the primitive oceans reached the consistency of hot dilute soup".

Common to both theories is the central concept that before the origin of life evolutionary processes were at work. Prebiological evolution was chemical in nature. The origin of life marked a turning point, when chemical evolution turned over to organic evolution. The origin of life itself is therefore part of

a continuous evolutionary process—continuous but not gradual. The British biochemist Norman Pirie in 1959 emphasized that during the period of chemical evolution, the hypothetical first-formed living organisms ("eobionts") produced would have been structurally very simple but chemically very diverse. Once life had originated, chemical diversity was replaced by biochemical uniformity. At the same time, the simple structures of the prebiological phase were replaced by more complex ones. Pirie expressed his ideas in a diagram which has become famous as the "hour-glass" model of the origin of life, which also suggests that the origin of life took place on Earth between 1,000 and 2,000 million years ago.

More recent geological evidence, summarized here, suggests that this time-scale is wrong. The origin of life probably took place very shortly after the origin of the Earth, perhaps as much as 4,000 million years ago. Moreover, the first kind of life that appeared on the Earth was not diversified in the same

Below An artist's impression of the landscape of the Earth some 4,000 million years ago when protolife is thought to have originated. There are basically two rival theories of the origin of life. One theory suggests that the formation of protolife was powered by the action of ultraviolet light from the sun acting on simple chemicals (water, carbon dioxide and ammonia) in the oceans, forming a hot dilute soup of organic substances. The other theory supposes that prebiological synthesis was powered by volcanic activity.

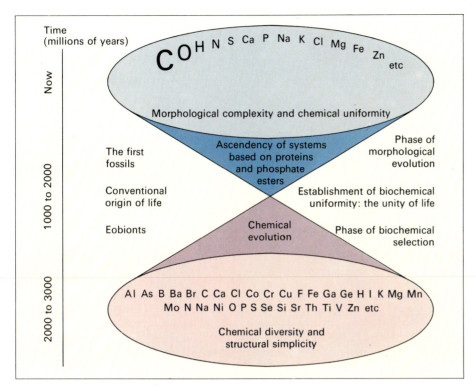

A reconstruction of Norman Pirie's hourglass model of the origin of life. At each level of time the width of the cone represents the number of ways in which living or life-like systems worked. The size of the atomic symbols is an indication of the contribution each element may have made to the process of the time.

way that life is today. For perhaps 3,000 million years, life was limited to prokaryotic organisms, ie organisms composed of cells without a nucleus, reproducing in a more primitive way than that which evolved later. Only during the last 1,000 million years have eukaryotic organisms appeared (ie those possessing a nucleus containing chromosomes), with full sexual reproduction and the consequent taxonomic diversity characteristic of all later times. These new discoveries suggest that the origin of life can better be represented by a "dumb-bell" model in which a long period of prokaryotic uniformity (3,000 million years) separates the early chemical diversity from the later taxonomic diversity. The three periods of biological history are separated by two events in which changes were very rapid: the first was the origin of life, 4,000 million years ago; the second the origin of sex, 1,000 million years ago.

Experimental Evidence for the Origin of Life

In 1953 the American biochemist Stanley Miller performed the first laboratory synthesis of organic compounds in an environment designed to simulate prebiological conditions. Miller exposed a mixture of methane, ammonia, hydrogen and water to an electric spark discharge, and kept the materials circulating in a closed system which included boiling water and a condensing tube. He was able to demonstrate the synthesis of a whole range of organic compounds including amino acids, sugars, urea and urea derivatives. Since then, many laboratories have repeated his experiments

with almost embarrassing success. Many alternative energy sources (including ultraviolet radiation) and many alternative gas mixtures have proved productive (provided that they have contained no free oxygen, since life originated at a time when oxygen was lacking). The yield of organic compounds, however, has varied in both quantity and composition. Not only simple molecules have been produced, but also polymers (larger molecules that contain a repeating sequence of the original molecules). In some experiments the polymers are red, and this lends weight to the suggestion that the red color in the atmosphere of Jupiter (particularly evident in the Great Red Spot) may be due to the synthesis of similar red polymers. More recently, electron-microscopic investigation of the insoluble polymers produced in such a "Miller experiment" has revealed the existence of cell-like entities which might be considered as the experimental analogue of prebiological "protocells".

The class of experiment initiated by Miller gives strong support to the theory that the origin of prebiological material took place in the atmosphere of the primitive Earth, and was from there washed down by rain into the early oceans, and so formed the "hot dilute soup" postulated by Haldane. This concept of a "prebiotic soup" forms the basis of the theories championed by Miller and Orgel.

Another class of experiment was reported by the American biochemist Sidney Fox in 1960. In these experiments mixtures of amino acids were heated to melting point, and the melts quenched with cold water. The resulting product has been shown to lead not only to the thermal polymerization of the amino acids to produce what Fox has called "proteinoids", but also to the aggregation of

such proteinoids into cell-like units approximately 2μm in diameter. These "proteinoid microspheres" have varying degrees of complication in their structure. Some have a double-walled membrane as an outer boundary, and this can be demonstrated by electron microscopy. They can also be made to react with their surrounding solution (by altering its acidity or by varying the pressure exerted by the cover-slip on the slide); they can grow, or divide by fission, or produce buds, or disintegrate.

Fox's experiments suggest a volcanic scenario for the origin of life. A volcano in eruption is powered by the expansion of the volatile constituents, and these include water and gases rich in carbon (carbon dioxide, carbon monoxide and methane), all of which are turbulently ejected into the atmosphere together with a spray of lava and the dust and solid fragments of more refractory material. A volcano in eruption is invariably accompanied by violent thunderstorms and torrential rain. If the atmosphere were to be free of oxygen, volcanic clouds would provide a perfect location for Miller-type organic synthesis. The amino acids thus produced would have been washed down by rain in solution on to the hot lava below. These amino acid solutions would have been repeatedly evaporated to dryness, melted and quenched by rain. Fox-type organic syn-

A reconstruction of the dumb-bell model of the origin of life.

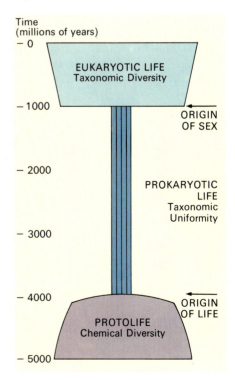

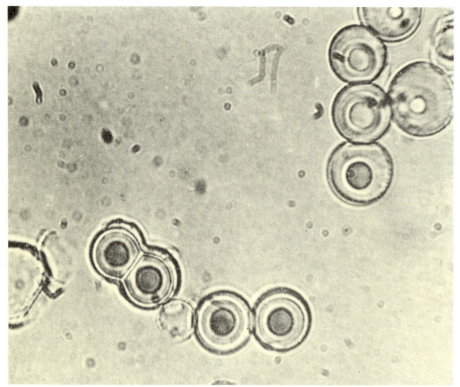

These cell-like structures, called "proteinoid microspheres", were produced experimentally by Sidney Fox by heating organic chemicals to melting point and quenching them with cold water. They resemble what may have been the first living organisms. These examples clearly show the double-walled membrane as the outer boundary.

thesis must have occurred, and the resulting proteinoid microspheres would have been washed into volcanic pools.

The weakest point of the volcanic theory for the origin of life concerns the subsequent history of the postulated protocells; for how did they come to ingest nucleic acids, and how did these link up with the protein sequence already in the cells? However, the alternative theory, in which the oceans formed a prebiotic broth, has produced no more satisfactory answer to the problem. One

suggests that the proteins came first, and that these ingested nucleic acids; the other suggests that the nucleic acids came first, and that these linked up with specific proteins. Perhaps neither is true. Maybe the first link was made before polymerization between a nucleotide and an amino acid, but such a possibility has as yet no experimental evidence to support it.

According to the volcanic theory, the most likely site for the origin of polynucleotides would be the thermal pools which invariably

accompany other late-stage volcanic effects. In these pools the phosphates, sugars and nucleosides would, it is supposed, be concentrated, and would combine to form nucleotides. Polymerization would follow.

Geological Evidence for the Sequence of Events

If the theories for the origin of life suggested by experimental evidence are correct, it should be possible to demonstrate in the geological history of the Earth some contrast between the conditions which existed before the event, with those which succeeded it. Moreover, it might still be possible, under certain conditions, to demonstrate that the natural, nonbiological synthesis of organic compounds has continued to the present day in certain environments. For example, despite the presence of free oxygen in the atmosphere, amino acids have been reported from some of the hot volcanic pools of the Kamchatka province in the USSR.

The Earth was formed about 4,600 million years ago. The oldest known rocks are sediments which were metamorphosed about 3,760 million years ago. They were formed as water-laid sediments some unknown time before that. The first rocks in which fossils occur abundantly are only 600 million years old; rocks younger than that are called Phanerozoic. The first 4,000 million years of the Earth's history is called Precambrian, and this can be divided into two parts: the earlier Archaean (before 2,600 million years ago) and the later Proterozoic (600 to 2,600 million years ago).

Although fossils are very much rarer in Precambrian rocks than they are in the Phanerozoic, they do occur, and they are of three kinds. Firstly, there are microscopic spheres and filaments, which appear to be the remains of bacteria and blue-green algae (ie prokaryotic organisms). These are better preserved in Proterozoic rocks than they are in the Archaean, but they occur in both. The oldest so far discovered come from near the bottom of the Swaziland Series of South Africa, and are more than 3,300 million years old.

The second kind of fossil is found as massive calcareous reefs; these are known as stromatolites and are believed to be laid down

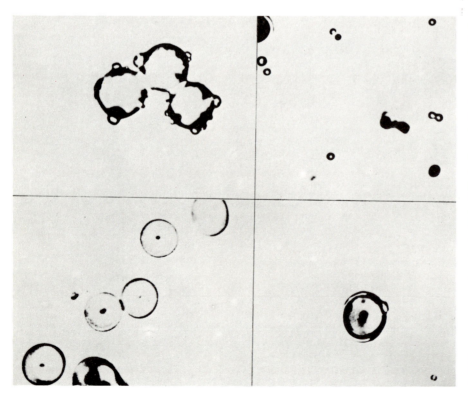

Electron micrograph pictures of Fox's microspheres showing spheres "growing" and "reproducing" by budding. **Top left** buds on parental microspheres; **top right** separated buds; **bottom left** microspheres grown from buds; **bottom right** second-generation microsphere with bud.

A volcanic cloud from Surtsey, a new island formed southwest of the Westman Islands. Iceland, in 1963. Such volcanic clouds, with their high content of water and gases rich in carbon, in conjunction with the violent thunderstorms that frequently occur in such circumstances, provide the perfect location for the Miller-type synthesis of protolife.

by blue-green algae. The oldest known (about 2,900 million years) form the Bulawayan stromatolites of South Africa.

Finally, there are fossils which are believed to be the remains of metazoa (multicellular animals with differentiated tissues and organs, ie eukaryotic organisms). The best known are those which occur at Ediacara in South Australia, and these are about 700 million years old, but others of about the same age occur in other parts of the world. Trails believed to have been made by metazoans have been reported from somewhat older rocks (1,000 million years) in South Africa, and microscopic fossils about 900 million years old (from the Bitter Springs chert of Australia) have been interpreted as representing the remains of eukaryotic cells. However, this interpretation has been disputed, and they may still be prokaryotes.

The interpretation of the earliest fossils and the earliest rocks is still controversial. The very earliest fossils, from near the bottom of the Swaziland System, are about half the size of those that occur later in the same system. Some would interpret the earlier and smaller fossils as the remains of protocells, and would therefore date the origin of life at about 3,300 million years ago, only the later Swaziland fossils being the remains of truly living organisms. But the occurrence of the spheres and filaments, taken in conjunction with the evidence considered below of the biogenic nature of still older rocks, lends support to the contention that these very old fossils are truly those of living organisms.

The oldest rocks accurately dated so far come from Greenland. They include three kinds of rock which have been interpreted as biogenic, ie produced by living organisms. These include graphitic schists, banded ironstones and marbles. The graphitic schists and the marbles contain carbon which the biogenic theory interprets as the metamorphic equivalent of bituminous shale and stromatolite. The alternative theory regards them as the remains of prebiological carbonaceous sediments. The evidence in support of this contention is that the carbon isotope ratio in the graphitic schists differs from that characterizing bituminous shales of known biological origin. The biogenic theory believes that the difference is due to the effect of metamorphism (ie changes produced in the rocks by temperature and pressure).

If the biogenic theory of these early rocks is correct, the origin of life must have occurred some time before the rocks were metamor-

A geological column (**center**) showing the probable ages of various dated horizons (**left**) relevant to the present theories as to the dates of various biological events. Dates are given in millions of years before the present.

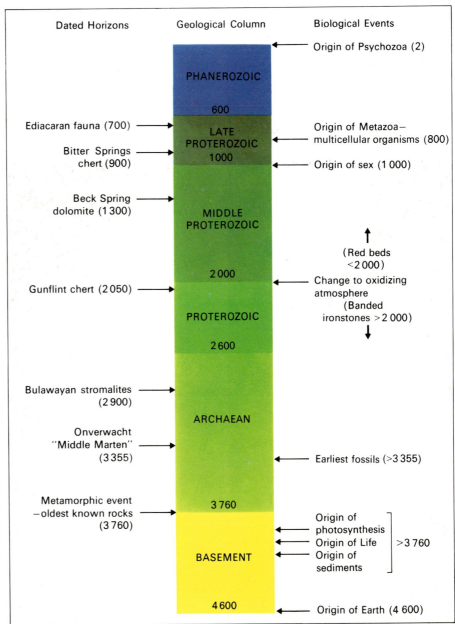

phosed. The event cannot have occurred long after 4,000 million years ago. This conclusion would mean that no known terrestrial rocks antedate the origin of life, which must have occurred during the first 600 million years of the Earth's existence.

This evidence suggests that the origin of life led to the rapid establishment of various kinds of prokaryotic life, including reef-forming blue-green algae, and various kinds of bacteria. Thereafter and for the next 3,000 million years evolution was almost at a standstill. Then, somewhere between 1,000 and 700 million years ago, the nucleus was invented, sexual reproduction became sophisticated, and diversity blossomed at an astonishing rate. Almost all the phyla we now know in the plant and animal kingdoms became established, and almost all aqueous environments became colonized with new ecosystems.

There remains one mystery. Geological evidence suggests that 2,000 million years ago there was a sudden change in the composition of the atmosphere. Before that time, there was no free oxygen. The rocks that occur include sediments which, if oxygen had been present, would have been oxidized (pyritic sands and uraninites). The banded iron formation also occurs before this time but is not found later. The banded irons include layers containing ferric (oxidized) iron. Rocks older than 2,000 million years do not include any red sandstones, but these become common at later times, and are

thought to owe their red color to an oxidizing atmosphere. One would have expected that the changeover from an anoxygenic to an oxygenic atmosphere would have been accompanied by a profound effect on the living organisms present. Yet no observations so far made support such a proposition.

Evidence for the Origin of Protolife

If the evidence for the origin of life before 4,000 million years ago is accepted, we can expect no evidence from terrestrial rocks for the existence of protolife, for life seems to be as old as our oldest rocks. But experimental evidence shows that environments suitable for the synthesis of organic compounds are not likely to have been limited to the planet Earth. Indeed the American chemist Edward Anders and his colleagues have performed experiments which suggest that organic synthesis occurred very early during the origin of the solar system, even before any of the planets themselves were formed. Moreover, the products of this occurrence are believed to have survived in the form of carbonaceous meteorites, and several of these have fallen onto the Earth in recent times, and are available for study.

Anders' experiments, using inorganic catalysts designed to simulate meteoritic dust, were constructed to test the hypothesis

that carbonaceous meteorites could form in the solar nebula as a result of thermal reactions. He was able to demonstrate that if gases including carbon dioxide, carbon monoxide, methane, ammonia, hydrogen and water are exposed at high temperatures to such catalysts, the organic compounds produced include almost all those which characterize the meteorites in question.

One of the most famous carbonaceous meteorites was that which fell at Orgeuil in France in 1864. Unfortunately, since it fell it has become heavily contaminated with biological material of terrestrial origin. However, in 1969, two large carbonaceous meteorites fell, one at Allende in Mexico, the other at Murchison in Australia. Great care was taken to collect uncontaminated samples, and it was soon demonstrated that the Murchison meteorite contained amino acids which were indigenous to the meteorite and could not have been contaminants. Subsequently, other carbonaceous meteorites were re-examined, and techniques were devised to distinguish between indigenous material and that which could be the result of contamination. It now seems clear that there is a great variety of original carbon compounds. By far the greater proportion of these are random insoluble polymers difficult to analyze, but not necessarily dissimilar from the "kerogen" or "sporopollenin" which characterizes terrestrial carbonaceous rocks, and is equally difficult to analyze. Among the soluble contents, the amino acids can often be recognized as indigenous. Among the most controversial of the early finds in the Orgeuil meteorite were fossil-like organisms termed "organized elements". Some of these have been shown without doubt to be contaminants, but others have been confirmed as both organic and as indigenous. Two classes in particular appear to be genuine. One class is hexagonal in shape, and these hexagonal elements appear to be organic coatings of pseudomorphs derived from some hexagonal mineral, probably troilite, an iron sulfide (pseudomorphs are alteration products which have a different composition but the same shape as the mineral they replace). The other class seem to be hollow spheres, and these could well have the nature of the "microspheres" or "protocells" produced in Miller-type or Fox-type syntheses.

It seems, then, that although it is not possible to demonstrate the existence of protolife on the Earth, it certainly does occur in the solar system. Indeed, it has been suggested by the astronomers Fred Hoyle and Chandra Wickramasinghe that life on Earth began when forms of protolife were brought to this planet by means of meteorites.

The red spot on Jupiter, which may be caused by the synthesis of polymers similar to those produced in laboratory experiments attempting to re-create the conditions in which "protolife" may have developed.

The Evolution of Life on Earth

Once life established itself on the Earth, as a primordial unicellular organism, Darwinian evolutionary forces are assumed to have taken over and, as a result of mutation and natural selection, a slow but steady increase in complexity of life-forms developed (see Evolution and Natural Selection, p. 89). As Nobel prize winner Albert Szent-Gyorgi has written, "Life has developed its processes gradually, never rejecting what it has built, but building over what has already taken place. As a result, the cell resembles the site of an archaeological excavation, with the successive strata on top of one another, the oldest one the deepest". One of the key events must have been the development of the eukaryote organism with a discrete nucleus, as distinct from the more primitive pro-karyotes. Subsequent events in evolution are more familiar to us: the development of aerobic respiration; the evolution of marine life-forms; the establishment of life on land; the age of dinosaurs; and eventually in very recent times, the emergence of Man.

In most theories of the origin of life, much attention is given to the nature of the first living organism (or organisms). While it is clearly very unlikely that it is still living, consideration has been given to the question: which present-day organisms most closely resemble this earliest form of life? One possible candidate, and one of the most ancient of present-day bacteria, is *Kakebekia*. This organism was first described as a fossil from the Gunflint chert, dating back 2,000 million years. Startlingly, a living form of the same organism was discovered quite recently growing in the soil at the base of one of the walls at Harlech Castle in North Wales. Even more remarkably, the soil was rich in ammoniacal compounds and experiments showed that the bacterium grew much better under anaerobic conditions (in the presence of ammonia and methane) than in air—truly, a "living fossil", if there ever was one.

In considering the origin of life, the relationship of primitive life-forms to higher plants has always excited the interest of botanists, in particular the contrast between the apparently simple unicellular bacterium and the highly complex, highly developed multicellular angiosperm plant. It has, indeed, been suggested that some parts of the complex cellular machinery of higher plants originated from the introduction of (or infection by) simpler organisms. One theory (the so-called endosymbiotic view of chloroplast origin) proposes that the chloroplasts of higher plants are modified blue-green algae which have at some stage invaded their cells. Another current theory proposes that the mitochondrion (the organ of respiration) of higher plants is of bacterial origin. There is recent biochemical evidence (from studying DNA compositions and enzyme activities) for these theories.

Much of our information on the origins of plants, however, is based on examination of plant fossils, collected from rock deposits of varying ages. Dating of these fossil remains is determined by their inorganic chemistry: on measuring the degree of decay of the radioactive elements present, the rate of decay being constant with time. The most important elements used are uranium, thorium, rubidium and potassium.

Until recently, the organic chemistry of fossil plants was not seriously considered, but recent progress in analytical methods has provided us with the means of measuring the minute amounts of organic matter present in these fossils. We can actually obtain evidence that particular life processes were taking place in these plants many millions of years ago. A key process in primitive plants was the development of photosynthesis. Evidence that this process occurred in the living fossils can now be obtained by comparing the ratio of carbon-12 (^{12}C) to carbon-13 (^{13}C) in the organic matter of the rock to the corresponding ratio in the carbon content of the rock itself. While the ratio of ^{12}C to ^{13}C in the atmosphere is constant at 99:1, any biological activity such as photosynthesis, which uses only $^{12}CO_2$, changes this ratio by enrichment of ^{12}C. Thus, by studying suitable fossils, it has proved possible to show that photosynthesis took place in aquatic, anaerobic, bacteria-like organisms 3,100 million years ago.

Organic analyses of fossil plants have shown that other processes known in plants today were taking place in plants many millions of years ago. For example, comparisons have been made between the present-day horsetail, *Equisetum arvense*, and *Equisetum* fossil deposits of 200 million years ago in the Triassic system. A mixture of four long-chain hydrocarbons was isolated from the fossil and, remarkably, the same four hydrocarbons in the same proportions were found in the lipid fraction of the living plant. These and other experiments show that much of the biochemistry of the plant kingdom has remained essentially unchanged during countless years of evolutionary modification of life-forms. Furthermore, they add support to the view that life originated only once on this Earth and that the immense range of living organisms known today all developed from one or a few primitive organisms. P.C.S.-B./J.B.H.

Evolution and Natural Selection

As we have just seen, knowledge of the origin and early history of life has increased significantly only as a result of studies in relatively recent years. The theory of evolution by natural selection was formulated well over a century ago and consequently, as we shall see, it had to develop in the absence of clear ideas about where and how living matter originated.

Before evolutionary relationships between plants could be discerned, an effective method of classifying existing plant species was essential. This was provided by Linnaeus, when he introduced his classification system based on their sexual organs. This system was rapidly accepted by botanists, since it had produced order where previously there had been chaos. There was no accepted nomenclature before Linnaeus and his two-volume *Species plantarum* (1753) introduced the binomial system which is still in use (see Nomenclature, p. 29). Linnaeus' classification purported to reveal the order established by the Creator. From the fact that a scale of nature could be readily erected, it was but a short step to the concept that such a hierarchical scheme could well be explained as a consequence of relationship through descent from a common ancestor. Linnaeus incidentally included Man with the apes in his classification, which caused great consternation at the time.

The greatest impetus to the establishment of the idea of evolution came with the work of the French naturalist Jean Baptiste Lamarck (1744–1829) and his development hypothesis. He proposed that characteristics acquired during the life of an organism could be passed on to its progeny, a theory caricatured by many detractors in drawings showing giraffes obtaining their long necks as a result of continued stretching to reach succulent leaves on trees. Interestingly, he had no compunction in suggesting that Man was derived from apes. Instead of the static scheme of Linnaeus, there was now a dynamic one.

By the early part of the 19th century the idea of evolution was becoming firmly implanted. In 1844 Robert Chambers published anonymously *Vestiges of Creation*, an extremely popular book arguing that all animals and plants have changed through time, gradually evolving as the conditions changed, and that all living things have evolved from simpler forms of life. Numerous scientists had written in favor of evolution and some had described natural selection as the possible mechanism. William Wells (1757–1817), in a paper presented to the Royal Society of London in 1813 (published in 1818), outlined the theory of natural selection acting on random variation and discussed the origin of the Negro race. However, all this was still only speculation, although eminently plausible speculation. Before it could be finally established it was necessary for factual evidence to be accumulated. This was Darwin's great contribution.

Darwin and the Theory of Evolution

At the start of the world voyage of the *Beagle* in 1831, Charles Darwin was vehemently opposed to the ideas of Lamarck and certainly to those of his evolutionist grandfather Erasmus Darwin (1731–1802). His conversion·stemmed initially from Hooker's gift to Darwin of Lyell's *Principles of Geology*. The notion of Uniformitarianism, that the geological history of an area was the consequence of processes that are still in progress today, had a profound effect on the young Darwin.

In South America he discovered remains of giant ground sloths (*Megatherium*) and giant armadilloes (*Glyptodon*). The fact that these two groups with their living relatives were confined to South America suggested that the extinct fossil forms had been replaced by similar related forms. Although this suggested to Darwin that species had become modified through time, until he could explain how the adaptation of animals and plants to their respective ways of life arose, he felt that it was not worthwhile trying to establish the modification of species by such indirect evidence.

It was during his visit to the Galápagos Islands during September and October 1835 that his belief in the immutability of species began to founder; as he recorded in his ornithological notebooks: "When I see these Islands in sight of each other and possessed of but a scanty stock of animals, tenanted by these birds but slightly differing in structure and filling the same place in Nature, I must suspect they are only varieties ... If there is the slightest foundation for these remarks, the Zoology of the Archipelago will be well worth examining; for such facts would undermine the stability of species". Two years later he began his first notebook on "Transmutation of species".

In October 1838 Darwin read Malthus' 1798 essay on population, the theme of which is the survival of the fittest in human societies: "and being well prepared to appreciate the struggle for existence which everywhere goes on from long continued observation of the habits of animals and plants, it at once struck me that under these circumstances favourable variations would tend to be preserved, and unfavorable ones to be destroyed. The result of this would be the formation of a new species. Here then I had at last a theory by which to work". (Darwin, *Autobiography*.)

Darwin wrote a sketch of his "new" theory in 1842, and in 1844 a 250-page essay. Lyell and Hooker were privy to this but no hint of it was published, although Darwin continued for many years to sift and put together the evidence for his theory. Then without warning in 1858 Alfred Wallace sent Darwin a short manuscript entitled, "On the tendency of varieties to depart indefinitely from the original type". Darwin was in the process of producing a monumental work on the origin of species and suddenly here was the key to natural selection already spelt out independently by Wallace (no one remembered Wells). Wallace's letter and essay "came like a bolt from the blue". Lyell and Hooker submitted an abstract of Darwin's 1844 essay, a letter from Darwin to Asa Gray dated 1857 (to prove that Darwin had not changed his views since writing the essay) and Wallace's article to the Linnean Society

on 1st July, 1858, and these were published in the same year as a joint paper by Darwin and Wallace. The following year Darwin published his book *On the origin of species by means of natural selection or the preservation of favoured races in the struggle for life*.

Darwin's thesis can be summarized as follows. It could be observed that organisms produced many more offspring than themselves and that this type of geometrical increase was not reflected in the size of subsequent populations. This meant that most offspring did not survive, only the fittest. This was a fact that anyone could understand, and it explained the mechanism by which evolution could have taken place. Granted the existence of variation within a population, it was easy to see how individuals with certain advantageous traits would stand a better chance at the expense of other individuals. With changing conditions the distribution of traits within a population would change.

The Genetic Basis of Evolution

As we have seen, Darwin's evolutionary theory depended upon the variation present in populations which allowed differential inheritance of traits from one generation to the next. However, there was no satisfactory theory to explain the basis of this variation or how new variation arose. Darwin tried to solve the problem by putting forward his theory of *pangenesis*. This suggested that cells of the plant produce minute particles ("gemmules" of Darwin) which circulate freely within the plant and are transmitted from parent to offspring. As a consequence of this, blending of gemmules occurs in the offspring which would thus be of intermediate characters between their parents. The reappearance of parental characters in grandchildren was considered to result from the transmission of gemmules in a dormant state. Furthermore, Darwin also expressed a belief in Lamarck's idea of the inheritance of acquired characteristics—if, for example, a normally tall plant is dwarfed by growth in unfavorable conditions, it would produce dwarf offspring. No experimental evidence was produced to support these beliefs. The English biologist Sir Francis Galton (1822–1911), for example, who intertransfused blood of different colored rabbits and studied the color of their offspring, found no evidence that Darwin's gemmules could explain his results.

This unsatisfactory state of affairs came to an end in 1900 with the rediscovery of Mendel's work of 1865. By careful analysis of the progeny of crosses between purebreeding lines of peas differing in certain distinctive characters the Austrian biologist Gregor Mendel was able to postulate the existence of physical factors (later called genes by the Swedish biologist Ivar Johanssen) which were inherited intact from

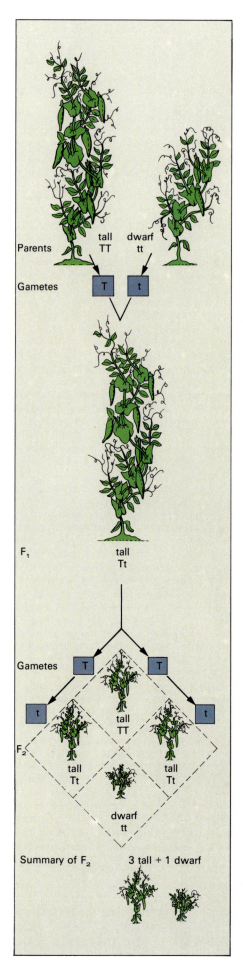

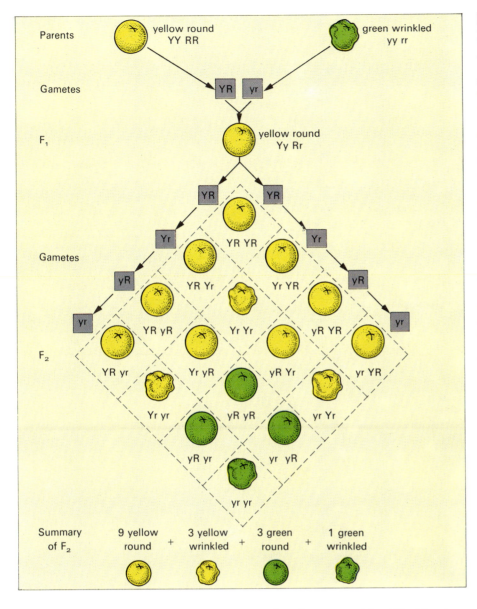

| Parents | yellow round YY RR | green wrinkled yy rr |

Summary of F₂: 9 yellow round + 3 yellow wrinkled + 3 green round + 1 green wrinkled

proximity, but parental combinations of such linked genes were always more common than new combinations (recombinants) among the progeny.

Although the first half of the 20th century saw an immense increase in knowledge of the inheritance, behavior and mode of operation of genes, of the rates at which they somehow changed their structure and effects (mutation) and the ways in which this mutation could be induced artificially, and of the behavior and importance of chromosomes and chromosomal change to the viability, variation and evolution of plants, the fundamental nature of the genetic material was not resolved until 1953. In that year James Watson and Francis Crick provided evidence that deoxyribonucleic acid (DNA) was the virtually ubiquitous genetic material among plants and animals. They demonstrated that the structure of its molecule, a double helix of two sugar-phosphate "backbones" joined by pairs of purine (adenine, guanine)-pyrimidine (cytosine, thymine) bases provided an explanation of how the exact replication of genes might occur. It also suggested, as was subsequently proved, that the sequence of bases on the strand might provide a code (later shown to be based on triplets of bases) which could specify the amino acids, and hence proteins which control most of the processes in, and characteristics of, the cell and organism. Once this "Genetic Code" was understood it became possible to determine how certain chemicals (mutagens) could alter it by affecting the occurrence of particular bases and thus how mutation, both artificial and spontaneous, was achieved. Although many problems still remain, the age-old question of the fundamental source of the variation on which evolution depends was at last made clear.

Evolutionary Theory since Darwin

It might be thought that after Darwin's theory, documented by a vast number of observations, of evolution by natural selection acting upon the variation present in populations, and the gradual clarification of the physical basis of that variation, there could have been no further grounds for doubt and controversy. This was far from the truth.

Darwin argued that species were ever-changing entities, the products of natural selection, with evolution being the result of continual and gradual change. However, around the turn of the 20th century, people such as Francis Galton, Karl Pearson and Hugo De Vries became interested in the statistical study of variation in natural populations of plants and animals. It was found that some features, such as the stature of plants or soldiers, showed a continuous variation within a population while others, such as the occurrence of red or white flowers, hairy or glabrous stems in foxgloves

The genetic basis of evolution. The individual plant or animal that we see, comprising the full set of characters, such as leaf-shape, flower-color, flowering time, soil-preference etc. is termed the **phenotype**. The phenotype results from the interaction of the prevailing environmental conditions and the **genotype**, which comprises all the genes inherited from previous generations. The **genes**, composed of the nucleic acid, DNA, may exist in two or more forms, known as **alleles**, that differ in their structure and effects as a result of the structural or chemical changes constituting **mutation**. Since any individual receives half of its genotype from each parent, each gene is represented at least twice; when the maternal and paternal representatives of each gene are of the same allelic form, the individual is said to be **homozygous** for that gene; where the alleles differ the term **heterozygous** is used.

The existence and inheritance of genes was first deduced by Gregor Mendel working with garden peas. In one experiment (**left**) he crossed pure breeding homozygous strains of peas differing in stature. Since all first generation (F₁) hybrid progeny were tall he deduced that tall alleles (T) are dominant to dwarf alleles (t). By intercrossing the F₁ plants he obtained tall and dwarf F₂ progeny in proportions which indicated that maternal and paternal genes segregated, were brought together at random during fertilization, and that genes such as that for dwarfness retained their identity from generation to generation. In a

second set of experiments (**above**) Mendel used homozygous lines of plants differing in two characters, the color and shape of the peas, with yellow and round dominant to green and wrinkled as shown in the F₁ progeny. The relative numbers of the different phenotypes in the F₂ progeny showed that alleles of different genes segregate independently.

generation to generation and which were the determinants of the characters seen in the plants. His further demonstration of the independent segregation and recombination of these genes was amplified during the early part of the 20th century when the genes were shown to be located on the chromosomes which provide a physical continuity between cell and organism generations. Genes present on different chromosomes showed Mendel's independent segregation and recombination, while those present on the same chromosome were linked by their presence on one structure; the extent of their recombination depended upon their physical

(*Digitalis purpurea*), showed discontinuous variation with abrupt changes from one type to the other. These different patterns of variation served to support two opposing views of how evolution came about. Continuous variation was used to support the gradual change advocated by Darwin (even though he was aware of the occurrence and perpetuation of abrupt changes—"sports"), while discontinuous variation supported the views of the "mutationalists", spearheaded by De Vries and Bateson, who considered that evolution proceeded by means of abrupt changes and that all continuous variation was environmentally induced. The discontinuous pattern of variation shown by the characters studied by Mendel in his fundamental work on the nature of inheritance was considered to be strong support for this view, while the Dutch botanist Hugo De Vries, in his classic studies of the evening primrose (*Oenothera lamarckiana*), recognized the appearance of a number of new types, which appeared suddenly by a

process he termed "mutation", and differed from their parents in several features, and bred true.

It was not until the 1930s that Darwin's ideas, now based on sound genetical precepts about the basis of variation, triumphed over the "mutation theory". Intensive cytogenetical studies of many plant groups showed that in working with species of *Oenothera* De Vries had chanced upon a particularly

Below The cytological basis of inheritance. Genes are located in linear order on the thread-like **chromosomes**, present in the cell-nucleus and composed of nucleic acids and proteins. Each chromosome is split longitudinally into two identical **chromatids**, held together at one point, the **centromere**. During the growth and development of an individual, cell-division is extremely important. In order that all the cells should be genetically identical the chromosomes undergo **mitosis** each time a cell divides. These photographs trace the course of mitosis in the root apex of *Crocus balansae*.

At the onset of mitosis, during early prophase (a), the chromosomes appear as long thin threads each composed of two chromatids. The

chromosomes become shorter and thicker by internal coiling during late **prophase** (b) and **metaphase** (c), by which stage the chromosomes have attached to longitudinal spindle microtubules (invisible in light microscope preparations such as these) and become orientated on the equator of the cell. During **anaphase** (d–g) the centromeres divide into their sister centromeres which then move apart towards opposite poles of the cell, dragging their attached chromatid-arms (now constituting a daughter chromosome) behind them. During **telophase** (h) the two polar groups of daughter chromosomes uncoil and a new nuclear membrane reforms around them, to give two cells which have the full complement of genes and, during **interphase** (i), each daughter synthesizes new material to make a 2–chromatid structure.

Opposite The cytological basis of inheritance (continued). At some time prior to sexual reproduction, cell-division (known as **meiosis**) takes place, during which the number of chromosomes is halved. At fertilization the nuclei of the male and female sex-cells (gametes) fuse in producing the zygote, which then develops into the adult organism; without meiosis the chromosome number would double each generation, thus upsetting the physical stability upon which the orderly inheritance of genes depends. In higher plants (angiosperms and gymnosperms) meiosis gives rise to spores (pollen in the male and megaspores in the female) which divide by mitosis to give the tiny 2- or 8-celled gametophytes that produce, respectively, the sperms and eggs (gametes).

When meiosis begins there are two sets of chromosomes, one derived from each parent; thus, for any chromosome contributed by one parent there is an identical one (or nearly so) from the other parent, the two being said to be **homologous**. Meiosis comprises 2 consecutive divisions, each of 4 stages—prophase, metaphase, anaphase and telophase, with the first prophase divided into 5 stages—as shown here in the meadow grasshopper (*Chorthippus parallelus*).

At the onset of meiosis, **leptotene** (a), the chromosomes are separate and appear single. During **zygotene** (b) homologous chromosomes pair exactly to form 8 bivalents, the single sex (X) chromosome having no partner. In **pachytene** (c) chromosome-contraction, which began at the onset of meiosis, is pronounced so that the chromosomes are much shortened and thickened. The two chromatids of each chromosome may be seen but they become much clearer during **diplotene** (d) as the homologues move apart to reveal the criss-cross of chromatids between them at one or more points; these **chiasmata** reveal the points at which homologous chromosomes can exchange segments and the allelic forms of the genes they carry (**recombination**). At **diakinesis** (e) the chromosomes are almost fully contracted and the centromere (<) may become visible. Later the nuclear membrane breaks down, the spindle forms between the cell's poles and **prophase-I** is completed. By **metaphase-I** (f) the bivalents have attached to the spindle-fibers on each side of the equator; during **anaphase-I** (g) the centromeres of each bivalent are pulled towards opposite poles, trailing the chromosome arms in which sister-chromatids are widely separated, to give two polar groups of chromosomes, each having half the number of the original cell; at **telophase-I** the X univalent moves to one pole. There may or may not be any interphase before the second meiotic division begins. At **prophase-II** (h, i) the sister chromatids are widely separated and attached only at the centromere. By **metaphase-II** (j, k) the centromeres of chromosomes in each group become aligned on the equators of the two spindles and then divide before the daughter centromeres of each group move to opposite poles, dragging the single chromatid (now daughter chromosome) arms behind them during **anaphase-II** (l, m) to give a total of four groups at **telophase-II**, two of which do and two do not contain the X-chromosome.

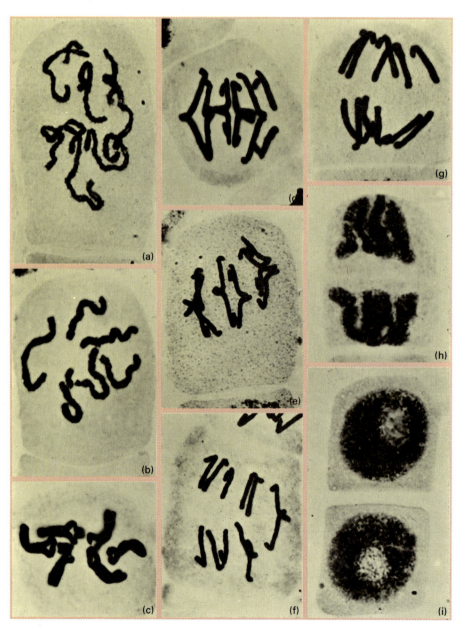

(a)
(b)
(c)
(d)
(e)
(f)
(g)
(h)
(i)

complex situation. While some of the variants he observed were due to fundamental changes in the structure of the genes (mutations as we now know them), others were due to major rearrangements of chromosome segments and groups of chromosomes, involving large complexes of genes. Ronald Fisher, Sewall Wright, J. B. S. Haldane and others provided mathematical models which showed how changes in the variation in natural populations resulted from the interaction between gene mutation (a relatively slow process), reassortment and recombination of genes during meiosis (reductive division of the nuclei of sex cells), selection of features controlled by particular genes or combinations of genes, chance fixation of genes, dependent on the size of the population, and the degree of reproductive isolation (the extent of gene migration into and out of the population). These models have been substantiated by an ever-increasing number of studies on various groups of plants for which the chromosomes, as both carriers and markers of the genes, have made a great contribution.

Not all the processes are equally important in all groups of plants, or animals, but all play their part in the evolution of species. Changes in chromosome number, chromosome structure and population size (see Genetic Drift, p. 104) and the degree of reproductive isolation (see Isolation, p. 108, Vicariance, p. 222 and Endemism, p. 227) and the different breeding systems, along with mutation and recombination of genes and changes in the selection pressures imposed by the changing environment (see Genecology, p. 36) have all been shown to be means by which the genetic structure and evolutionary status of populations is changed naturally or, by the interference of Man, artificially (see also Types of Selection, p. 99).

Acquired Characters

Whilst the mainstream of thought concerning evolution and selection has been based on the facts of Mendelian genetics, there have been periodic echoes of Lamarckian ideas on acquired characteristics. During the 1940s and 1950s the Soviet worker Trofim Lysenko upheld the view that environmental factors were of paramount importance in molding the evolutionary potential of species. Although most of the claims have been subsequently disproved it has become clear in recent years that the genetic material is not all included in the chromosomes and therefore subject to "neo-Mendelian" rules. It has been shown that non-nuclear DNA, such as that present in the mitochondria, for example, may provide an explanation for the inheritance of some characters apparently "acquired" as a result of exposure to certain environments. The character that we see in the plant results from the effect of the message promulgated by the DNA and modified by the environment, and it is clear that the balance between these two factors is not the same in all cases.

Many plants can show considerable variation in a particular character. Amphibious plants, for example the common water crowfoot (*Ranunculus aquatilis*), may have leaves which grow both in and out of the water, and the morphological difference in outline between the submerged and aerial leaves is quite considerable. The lobes of the submerged leaves are more or less threadlike and the lobes of the terrestrial ones are much larger. A completely submerged plant possessing only the slender-leaf form can develop the aerial-type leaf when grown in terrestrial conditions, and of course a broad-leaved terrestrial plant can develop threadlike leaves when grown under water.

If the theory of Lamarck were true, then such characteristics would be inherited in succeeding generations. The offspring of submerged plants derived from plants acquiring the threadlike leaf should only have threadlike leaves, and similarly plants with broad leaves should only occur in the succeeding generations of the terrestrial plant. We now know that this is not true and that environmental modification is not a major factor in evolution.

This does not mean that the hereditary material cannot be altered by the environment. On the contrary, the effect on the germ plasm of such agencies as heat, cold, chemicals and radiation is well known. However, at least in higher organisms, such changes are produced either at random as in spontaneous mutations or as chromosome rearrangements, both types being changes which do not allow the organism to become better adapted to that environment.

Determinism

Another argument which has continued into recent times is that concerning the "determinism" of evolution. For example, if all evolutionary change is the result of selection by the total environment at a particular time acting upon the variation available then it will inevitably show numerous changes of direction. If, on the other hand, evolutionary change shows rigidly undeviating trends that are not orientated by current adaptations ("orthogenesis"), it may be said to show determinism, implying that it is controlled by factors other than those of the immediate environment in any given period of time. Any study of fossil groups shows that such trends do, indeed, seem to have occurred. Current views, however, would suggest that the gradual evolution of a complex structure in response to continuing selection of small variations by a changing environment would result in a structure so intregrated into the life and economy of the plant or animal that only a profound reduction in its selective value would lead to the extinction of individuals which retained it. In most cases the structure would be retained long after its value was gone, so long as it was not positively detrimental to the success of the organism. Thus, for example, the orchid genus *Ophrys* has built up a most complex floral structure adapted to cross-pollination by certain bees, but this structure is still maintained in those species which have adopted self-pollination, presumably because it is less disadvantageous to have a non-functioning complex flower than no flower at all!

Darwin first educed the theory of evolution by selection acting upon small differences between individuals in a population, between populations, and between species. The basis of these differences has been discussed, and an account given of the ways in which such evolution at the level of the population and species has been observed and experimentally verified by naturalists, geneticists

Fossils provide important evidence for the course of evolution in various groups. The fossilized surface of a large branch or trunk of *Lepidodendron aculeatum* makes a characteristic pattern on this piece of coal from the Upper Carboniferous period, found in Derbyshire, England. The species is an early ancestor of our much smaller modern club mosses of the genus *Lycopodium*.

and plant breeders. The data available to us today are derived from studies of the genetical and morphological affinities between plants alive on Earth today, allied to morphological studies of fossil plants. The direct genetical approach can be applied when it is possible to hybridize two plants and study the inheritance of features in their progeny; this approach is thus most useful in studying the evolutionary relationships of species and groups of species. Where this approach is impossible, as in assessing the relationships of living genera and families which do not interbreed, similarity of structure, whether chemical or morphological, gives a clue to similarity of ancestry—though with many reservations.

Once the fossil record is considered, the problems become even greater. Does, for example, the apparently sudden appearance of the flowering plants (angiosperms) in the Cretaceous period result from their sudden evolution at that time or from an incomplete fossil record? Opinion is still divided on these issues. It is certainly clear that rapidly changing environmental conditions stimulate the origin and extinction of groups of organisms (see Endemism, p. 227), but this is simply a reflection of a changed balance in the basic evolutionarily important factors mentioned earlier. In consequence, most modern workers hold, to varying degrees, to the concept of "genetic uniformitarianism"—that the processes we can now see and simulate in the evolution of populations and species are the same as those which, in the past, gave rise to the groups we now recognize as genera, families and orders. Only continuing experimentation, observation and exploration will show whether the processes now seen to be important in the evolution of plants currently on Earth were equally important in the evolution of the plants we now know only as fossils.

L.B.H.

Evolution and the Fossil Record

The preserved record of fossil plants is very considerable but is beset with enormous problems which do not apply to the animal kingdom to anything like the same extent. Darwin long ago recognized that the proof of the theory of evolution would come from the fossil record, although such evidence was not available during his time. Detailed family trees of the horse and elephant and many groups of invertebrates such as the ammonites (shelled relatives of the squids) became classics, since detailed successions of forms preserved in successive layers of strata provided clear proof of gradual evolution through time of particular lineages. In the case of the horse, for example, the reasons for the changes could be inferred as gradual adaptation to changing environments, in particular the spread of grasslands and the need to be able to cope with more siliceous

plant material. Hence, the complex ridges of enamel on the crowns of the teeth and the need to be fleet of foot to escape from predators on the open plains. Such examples helped establish the fact that evolution had taken place.

In the case of the plant world there are few comparable studies. One of the classics was the English botanist Margorie Chandler's study of the seeds of the herbaceous water plant *Stratiolites*. In the Eocene the seeds were small, broad and heavily sculptured, but successive deposits showed a gradual change so that by the Upper Pliocene they had become larger, narrower and smoother, in fact indistinguishable from those of the living species, *Stratiotes*.

One of the great difficulties in tracing the details of plant evolution is that with fossils there is often no way of relating spores, pollen, leaves, stems, seeds and fruits to particular species. In fact the different parts have their own Linnaean binomial nomenclature; thus there are "form" genera and

species relating to leaves, and different form genera for other parts. The giant horsetail *Calamites* refers to fossils of the stem, the genus *Annularia* refers to the leaves of the same plant.

A detailed evolution of seeds or leaves can often be traced. An excellent example of this is the record of oak leaves from the Cretaceous and Tertiary rocks of North America. The oldest leaf form genus, *Quercophyllum*, comes from the early Cretaceous of Virginia and Wyoming. By late Cretaceous times there are leaf impressions from Greenland that are attributed to the living genus *Quercus*. Also by Late Cretaceous times there is evidence that the temperate deciduous white oaks and the evergreen black oaks were in existence. Changes in leaf form can be traced up to Miocene times, since when there seems to have been little change. In the western part of North America there has been a reduction in the number of white oak species whereas in the eastern part the Miocene variety has been retained. It is possible to draw up a fossil history of oak leaves, but until a history of acorns is known or even a single Cretaceous acorn, it is not possible to erect a true evolutionary history of the oaks.

In this context, a cautionary tale is provided by the fossil fruits from the Eocene that seem to be intermediate in character between walnut (*Juglans*) and hickory (*Carya*) and are given the form genus *Juglandicarya*. This indicates that the walnut and hickory were not yet differentiated from one another. However, Cretaceous leaves named *Juglandites* and Eocene *Carya* leaves are known, suggesting that in the structure of the leaves such differentiation had already occurred.

The difficulties of drawing up detailed

Fossilized stem (**left**) and leaves (**below**) of the giant horsetail *Calamites*. Leaves of the same plant were found separately and assigned to a genus, *Annularia*, and only at a later date was it discovered that *Calamites* and *Annularia* were the same plant.

family trees which include fruits as well as leaves are legion, in view of the patchy nature of the fossil record of plants. In spite of this, the overall evolution of the plant kingdom in its broadest outline is well established. The major structural grades of organization from bacteria to angiosperms are preserved in the fossil record but there is a dearth of linking forms to indicate how each major grade evolved into the more advanced.

Evolution of Higher Plants

As was noted earlier (see The Origin and Early History of Life, p. 84), it is believed that, about 600 million years ago, the point was reached when oxygen formed 1% of the atmosphere and aerobic respiration became more efficient than anaerobic (the Pasteur point). By just over 400 million years ago, at

An artist's reconstruction of four psilophytes—the first primitive land plants, which appeared in the Lower Devonian, some 400 million years ago.

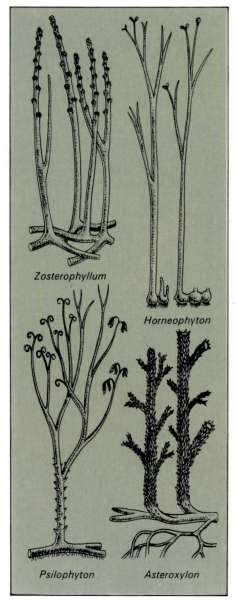

Zosterophyllum

Horneophyton

Psilophyton

Asteroxylon

the end of the Silurian, there occurred a further breakthrough in the evolution of the plant kingdom: the first vascular plants made their appearance—this was the beginning of the colonization of the land. The early form *Cooksonia* had no roots as far as is known, and bore no leaves or flowers but consisted merely of a branching stem, the ends of which carried spore capsules. From such simple beginnings there evolved during the Devonian period a large variety of plants. *Zosterophyllum*, a later and more advanced form, seems to have given rise to the lycopods or club mosses, while *Psilophyton* is a possible ancestral form of the ferns and horsetails. All these primitive plants reproduced by releasing spores (the male microspores, and female megaspores).

Living ferns, horsetails and some club mosses have complex life histories in that the released spores germinate to form a minute reproductive stage, the prothallus, which produces male and female cells that unite to form the embryo, which then develops into the fern plant. The next major stage in plant evolution was the development of the seed. It is not difficult to visualize how this came about, as some of the club mosses and horsetails represent a kind of halfway stage. In some living club mosses the prothalli develop within the spores, and the smaller spores then release the male cells. In some instances, the female spores are not released but remain attached to the parent plant so that the female prothallus receives a certain amount of protection. Fertilization and the formation of the embryo all take place in direct association with the parent plant. The oldest known seed is *Archaeosperma* from the Upper Devonian of North America, which is protected by expanded stem-like structures.

During the Carboniferous period, when the coal forests flourished, there were giant club mosses, such as *Lepidodendron*, giant horsetails, such as *Calamites*, giant ferns and seed ferns. The last, as the name implies, were not true ferns but were true seed-bearing plants related to the conifers. The first conifers appeared at the end of the Carboniferous, the seeds being borne on scales protected by woody cones. With the end of the coal swamps, the gymnosperms—the conifers and their allies—dominated the flora. In the ginkgos (maidenhair trees) and the palm-like cycads and their allies the female ovule is exposed directly so that the male pollen has direct access.

The final stage in the evolution of plants was the origin of the angiosperms or flowering plants. Here the ovules are enclosed within carpels. The pollen is caught on a stigma and a pollen tube germinates, through which the male nuclei are transported to the egg within the embryo sac of the ovule. In some of the most primitive angiosperms, such as *Drimys* (Winteraceae), a close relative of *Magnolia*, the carpels are not completely closed. When the embryo develops, the seed is contained within a fruit formed from the carpel. In Permian deposits of Kansas and Texas a cycad has recently

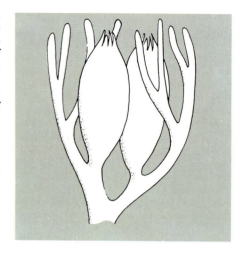

A reconstruction of the oldest known seed—*Archaeosperma* from the Upper Devonian. The seeds were apparently borne in pairs loosely enfolded in sterile appendages.

been discovered having two rows of ovules borne on a laminate frond, which seems to have been partially rolled over the ovules. This represents a possible intermediate stage, a kind of missing link which shows how the angiosperm carpel could have originated from the lamina of the sporophyll fusing and thus enclosing the ovules. The fact that in the more primitive flowering plants the ovule may not be completely enclosed by the carpel lends support to the theory of the probable cycad origin of the angiosperms.

One of the most remarkable advances in plant evolution is the relationship between the insects and the flowering plants, the latter relying on the former to ensure pollination. As well as the exploitation of the insect world (nectar and pollen is donated) for pollination, the development of fruits ensures a wide dispersal through other agents, especially birds and mammals attracted by the fruit which is consumed while the seeds are dispersed by passing through the digestive tract unaffected or by being thrown away.

Plants were probably the first form of life to become established on the land and therefore constituted the foundation for all subsequent animal life by providing nourishment. The most advanced plants, while still doing this, now have a more productive, creative working relationship with the animal world. They are still providers but receive in their turn the services of the animal kingdom.

L.B.H.

Adaptation

Adaptation, the process of hereditary modifications to suit the environment, is a universal feature of all living organisms. To suit a particular niche, every species possesses a whole complex of characters which cannot be shared by any other organism. Charles Darwin summed up the situation: "Can a more striking instance of adaptation be given, than that of a woodpecker for climbing trees and for seizing insects in the chinks of the bark?'' Extrapolating from this, we see that birds of prey possess strong talons and powerful visual capabilities for catching small animals, fish have fins and gills for life under water, plants have green leaves for autotrophic nutrition and mammals have forelimbs for grasping. Nearly every part of the Earth's surface is inhabited by some form of life. Living organisms are found at the bottom of the sea, on the sea shore and on the land, from the polar ice caps to the hottest deserts and a great distance into the air. At the farthest extremes, bare rocks are colonized by lichens, glaciers by red algae and even hot springs by sulfur bacteria.

Even casual glimpses of any place on the Earth immediately show that no one organism can live in all environments but each lives in a particular range of conditions. Thus, we can expect to find fish and seaweeds in the sea, birds in the air and cacti in deserts. The correlation between vegetation and climate is evident in the use of words like tundra, steppe and rain forest, which imply an impression of the total landscape. So, too, the diversity of plants, of which there are perhaps more than half a million different species, shows a discrete correlation between distribution and habitat, as expressed in such terms as rock plant, dune plant, calcicole and halophyte. The list of habitats and corresponding plant forms is almost endless, and the reason why each kind of organism is restricted in nature to its own special place or niche and not found elsewhere is that it is specialized or adapted for one set of conditions (see Life-form, p. 50, Structure of Vegetation, p. 47).

The major physical factors affecting plant structure and physiology are water availability, dryness, salinity, heat, cold exposure, toxicity and the biotic influences of competition. Some of the more obvious adaptive modifications in plants are those seen in extreme environments, because one or two of the physical factors become dominant in their influence and the others quite subsidiary. For the purposes of looking at the processes of adaptation in plants it is easier to see modifications of organs which have responded to extreme environments and to define the intensity of those modifications which can be exerted on the species.

Origin of Adaptations

A key question when considering the wide range of biological diversity that exists in the world is ''where does it all come from''. The answer lies in the fact that variation in populations provides the raw material on which the selection pressure of the environment exerts its influence.

All species of plants or animals exist in discrete groups of individuals. These populations consist of individuals of varying ages from the newly born to the aged and, since all organisms are mortal, with a comparatively short life span in comparison to the existence of a species, there is a continual influx of new individuals into the population by birth and a continual loss by death. Consequently, the population is always gradually changing in terms of the available individuals at any one time. A constant new supply of individuals into the population coupled with continual loss means that there is always an opportunity to have replacements better equipped for the rigors of the environment. If one considers an oak tree or an orchid flower, which both produce millions of seeds in their lifetime but need only replace themselves once to keep a constant population in existence, then one quickly realizes that such prolific supplies of seeds are far more than are required to fill the available environmental niches for that species. It is inevitable, therefore, that many, in fact most, of the seeds get killed off at a very early stage and usually these include the weak, nonviable and obscure genetic mutants, the so-called ''unfit'' individuals. It is the ''survival of the fittest'' selected by the prevailing environment which maintains the population.

The considerable range of variation in populations is brought about by the processes of mutation, recombination and isolation. In different and changing environments only a few of those genetic variants will ever be ideally suited to live there. Consequently, these are the individuals that have adapted to those conditions. To understand how plants evolve and become adapted to particular conditions, it is necessary to appreciate the power of selection and the range of variation. In very extreme environments the selective pressures on a plant population are severe, and only a small fraction of the variation in the population can survive the limitations set by the environment. This is particularly striking if we compare the cultivated products of a particular species, where the selective pressures are diminished or eliminated, with naturally occurring populations where selection is severe. Consider for example the cabbage, *Brassica oleracea*. The wild populations of this species, known as variety *oleracea*, a form commonly found on cliffs around the coast of the Mediterranean sea and the Atlantic coast of Europe, consist of a small-leaved, white-flowered biennial rosette plant with tough tap roots. By comparison, there are very many cultivated varieties of *B.*

(a)   (b)   (c)

A remarkable advance in plant evolution is the relationship between insects and the flowering plants. The peak of this form of evolution is found in the process of pseudocopulation by which many orchids are pollinated. For example, flowers of members of the orchid genus *Ophrys* closely resemble the form of a female bee (a). The male insect often attempts to copulate with the flower (b), and in the process the sticky pollen masses (pollinia) stick to its head (c). When returning to another flower of the same species and again attempting copulation, the pollinia will collide with the stigmatic surface of the flower, transferring pollen to it and bringing about pollination.

1

2

3

4

5

6

7

8

9

The species of the genus *Euphorbia* present what is probably the most striking range of adaptive evolution, and hence vegetative diversity, found within a single genus in the plant kingdom. These nine pictures are a small sample of representative species (over 2,000 in all) of this genus but they give some indication of the range of form there can be in a single genus. (1) One of the largest species is the semi-succulent dichotomously branching candelabra tree (*E. candelabrum*), seen here growing in Kenya. (2) Two species of succulent *Euphorbia*, growing on the volcanic coastal slopes of Tenerife, Canary Islands. *E. canariensis* resembles a columnar cactus, while *E. aphylla* grows in more compact cushions. These species are said to be sympatric, ie they are very closely related and live in the same habitat, but are unable to interbreed because they are reproductively isolated. (3) Shrubby forms are common in tropical grasslands, such as the single specimen of *E. kamerunensis* and numerous smaller specimens of *E. poissonii* shown here growing in the northern Cameroun. (4) The popular houseplant, poinsettia (*E. pulcherrima*), with its distinctive red bracts, develops into a large shrub when grown outdoors in subtropical and tropical regions. (5) Another species from the volcanic slopes of Tenerife is the woody, semi-succulent *E. regis-jubae*, which has distinctive rosettes of leaves at the stem tips. (6) At the opposite extreme to the tree and shrub forms of drier regions are the shade and moisture-loving spurges of temperate regions, such as this mesophytic wood spurge (*E. amygdaloides*). (7) Even the temperate spurges show modifications to suit their habitats. The sea spurge (*E. paralias*) is a halophyte living on sandy shores and sand dunes where salt and lack of water are a problem; its leaves are therefore succulent, holding reserves of water. (8) The uniting feature of all species of the genus *Euphorbia* is the specialized inflorescence called a cyathium. This comprises a single, simple, stalked, central female flower, with several simple male flowers, all enclosed in leaf-like bracts, which form the involucre. In the sun spurge (*E. helioscopia*) there are four cyathia and the bracts are green. However, in some species the bracts are coloured, as in the poinsettia and (9) *E. fulgens*, a leafy shrub from Mexico, where each cyathium and its colored bracts appear to be a single flower, not an inflorescence.

oleracea grown in gardens and commercially which agree with the wild species in flower, fruit and seed structure but vary widely in vegetative morphology and the form of their immature inflorescence. Similarly, ornamental chrysanthemums, of which there are reputed to be 7,000 or more varieties grown in the world today, are all derived from a single wild hybrid between *Dendranthema morifolia* and *D. sinensis*. Both examples serve to illustrate the powers of natural selection, whereby the potential range of variants which can be produced in succeeding generations outnumber by far the actual range produced in nature.

Types of Selection

Since natural selection results from the operation of the various factors which together constitute the environment faced by any population of plants, its results can be complex. Nevertheless, the variation patterns produced can be interpreted in terms of three principal kinds of selection—directional, stabilizing and disruptive.

Directional Selection
In any population of plants in a particular environment there will be an optimum group of characters (phenotype), out of the many available in the population, which will be most successful. Individuals with characters which diverge considerably from this optimum have a markedly lower chance of survival to reproductive age, and are said to be under strong selective pressure. In time, selection will shift the average phenotype of the population towards the optimum, that is selection acts in one direction away from the original genetical characteristics of the population. This is termed "directional selection". In order to respond to directional selection, the population must contain enough genetical variation for such a change to be possible. This is the main type of selection used in plant breeding.

Stabilizing Selection
As the optimum is approached, selection will tend to oppose deviation from it in either direction and is then said to be "stabilizing". The total variation in the population will tend to be decreased by this process to give an increasingly narrow range of phenotypes centered around that which is best suited to the prevailing conditions. Wherever directional selection is tending to an average phenotype not outside the limits of the present variation in the population there will be some element of stabilizing selection going on.

Disruptive Selection
Under certain circumstances, there may be more than one optimal phenotype for a population. For example, in areas where the environment shows a gradual or clinal change over the area occupied by the population (see Genecology, p. 36), or where

it comprises a mosaic of ecological niches, there may be more than one combination of characters conferring equal success. Certain populations have intrinsically more than one optimum, the functioning of an individual depending upon other individuals acting in a different though related way. For example, dioecious species have two optimum phenotypes, male and female. Certain plants that are self-incompatible (cannot be fertilized with their own pollen), require more than one optimal phenotype. A well-known example is the primrose (*Primula vulgaris*) which has two kinds of flowers: pin-eyed, with the style longer than the anthers, and thrum-eyed, with the anthers longer than the style.

Disruptive selection leads to selection towards two or more optima, and if it is so severe that there is genetical isolation between members of the different phenotypic groups, it can lead to speciation, especially where the disruptive selection is due to environmental features (see Speciation, p. 105).

Adaptability

Having established that the physical or biotic environment eventually selects, in a fairly ruthless manner, which of the individuals survive in any particular environment, it is now possible to consider the various types of changes which organisms undergo in response to their environment. Changes in the hereditary characteristics of organisms constitute adaptability or adaptive evolution, and examination of widespread plant species occurring in different extreme environments illustrates the point that evolution and hence adaptation are invariably canalized by selection and reproductive or eco-geographical isolation.

Comparison of sea-cliff plants with related inland populations furnishes particularly good examples. Plants occurring on cliffs are characteristically prostrate, as in the buckthorn (*Rhamnus catharticus*) and sea plantains (*Plantago coronopus* and *P. maritima*), which even if they ever do grow erect, do so in compact clumps so that the separate shoots provide each other with protection. Species that occur both on cliffs and on inland habitats always have genetically distinctive populations on cliffs characterized by extreme dwarfness and compact growth. Classic examples here include *Achillea millefolium* and *Agrostis stolonifera*, which have such genetically distinct races that they frequently behave as different species.

Adaptive evolution is most impressive when species of a single widespread genus adapted to different environments are compared with one another. The cosmopolitan genus *Senecio* (family Compositae) is clearly remarkable in this respect since it contains some 1,500 species occurring from coasts to the highest mountains in a wide range of ecological conditions. Amongst some of the widely different species are the thick

stemmed candelabra-shaped trees of the subgenus *Dendrosenecio*, from the Afro-alpine flora of the East African mountains, which can grow up to 10m (33ft) or more in height, the succulent cactus-like *Senecio kleinia*, from the xerophytic zone of the Atlantic islands, and the well-known annual groundsel (*Senecio vulgaris*), a common European weed of cultivated ground (see also Adaptive Radiation, p. 103).

Main Adaptations in Plants

The most important environmental factors to which plants must adapt themselves, both morphologically and physiologically, are water availability, temperature changes, light regimes, soil conditions and biotic competition. For any species, each of these factors has a minimum, optimal and maximum value. Consequently, there are species with very wide or narrow ecological adaptations. Species adapted to extreme environments have undergone quite irreversible modifications to adapt to their particular and often narrow ecological conditions, and a description of some of these follows.

Hydrophytes
Hydrophytes are specialized flowering plants that actually live in aqueous habitats. They show certain structural features peculiar to their environment and also possess a conspicuous number of characters which indicate their terrestrial ancestry, modifications which suggest that the ability to live in an aquatic environment is distinctly adaptive. Such development of similar features in taxonomically unrelated plants is known as *convergent evolution*.

Typical adaptations of wholly or partially submerged plants are conspicuous air chambers in tissues of stem roots and petioles. This is because the necessary available gases for photosynthesis and respiration are in quite different concentrations in water and in the atmosphere. The wide air spaces not only provide storage and active diffusion of gases to the plant tissues but also increase the buoyancy of the plant so as to keep the stem and leaves away from the river or sea bottom.

Anatomically, plants living in aqueous environments are quite reduced in structure when compared to related species on dry land. The vascular system is generally less well developed, the water-conducting vessels (xylem) often being completely lost and the stem consisting of normal cortical tissue (parenchyma), there being little evidence of heavily thickened support tissues (sclerenchyma). The submerged leaves and stems are usually very delicate, thin and rich in sap. Submerged leaves are often divided into strap-shaped or linear filaments. Amphibious and partially submerged plants frequently exhibit leaf hetermorphy with broad, thick floating or emergent leaves, more like those of land plants, and threadlike submerged leaves. The outer walls of the epidermis of submerged shoots are often very thin and the very delicate cuticle offers hardly any impediment to the entry of dissolved gases, water and salts.

Morphological clues to the terrestrial ancestry of water plants are the presence in some species of thin cuticles, functionless pores (stomata) on the leaves, occasional strengthened tissues and the widespread occurrence of aerial wind and insect pollination systems. In primitive aquatic plants, particularly marine algae, sexual reproduction is brought about by the fusion of swimming motile male gametes with stationary female spores. However, even in flowering plants with submerged flowers, despite various minor structural modifications of thin walls and slimy envelopes which can be found on the sexual parts, the male gametes are in the pollen grains and pollen germination still occurs on the stigma—both originally adaptations for sexual reproduction in drier terrestrial conditions.

Xerophytes
Xerophytes, plants adapted to dry climates, have characteristic structural modifications associated with methods of water conservation. Interestingly, plants of diverse origin have developed similar types of modification. The most common type of structural modification is a thick, cuticular layer all

Right A light micrograph of a transverse section of the stem of a rush (*Juncus* sp.), showing the aerenchyma (× 120). This tissue gives buoyancy to tissues of aquatic plants and allows gaseous exchange between aerial and submerged organs.

Below Plants that live in water (hydrophytes) possess a range of adaptations for this environment. Shown here is the river margin vegetation of Shire river in Malawi with (**from left to right**) submerged aquatics that have small, finely cut leaves, floating leaves of the rooted water lilies (*Nymphaea* sp.), free-floating plants of the water fern. *Azolla nilotica*, free-floating plants of *Pistia stratiotes* (**bottom right corner**) and aerial arrow-shaped leaves of a species of *Ipomoea* spreading from the bank.

over the surfaces of the aerial parts. This waxy layer cuts down water loss through the epidermis when the breathing pores, the stomata, are closed. Even xerophytes have to photosynthesize, however, and obviously the stomata need to stay open during these times. Naturally they are sites of potential water loss and so the best-adapted xerophytes have stomata depressed into pits or furrows, depressions which serve as pockets to trap water vapor so as to make the diffusion gradient from the inside to the outside of the plant much less severe.

The most obvious adaptations for growth in dry climates are found in those xerophytes having some means of storing necessary water for metabolic processes during unfavorable dry periods. Succulent plants, such as cacti and *Euphorbia* species, retain enormous quantities of water within the plant tissues. Adaptations associated with water storage often involve the loss or reduction of plant parts that have conspicuous water loss.

A number of plants that grow in normally arid environments are able to complete their

life cycle in a very short time. These species undergo long periods of dormancy either in the form of seeds or as curled up vegetative balls during the months of dry weather. As soon as the rains come, the seeds germinate, the plants grow and produce flowers and seeds long before the brief wet season comes to an end.

These examples of xerophytic modifications clearly demonstrate the concept of adaptation to the environment. Plants from different, quite unrelated families become modified in one or more of the above ways, and produce surprisingly similar-looking individuals. The convergence of structural and functional adaptation can be so striking in some cases that occasionally it is impossible to identify the individual species.

Adaptations to Temperature and Seasonal Differences
The effects of temperature on plants are often difficult to assess because they are so closely dependent on the available water supply. Nevertheless, even in places of high water availability, high air and ground temperatures can prevail beyond the normal capabilities of plants. Apart from a few bacteria and heat-loving algae, most plants are unable to withstand temperatures beyond 55°C (131°F). However, various adaptations to overcome very high temperatures include the development of very small, vertically arranged narrow leaflets, as in *Eucalyptus*, the drooping of leaves at high light intensities as in various scrub *Acacia* trees, thick white felt-like clothings of hairs, as in many dune plants, and the thick, leathery "lacquered" leaves typical of laurels.

Woody plants and, particularly, deciduous trees provide conspicuous examples of seasonal adaptation in Northern and Southern Hemisphere temperate floras which experience warm, wet summers and cold, dry winters. Most species protect their apical meristems during cold and dry winter periods with bud-scales derived from stipules. These are usually brown, leathery, overlapping scales covered with a sticky suberin resin or hairs to form a waterproof sheath over the delicate apex. In the spring, new leaves burst forth from the bud, and the scales, now having completed their task, are shed to leave distinctive girdle scars on the stem. The leaves themselves provide the necessary food store for the flowers and fruits as well as for storage in the stem over winter. Having completed their task in the spring and summer, they too are shed in the autumn.

Perennial herbs are also particularly conspicuous examples of adaptation to seasonal climates. By definition they are those plants which sacrifice not only their leaves in the autumn but most of their aerial parts as well. The new season's resting buds which arise in the following spring arise from aerial stems which grow very near to the ground or from subterranean stems occurring just below it. The origins of perennial herbs are diverse and the methods of overwintering are similarly variable.

Essential foods provided by the aerial leaves are transferred to the storage organs during the summer months to facilitate the emergence of the overwintering resting buds as new shoots in the following spring. During the winter, protection is afforded to the delicate resting bud either by a covering of snow, as in alpine plants, or by virtue of the fact that they are underground.

The Struggle for Light

The amount of available light beneath tall trees, especially in major tree environments such as rain forest, can often be very small. Since most terrestrial autotrophic plants need a reasonable amount of light for normal photosynthesis to take place, there are various modifications, frequently associated with the forest environment, for plants to reach the sun. Lianas and epiphytic growth-forms represent two contrasting syndromes of characters in plants for obtaining light. Climbing plants are rooted in the ground but have developed the ability to climb up other plants, and also along rocks and walls when necessary. Their stems are thin, without any major support system but their foliage is rapidly raised from the forest floor into

Left Plants that live in dry habitats (xerophytes) show numerous adaptations to survive periods of water shortage. The *Opuntia* species shown here has no leaves, the stems are succulent (water-storing) and green, having taken over the photosynthetic function normally carried out in the leaves, and all parts of the plant are covered by a thick cuticle that prevents water loss.

Below Adaptations to seasonal differences in climatic factors are important for the ability of plants to survive in different parts of the world. This diagram illustrates how ten important trees of the West African forest have adapted in various ways to the annual cycle of dry and rainy seasons.

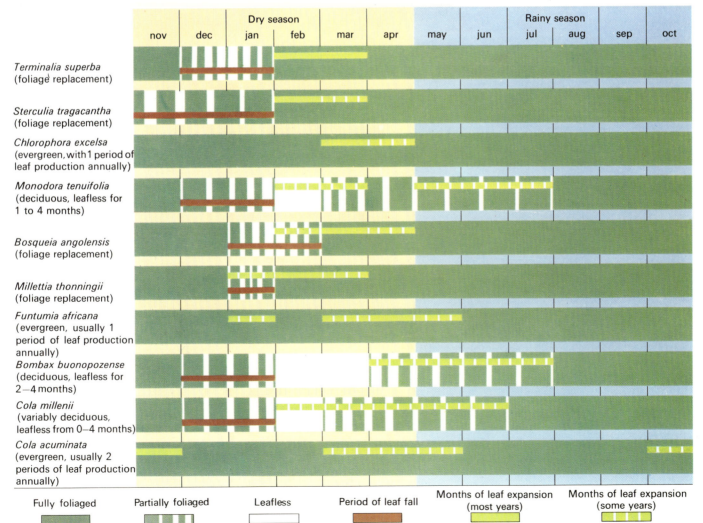

sunlight. Climbing is accomplished by a variety of methods.

By contrast to the climbing plants, which are invariably rooted in the soil, epiphytes are always attached to other plants well above the ground, to acquire well-illuminated habitats. Since they are rarely parasitic, the principal problems of epiphytes are related to the acquisition of essential minerals and water and attachment to their hosts. Because of this, the most favorable habitats are found in regions of appreciable daily rainfall and high constant humidity.

Halophytes, Parasites and Carnivores

Habitats of extremely high or extremely low mineral supply present considerable adaptive problems to those plants which grow there. Plants of salt deserts, parasites, and plants of nitrogen-deficient areas represent three very different extreme environments.

Although the oceans of the world have a salt concentration of about 3.5%, plants dwelling on sea coasts and on the salt pans of alkaline steppe deserts often have to endure

salt concentrations of up to or even beyond 10%. Coastal and desert halophytes all tend to belong to the same families of plants, notably the Chenopodiaceae, which have adapted to these environments. They not only require the means of disposing of high, often injurious concentrations of salt ions but also require the ability to deal with wide fluctuations in osmotic pressure and the occasional onset of drought conditions.

Above These pitcher plants (*Nepenthes pervillei*), which are endemic to the Seychelles, are quite capable of producing some of their food materials by photosynthesis, but because their environment lacks certain nutrients (mainly nitrogen compounds), they have become adapted to catch small animals, such as insects, to compensate for this nutrient imbalance.

The parasitic existence is quite a contrast, where special adaptations have developed to obtain nutrients from other organisms. Parasites have generally lost the capacity for autotrophic nutrition and as a consequence obtain their nutrients at the expense of other plants. Their leaves become quite superfluous and thus are often inconspicuous and never green. Consequently their transpiration rates are all markedly reduced; so too are the roots. In fact, since they obtain their nutrient supplies directly from the phloem of xylem vessels of their host plant, root-systems are exchanged for special absorbing organs known as haustoria which allow the parasite to penetrate its host and draw nutrients from the conducting elements. A major adaptation of all parasites is the development of various biochemical mechanisms to overcome the natural immunity of the host.

Perhaps the most bizarre form of heterotrophic nutrition is that of carnivorous plants. At quite the opposite extreme to halophytes and parasites which live in food- and mineral-rich environments, carnivores are adapted to live in environments such as peat bogs and volcanic ash, which are extremely poorly supplied with plant nutrients such as nitrogen. Although all of them are capable of quite normal autotrophic nutrition by photosynthesis in the leaves, to overcome nitrogen deficiencies, they have adapted a range of secondary trapping mechanisms, which are usually modifications of the leaves, for catching insects.

C.J.H.

Below The main adaptations to assist climbing. (a) The sharp down-curved spines of the blackberry (*Rubus fruticosus*) hook on to the surrounding vegetation. (b) Virginia creeper (*Parthenocissus quinquefolia*) produces modified shoots with sucker-like pads that stick to walls etc. (c) In bryony (*Bryonia dioica*) the modified shoots take the form of tendrils. (d) Lesser bindweed (*Convolvulus arvensis*) climbs by twining stems. (e) Ivy (*Hedera helix*) adheres to tree trunks, rocks or walls by means of modified roots.

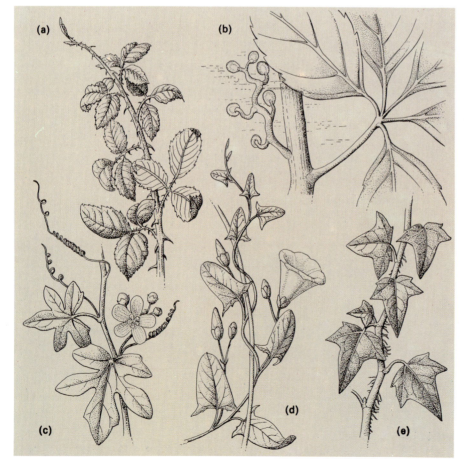

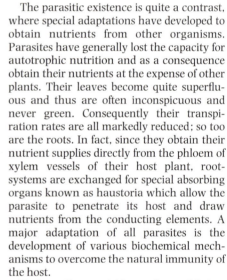

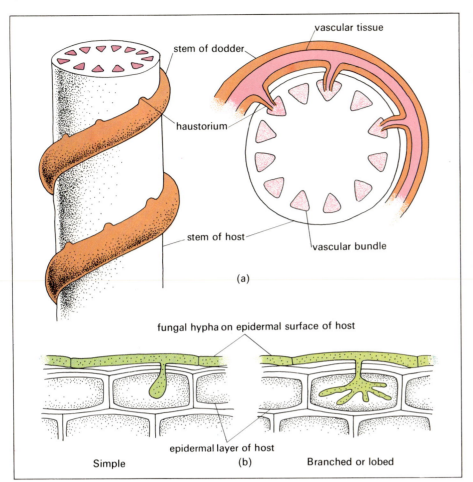

stem of dodder

vascular tissue

haustorium

stem of host

vascular bundle

(a)

fungal hypha on epidermal surface of host

epidermal layer of host

Simple　(b)　Branched or lobed

Parasites have developed special adaptations to obtain nutrients from other organisms. One of the most widespread adaptations is the possession of special absorbing organs, haustoria, which penetrate the host plant's tissues, and these are produced by organisms as widely different as (a) the parasitic angiosperm, dodder (*Cuscuta*) and (b) the fungal pathogen *Erysiphe* (a powdery mildew).

Adaptive Radiation

We have established that one of the most important methods of evolutionary change is by the action of natural selection favoring those organisms most suited to the conditions under which they live. By repeated selection, generation after generation, a group of animals or plants can become increasingly able to cope with its environment—it shows increased adaptation. Adaptive radiation is a particular aspect of adaptation which involves the evolution from one kind of organism of several divergent forms which are specialized to fit distinct and diverse ways of life.

In its rudimentary form, adaptive radiation can be observed in species within which ecotypic differentiation (see Genecology, p. 36) has been demonstrated. Thus, genecological studies have shown that many species have genetically determined ecotypes, each adapted to a particular habitat and differing in various morpholog-ical and physiological features. In these cases, the ecotypes are demonstrably variants on a common theme, derived from a group of plants which are closely related, capable of exchanging genes and, indeed, included in a single species.

A further stage of evolutionary divergence is shown by genera and groups of species within which there is evidence of adaptive radiation. In Australia, to which it is largely confined, the genus *Eucalyptus* has been able to adapt to a wide range of environments, from the subalpine zone to lowland tropics, and there are species adapted to almost every range of moisture conditions from rain forests to desert margins. There are many similar examples of genera showing adaptive radiation in response to environmental diversity and there is good evidence for this even at the family level. In the largely Southern Hemisphere family Restionaceae, for example, some Australian species from wet habitats have the simplest form of culm (stem) anatomy found in the family. Most other Restionaceae, however, grow in drier habitats to give xeromorphic features (ie adaptations to conditions of water shortage). Interestingly, in this case, the adaptive radiation involves different modifications to cope with similar environmental conditions. In South Africa, excessive water loss from the stomata of most species is prevented by thick-walled "protective-cells", which surround each substomatal cavity and are modified cells of the chlorenchyma. In the Australian species subject to water stress, a very similar structure has been evolved, but by modification of the epidermal cells.

The critical importance of seed production for the evolutionary success of plants is shown by the variety of ways in which effective pollination is achieved, and a high proportion of the morphological characters which differentiate species are concerned with parts of the flower adapting it to efficient pollination by specific pollen vectors. Consequently, it is not surprising that many examples of adaptive radiation in plants concern pollinating mechanisms. Such a situation is shown by a closely related group of columbines (*Aquilegia*) in the southwestern United States, which are pollinated by hawk moths inserting their prosboces into the corolla spurs in search of nectar. *Aquilegia pubescens*, with a spur 3–4cm (1–1.5in) long, is pollinated by moths whose proboscis is 3–4.5cm (1–1.8in) long, while the pollinators of *A. chrystantha* and *A. longissima*, with spurs respectively 4–7cm (1.5–2.8in) and 9–13cm (3.5–5in) long, have prosboces respectively 3–10cm (1–4in) and 8.5–12cm (3–4.5in) in length.

The phenomenon of pseudocopulation in the orchid genus *Ophrys* (see also p. 97) also provides a striking example of adaptive radiation. The flowers of many *Ophrys* species resemble the females of different species of pollinating bees. The male bees are attracted to the *Ophrys* species whose flowers resemble the appropriate female bee and since each species of *Ophrys* usually attracts only one species of bee, the basic flower form shows a variety of adaptations in response to the variety of pollinating bees.

The genus *Ipomopsis* (family Polemoniaceae) shows a wide diversity of flower forms and modes of pollination. Some of the 27 species have blue, salver-form flowers pollinated by bees, others have red, trumpet-shaped flowers pollinated by hummingbirds, while other species have long, tubular, violet or whitish hawkmoth-pollinated flowers. Yet others have small, whitish, beetle-pollinated flowers and others have small, yellowish flowers with a cloying nitrogenous odor attractive to scavenger flies, while the small, inconspicuous flowers of other species do not attract pollinators and are predominantly self-pollinating. On a large scale, adaptive radiation in response to this range of pollinators is considered to have occurred throughout the whole family Polemoniaceae, with successive divergences from the original bee-pollination being accompanied by a series of adaptations involving characteristic modifications to the color, form and odor of the flower.

While, as we have seen, some organisms become better adapted by means of natural selection, this is at the expense of those which are not so fitted to the prevailing conditions and which, in consequence, may become extinct. Such extinction can, consequently, frequently obscure the sequence of events leading to the present adaptation of a group of organisms. Because of this, many of the most prominent examples of adaptive radiation come from islands (see p. 231), which, because of their isolation, can have a low level of predators, parasites and other competitors, so that the occasional immigrants can often occupy a much wider range of niches and habitats than is possible in continental areas where selection (and extinction) from these sources is much more intense. For this reason it is not surprising that the classical example of adaptive radiation comes from the Galápagos Islands, where the 13 species of Darwin's finches, adapted to different habitats and food supplies not previously available, have obviously evolved from one ground-based species in South America, some 965km (600mi) distant. Some of the most conspicuous examples of adaptive radiation amongst plants also come from such isolated archipelagos as Hawaii, the Canary Islands, the Juan Fernandez Islands and the Galápagos Islands.

In Hawaii the endemic genus *Cynanea* (family Campanulaceae) shows a wide range of growth-forms and leaf shapes related to habitat. *Cynanea pilosa* and related species are unbranched rosette plants with simple, broad leaves, and grow in wet forest. Also in this habitat are taller plants with larger leaves, such as *C. gayana* and its relatives. In drier habitats the plants are taller, with branches, stems and small leaves. In tall forest the palm-like *C. leptostegia*, with long strap-shaped leathery leaves, attains a height of 14m (46ft), depending upon the height of the surrounding forest. The genus also shows a wide range of flower color, and shape of corolla and calyx lobes, indicating a diversity of pollinators which, together with the habitat, have undoubtedly been involved in producing the conspicuous adaptive radiation shown by this genus.

The flora of the Canary Islands is famous for the number of endemics (see p. 227), relicts of groups which were formerly present in the Mediterranean region during the Tertiary epoch. Since their isolation in the archipelago, several genera or distinctive parts of genera have evolved numerous species and subspecies in response to the wide range of climatic and ecological conditions available, from hot, arid desert scrub to cool, wet evergreen forest and high altitude desert with a marked continental climate. The striking, woody, Macaronesian sowthistles (*Sonchus* spp.), belonging to the subgenus *Dendrosonchus*, demonstrate adaptive radiation in habit, leaf shape and capitulum size. Starting from widespread species such as

Sonchus pinnatus, which are woody, broad-leaved, tree-like plants up to 4m (13ft) high, with large capitula, it is possible to discern several evolutionary trends which are closely correlated with the different habitats available. With increasingly dry conditions the stems and branches become thinner and the capitula smaller, while the area of the leaves is reduced by greater dissection, so that in the driest zones the leaf-segments are linear (eg *S. leptocephalus*). In moist forest habitats, species such as *S. jacquini* have large capitula and broad leaves, and rarely exceed 2m (6.5ft) while in the high mountain forests the trend continues to give forms such as *S. acaulis*, which has a very short woody stem, a rosette of very large leaves and extremely large capitula. The distinctive habitats provided by the cliffs at lower elevations in the Canary Islands result in forms, such as *S. radicatus*, which have a very short woody stem, a dense rosette of small leaves with rounded, often overlapping lobes, and few large capitula in each inflorescence. Similar evolution in response to this range of habitats is shown by Canarian members of, for example, *Aeonium*, *Echium* and *Argyranthemum*.

Not all the evolution of diversity in such island groups can, however, be ascribed to adaptive radiation. Although in many instances it is clear that the diversity results from the positive selection pressures exerted by the conditions of the different habitats, in others the evolution has proceeded as a result of reduced gene flow between populations geographically isolated on different islands in the archipelago, or even in different valleys separated by unfavorable terrain on the same island. This latter process, which gives rise to vicariance (see p. 222), the occurrence of closely similar species in widely separated areas, is often difficult to distinguish from adaptive radiation without considerable study utilizing many sorts of information, so that great care must be exercised not to

overestimate the importance of adaptive radiation when interpreting the often bewildering variety of related forms in many island floras. If this care is exercised then knowledge of these groups gives a clear indication of the important role of adaptive radiation in the evolution of species and genera in areas where extinction has obscured some of the intermediate stages leading to the present diversity (see Isolation, p. 108).

It is frequently difficult to distinguish in practice between variation patterns attributable to adaptive radiation, just described, and those resulting from the much-debated process known as genetic drift, in which selection plays no part.   D.M.M.

Genetic Drift

The patterns of variation shown by plants occurring in small populations, especially on islands, have been shown to result from adaptive radiation, as we have just seen. However, under the same circumstances, the variation can, alternatively, be shown to result from the oft-disputed process known as genetic drift.

In 1908 G. H. Hardy and W. Weinberg independently derived the mathematical expression of the relationship between the relative frequencies of two or more alleles (alternative forms of a gene, the unit of inheritance) in a population. The resultant Hardy-Weinberg Law states that an allele which is neither advantageous nor deleterious, ie selectively neutral, will tend to remain at a constant frequency from one generation to the next. The constancy is, however, based on statistical probability and the frequency of the allele will, in fact, fluctuate somewhat from generation to generation owing to purely chance variations in the reproduction of plants having different genotypes. Similarly, a coin flipped repeatedly will be expected

Adaptive radiation in the giant hawkweeds (*Sonchus*) of the Canary Islands. Beginning with the arborescent *S. pinnatus* of moist valleys, it is possible to demonstrate changes in the growth-form, leaf-shape and capitulum-size in response to the different environments provided by the lowland xerophytic zone (*S. regis-jubae*, *S. leptocephalus*), humid lower altitude forests (*S. jacquini*, *S. abbreviatus*), drier upper forests and montane zone (*S. acaulis*, *S. platylepis*) and exposed low altitude cliffs (*S. radicatus*).

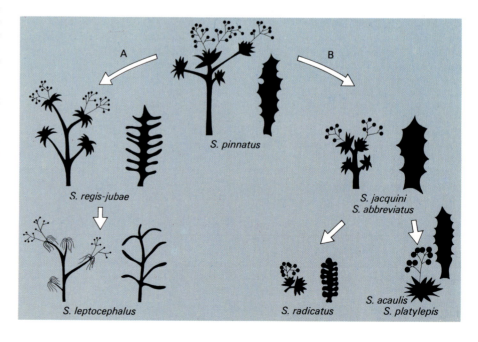

On the Canary Islands, endemic species of very limited distribution occur as relatively small populations in which genetic drift may have been an important evolutionary force. *Senecio heritieri*, pictured here, is restricted to a small area of rocky slopes on the south coast of the island of Tenerife. Related endemics occur elsewhere in the archipelago.

characteristics, occur on different islands, or even in different localities on the same island, and are thought to have arisen as a result of genetic drift. Similar variation patterns are seen in plants in scattered populations isolated on mountain tops. In many of these cases, not only are the populations small at present but the plants reached these often isolated areas as a result of the long-distance dispersal of one or a few propagules (seeds, spores etc.), which represented an unbalanced sample of the populations from which they came. The "founder principle", proposed by the American biologist Ernst Mayr, is used to account for the unusual characteristics of populations derived from such unbalanced samples of their parent population.

New populations produced from one or a few founders have a similar appearance to those caused by genetic drift. Populations at the margin of a species' range tend to be more isolated than the others and, since they are often subject to new environmental stress, can be subject to extinction or near extinction in unfavorable seasons. Near extinction followed by regeneration of the population from one or a few genetically aberrant survivors can lead to the rapid development of populations with new characteristics. Many marginal populations reflect this in their differing variation patterns. Neither in these nor in island populations can genetic drift be shown to be wholly responsible for the variation but it can be argued that they result from the joint action of selection and genetic drift more obviously than in large populations, in which selection is almost entirely responsible for the variation.

D.M.M.

to come up heads 50% of the time; in any trial, however, we tend to get rather more or rather less than half the throws giving heads. Such chance fluctuations in frequencies of an allele within a population of plants or animals are known as genetic drift, a term coined by the noted American mathematical geneticist Sewall Wright.

Genetic drift will not greatly affect the frequencies of particular alleles in a large population. In a small population, however, a chance excess or deficit of one genotype may cause a comparatively large change in the frequency of a particular allele, to the extent that it may be either completely eliminated or become present throughout the population. Since evolution is effectively the difference in gene frequencies between generations, it follows that genetic drift is of comparatively minor evolutionary importance in large populations, in which most changes in gene frequency result from selection, but it may be expected to be a

major factor in the evolution of small populations, as was pointed out by Sewall Wright in the 1930s.

Whether or not genetic drift as an evolutionary factor does indeed accord with these predictions when natural populations are studied has been much debated during the past 40 years. It can be shown mathematically that whether genetic drift will occur or not in a real population depends not only upon the size of the population but also upon the selective value of the alleles present, the rates at which the genes mutate and the extent to which there is immigration or emigration of genes by crossing with other populations.

There is certainly considerable evidence that populations of plants which are permanently or periodically reduced in numbers exhibit distinctive patterns of variation. Many island plants, which occur in small populations on separate islands or in different valleys of the same island, as in Hawaii and the Canary Islands, for example, show morphological variation between populations in characters which are not always obviously of adaptive value. The Greek Islands of the Cyclades in the Aegean sea have many similar habitats. Sporadic forms of *Nigella arvensis* (family Ranunculaceae), differing greatly in their fruit and flower

Speciation

The general outline of evolution and natural selection is clear and the role of adaptation and related processes is understood. However, much of our knowledge of them is inferred from observing the diversity of fossil and modern plants in relation to environmental and other factors presumed to have been important in their evolution. The techniques currently available do not permit direct study of the evolution of families and genera, for example, so that informed speculation must continue to be used in discussing their origins and development. We are on much surer ground when discussing the origin of new species, the process known as speciation, since it is possible to demonstrate this practically. Consequently, like Darwin, we emphasize the role played by speciation as a particular feature of evolution in this chapter.

Species normally remain distinct from one another because there are one or more isolating barriers to interbreeding. These may prevent the formation of hybrids or result in the production of partially or completely sterile hybrids. Speciation therefore involves the establishment of these barriers, and the processes by which they arise can be divided into two groups. In one, small genetic changes take place over a considerable length of time, often in response to environmental factors, to give new species; this is frequently termed gradual speciation. In the second group we have rapid speciation, resulting from reorganization of the chromosomes, which may involve changes in their number.

Gradual Speciation

In gradual speciation we may envisage a situation whereby an existing species gradually extends its range. In so doing, it may colonize areas which differ from one another environmentally, particularly in soil and climate. Colonization takes place by individuals with phenotypes and hence genotypes which allow them to survive and reproduce in the new habitat. Over a number of generations natural selection will take place to produce individuals which are progressively more adapted to the environment.

One especially clear example of this gradual adaptation is the formation of races of a number of species tolerant to high concentrations of normally toxic heavy metals on old spoil heaps. One species that has been studied in particular is the common bent grass (*Agrostis tenuis*) which, in North Wales, is known to have races tolerant to high concentrations of lead, copper and zinc. It has been shown that populations of this grass growing on normal soils possess occasional individuals with some tolerance to one or other of these metals, which presumably arose by mutation. When a new spoil heap is made, partially tolerant plants will be able to colonize the less toxic parts and in later generations selection will take place for more tolerant individuals, until plants are produced which are able to colonize the whole heap. Tolerant populations are, of course, not isolated, usually being surrounded by non-tolerant populations.

Genetic adaptation of species to varying habitats was first studied in detail by the Swedish botanist Göte Turesson, who called races adapted to particular habitats *ecotypes* (see Genecology, p. 36). Over long periods of time, races may become isolated from one another owing to geological or climatic changes, eg the formation of islands or the effects of glaciations. Under these conditions, when no exchange of genes (gene flow) takes place, adaptation to changing conditions may give forms which are so distinct as to be

Only three species of *Ranunculus* occur in alpine areas of the North Island of New Zealand. *Ranunculus nivicola* has twice as many chromosomes (2n = 96) as *R. verticillata* and *R. insignis* and is thought to have evolved by the doubling of the chromosome number (allopolyploidy) in a hybrid between the two diploid species.

distinguished as species. This can be illustrated by the plane trees (*Platanus*). Apart from several species in the southwest United States there are three species which have very distinct geographic areas and differ in both vegetative and fruiting characters. These are *Platanus kerrii* in Indochina, *P. occidentalis* in the eastern United States and *P. orientalis*, ranging from southeast Europe to Iran. Species of this kind which are geographically separate are known as *allopatric* species. Fossil leaves of this genus, from the Tertiary period, are known from localities across Eurasia and the present-day species have presumably evolved as the result of a breakup by the recent glacial periods of a more or less continuous distribution of the plants through Asia some millions of years ago. Presumed hybrids of garden origin between *P. occidentalis* and *P. orientalis* (London plane) are vigorous and fertile, indicating that the morphological changes have not produced genetic isolating barriers.

In many cases, although genetic divergence takes place when populations are geographically isolated, further migration results in them coming into contact again, that is they become *sympatric*, over at least part of their areas. If the genetic changes which have arisen in isolation include isolating barriers, then the populations will remain distinct and be regarded as species. The evolution of such barriers thus plays an important role in the evolution of separate species (see Isolation, p. 108).

Rapid Speciation

Rapid speciation can take place in a number of ways. The most common method is through *polyploidy*, and it is estimated that about half of the species of flowering plants are polyploid. The most common change is for a plant with two sets of chromosomes (diploid) to give rise, by irregular cell-division, to a plant with four sets of chromosomes (tetraploid). Hybridization between these diploid and tetraploid plants means that the resultant offspring will be triploid, with three sets of chromosomes. Whether the diploid and tetraploid parents had similar or different chromosomal configurations, it is clear that there will be difficulty in the pairing of their chromosomes in the hybrid, so that the equivalent of one set of chromosomes, at least, will not be included in the regular division of meiosis. Consequently, the triploid hybrid plant cannot produce balanced gametes and will be largely sterile, which means that the diploid and tetraploid parents are isolated from each other and unable to exchange genes. Because polyploids arise from their parental diploids as a result of a chance process which occurs within a short period of time, the consequent genetical isolation between parents and offspring takes place rather quickly, so that differentiation between them is an immediate result of polyploidy. For this reason, the change from diploid to polyploid plants can

R. insignis
2n = 48

R. verticillatus
2n = 48

R. nivicola
2n = 96

Platanus orientalis

Platanus occidentalis

■ *Platanus occidentalis*
■ *Platanus orientalis*

Platanus acerifolia

The genus *Platanus* provides a good example of how geographical isolation prevents gene flow, with the result that the different forms become recognized as species. Natural populations of *Platanus occidentalis* and *P. orientalis* are isolated geographically. The two species obviously cannot hybridize unless deliberately grown together. Where this has been done the result is a fully fertile and vigorous hybrid *P. acerifolia* (*P. hybrida*), the familiar London plane.

result in an abrupt severance in their genetical connections, which can lead to an immediate break in their evolutionary futures—and hence rapid speciation.

Nearly all polyploid species seem to be allopolyploid, ie arise through hybridization between different parental species, followed by chromosome doubling. In most cases it has only been possible to identify the parents by hybridization experiments and chromosome studies. In a very few other examples, evolution of a new species has been observed to take place in recent historical time. An outstanding example is that of the cord grass now known as *Spartina anglica*. The only native British *Spartina* is *S. maritima*, which is found in salt marshes in south and east Britain. In the early 1800s, however, *S. alterniflora*, native to northeast America, became accidentally introduced and established in Southampton Water. About 1890 a vigorous new and fertile plant was recorded in Southampton Water, which has been shown to have arisen through hybridization of the two existing species to form a sterile hybrid, followed by doubling of the chromosome number in the hybrid, which made it fertile. This is a new species, *S. anglica*, which today is extensively used for reclamation of mudflats in Europe, together with the sterile hybrid (*S. × townsendii*), which still grows with the parents and *S. anglica* in Southampton Water. *Spartina*

anglica is effectively genetically isolated from its parents since any cross between it—a tetraploid—and its parents—diploids—will produce a sterile triploid. It also cannot backbreed with *S. × townsendii*, since the latter is also sterile.

In polyploids, the changes in chromosome number result in immediate isolation from the parental species, usually because of various postzygotic mechanisms. Speciation can, however, take place via hybridization alone, if other isolating barriers arise. This may, for example, happen if the hybrids are adopted by a pollinator which does not visit the parents. In California there is a group of three partially interfertile *Penstemon* species: *Penstemon centranthifolius* has red, trumpet-shaped flowers, pollinated by hummingbirds, *P. grinnellii* has blue, broad-throated flowers, pollinated by carpenter bees, while *P. spectabilis* has intermediate-shaped flowers, which are purple and pollinated by solitary wasps. *P. spectabilis* is presumed to be of hybrid origin, intermediate in morphology between the other two species and adopted by a different pollinator, which together with the availability of an intermediate, open habitat, was essential for its successful establishment.

Another mechanism by which rapid evolution of new species may take place involves reorganization of the structure of the chromosomes. The genus *Clarkia* has been particularly well studied from this point of view. Similar mechanisms are known in other genera, all of which involve groups of annual or biennial species. It seems clear that a necessary prerequisite for successful establishment of a new species by this mechanism is that small populations or single individuals must become isolated from the parental populations. This may take place by dispersal to a new site or by elimination of the

majority of the population by some drastic event, eg drought, a process which has been called *catastrophic selection*. Such a small derivative population will need to inbreed in order to survive and there is evidence that such a process, especially if imposed on a normally outbreeding species, may cause chromosome breakage. This may lead to various consequences, including the exchange of material from different chromosomes (translocations), the reversal of sections of chromosomes (inversions) and increase or decrease in chromosome number. Although individuals will initially be heterozygous for such changes, inbreeding will soon result in the formation of homozygous individuals, which will be reproductively isolated from the parents, because hybrids will be heterozygous and infertile. In one species pair which has been carefully studied, *Clarkia lingulata* differs morphologically from *C. biloba* only by lacking notches on the petals; *C. lingulata* has a gametic chromosome number of 9, compared with a chromosome number of 8 in *C. biloba*, the extra chromosome having been derived from parts of two different *C. biloba* chromosomes.

Apomixis—Reproduction without Sex

Although the species concept has been developed from a study of sexually reproducing organisms, it is also applied to groups which reproduce asexually. Apomixis is the phenomenon, shown by many higher plants, whereby reproduction is not dependent upon fertilization, and the sexual process is wholly or partly superseded. It is normally subdivided into *vegetative reproduction* and *agamospermy*. Vegetative reproduction takes place by simple fragmentation of the plant or by the production of rhizomes, stolons or tubers which, when detached, give rise to "daughter" plants. In other cases, particularly in a range of species from mountains and arctic environments, the inflorescences are modified to produce small plantlets (pseudovivipary) or small bulbils are produced, either in leaf axils or instead of flowers (bulbifery). Plants reproducing by agamospermy produce embryos and seeds by normal division of cells in or surrounding the embryo sac, with no fertilization involved. In all these cases the progeny are identical with the parent from which they are derived. With apomixis, therefore, the identity between parents and progeny gives rise to pure-breeding lines which maintain their differences and so appear similar to normal sexual species. However, sexual reproduction can occur occasionally in normally apomictic groups. When this happens the progeny can be genetically different from their parents and if perpetuated by subsequent apomixis, become recognized as new apomictic species, which have arisen within a very short period of time. The numerous true-breeding variants in dandelions (*Taraxacum*), blackberries (*Rubus*) and hawkweeds (*Hieracium*), for example, result from apomixis.

T.T.E.

Isolation

The reduction of interbreeding between populations is one of the important steps in speciation, and the study of the factors, termed *isolating mechanisms*, which bring this about, is essential for understanding how evolution occurs. Indeed, isolation in this wide sense is a universal phenomenon, practically synonymous with evolution. The

Forms of vegetative reproduction. (1) Crocus corm; (2) daffodil bulb; (3) iris rhizome; (4) blackberry "tip layering"; (5) strawberry runner; (6) mint stolon; (7) potato tuber.

attention of the botanist is focused on those examples where an isolating mechanism can be seen at work, preventing the interbreeding of potentially interfertile populations and thus contributing eventually to the evolution of new species.

Isolating mechanisms can be grouped into three main classes: spatial, ecological and reproductive. Spatial isolation, ie geographical or eco-geographical isolation, results from the differentiation of recognizable populations and species in topographically separate areas. Ecological isolation, by contrast, is the separation of different populations and species by climatic or soil conditions or by competition from other organisms. Reproductive isolation is by far the biggest class and refers to all of those genetically controlled processes which prevent successful hybridization between different species. These include either premating

(external) blocks to gene exchange such as mechanical, ethological and temporal differences in the floral mechanism, or postmating (internal) blocks operating after pollen or spore transfer between one species and another. Here we can include pollen incompatibility, hybrid inviability, hybrid sterility and hybrid breakdown.

The different classes summarized above demonstrate the real extremes of isolation to be found in the plant kingdom. It is important to emphasize, however, that rarely does one isolating mechanism operate on its own, but usually there is a combination of two or more systems which reinforce population or species differences.

Spatial Isolation

Spatial or *allopatric* isolation differs from all other types of isolating mechanisms because it is a completely fortuitous event and independent of genetical changes in the plant species. In other words, it is the spatial separation of different populations sufficient to prevent cross-pollination or gene flow between them, thus allowing them to continue through time as two distinct entities. Spatial isolation is brought about by two types of event: either by long-distance dispersal or, more commonly, through the fragmentation of a formerly continuous distribution as a result of historical changes in the climate or geography (eg drought, continental drift and glaciation). There are many degrees of geographical isolation ranging from one long continuous population effectively isolated at either end, right through to transcontinental or transoceanic disjunctions where gene exchange is virtually, if not totally, prevented by the massive distances involved. (See also Restricted Distributions, p. 226).

If two widely separated populations occupy identical habitats, it is common for both of them to belong to the same species even when isolated for hundreds, thousands, or even millions of years. A good example here is the fern species, *Adiantum reniforme*, which is now found only in the Canary and Cape Verde Islands, and the Malagasy Republic (Madagascar) off the east coast of Africa. This species is said to be a living relict of the Tertiary period, when it enjoyed a much wider distribution throughout wet temperate areas in Africa, but today only occupies this bizarre, disjunct distribution.

More often than not, ecological conditions are not the same in the geographically separated habitats, so there has been divergence between the isolated populations as a result of natural selection for different genotypes. An interesting feature of geographically isolated, or allopatric populations, is the fact that even though divergence may have taken place they still retain the ability to exchange genes when brought together into cultivation. The degree of divergence which might occur between two populations depends almost entirely upon

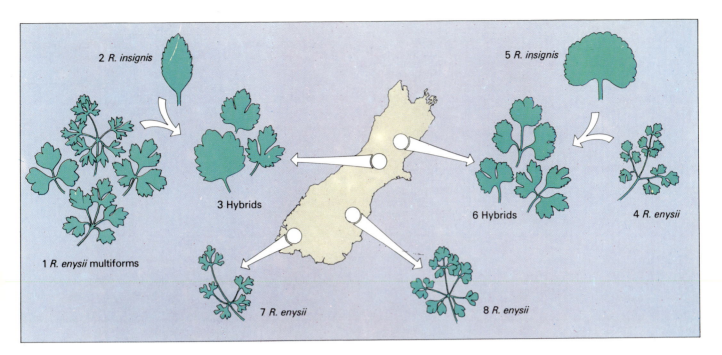

2 *R. insignis*

3 Hybrids

1 *R. enysii* multiforms

5 *R. insignis*

6 Hybrids

4 *R. enysii*

7 *R. enysii*

8 *R. enysii*

Introgressive hybridization (or introgression) is the transfer of genes from one species to another by the formation of hybrids and the back-crossing of these with the parents, producing individuals that closely resemble the predominant parent, but which possess some of the characters of the other parent which formed the hybrid. Through this process, new gene combinations may become incorporated into a species, allowing it to extend its range. Shown here is introgression in *Ranunculus enysii*. At the southern end of South Island, New Zealand, populations of *R. enysii* are rather uniform with finely divided leaves (7 and 8). Further north the populations are of very variable leaf shape (1) due to introgression of genes from *R. insignis* (2), with which it readily forms hybrids (3). Further north still, uniform populations of *R. enysii* (4) are in slight contact with a reniform-leaved form of *R. insignis* (5) and hybrids (6) are produced locally.

the difference in the habitat conditions, the size of the populations and the length of time for isolation (see Speciation, p. 105).

By far the commonest type of geographical or eco-geographical isolation in plants, and animals too for that matter, is that in which populations are widely separated. An important observation to establish here is the idea that ecological factors are just as important as geographical separation in preventing overlap of range under natural conditions.

When there are different habitat conditions for two separated populations it is often not possible to make a clear distinction between purely geographical and purely ecological isolation. It was originally suggested by the American biologist Ledyard Stebbins that in the evolution of geographically isolated species or subspecies, initial separation of an ancestral population was just as likely to be caused by different selection pressures in the environment, resulting in ecological differentiation of two populations, as by geographical separation. Thus, in this situation, geographical races may have arisen by ecological adaptation, as ecotypes.

Ecological differentiation can be extremely subtle, in the sense that even the slightest differences in climate or soil, or competition from other plants or animals may be sufficient to initiate divergence and adaptation. Transplant studies, initiated to determine the ecological responses of species to different environmental conditions have shown that most species are ecologically variable. More surprisingly, differences can occur between even two cross-breeding populations standing only a few feet apart. Thus, divergence can take place as a "mosaic" process just as soon as there is *any* form of separation between groups of individuals in a population. (See also Genecology, p. 36.)

Between the two extremes of population separation and conspicuous species divergence, there occurs a wide array of intermediate situations producing populations, which can variously be described taxonomically as races, varieties or subspecies. Many examples of species comprising several different races are known. A particularly good example is the columbine (*Aquilegia vulgaris*) which has many different montane and lowland variants, as a result of obvious ecological and geographical isolation. Similarly, there are many species with such distinctive eco-geographically isolated populations that they are recognized as subspecies. The kidney vetch (*Anthyllis vulneraria*), for example, is remarkable in that it is a widespread species with no fewer than 24 subspecies in Europe alone.

It seems in the great majority of situations that geographical isolation, when divorced from other factors, results only in the disjunction of populations and does not lead to the evolution of species. In very small populations, however, especially those formed by chance events such as long-distance dispersal, rapid divergence and thus recognizable geographical isolation may occur by means of genetic drift (see p. 104).

Ecological Isolation

So far, the emphasis on extrinsic isolation has revolved around *allopatry*, the actual physical separation of populations with and without the involvement of an ecological component. Brief mention has already been made of the fact that closely adjacent cross-breeding populations can diverge under intense environmental selection pressure, and it is here that the process of *sympatric* ecological isolation in overlapping distributions can be considered. The ecological separation of sympatric populations is considered common in plants and several good examples are known. One of the earliest reported examples of ecological isolation is that of the golden rods *Solidago rugosa* and *S. sempervivens* in the United States. The former species is characteristic over dry, poor soils throughout the United States, but *S. sempervivens* occurs in salt marshes around much of the coast. For most of the time they remain very distinct but they do overlap considerably in general distribution, and the hybrid between them, *S. × asperula*, is not uncommon. It is interesting to note that the hybrid is successful when there are intermediate, intergrading habitats, but absent when there are sharp discontinuities between the two soil types. Thus, the two species are kept apart by ecological isolation.

The fact that many closely related sympatric species retain their identity as a result of ecological isolation in no way implies that they necessarily evolved as a result of ecological divergence. It is just as possible that two species might also have originally evolved eco-geographically and then come together again after genetical divergence and changing habitat conditions.

The most convincing evidence for ecological isolation has come from statistical studies of morphological and physiological character expression in relation to the environment. The sea plantain (*Plantago maritima*), for example, consists of outbreeding colonies of varying size. Although differences in small populations might well have been caused by genetic drift, for large outbreeding populations it has been shown that even the subtlest changes in the environment (eg changes in the overall water content, salinity, acidity etc) are mirrored in changes of leaf-shape, flower-size and general habit. This phenomenon of continuous graded variation, which can be correlated with observable gradients in environmental conditions, is known as an *ecocline*. Where there are sharp discontinuities in the environmental conditions, sharp discontinuities also occur in the morphological and physiological features of *Plantago maritima*. Thus, with time, in just the same way that geographically isolated populations can eventually diverge into different species, ecotypes can do the same (see also Genecology, p. 36).

Reproductive Isolation

The third group of isolating mechanisms, which often occur in conjunction with those just described, are the internal (intrinsic) systems that have evolved genetically to maintain population and species differences.

External Reproductive Isolation

Some of the most obvious systems of reproductive isolation are those which have evolved to prevent pollen and spores actually reaching different species at all. These take place by ensuring pollination specificity as in mechanical and ethological isolation, or simply by the prevention of overlap in the timing of flowering.

Mechanical isolation results from structural differences in the flowers of two or more species which normally prevent pollen trans-

Ecological isolation and introgressive hybridization of the spiderworts. Normally, *Tradescantia canaliculata* grows on cliff tops and *T. occidentalis* at the foot of cliffs. However, if the slopes of the cliff are gentle the habitats of the two species overlap and the species interbreed (arrows). The resulting hybrids can back-cross with the parent species.

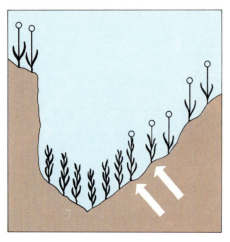

Pollen incompatibility is an effective form of internal reproductive isolation. In the radish (*Raphanus*) the incompatibility mechanism is now believed to be controlled by a coating called "tryphine", which is applied to the pollen grain late in its development. The reaction between tryphine and the stigmatic wall after pollination prevents pollen tube growth of incompatible pollen. Thus in many cases incompatible tubes fail to enter at all. If they do, a second part of the system is brought into play (**right**). As soon as the stigmatic surface is ruptured (arrows), the cytoplasm (Cl) first becomes highly active (× 10,000). **Below right** Immediately underneath the point of penetration (arrow) this cytoplasm lays down layers of callose (Cl), a structural carbohydrate, which results in the final arrest of pollen tube (T) development (× 6,500). **Bottom right** Incompatible pollen tube (T) of radish, 10 hours after pollination. Despite its having penetrated the stigmatic cuticle (arrows), the tube has finally been stopped at the layer of carbohydrate formed under it (C) (× 4,000).

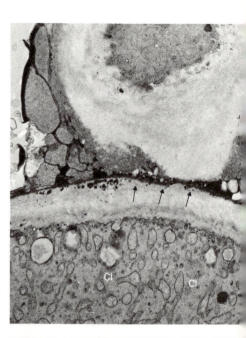

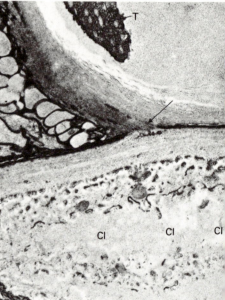

fer from one species to another. Plant families with complex flowering mechanisms such as the Orchidaceae, the Leguminosae, the Asclepiadaceae, the Scrophulariaceae and the Labiatae are most frequently the ones to have evolved mechanical isolation. Orchids are the most remarkable family in this respect and *Ophrys* a particularly good example. The pollen of *Ophrys* (and most orchids for that matter) does not occur in single grains within an anther but is agglomerated into sticky balls called pollinia. The pollinia must be transferred from one flower to another by insect pollinators and cannot be transported by wind. The relative positions of the hanging pollinia and stigmas to one another make self-fertilization virtually impossible. The flowers bear a superficial resemblance to the female of a particular insect species and the males of the species are attracted by the appearance of the flower and attempt to copulate with it. During this pseudocopulation the pollinia become detached and stick to the abdomen of the insect. When the male insect visits another flower of the same species the pollinia will come into contact with the stigmatic surface and bring about pollination. The pollinia are shaped and positioned on the insect's body in such a way that they can only come into contact with the stigmatic surface of the orchid species first visited and not any other species. In the genus *Ophrys* there are many interfertile species but interbreeding is prevented by two mechanical mechanisms, ie each orchid species mimics the female of only one insect species and there is a sophisticated "lock-and-key" mechanism between pollinia and stigmatic surfaces which prevents pollination should the insect visit the wrong species.

Ethological or *behavioral isolation* occurs quite frequently in plants. Bees, hawkmoths and various other insects are able to perceive and differentiate between the different shapes and colors of many different flowers, particularly in such genera as *Brassica*, *Pedicularis* and, as already mentioned above, *Ophrys*. This gives them the opportunity to visit and feed preferentially on one species of flower during a succession of feeding visits.

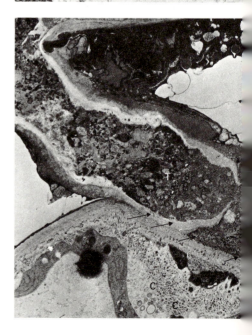

Such "flower constant" behavior strengthens existing species differences and tends to keep cross-pollination to within the limits of one species.

Temporal isolation maintains specific integrity by flowering and pollination taking place at different times of the year or even different times of the day. Familiar examples of this phenomenon can be found in the snowdrop genus, *Galanthus*, the campion genus, *Silene*, and in *Crocus* and the autumn crocus, *Colchicum*. In the common snowdrop (*Galanthus nivalis*) the subspecies *reginae-olgae* flowers in the autumn, subspecies *cilicicus* in midwinter and subspecies *nivalis* in the spring. Hybrids between the subspecies are rare, even when cultivated in the same area, and absence of hybrids in the wild results from geographical separation and minor morphological differences reinforcing the seasonal flowering times. More frequently, as in the parallel genera *Crocus* (family Iridaceae) and *Colchicum* (family Liliaceae), not only do species pairs show seasonal isolation but whole groups of species are either spring or autumn flowering.

Internal Reproductive Isolation
Some of the most effective isolating systems are those which prevent hybridization. The chief mechanisms involved depend on the prevention of pollen germination by the stigma, the prevention of entry of the pollen tube into the ovule, or weakness, reduced fertility or sterility among hybrid offspring.

POLLEN INCOMPATIBILITY. The first effective carrier to operate in the reproductive phase is the failure or ineffectiveness of pollen tube growth after the pollen has come into contact with the stigma. For successful fertilization to take place, the pollen grain must germinate on a receptive stigma, penetrate the style and ultimately fertilize the embryo and endosperm. Incompatible pollen grains can fail to reach the embryo for a variety of reasons; mainly by protein interaction through genetic control by the stigma or the pollen grain, but also by physical differences such as differences in style length or thickness. Although examples of this phenomenon are rarely documented, it is perhaps pollen incompatibility more than any other mechanism which initially prevents wide crossing taking place in flowering plants.

HYBRID INVIABILITY OR WEAKNESS. Frequently,

interspecific first generation (F_1) hybrids are very vigorous, but they often display a number of structural weaknesses in the vegetative phase as a result of incompatible genetic and cytoplasmic interaction between the two parents, an unfavorable reaction which leads to early death. Characteristic features of inviable F_1 hybrids include chlorosis, ie a complete lack of chlorophyll, preventing any plant development beyond the seedling stage, genetic "tumors" and various degrees of stunted or twisted growth. Chlorotic seedlings and distorted leaves are commonly found in *Epilobium* and *Chrysanthemum* hybrids, and genetic, virus-like tumors have been reported in poppy hybrids involving *Papaver dubium* and *P. rhoeas* as well as in radish (*Raphanus*) and cabbage (*Brassica*) hybrids.

HYBRID STERILITY. Like incompatibility, this embraces a wide variety of phenomena. There seem to be two basic causes of hybrid sterility in F_1 plants as a result of parental differences, namely incompatible gene interaction or malfunction of chromosome pairing at meiosis (see p. 93). Hybrid failure of this type can take place before, during or after meiosis takes place in the anthers and ovules. Pre-meiotic genetic events include abnormal flower development, such as the conversion of stamens to petal-like organs, a complete

loss of anther material, or the degeneration of pollen mother cells before pollen is produced. All three events lead to male sterility and cessation of the male germ line. Meiotic events leading to complete sterility in both male and female germ lines are various but usually occur as a result of non-pairing or irregular pairing of the chromosomes, resulting in the formation of useless pollen and ovules. Post-meiotic effects are equally numerous and again involve all stages of development from the degeneration of the pollen or ovules before utilization in the production of the next generation through to complete abortion of the second-generation embryo.

HYBRID BREAKDOWN. Finally, there are a number of examples demonstrated in the plant kingdom where the individuals of the second or later generations show general weakness or markedly reduced fertility even when the first-generation hybrids are quite vigorous and fully fertile. It seems that the same factors as those associated with first-generation inviability or sterility are in operation but that they are manifested in a later generation. Well-documented examples of such isolating mechanisms can be found in the elder hybrid *Sambucus nigra* × *S. racemosa* and the larch hybrid *Larix gmelinii* × *L. kaempferi*.   C.J.H.

Right Incompatibility mechanism involving the interaction of the pollen grain with the style of a flowering plant, eg the primrose (*Primula vulgaris*). The pollen grain is haploid, ie carries a single S allele (shown here as S_1) while the style is diploid and carries two S alleles (shown here as S_2). Pollen tubes with the same allele as one of the styles will be unable to penetrate the style tissue and are therefore unable to bring about fertilization. Thus if the pollen grains are either S_1 or S_2 no fertilization will take place (**left**) or if the pollen is S_1 or S_3 only the S_3 pollen tube will reach the ovary to fertilize the ovules, giving two types of progeny, S_1S_3, S_2S_3 (**center**). Only if pollen grains have completely different S alleles (eg S_3S_4) will complete fertilization occur, the progeny being S_1S_3, S_1S_4, S_2S_3 or S_2S_4 (**right**).

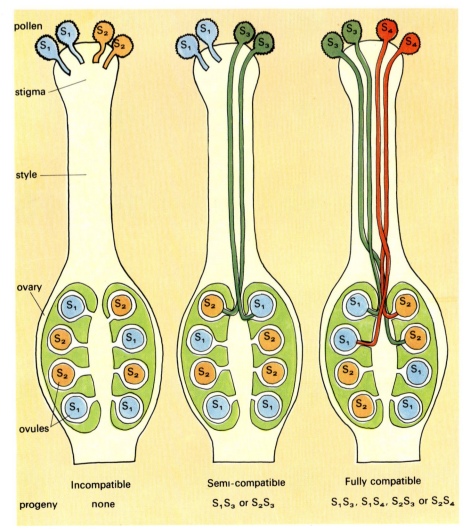

Incompatible	Semi-compatible	Fully compatible
progeny none	S_1S_3 or S_2S_3	S_1S_3, S_1S_4, S_2S_3 or S_2S_4

CHAPTER FOUR

Environmental Factors

The Background to Plant Distribution Today

The plants occurring in any particular region, whether considered as individual species or as members of the communities which make up vegetation, do so as a result of the historical components, dealt with in the previous chapter, upon which are superimposed the effects of the present-day environment. It is now generally accepted that the environment comprises not only the physical features with which plants have to contend but also the other plants with which they grow to form populations and communities. The environment thus constitutes a variable complex of factors which can interact in numerous ways. In the present state of knowledge we have little or no idea what the consequences of many of these interactions are for the world's plant cover. Certain factors seem particularly important, and have consequently received a good deal of attention, but too often in isolation from other features of the environment. This chapter, therefore, is necessarily an over-simplified introduction to a complex subject that still requires considerable research.

While it is clear that the physical features of the environment cannot be truly separated from the biological characteristics of the plants in determining where they can grow, some attempt has been made here to do so for convenience of presentation. Thus, the ability of a plant species to cope with various physical features, such as temperature and exposure, is the result of selection and adaptation acting on its genetic endowment over the generations to give the tolerance shown by its present-day populations. Similarly, the capacity of plant species to maintain and extend their range depends in part on their inherited ability to disperse to and colonize new areas, although chance also plays an important part, especially in the early stages of migration. Competition, which is important for determining the relative importance of different species in plant communities, also results from

natural selection acting upon those features which govern the fitness of plants.

Of the physical factors involved in determining the environments occupied by plants, some may be considered as operating on a large scale, providing the framework for the worldwide distribution of plants. Thus, the principal climatic zones of the world, which are largely delimited by temperature and rainfall regimes that result to a considerable extent from the facts of the Earth's orientation, rotation and major topographic features, determine the broad outline of the important vegetation types seen on the Earth. Similarly, the variation in day length and seasonality between polar and equatorial regions imposes considerable constraints on the plants. Another very important large-scale factor exerting considerable influence on the plant cover is, of course, soil, the major classes of which are determined in large part by the features of the world's principal climatic zones.

The above factors can also operate on a much more restricted scale. The often complex regional variation in the underlying rocks, which are clearly important in determining the properties of the soils derived from them, together with features of regional hydrology and temperature, belong to the group of factors which affect distribution of plants and the communities they constitute within relatively small areas. It is important, however, to remember that at this more local level, and undoubtedly on the larger scale as well, the plant cover is itself important in modifying the factors referred to above. For example, the properties of soils depend to a marked extent on their organic content, and this is derived from the plants that they support, the dead leaves, stems etc of which are incorporated on decomposition, while the fungi and other organisms living and dying below the surface are also clearly important. Furthermore, the plant cover alters the physical conditions near the ground so that, for example, wind effects are dramatically reduced within a forest canopy, as is the amount of light penetrating to the ground surface, while water loss by evaporation etc is profoundly altered. Thus, within complex vegetation such as high forest a whole series of microclimates is produced under the primary influence of the dominant trees which consequently provide the conditions governing the lives of innumerable plants (and animals) growing within it.

Opposite The fronds of one of the giant kelps, *Durvillea antarctica*, form a vast floating canopy up to 15m (50ft) across. Without major strengthening tissues this alga can only attain such a great size because of the support afforded by its marine environment.

What is the Environment?

When describing and discussing the factors affecting the distribution of plants as individuals, or as members of populations, communities or different kinds of vegetation, the terms environment, habitat and niche are all used in referring to the places in which plants can live and the conditions governing them.

Environment

The environment comprises the whole complex of factors which affect the occurrence of organisms and which interact with each other as well as with the plants and animals. It is convenient, although artificial, to distinguish between physical components, such as climate, and those arising from the activities of other organisms (biotic factors); artificial because many of the ways in which organisms act upon each other are indirect, through their modification of physical factors, but convenient because it provides a framework upon which to build hypotheses.

The complexity of factors impinging on any individual plant can be appreciated by considering plants growing on the steep north- and south-facing slopes of a mountain, situated in a temperate region of the Northern Hemisphere. It is obvious that on any one day there will be differences in the environments experienced by plants on these two slopes. The south slope will receive solar radiation for longer and at a steeper angle. The light intensity and temperature will therefore be greater. Evaporation will be

A diagram illustrating the sources of loss and gain of environmental resources and their cycling through an ecosystem.

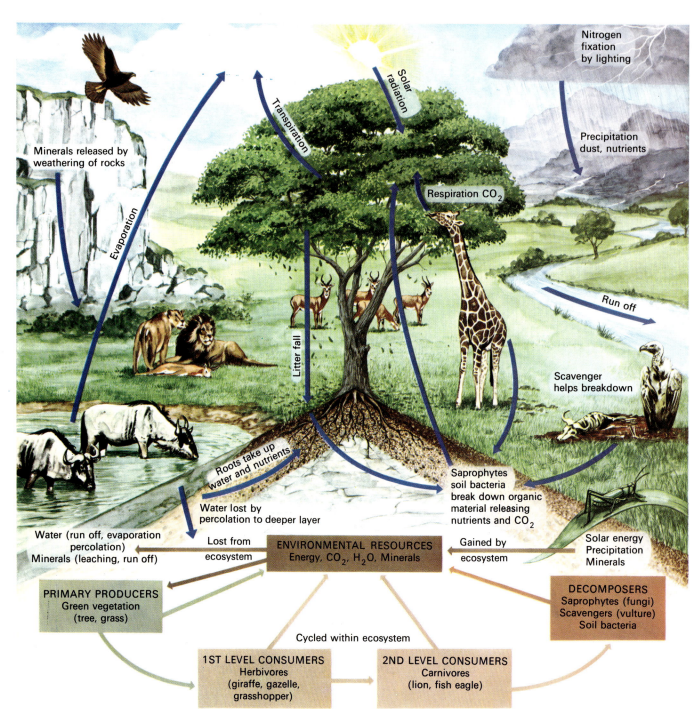

Nitrogen fixation by lighting

Solar radiation

Transpiration

Precipitation dust, nutrients

Minerals released by weathering of rocks

Evaporation

Respiration CO_2

Run off

Litter fall

Scavenger helps breakdown

Roots take up water and nutrients

Saprophytes soil bacteria break down organic material releasing nutrients and CO_2

Water lost by percolation to deeper layer

Solar energy Precipitation Minerals

Water (run off, evaporation percolation) Minerals (leaching, run off)

Lost from ecosystem

ENVIRONMENTAL RESOURCES Energy, CO_2, H_2O, Minerals

Gained by ecosystem

DECOMPOSERS Saprophytes (fungi) Scavengers (vulture) Soil bacteria

PRIMARY PRODUCERS Green vegetation (tree, grass)

Cycled within ecosystem

1ST LEVEL CONSUMERS Herbivores (giraffe, gazelle, grasshopper)

2ND LEVEL CONSUMERS Carnivores (lion, fish eagle)

more rapid, as will the metabolism of the organisms in the soil and surface litter. Thus there will be less organic matter in and on the soil and the soil will be drier. The north slope will be cooler, more shaded, damper and perhaps have more litter. In winter, frost will be more frequent and prolonged and the soil will consequently be more unstable. These are only a few of the differences to be encountered, but it will be clear that they will have a profound effect on the plants growing on the two slopes. The plants on the south slope will be warmth-loving, drought-resistant species and will possibly have a longer growing season; those on the north slope will be likely to be less drought-tolerant, and require cooler, damper conditions. Even in lowland areas one can find very striking differences between the vegetation of north and south slopes.

It is important to realize that while at any one time differences will be easily detectable, each and every environment is fluctuating, daily, annually and (more slowly) over longer periods. Environmental differences are not static, and every organism has to be adapted to the whole range of environmental fluctuations it will experience throughout its lifetime.

It is impossible to give details of all the environmental factors involved, but some attempt will be made to define the importance of each.

Solar radiation accounts for almost all the energy entering the biosphere, and light energy provides the basis for the functioning of all ecosystems by means of photosynthesis. Only a small amount of ultraviolet radiation reaches the Earth's surface (fortunately, because it damages cells), but it may be crucial in some plant-pollinator relationships. Plants respond in many ways to variations in the quality, intensity and duration of light.

Temperature controls the rate of biological processes and imposes considerable constraint on the ability of plants to occupy particular areas—some being adapted to very high and often diurnally greatly fluctuating temperatures, as in deserts, while at the other extreme some plants can metabolize at very low temperatures and endure long periods below freezing. Vegetational zonation, latitudinal and altitudinal, is largely under the control of temperature.

Water is an essential component of all living matter. It becomes available to most animals and plants almost solely as precipitation and, as the amount and annual distribution of water is uneven over the Earth's surface, water availability is another major factor in determining plant distribution. Some plants can tap water deep below the ground surface, but these are few. The uneven distribution of water is perhaps one of the greatest barriers to successful increase of world food production.

The *atmosphere* contains the essential gases oxygen and carbon dioxide and, except for altitudinal differences, does not vary much over the Earth. Its movement, manifes-

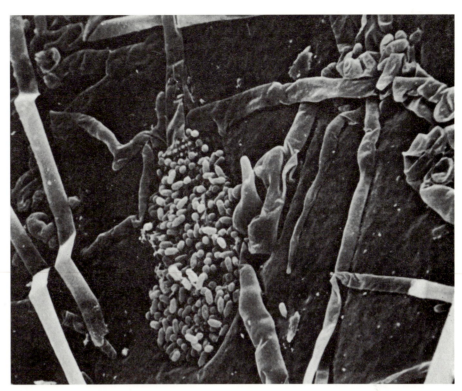

Unseen to the naked eye, the surface of most leaves is covered by populations of various microorganisms (phyllophane) which here find an environment in which they live off the nutrients secreted by the leaf. This scanning electron micrograph shows part of a cucumber leaf supporting the ribbon-like hyphae of a fungus and a mass of oval bacteria (× 2,000).

ted as wind, is responsible for circulating these gases and water vapor—hence rainfall. Wind may also be responsible for the dispersal of pollen, seeds and fruits over sometimes very long distances.

The *soil* is a complex system derived from the interaction of the substratum with water, climate, decaying organic matter, microorganisms, living roots and larger animals. Consequently, soils display an immense range of variation. Soil-plant relationships are of crucial importance, since soils provide the reservoir of water and minerals for the ecosystem, and they combine the chemical, physical and biological activity that makes these available and controls their cycling through the system.

The importance of *fire* as a "natural" factor has only been fully realized in recent years. While Man has been responsible for many fires, natural fire, mainly caused by lightning, has been a molding force in many ecosystems, and it is so frequent in some, eg chaparral (see p. 167), that it is the controlling factor. Such ecosystems are composed of fire-resistant organisms and there is a surprisingly large number of plants that are adapted to tolerate fire.

Biological factors are part of every plant's environment, except perhaps a few pioneers of newly available substrata. Green plants modify each other's environments by using water and minerals and returning their waste products to the soil, and by creating shade. Plants growing in close proximity compete for light, water, minerals and space. Most competition (see p. 146) is indirect, through modification of other organisms' physical environment. Some is direct, by the release of substances which affect other organisms adversely (see Allelopathy, p. 147). Competition is a potent factor in producing and maintaining diversity in natural ecosystems.

Fungi and *bacteria* are responsible for breaking down dead plant and animal remains, as are a whole range of animals which take part in the decomposer system, thus making nutrients available. Animals are also responsible, as herbivores, for the destruction of plants, and many plants have evolved mechanisms protecting them from grazing to some extent. Plant-animal relationships are important at many levels of ecosystem organization, in pollination and seed dispersal, for instance; plants and animals also create microhabitats for each other, thus increasing each other's diversity. Amongst animals, it should be noted that over recent centuries Man has probably exerted the major influence on the environment, both of plants and other animals (see Man and the Green Planet, p. 237), and much current interest in our green planet results from such human activities.

Having briefly considered the several components which together make up the environment, it must be re-emphasized that it is as a complex of all these factors that the environment acts upon a plant. Nevertheless it is possible sometimes to identify a single factor which may be "limiting" to a particular plant in a particular place. By using these limiting factors as foci of attention in investigating the ecology of a plant, much can be learned about what enables it to grow

A conspicuous role of plants in the biological environment is demonstrated by their relationship with pollinating animals, such as bats, birds and moths. Here a glossophagine bat (*Leptonycteris sanborni*) is feeding on the pollen and nectar of a century plant (*Agave palmeri*) in southeastern Arizona, United States.

in some places and prevents it from growing in others.

It must be remembered that if one factor is altered there is often a chain reaction, ending in results that were never anticipated. This is one of the reasons why human influence can be so destructive: what seems at first to be a harmless modification precipitates a whole series of changes. It is the "wholeness" of environments that makes them so difficult and at the same time so fascinating to study, and which also makes them so susceptible to external influences. It should be pointed out, however, that many ecosystems have a considerable "buffering" potential which enables them to respond to some change without damage. This is especially true of more complex systems, some of which can tolerate a surprisingly large amount of interference without being fatally damaged. Herein lies one of man's greatest challenges—to learn to live with, and in, his environment without destroying it.

s.w.

Habitat

Although the two terms, environment and habitat, are often used in a similar context, the habitat of any organism is the localized environment in which it lives. As such the term is usually restricted to the conditions favored or tolerated by a particular species within a particular region. The same species may tolerate different conditions in different regions; for example, within the environmental conditions available in western Europe, *Pulsatilla vulgaris* occupies habitats with alkaline soils in Britain and acid habitats in mainland Europe.

D.M.M.

Niche

Few, if any, plant species can live in the full range of habitats that have been colonized by vegetation. All plants are more or less specialized in the conditions they require for successful growth and reproduction, and therefore more or less rigidly limited to specialized niches. A plant specializes both in relation to local physical and soil conditions and in relation to its neighbors with whom it must compete.

Specific adaptation to local physical and soil conditions limits the range of habitats

that species can occupy even if other vegetation is absent. Important examples of niche differentiation at this level are between calcicole (restricted to base-rich soils) and calcifuge (intolerant of alkaline soils) species, wetland (also aquatic) species and those of dry land, frost-tolerant and intolerant species. Differences of this type determine that the vegetation which can potentially occupy an area is restricted and that some species are excluded from the habitats of others without any biological interaction. It is clear, however, that this level of niche differentiation is rather coarse, because most botanic gardens can maintain in a healthy condition a range of species far in excess of what would survive if competition occurred between species and natural succession was allowed to proceed.

The competitive interactions between species may narrow, or occasionally extend, the range of habitats that a species might occupy if other species were absent. Thus the presence of a vigorous competitor (or of a predator or an agent causing disease) may make it impossible for a population of a species to survive and reproduce itself in some localities. Species that make demands on the same resources at the same time tend to enter into a remorseless struggle for existence in which some are eliminated. The "law of competitive exclusion" has been variously phrased but essentially states that two species with the same ecologies cannot persist for long periods together—ultimately one succeeds at the expense of the other. A classic example comes from experiments with mixed cropping of oats and barley. The number of grains produced by pure populations of either species is unaffected when grown in soils over a wide range of soil pH. However, in mixtures, the oat is the more

Typical calcifuge vegetation including heathers (*Erica*) and gorse (*Ulex*) in Serra da Estrela, Portugal. Such calcifuge plants cannot tolerate any lime in the soil. They are therefore specialized to occupy a particular local environmental niche with acid soil.

A limestone pavement on the Gower Peninsula, South Wales. Crevices in the rock provide some soil for plants to grow in. However, this soil has a very high lime content and can only be colonized by lime-loving (calcicole) plants, such as the yellow flowered hoary rockrose (*Helianthemum canum*) seen here. Such restricted habitats provide a niche for calcicole species.

Physical Features of the Environment

Climate

It is an evident fact that climatic variation over the surface of the Earth is a major influence on the distribution patterns of plants. The variation is due largely to the different penetration of sunlight energy; in equatorial latitudes, at mid-day, energy penetrates the atmosphere at approximately 90°, so less is absorbed in transit than at the poles, where the angle is considerably more oblique. Similarly, the surface area upon which the radiant energy impinges is much greater at the poles, hence the sunlight is more dissipated. Thus equatorial regions become heated to a greater degree than the high latitudes and this results in an instability developing within the air masses, which rise over the equator and fall again at higher latitudes to give high pressure belts around the Tropics of Cancer and Capricorn, in the Northern and Southern Hemispheres respectively. High pressure also develops over the poles, leading to the movement of polar air masses towards the equator. Between these two regions are the unstable temperate regions, in which there is global circulation of air masses resulting from the spin of the Earth.

These air movements, modified by the effects of oceans, land masses and mountain ranges, lead to the development of global climatic patterns. But such patterns are not stable; they vary from season to season and they also drift from decade to decade and century to century because of a variety of

factors both within the Earth's atmosphere and in other parts of the solar system. Variations in climate over millennia can be very considerable, as can be seen from the succession of ice advances into temperate latitudes which our planet has experienced over the past two million years.

The distributional limits of vegetation are largely controlled by climate, because such vital factors as temperature and water availability are intricately bound up with the pattern of global climates. Each species has certain requirements of these factors and, collectively, these species join in communities to form vegetation types. For each species there are climatic limits, beyond which it is unable to survive, and a climatic optimum at which it grows and reproduces most efficiently (although these limits and optima may be modified under field conditions by the competition from neighboring plants or by predation from herbivores).

Climate varies not only on a global scale, but also locally. Within relatively small areas this variation is caused by such features as topography and aspect. A crevice in a rock receives less direct sunlight and therefore experiences smaller variations in temperature than a rock surface; wind speeds would also be lower in such a crevice and humidity would therefore be higher. Furthermore, within and up to about 2m (6.5ft) above stands of vegetation the climate is distinct from that described for the region in general terms. This small-scale climate is known as the *microclimate*.

The solar radiation incident upon a plant or an area of vegetation will in part be reflected but the remainder will either be trapped during the processes of photosynthesis or be utilized in the evaporation of water and the heating of plant tissue. The amount of energy reflected will vary greatly, being greatest over short grass communities such as lawns, and least over forests, because of the greater potential for internal scattering in the latter.

The rate of evaporation of water will be determined both by levels of solar radiation and by the hydrological status of the site, and will have a profound influence on the microclimate of the site. The amount of water available at the roots of the plant for passage upwards to the leaves and subsequent transpiration will vary, and so too will the relative humidity generated within the plant canopy, which will in turn affect the rate of transpiration.

The absorption of radiant energy by plants results in considerable differences between the temperature of the air and that within the tissues of the plant. This is of especial importance to plants growing in cold climates: on days when the air temperature may be below freezing, photosynthetic tissues may be warmed by radiant heating to a temperature at which metabolism can proceed. Local differences in topography can lead to great differences in temperature. As we have noted, in the Northern Hemisphere a south-facing slope will generally be warmer

---

aggressive competitor on acid soils and barley on the more neutral. Over several generations, barley would exclude oats on the neutral soils and the reverse would happen on the acid soils. This is the competitive exclusion principle in action. If the soils are very acid, barley fails whether oats are present or not; under these conditions the tolerance of barley is not wide enough to permit its growth and reproduction, and its exclusion occurs regardless of competition.

As a result of a competitive interaction between populations of two species, both may be changed genetically. Over periods of evolutionary time, forms that fail in a struggle for existence will be eliminated and selection may favor differences which enable species to avoid the full intensity of the interspecific struggle for existence. There is probably powerful selection pressure resulting from competitive exclusion and the differentiation of species into different niches may result.

These two components of niche specialization can be identified by experiment. If one of two species fails to grow even when other vegetation (and pests) are removed, it is presumably adapted to different physical and soil conditions. If, however, both species will grow in the absence of the other but only one survives when both are grown together, there is presumably proximal interaction and competitive exclusion. If the two species persist together in equilibrium it is likely that they have evolved differences in biology which minimize the interference each exerts on the activities of the other.

J.L.H.

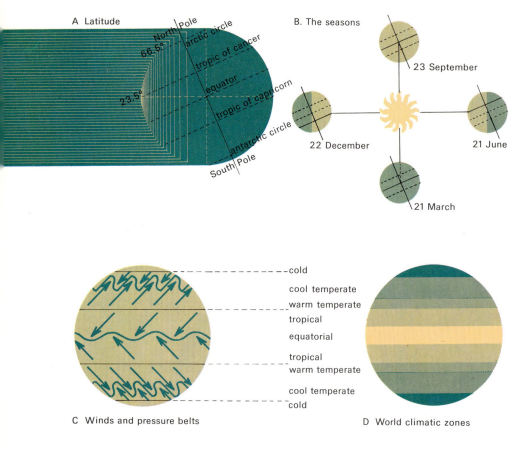

A Latitude

B. The seasons

23 September

22 December

21 June

21 March

C Winds and pressure belts

cold
cool temperate
warm temperate
tropical
equatorial
tropical
warm temperate
cool temperate
cold

D World climatic zones

Summary of main climate types.

Equatorial. Hot and wet all the year, with no marked seasons.

Tropical. Typically with a winter dry season and a summer wet season and often affected by monsoons. Areas on east coasts without a winter off-shore monsoon have winter rain from the easterly trade winds. Farthest away from the Equator, some areas are not reached by the belt of equatorial rains in summer and have a dry season. Some of the great deserts of the world are found in tropical areas in the centers and towards the west coasts of continents.

Warm temperate. Affected by westerly winds in winter and easterly winds in summer. West coasts have a typically Mediterranean climate, having warm winters with some rain, and dry summers. East coasts have drier winters and wetter summers, while central areas are often very dry, with summer convectional rain.

Cool temperate. Affected by westerly winds all the year round, west coasts and east coasts have changeable weather with rain falling throughout the year. Temperatures may fall well below freezing in winter without the influence of warming currents. Central areas are drier and often very cold in winter and very hot in summer, with rain coming mainly from convectional summer thunderstorms.

Cold. Long cold winters are followed by short summers with insufficient heat to unfreeze anything except the top layers of the soil, although in the height of summer the sun may shine for up to 24 hours.

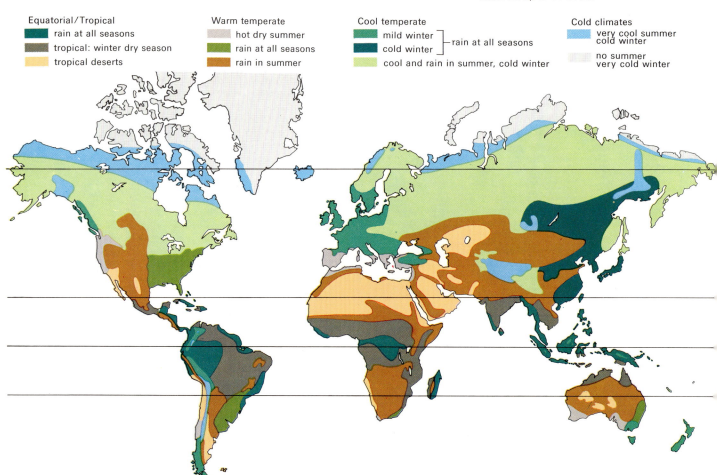

Equatorial/Tropical
- rain at all seasons
- tropical: winter dry season
- tropical deserts

Warm temperate
- hot dry summer
- rain at all seasons
- rain in summer

Cool temperate
- mild winter
- cold winter } rain at all seasons
- cool and rain in summer, cold winter

Cold climates
- very cool summer cold winter
- no summer very cold winter

Top The thicker main veins on these leaves of a *Mahonia* species are clearly outlined by the ice crystals which have formed on the thinner parts of the leaves. Microclimatic differences, such as this in the phyllosphere (air close to the leaf surface), may have significant effects on photosynthesis and water transport in the leaves.

Above Ice crystals still encrust the upright seta of a moss though they have melted from nearby horizontal surfaces. This is due to the variation in thermal capacities of the different surfaces. The main green parts are therefore able to carry out photosynthesis even though much of the plant is encrusted in ice and the temperature is below freezing point.

than a north-facing slope. Because of prolonged snow lie, growth will begin later in spring on the north-facing slope. Such differences may allow species to survive outside their main range, with southern species further north than might be expected and *vice versa*.

The closer that a surface is approached, whether it is the ground, the vegetation or, on a smaller scale, the surface of a leaf, the more the speed of wind is reduced by friction. Within the plant canopy itself, the air may be very still, so that while a gale might be blowing above a wood, for example, the vegetation of the forest floor may be in calm air.

Opposite Climatic variation depends ultimately on the way in which the sun's heat reaches different parts of the globe, variations being due to latitude (A) or the seasons (B). The resulting wind and pressure belts (C) and general climatic zones (D) interact with surface features (land distribution, mountains etc) to produce the complex pattern of climates shown in the map of the world.

It can be seen that any particular attribute of the microclimate measured at a certain point in time is the product of a dynamic system, dependent to a great extent on feedback. Variations in microclimate are not always a direct reflection of seasonal changes in incoming radiation. In deciduous woodland, for example, the microclimate of the plants of the forest floor is most drastically altered by the development of leaves in spring and their shedding in the autumn from the tree canopy above.

P.D.M.

Temperature

The maintenance of all living organisms demands a constant succession of biochemical reactions within their bodies, usually termed metabolism. Because of this chemical basis for life, living things are influenced, if not totally governed, by the laws of physical chemistry.

One such law states that the rate of many chemical reactions will increase by a factor of two to three times when the temperature is raised by 10°C (18°F). In biological systems one cannot apply this principle without certain qualifications, because the majority of biochemical reactions taking place within organisms require the catalytic activity of complex protein molecules, the enzymes. These depend for their activity upon a certain structural arrangement which is destroyed when the temperature rises too high, that is above 40°–50°C (104°–122°F). This denaturation of enzymes at high temperatures means that life becomes impossible and the rate of biochemical reactions declines in this upper temperature range. At temperatures between 0°C and 30°C (32°–86°F), however, many biochemical reactions are speeded up very considerably by increased temperature. For example, seed germination of some plants may increase by a factor of 12 times for a 10°C (18°F) rise.

Below 0°C (32°F) new problems arise for plants, because water and dilute salt solutions within the plant tissue may freeze and the resulting expansion may cause disruption of cellular structure, particularly membrane systems. Often it is the rapidity of freezing and thawing which is responsible for such damage, and the events preceding this stress are also important. If temperatures fall gradually and the period and intensity of frost increases gradually, then many plants are able to adapt to the new stress. Some plant species show seasonal adaptation which allows them to tolerate cold in midwinter but not in the summer. Such temporary acclimatization appears to be related to increased quantities of unsaturated fatty acids in the cell membranes which protect them from damage caused by freezing.

These physiological characteristics mean that each species has certain limits of temperature tolerance, outside of which it is killed, and this in turn means that the

geographical distribution of plant species must be within its extremes of tolerance. Many plants found in arctic tundra conditions are able to withstand very low temperatures during their winter-cold-adapted state. Laboratory testing has shown that many species can survive −80°C (−112°F) for several days. These species are generally rather limited in distribution as one moves southwards into temperate latitudes and many explanations have been put forward to try to explain this. One theory, for which there is now considerable circumstantial evidence, is that these arctic plants are limited in the south of their range by high summer temperature. For example, the rush *Juncus trifidus* occurs only in the Scottish Highlands in Britain and it is found there only where the mean annual maximum temperature is less than 22°C (72°F); the same applies to the dwarf birch (*Betula nana*).

Temperatures of over 22°C are unlikely to damage enzymes in these plants, but they could have adverse effects upon their energy balance. High temperature can have the effect of increasing respiration more than it increases photosynthesis. If high temperatures are maintained through the night then this becomes particularly severe and a plant may experience difficulty in maintaining growth. This is one reason why the potato (*Solanum tuberosum*), which originates from high tropical mountains in South America, is an unsuccessful vegetable in low altitudes in the tropics; its night-time respiratory rate is too high to allow survival. Nevertheless, it is well adapted to the normal temperature regimes of temperate latitudes.

The interpretation of distributions that have limits in common with certain isotherms (lines on a map connecting places with the same temperature at given times or on average over a certain period) is fraught with difficulties, since the true cause of

Present and Pleistocene distributions of *Dryas octopetala* in Europe. Present occurrences are shown in green, while fossil occurrences of the last glaciation and during its retreat are shown as red squares. Fossil finds other than those of the last glaciation are shown as blue triangles. These fossil finds indicate a wider distribution than at present, this restriction in range being probably caused in part by rising summer temperatures.

The importance of temperature in determining plant distribution is shown by many crops. The potato (*Solanum tuberosum*), for example, a widespread crop in temperate regions of the world, is a native of regions in subtropical latitudes that have temperatures lowered by their high altitude. The potato field in eastern England (**opposite**) is only possible because of the low temperatures experienced in the Bolivian Andes (**above**), at an altitude of about 3,500m (11,500ft). Here, potatoes are being harvested in an area where the crop originated perhaps as long as 5,000 years ago.

limitation may be competition or some other factor. Laboratory experimentation may assist interpretation, but there are times when it can complicate the issue yet further. For example, in the Northern Hemisphere, the annual, arctic-alpine species *Koenigia islandica* has a southern distributional limit which coincides with the 24°C (75°F) annual maximum isotherm. When tested in the laboratory, the lethal temperature for *Koenigia* is 45°C (113°F) (the lethal temperature is the temperature which, when maintained for 30 minutes, kills 50% of the experimental plants). Isotherms are based upon air temperatures, whereas the critical feature for small herbs like *Koenigia* is the soil surface temperature, which may be considerably warmer than the air in the type of wet, rocky areas where this plant is found. Thus experimental data are often difficult to apply to the field situation.

Frost sensitivity is one of the most important factors in determining the distribution of many oceanic species. For example, holly (*Ilex aquifolium*) is found in Denmark only where the mean temperature of the coldest month is above 0.5°C (33°F), provided that the mean temperature of the warmest month

is greater than 16°C (61°F). In situations where the summer temperatures are lower (down to a limit of 13°C or 55°F) holly can grow only if winters are milder (about 4°C or 39°F). As with *Koenigia*, it is difficult to discern the precise nature of the temperature sensitivity of *Ilex*, since, although its distribution seems closely correlated with the mean monthly temperatures quoted, it has also been observed to survive occasional cooling to temperatures of −17°C (0°F) after periods of cold which might have induced a hardening response.

As well as having extremes of temperature tolerance, each species has an optimum temperature at which its growth is maximal for any given light intensity. It is found that these optima vary with plants from different climatic regions, as a result of millions of years of evolutionary selection in response to the prevailing regional conditions. Thus, the Norway spruce (*Picea abies*) has a maximum net photosynthetic rate at 18°C (64°F), the European beech (*Fagus sylvatica*) at 21°C (70°F) and *Acacia craspedocarpa*, a tropical species, at 37°C (99°F). Consequently, each species is well adapted to the climatic regime within which it is found.

The concept of the temperature growth optimum takes into account the largest number of biochemical processes involved in the accumulation of matter in the plant. It is possible to isolate some of these processes and look at them individually; for example, one can isolate a specific enzyme system and determine its temperature optimum. This has been done for the tufted hair grass (*Deschampsia caespitosa*), which has a very wide distribution both in temperate regions and on mountains in the tropics. One of the enzyme systems involved in photosynthesis has been studied from a temperate and a

tropical population of this grass and the temperature for optimum activity was found to vary, being higher (30°C or 86°F) in the tropical population than in the arctic one (18°C or 64°F). Thus, the optimum temperature for growth in a plant species reflects the sum of individual enzyme optima. At temperatures close to its optimum the species will best be able to compete with other species of similar ecological requirements.

Another process in the life of a plant that is strongly influenced by temperature and that can affect distribution is seed-germination. Each species has temperature limits within which it is able to germinate and it also has an optimum. Measurements of the germination requirements of a number of species have shown that these are closely connected with the plant's requirements for survival under different climates. For example, a study was made of three species within the family Caryophyllaceae: *Silene secundiflora* (a Mediterranean species), *Lychnis flos-cuculi* (from the deciduous forest zone of Europe) and *Silene viscosa* (a steppe species of eastern Europe). The Mediterranean species germinated between 8°C and 14°C (46°–57°F), which was the lowest temperature requirement of the three. In Mediterranean areas the summer period is hot and dry and is the least

Tropical acacias, such as this one growing in the northern Sahara, have, over millions of years, not only become adapted to tolerate hot, dry conditions, but their physiological processes also now operate much more efficiently at high temperatures. For example, the maximum net photosynthetic rate occurs at 37°C (99°F) for some tropical acacias compared to a temperature of 18°C (64°F) for a cool-temperate plant such as the spruce.

suitable for plant growth. A low temperature germination requirement results in autumn germination and establishment can occur during the mild, damp winter. *Lychnis flos-cuculi*, the deciduous forest zone species, germinated between 17°C and 28°C (63°–82°F). This fairly high temperature requirement can be explained by the fact that in this region the cold winter is the most unfavorable period for growth and germination is therefore most effective if it takes place in spring. A high temperature requirement for germination ensures that this occurs. In the steppe species (*Silene viscosa*), germination occurred between 12°C and 29°C (54°–84°F), which is a very broad response. In steppe situations water supply is a limiting factor for growth, and germination can take place only when water is available. It is essential that the plant should be capable of taking advantage of wet periods and not be limited by a further requirement for a specific temperature regime.

The germination requirements of a species are therefore likely to be of considerable importance in determining its distributional range. If the temperature demand of a species is not met at a site, or if it results in germination during a period unsuitable for growth, then the species cannot survive in that location. The gray hair grass (*Corynephorous canescens*) reaches the northwestern limit of its European distribution in the northern parts of the British Isles. It is a plant of open habitats, particularly sand dunes in Britain, and has a rather short life span of two to six years. This being so, reproduction by seed is very necessary for the maintenance of viable populations and in this respect the species is well adapted, for a vigorous tussock can produce up to 36,000 seeds. The optimum temperature for germination has been found to be 15°C (59°F) and this normally takes place soon after the seed is set, provided water is available. In northern parts of Britain ripe seed is not formed until late in the summer and by this time temperatures are becoming too low for effective germination and establishment of the species. Only isolated populations are able to maintain themselves in this region. Thus the precise mechanism whereby temperature can make its effect felt may be complex and may be associated with seed germination and seedling establishment.

In desert regions, high daytime temperatures, coupled with low water availability, act as limiting factors for many plant species. Under these circumstances most of the plant species which survive do so by restricting active photosynthesis and growth to those occasions when temperatures are lower and water is available. There are, however, some species which are able to grow and, indeed, which grow best when temperatures and light intensities are at their greatest. Such an ability obviously places these plants at a considerable advantage when in competition with less well adapted species. One such plant, *Tidestromia oblongifolia*, is a native of Death Valley, California,

and it has been found to photosynthesize most efficiently in the middle of the summer day, when its leaf temperature may be between 46°C and 50°C (115°F to 122°F). In fact, reduction of the leaf temperature to 44°C (111°F) was found to impede photosynthesis, and transplantation to less extreme sites resulted in its failure and death. Here, then, is a species in which adaptation to high temperature has resulted in such a high degree of physiological specialization that its distribution is necessarily restricted.

Under high temperature conditions it is often difficult to separate the direct effects of heat from those associated with water scarcity. Many of the adaptations of desert species, both morphological and physiological, are more closely related to this stress. As a result, any consideration of the factors limiting plant distribution in tropical and subtropical latitudes must take into account precipitation as well as temperature. Many temperate species appear to be limited by low temperature in high latitudes and by water stress, often exacerbated by high temperatures in lower latitudes. It is probably fair to say, therefore, that temperature, both in its direct and indirect effects, is the most critical of climatic factors in the determination of the geographical limits of plant distribution.

P.D.M.

Light

Light from the sun is the ultimate origin of all biological energy. Plants and the photosynthetic bacteria can use light energy directly to obtain and store chemical energy for themselves and are ultimately the source of energy for all nonphotosynthetic organisms (animals, fungi, etc). In recognizing the prime importance amongst biological processes of photosynthesis, we should not forget that plants do not use light solely to obtain energy. The detection of light provides an important mechanism by which plants can acquire essential information about the environment in which they are growing; this is important for organisms which, lacking the power of locomotion, must adapt to prevailing conditions by using the sun's energy to the best effect in order to survive.

What is light?

Light comprises a small portion of the spectrum of electromagnetic radiation (energy which can be transmitted through a vacuum), which is most readily described in terms of its wavelength. Such radiation extends from gamma rays (wavelengths as short as 10^{-4}nm) to radio waves (wavelengths up to 10km). According to the quantum theory, electromagnetic radiation is composed of discrete packets of energy called *quanta*, the energy present in each quantum depending on the wavelength of the radiation. What we know as light encompasses that region of electromagnetic radiation where the quanta have just the right amount of energy to interact specifi-

Plants of the herb layer of temperate woodlands are heavily shaded in the summer by the leaf canopy. Many of these plants, such as the carpet of dog's mercury (*Mercurialis perennis*) shown here, avoid this competition for light by growing rapidly and flowering in the spring months before the leaf canopy forms.

cally with biological molecules; in practice this is the range between wavelengths of about 300nm and 1,100nm. The quanta of radiation of wavelengths longer than 1,100nm contain too little energy to promote biological reactions: those of wavelengths lower than 300nm contain too much energy, so that they cause indiscriminate ionization of molecules (which is why ultraviolet radiation is so dangerous). The infrared region (800–1,100nm) is utilized only by photosynthetic bacteria, being harmful to plants and animals because it is strongly absorbed by water and causes excessive heating. Thus, the useful spectrum for plants extends from the near ultraviolet (about 300nm) to the "far-red" (700–800nm), a range slightly greater than that of visible radiation (380–700nm).

Natural Variation in the Radiation Environment

There are four aspects of the natural radiation environment which plants might detect through their photoreceptors and exploit. These are:

1. The irradiance (often wrongly called intensity), which is the number of quanta reaching them per unit time and area.
2. The spectral quality of the radiation, at various times of the day.
3. The duration of light in a day (and, possibly no less important, the duration of darkness).
4. The direction from which the light comes.

All these are profoundly influenced by the geographical location of the plants, as well as by their local environments. The irradiance of full sunlight in temperate regions is about

1,000 μmol per square metre per second (1 mol = 6 × 10²³ quanta), somewhat more in the tropics, but substantially less in, for example, woodland (typically as low as 10 μmol per square metre per second) or under water. Passage through the Earth's at-

Many plants of temperate zones undergo developmental cycles which are synchronized with the yearly cycle of seasons. These cycles represent physiological adaptations which enable the plant to withstand the extreme conditions of winter. Many different aspects of development are related to seasons but the most spectacular and most intensely investigated are flowering and dormancy. The diagram below shows the development cycle of four species particularly with reference to flowering and dormancy. Length of light period (photoperiodism) is an important factor governing flower initiation. Long-day plants, such as *Epilobium hirsutum* and winter wheat, require long days for flower initiation, while short-day plants such as *Salvia splendens* and chrysanthemums require the shortening days of late summer. Another factor is overwinter chilling (vernalization), winter wheat and chrysanthemums requiring at least one period of low temperature (−2°C to +7°C)(28°F to 45°F) for flower initiation to occur.

mosphere both markedly reduces the irradiance of sunlight, and substantially alters the spectral distribution. Ultraviolet and visible radiation is absorbed by ozone, red and infrared radiation by water vapor. The absorbance of ultraviolet light is particularly important since, for example, an exposure for 1 second to unmodified sunlight would be lethal for humans and, indeed, most plants.

The distance within the atmosphere through which light travels before it reaches the Earth's surface obviously depends on the solar angle. As the sun approaches the horizon, the irradiance decreases, and there are also changes in the spectral composition of the light: the relative proportion of red is smaller and blue and far-red greater as the atmosphere to be traversed increases. This means that the spectral quality is altered at higher northern and southern latitudes, and that it alters twice daily—as the sun rises and sets.

Greater changes in spectral quality result from the interception of radiation by vegetation. The green leaf acts as a strong visible-light-absorbing filter but transmits

most of the light above 700nm; there is also some transmission in the green, which is why leaves look green to us. A single leaf typically absorbs more than 95% of the blue and red light reaching its surface. Thus, plants living in the shade of other vegetation receive only a small fraction of the radiation of those in direct sunlight and they are also presented with light of strikingly different spectral characteristics. The same goes for plants in aquatic habitats. While light may penetrate to considerable depths in clear waters, it becomes progressively depleted at wavelengths longer than about 650nm. Additionally, in nutrient-rich estuaries and lakes the irradiance falls very rapidly (by up to 80% at 0.5m depth) due to absorption by marine vegetation, and it again becomes depleted in the red and blue regions.

The duration of day and night is potentially another source of valuable information for plants. The marked seasonal changes in day-length which occur in temperate and polar latitudes confer on those plants with the ability to detect and respond to day-length the opportunity of avoiding or adapting to

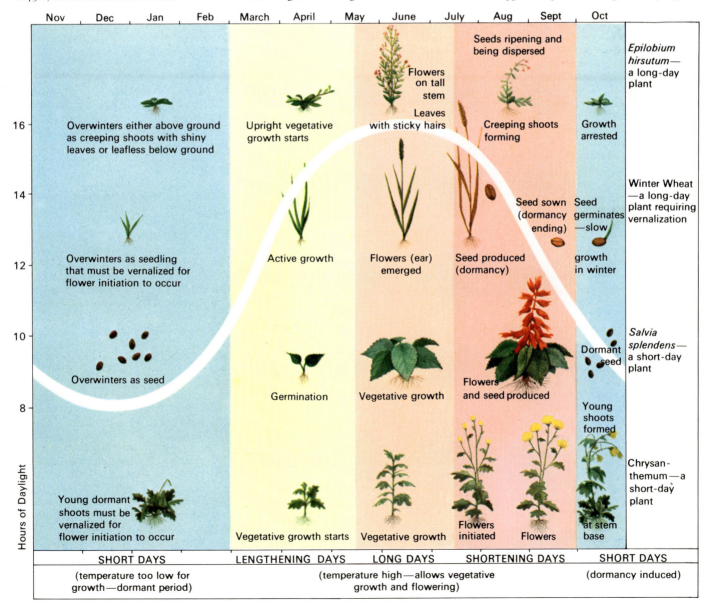

impending unfavorable conditions (eg extreme heat or cold) and exploiting an approaching favorable season. The plants which respond to such variation in day-length are said to exhibit *photoperiodism*.

Finally, the direction from which light comes is important. Plants often exhibit directional responses to light, collectively called *phototropism*.

Photoreceptors
In order that the radiation environment can be exploited, the light must first be absorbed by specific photoreceptive molecules. The energy of light, once absorbed, is used to drive photochemical reactions within the plant. There is a range of plant photoreceptors which between them are capable of utilizing most of the radiation that is potentially active biologically.

The light-absorbing pigments involved in photosynthesis are the chlorophylls, carotenoids and phycobilins. None of these pigments, except for a tiny proportion of the chlorophyll molecules, is involved directly in the light reactions of photosynthesis but trap the energy of sunlight and pass it on to the photosynthetic "reaction centers". In higher plants and green algae the light-trapping pigments are carotenoids and chlorophylls, which between them absorb blue and red light but reflect or transmit green light. Blue-green and red algae also contain various phycobilins, pigments which absorb green-orange light (500–600nm) well, enabling them to exploit the light penetrating to deeper waters in which the red and blue light is filtered out. Brown algae and diatoms use a different pigment, fucoxanthin (a carotenoid), to utilize the green-yellow radiation in the waters they inhabit.

The pigments involved in the developmental responses of higher plants to light absorb primarily in the blue and red (extending into the far-red) parts of the spectrum. The best known of these pigments is phytochrome, a blue-green chromoprotein with a chromophore (that part which actually absorbs the light) composed of the linear tetrapyrrole, porphyrin. The principal criterion for recognizing phytochrome involvement in a response is the promotion of the response by a low dose of red light (a few seconds is often enough), the effect being reversed by a subsequent pulse of far-red light. The ultimate response depends only on the last light treatment: in one experiment several hundred consecutive reversals were given with the response still dependent only on the final irradiation. Many aspects of plant development are controlled in this way. The explanation for this phenomenon is that there are two forms of phytochrome, one of which, designated Pr, absorbs red light strongly, whereupon it is converted to the other, Pfr, which absorbs far-red light more strongly, leading to conversion back to Pr. The Pfr form is thought to be biologically active, Pr inactive—hence the antagonistic effects of red and far-red light. Other pigment systems exhibiting similar properties are known in algae and fungi. All of these photoreceptors are probably involved in the detection of changes in light quality, which affects the proportion of the photoreceptor existing in each of the two photoconvertible forms.

It is known that phytochrome also absorbs blue light, and plants respond markedly to blue light, at least partly through phytochrome. They also contain another blue light-absorbing receptor. Indeed, blue-light responses are universal amongst living organisms from fungi to fruit-flies and exhibit remarkably similar characteristics throughout, suggesting a ubiquitous photoreceptor, the nature of which has proved surprisingly elusive. The best evidence now available suggests that it is a flavin (vitamin B_2 or a derivative) associated with a protein, or else a number of different flavoproteins. The most widespread responses to blue light are *phototropic*, where the plant grows towards (or away from) the light, and its photoreceptors are therefore responding to differences in the quantity of light reaching different parts of the plant, eg two sides of a stem. Whilst it is an over-generalization to say that the function of phytochrome is to detect light quality and that of the blue-light receptor to perceive quantity, it is clear that all plants contain, in addition to their photosynthetic pigments, photoreceptors which enable them to detect and respond to changes in light quality and quantity, and hence to measure the duration of daylight and the direction from which it comes.

Developmental Responses to Light
The radiation environment affects every stage of the plant's development, beginning with germination itself. For many seeds, especially those of plants which grow in waste places (ruderal species), germination is under phytochrome control, and they will not germinate, even at the soil surface, if they are shaded by the leaves of other plants. For example, seeds of the common plantain (*Plantago major*) sown under a canopy of plants growing at different densities germinate at a rate which is a function of the extent of shade cover. This cover is detected by the decrease in the red/far-red ratio produced by the vegetation, which in turn leads to an alteration of the proportion of phytochrome present as the active Pfr form.

Of course, many seeds germinate below the soil surface. If a seedling is growing beneath the soil surface (or in the dark), leaf development is inhibited and the stem extension rate enhanced until the seedling protrudes through the surface. Then normal leaf development, inhibition of stem extension and greening up of the tissues occur.

Summary diagram of photosynthesis in a cell. Energy, in the form of potential chemical energy bound in adenosine triphosphate (ATP) and reduced nicotinamide adenine dinucleotide (NADPH₂), is produced by the splitting of water in the presence of light, oxygen being a by-product. Carbon from carbon dioxide in the atmosphere is fixed (without the need for light), with ribulose diphosphate (RuDP) to form phosphoglyceric acid (PGA). PGA, ATP and NADP then enter the Calvin-Benson cycle (another process not requiring light), where fats, proteins, carbohydrates and other compounds required for "living" processes of the cell are produced.

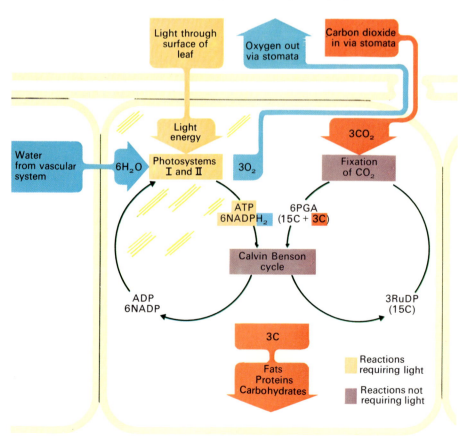

The long-day plant *Petunia* 'Orange Bells' flowers in response to a night-break or to long days. The sequence shows (**right to left**) increasing durations of red light (from $\frac{1}{2}$ hour to 16 hours) given in the middle of a 16-hour dark period. Although there are some differences in growth habit, all plants flowered about the same time; only the short-day control (**far right**), which had no light during the 16-hour dark period, remained vegetative.

This process of de-etiolation is again controlled by phytochrome. For example, a mustard seedling germinated in far-red light (which is inactive photosynthetically) develops much like one grown in daylight except that it is not green. However, some other plants, eg lettuce also require blue light in order for de-etiolation to occur.

Detecting light quality is of importance not only for seedlings. Throughout plant development, enzyme levels, chlorophyll content, leaf area and the dry weights of leaves and stems, for example, have been shown to be a function of the phytochrome photoequilibrium as affected by the spectral quality of the light. A striking example is provided by the arable weed *Chenopodium album*, in which the rate of stem-elongation increases in shade. Such a response can readily be explained ecologically as an adaptation for avoiding shade literally by growing out of it. Such weeds and other shade-avoiders generally show more marked responses to shade than, for example, woodland species which have adapted to low growth rates and maximal use of the restricted radiation they can obtain. A rather similar response to shade, controlled by phytochrome, is one which brings about a change in the ratio of the two forms of chlorophyll found in higher plants. Of the two forms, chlorophyll b absorbs light of shorter wavelengths than chlorophyll a. Hence, by increasing the proportion of chlorophyll b, a plant growing in the shade of another can make better use of the light not absorbed by the leaves above. Blue-green and red algae have comparable systems.

The development of many plants is markedly affected by the duration of the day or night. Some plants, termed "long-day plants" (eg *Fuchsia*), will only flower once the day-length exceeds a critical period; others (short-day plants), eg *Xanthium*, require days shorter than a critical length. Yet others are insensitive to day-length. In *Xanthium* the day-length perceiving mechanism is

extremely sensitive, and a single day shorter than $15\frac{1}{2}$ hours provides the signal for flowering, which then continues for up to 12 months even in continuous light. It has been shown that the length of the night controls this response. A short pulse of white (or red) light given in the middle of an otherwise long enough night can prevent flowering. The effect of such a pulse is far-red reversible, showing that phytochrome is the photoreceptor. Photoperiodic phenomena are of widespread occurrence in plants, especially those of regions where variation in day-length is significant.

It is clear from what has been said above that the quality, quantity and duration of light are of prime importance in the life of plants. Their abilities to exploit the different light conditions of the world are largely dependent upon their genetic constitution (genotype). The light regimes encountered by plants can therefore be crucial in determining what species are able to survive successfully at low or high latitudes, in open or closed vegetation, or in shallow or deep water, all features of fundamental concern in the ecology and geography of plants.

C.B.J.

Water

No living organisms that we know could exist without water. Life began in water (see p. 84) and has evolved in such ways that it has remained completely dependent upon it. Water is an essential constituent of all living tissues and may account for over 90% of the tissue's weight. Water is the medium in which many substances are dissolved and in which they react chemically, as well as taking part itself in many protoplasmic chemical reactions. It also provides the hydrogen for the reduction of carbon dioxide in photosynthesis.

Plants possess some structures providing rigidity to the tissues, but in nonwoody plants the bulk of the rigidity is the consequence of turgidity, maintained by the water in the cell vacuoles. Water forms a surface film on, and through, the pores in cell walls; dissolved substances pass in and out in it. The water in the vascular tissue of plants also transports dissolved substances.

Because the evolution of life took place in water, the transition, when it took place, from aquatic to terrestrial existence posed many problems. Terrestrial organisms require some way of obtaining water, and having done so, of reducing its loss. In green plants the problem is compounded by the fact that to carry out photosynthesis plants have to take in carbon dioxide from the atmosphere. This gas is taken in through minute pores—the stomata—and in the course of the exchange between the atmospheres within and outside the plant, water vapor is lost. For all terrestrial plants, the result of this is essentially a compromise between obtaining enough carbon in the form of carbon dioxide to maintain an adequate assimilation rate, and not losing so much water that the plant suffers damage.

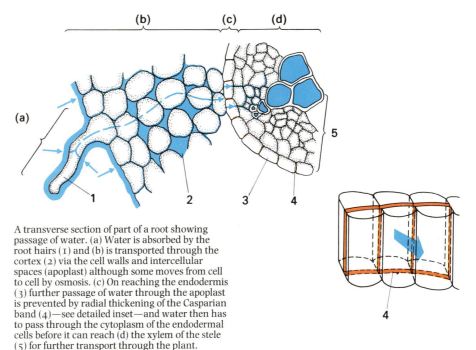

A transverse section of part of a root showing passage of water. (a) Water is absorbed by the root hairs (1) and (b) is transported through the cortex (2) via the cell walls and intercellular spaces (apoplast) although some moves from cell to cell by osmosis. (c) On reaching the endodermis (3) further passage of water through the apoplast is prevented by radial thickening of the Casparian band (4)—see detailed inset—and water then has to pass through the cytoplasm of the endodermal cells before it can reach (d) the xylem of the stele (5) for further transport through the plant.

In climates where rainfall and atmospheric humidity are high throughout the year, plants rooted in the soil can obtain sufficient water through their roots. Indeed, there is often a superabundance of water in the soil; waterlogged soils pose special difficulties, since they may become anaerobic (see Peat, p. 138). Plants in such soils have developed special structural and physiological mechanisms which enable them to deal with the oxygen deficit which may occur.

In areas where rainfall is less abundant, or is unevenly distributed throughout the year, or is unavailable for part of the year because it is frozen, then the balance between carbon gain and water loss becomes increasingly difficult to maintain. This has important

The dependence of plants on water is starkly illustrated by this photograph of dead vegetation in drought-stricken Niger in central Africa.

effects. Firstly, the amount of vegetation a given area can support is less. Secondly, the growth of the vegetation may become seasonal, being confined to the parts of the year when water is available. Thirdly, there is a reduction in the different types of plants that can exist. Fourthly, groups of plants which have special means of preventing water loss, or of more efficiently using the limited water they have, become more prominent.

There are two main types of very dry climate. The best-known, occupying the largest area, are the hot deserts and semi-deserts (see Deserts, p. 168) north and south of the equatorial zone. These areas are characterized by low rainfall and by unpredictability in its amount, duration and annual distribution. The Arctic (p. 214) and Antarctic (p. 218), despite their vast snow-fields and ice-sheets, in fact have very low rainfall, and what does fall is immobilized for much of the year as ice. This stress, added to

Above Light micrograph of a transverse section of the stem of a water lily (*Nymphaea*) showing the large air spaces. Such stems are submerged in water and the air passages aid buoyancy and assist in the movement of gases, particularly oxygen, around the plant.

low temperature, makes these areas highly unproductive.

Even within nondesert vegetation, local topography can have marked effects on the plant-cover, often through the influence of water availability—well-drained or poorly drained soils, for instance. There may be a combination of temperature- and water-effects, as in the differences between north- and south-facing slopes in temperate regions (see Exposure, p. 128). These can be quite striking, especially in areas of low rainfall. Even in Britain, water-dependent bryophytes are much more abundant on the north side of hills compared with the south. There are many factors which can locally determine large differences in water availability.

Drought injury can be mechanical or metabolic. Mechanical injury normally happens quickly, and often plants are killed before metabolic injury can occur. When cells dry out, the cell vacuole shrinks, the protoplasm lining the cells is thus subject to inward tension, and at the same time the cell-wall, which is tough, remains rigid. This may cause tearing of the protoplasm and the cells' death. In some plants, cell walls are soft and can collapse, thus minimizing such mechanical damage. Metabolic injury can include such problems as reduction in photosynthesis, respiration increase, various chemical changes in the cells, abnormalities of protein synthesis, loss of valuable metabolites from the roots to the soil, and so on.

Plants respond to drought stress in many ways. Even in moist climates, many plants have a limited ability to respond to drought. Many plants of arid regions evade drought by growing only during the brief periods follow-

Above A creosote bush (*Larrea divaricata*) growing in the gravel desert of the Upper Coachella Valley, southern California, United States. Such plants have become especially adapted to hot, dry conditions, and are known as xerophytes.

Below A major threat to plants undergoing water shortage is plasmolysis, during which the cell-contents shrink and break up. This diagram shows plasmolysis in epidermal cells containing colored cell sap. (a) The normal, fully expanded (turgid) state. When such cells are placed in a strong sugar solution, water passes out of the cells by osmosis, a physiological drought comparable to that produced in arid environments, the cell-contents contract away from the cell-wall (b) and become fragmented (c).

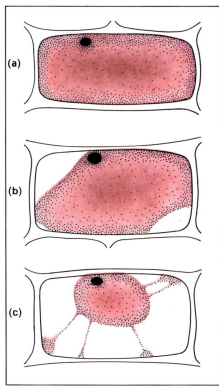

ing adequate rain. They are mainly ephemerals, eg many species of North American *Oenothera* (evening primroses), whose seeds may survive many years in the soil. These seeds will only germinate in the appropriate conditions of temperature and rainfall; the germination response is very precise, and the amount of rain sufficient to trigger germination is normally enough for them to complete their life cycle.

Some plants in arid areas rely on deep sources of ground water, and apart from their ability to draw water from great depths they have not necessarily any other xeromorphic characteristics. Much more characteristic are evergreen shrubs—such as the creosote bush (*Larrea divaricata*)—which usually have small leaves that may be deciduous in extreme drought, or they may have no leaves at all. The leaves may have thick cuticles, waxy surfaces, abundant light-reflecting hairs, and stomata sunk into

the surface, all of which may reduce water loss. Often their stems contain chlorophyll, so they can go on assimilating even if the leaves have fallen (eg the ocotillo, *Fouquieria splendens*). Many of them are extremely tolerant of long periods of drought, and can hold water very strongly in their tissues.

The other major groups of water-conserving plants are succulents, eg species of Crassulaceae, Cactaceae and Euphorbiaceae. These take up quantities of water after rain, and store it in their tissues. They utilize a special photosynthetic mechanism: dark carbon fixation (crassulacean acid metabolism), which involves reusing respiratory carbon. This means that they need open their stomata only at night, if at all, so water loss is minimized. Their carbon gain is slow, so they are slow growing, but they can survive in very dry conditions.

Another photosynthetic mechanism, the C_4 pathway, involves carbon dioxide being converted to malate in the mesophyll and its transport to special cells around the vascular bundle, where it is converted to carbohydrate. The process is similar to dark carbon fixation, but in the C_4 plants it takes place in different cells, whereas in the succulents it occurs in the same cells at different times of day, the malate being converted at night. C_4 plants are most efficient in very hot places with very bright light. Many of the drought evaders use this photosynthetic pathway.

Even in moist temperate regions, many

Below A fundamental property of all green plants is their ability to photosynthesize, utilizing the sun's energy to produce carbohydrates from carbon dioxide and water. Normally this involves the stomata of the leaves being open during daylight but many plants are able to withstand high temperatures and low water-availability by carrying out photosynthesis with the stomata closed during the hours of daylight, when water-loss from the leaves is greatest. This diagram shows how the "crassulacean acid metabolism" of such plants, which are frequent in hot parts of the world, differs from the sequence of events in "normal" plants.

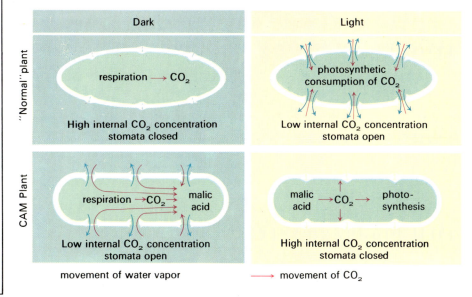

Eschscholzia minutiflora, a spring annual of open sandy or gravelly soils in California, United States, which is able to complete its life cycle during a relatively brief period following the winter rains that allow its seeds to germinate.

plants can be found which possess some of the water-conserving mechanisms found in arid areas; grasses with inrolled leaves and sunken stomata are common. Increased root growth and decreased shoots, and also smaller leaves, are all frequent in mesic plants exposed to drought. They enable these plants to survive brief periods of drought. Only the specially adapted plants can survive extremes. Though temperature may determine the main vegetation areas of the world, water is also crucial, and the two interact in ways which make it difficult to separate them or attribute particular distributions to either.

s.w.

Exposure

The type of vegetation found at any point across the world is largely determined by the various factors which contribute to what we call climate. A place is called exposed when one or more aspects of the prevailing climate are accentuated. The opposite of exposure is shelter. Both exposure and shelter are thus modifications of the regional climate and can be considered together. The main components of the climate that are affected by exposure are just those that are most important to the plant—temperature and moisture. These components are particularly changed by differences in wind speed and insolation.

The effect of exposure to a high wind speed may often be seen in trees on slopes or bluffs exposed to the prevailing wind. Shoots on the exposed side of the plant grow more poorly than those on the leeward side and in consequence the tree grows away from the wind. The parts of the tree protected from the wind may appear normal; from inside a wind-cut wood it may be difficult to tell whether or not the top is wind-pruned. The cutting of plants by wind may be seen on small undershrubs and herbs on mountain slopes, as well as on trees. The shoots of a mountain dwarf shrub, such as bearberry (*Arctostaphylos uva-ursi*), may be only a few centimetres high but they are often cut by the wind to give a smooth surface over uneven terrain.

The direct effect on plants of ordinary winds is to increase water loss and to bring the leaves nearer to air temperature. Damage to leaves may occur, either by rubbing together in strong winds or by abrasion from dust or sand particles carried in the wind. Near the sea coast, salt deposition may be damaging. Exposed leaves are almost always damaged more where the air is polluted (see Pollution, p. 253). Characteristically, the leaves of plants affected by wind feel more leathery, are smaller and have a more waxy sheen to the surface than do those leaves of the same species in more normal situations. Internally, the cells are usually smaller, and there are more vascular and strengthening tissues. All these characteristics enable the plant to control water loss more effectively and to hold their leaves at an angle for intercepting light efficiently, even under conditions of water stress.

On mountain tops, where the wind is particularly persistent, plants may be restricted to protected places, such as behind rocks. This protection is especially important if there is dust or sand in the wind, or when the drying effect of the wind is enhanced by the poor water retention of shallow or skeletal soils.

Wind speed is usually greatest on the brows of hills. However, the insolation received by the plants and the soil is affected by slope and aspect. This effect is particularly obvious in the middle latitudes where the pole-facing slopes receive least radiation and the equator-facing slopes most. On some limestone ridges in southern Europe the south-facing slopes are covered with a xerophyllous, small-leaved, open scrub, whereas the north-facing slopes are covered with forest. In a similar way, the southern limit of arctic plants is likely to be found on north-facing slopes, as it is for mountain sorrel (*Oxyria digyna*) in Britain, and warm south-facing slopes may provide the north-

ern outposts of southern species, as for bristle agrostis (*Agrostis curtisii*) in southwestern England.

The degree of exposure is affected by the duration of sunshine and the angle at which the sun's rays hit the ground (see Light, p. 122). Steep equator-facing slopes receive most, and north-facing slopes least radiation in the polar and temperate latitudes, whereas in the tropics all steep slopes receive less radiation than do horizontal surfaces. This effect is greatest on bright days, for on dull days the radiation received on all slopes is similar. For these reasons, the effect of slope and aspect is greatest in the sunny parts of the middle latitudes and least in the tropical and polar regions. A variation on this is seen in Jamaica where the trade winds blowing off the Atlantic Ocean are forced up by the mountains and cover the land over about 1,000m (3,300ft) in mist or fog which persists for most of the time. Sometimes, however, the mist disperses so that the full intensity of the sun reaches the vegetation and the heated leaves show the expected large increase in transpiration. Such a sudden change, if it persists, will kill plants not suitably adapted and all the plants here have thick xeromorphic leaves.

Differences in the receipt of solar radiation affect not only the photosynthesis of the plants but also—and this may be more important for plant distribution—the temperature and through this the soil-moisture content and the humidity of the air. The more exposed to the sun a slope is, the warmer it will be and the more prone it is to dry out. Drying out will be least on steep soils and where the bedding planes of the underlying rock bring water to the soil surface.

The difference in temperature between north- and south-facing slopes in the middle latitudes may be 5 or 6°C (9–10°F). This difference is reduced by an insulating cover of vegetation and increased by a dark soil. It has frequently been noted that in the middle latitudes seeds germinate earlier in the season on equator-facing slopes and, provided water shortage is not important, agricultural yields are also greater. Crops such as grapes and raspberries are therefore often planted only on south-facing slopes in northern Europe, to take advantage of this.

The effects of slope and aspect on exposure are modified by altitude. In turbulent wind conditions, the effect of altitude on temperature is of the order of 1°C per 200m of altitude. This, by itself, will have a profound effect both on individual species and on the vegetation. Thus, for example, in a year of plentiful hawthorn (*Crataegus monogyna*) blossom, a band of white may be seen ascending a British hillside during early May, and on a larger scale the change in vegetation one sees on climbing a mountain, passing through, maybe, forest then scrub and later grassy slopes to more or less bare rock, also results from exposure and altitude.

Allied to the influence of altitude are the other effects of topography. In the daytime, a sheltered valley may be hotter than a plain, as the still air in the valley is heated by the sun and by heat reflected by the valley sides. On a still, cloudy night the valley may remain warmer than the plain, but on a cloudless night it is likely to be cooler and indeed often much colder than the plain or the valley sides. At such times, radiation from the Earth cools soil and air. The cooled air flows, at a speed of about 1m (3.3ft)/sec, to the valley bottom, which then becomes cooler than the valley sides. If this "cold air drainage" is impeded by woods, hedges, or other structures then particularly cool patches may be formed.

Such "valley cooling" is greatest for depressions without outlets. In deep kettle holes in forested country in Europe the difference in climate may be such that no trees can grow in the depressions, and the vegetation approximates to that normally found much further north. Even a depression some 3–4m deep by 100m across may be sufficient to produce an effect. For example, a depression of this size on limestone pavement in northwest England, which is covered with a grass sward, lies among a dense stand of bracken fern (*Pteridium aquilinum*), which is

apparently attempting to colonize it. The fronds of bracken entering the upper slopes of the depression are killed by the occasional late frost, whereas those outside usually escape. Frost-sensitive species, such as grapes and citrus fruits, are cultivated most successfully on valley sides away from the frost pockets of the valley bottoms.

Exposure thus has an effect on the individual plant which may often affect the vegetation cover of a region. For example, on an isolated tree or shrub, the first part to bear leaves and flowers may be the sheltered or equator-facing side, while changes in the growth-rate of plants under exposure will obviously alter the competitive advantage of

different species and be a cause of differences in vegetation structure (see Mountain Vegetation, p. 183; Tundra, p. 173). R.M.W.

Soils

Soil is a mixture of mineral particles, finely divided organic matter, water, air and living organisms such as bacteria and earthworms. Soil should provide plants with water, inorganic nutrients (eg nitrogen, potassium and phosphate), a supply of oxygen for root respiration and an anchorage so that they are not easily blown over; soils vary in their ability to fulfill these requirements, which

The extreme climate of Tierra del Fuego provides many examples of the effects of exposure on vegetation and individual plants. **Right** With greater altitude the southern beech (*Nothofagus pumilio*) forest gives way to scrub, grassland and increasingly sparse vegetation as the exposed rocky summits are approached. The valley-bottom is too wet for trees and therefore supports a boggy grassland. **Top** The prevailing westerly winds have "cut" this *Nothofagus antarctica* tree on the shores of the Beagle Channel. **Above** In Tierra del Fuego, cushion plants, such as this *Benthamiella nordenskjoldi* (Solanaceae), flower first on their northern side, which receives most heat from the sun.

constitute the *edaphic* (soil-related) factors affecting plant growth and distribution. Plant roots often extend more than 1m (3.3ft) below the surface. A vertical section through the soil profile (see p. 136), reveals a succession of layers (horizons) creating far more diversity than mere inspection of topsoils suggests. In view of all these factors, it is not surprising that soils greatly affect both the distribution of species and the kind of natural vegetation.

Soil Formation (Pedogenesis)

The parent material of soils is derived by the fragmentation and decomposition of rock through weathering. Physical weathering results from a number of factors. Alternating heat and cold causes stresses in the rock because of differing expansion and contraction in its constituent minerals, which leads to fragmentation, giving irregularly sized, angular rock debris known as *breccia*. This is typical of deserts. Similarly, water freezing in fissures causes expansion and shattering; this is common in arctic and cold mountainous regions, the resultant debris often being termed *screes*. The action of glaciers, the sea, rivers and streams causes abrasion of the rock particles they carry and corrosion of the bed rock; erosion by sand-laden wind causes sand blast. Chemical weathering processes include hydrolysis or decomposition by water, which is accelerated under acid or alkaline conditions and high temperature; oxidation, where ferrous ion in ferromagnesium minerals, like

During the Pleistocene glaciation, wedges of ice formed in the soil leaving these characteristic patterns in unconsolidated gravel beds.

biotite, is oxidized to ferric oxide; and dissolution, for example when limestones dissolve in carbonated, acidic water.

Following chemical weathering, resistant minerals such as quartz remain. Cations such as calcium, magnesium, potassium and sodium, anions such as sulfate and chloride, and colloidally dispersible molecules such as silica may be removed by ground water. Clay minerals are synthesized from dissolved constituents, either at the seat of weathering, or elsewhere (eg in valleys). *Kaolin clays* are formed when the concentrations of silica, calcium and magnesium are low, ie when these are carried away by ground water and the parent rocks are dominated by alkali

metal ions, as in granite. *Montmorillonite clays* develop when the concentrations of silica, calcium and magnesium are high. They prevail where there is little movement of ground water, as in semi-arid conditions, valleys subject to dry seasons, and areas of basic rocks rich in ferromagnesium minerals. *Illites* are clay minerals similar to mica, and are features of areas of young soils; for example, they are found over much of the British Isles, where soils have developed since the last ice age. Iron and aluminum released on weathering may be incorporated in these clays or may occur as insoluble hydrous oxides.

To form soil, the parent material is acted on by external soil-forming factors and internal soil-forming processes. The external factors are climate (p. 117), biotic factors (p. 140), topography and age. Important internal processes are leaching and deposition of salts; colloidal dispersion of clay and humus in top soils, their leaching (elluviation) and deposition lower in the profile (illuviation); recycling of nutrient ions by vegetation; formation and mineralization of humus; and reduction and oxidation under fluctuating groundwater levels.

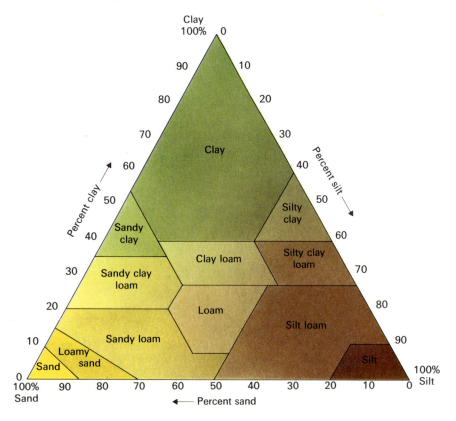

Left A soil-class triangle. Soils contain three broad fractions of particle (sand, clay, silt) and are named according to the proportion by weight of each type in a given sample. To use this triangle, only two of the fractions need be known, say, for example 30% clay and 30% silt. First a line is drawn parallel to the sand axis from the 30% point on the clay axis; the second line is drawn parallel to the clay axis from the 30% point on the silt axis. The name of the box in which the two lines intersect (clay loam in this case) is the class name of the soil in question.

Given adequate relief and time, the soils developed by these processes may be related to climate and vegetation.

Soil Composition

A typical humid, temperate, well-drained topsoil contains mineral particles of various sizes, intimately mixed with about 4% by weight of organic matter, mainly humus, derived from decomposition of plant and animal remains. These solids occupy about 60% of the soil volume. The pore space between them contains the soil atmosphere and the soil solution. In addition to the dead organic matter there is about 0.2% (dry weight) of living organic matter distributed in the top 15cm (6in) of soil. The principal living components are as follows: living roots 2,000kg/ha (1,800lb/acre), animals (mainly worms) 700kg/ha (625lb/acre), bacteria 500kg/ha (450lb/acre) and fungi 400kg/ha (350lb/acre).

Inorganic Fractions

The mineral particles are divided by size into coarse earth fractions (more than 2mm (0.08in) in diameter), which include gravel, stones and boulders, and fine earth fractions (less than 2mm), comprising sand, silt, and clay. The proportions among the fine earth size classes determine the texture of the soil.

Nearly all the particles larger than clay are residues from the weathering of rocks, the

Below A diagrammatic representation of the particles found in a "typical" soil aggregate. The structure of a particle from a particular soil type is determined by the aggregate size and shape. Sufficient clay is essential to form aggregates.

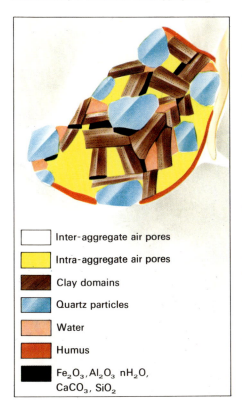

☐	Inter-aggregate air pores
◼ (yellow)	Intra-aggregate air pores
◼ (brown)	Clay domains
◼ (blue)	Quartz particles
◼ (pink)	Water
◼ (orange)	Humus
◼ (black)	$Fe_2O_3, Al_2O_3\ nH_2O,$ $CaCO_3, SiO_2$

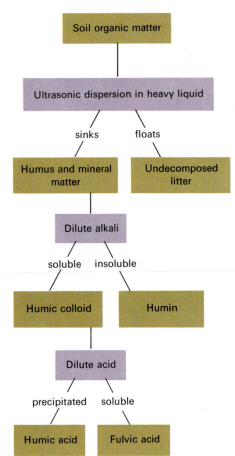

An important constituent of soil is its organic matter (humus) and this diagram summarizes the stages of its breakdown, an essential feature of a soil's fertility for plants.

sand and silt fractions usually being dominated by quartz with subsidiary felspars and micas. Some soils contain additional materials in the form of concretions of calcium carbonate, or iron, aluminum or manganese oxides. The clay fraction also contains some fine materials resistant to weathering, but most of it consists of newly formed crystalline and amorphous oxides and aluminosilicates. The clay is the most chemically reactive fraction; it is largely responsible for holding the supply of plant nutrient elements and for binding individual particles together to form structural aggregates.

Soil Organic Matter

Plant residues in the form of litter and roots consist of sugars, proteins, starch, fats and other cell contents that are rapidly decomposed by microorganisms, cellulose, pectin and other structural material that is more slowly decomposed and lignin which is fairly stable. Decomposition is aided by the soil fauna which fragment the litter, exposing large surface areas to attack. After the initial decomposition, a relatively stable mixture of resistant plant residues and microbial excretion products remains; this is humus. Humus is important for several reasons: it contains nearly all the soil's reserve of nitrogen and much of its phosphorus and sulfur, which are slowly released in inorganic forms when the humus mineralizes; it behaves as a polybasic acid and hence holds exchangeable cations;

it tends to stabilize soil structure; and it increases the available water, particularly in badly structured or sandy soils.

In well drained soils which are not very acid, most humus has passed through the gut of earthworms and is thoroughly mixed with mineral matter to form mild humus or *mull*. In very acid soils, as under coniferous or heath vegetation, only specialized organisms thrive, decomposition is slow and a raw humus called *mor* is formed on the surface of the mineral soil. It consists of an upper L layer of partially decomposed litter, an F (fermentation) layer and an H (humus) layer of black, greasy, amorphous humus. Under the anaerobic conditions of wet soils, decomposition of residues is only partial and peat is formed.

The Soil Pore Space

This is the home of plant roots. It holds the water they take into the transpiration stream and the oxygen needed for their respiration. The sizes of the pores and the connections between them determine the ease of drainage of the soil and its aeration—the diffusion of oxygen into the soil and of respired carbon dioxide out of it.

In most soils, the individual particles are bound together to form aggregates. The pore space may be roughly divided into large inter-aggregate pores and small intra-aggregate pores. Aggregates contain individual particles of silt and sand, ordered stacks of unit layers of clay (these are called domains) and humus. These elements are

This picture shows how the granitic rock at the base of the profile decomposes to form a soil consisting mainly of quartz, kaolin and iron oxides, with a shallow humus layer on the surface. The vegetation is savanna grassland.

Three sections of soil on a slope in the subtropics, showing the increase in depth of humus accumulation from upper slopes (**left**), through the mid slope (**center**) to the poorly drained valley bottom (**right**).

bonded by electrostatic forces. Aggregates containing 2:1 type clays, such as montmorillonite, swell when wet because water penetrates between domains (the water associated with the interlayer cations within domains does not change). Aggregates containing 1:1 type clays, such as kaolinite, swell little on wetting, partly because of hydrogen bonding between domains, and also because such aggregates usually contain free aluminum and iron oxides that form bridges between domains.

The structure of the soil is determined by the size and shape of its aggregates. These will only form if the soil has sufficient clay. Aggregation is promoted by local compression and drying of particles, which is achieved by roots, slow freezing, cultivation and earthworms. Roots, particularly the fine fibrous roots of grasses, densely permeate the soil, producing fine aggregates. Slow freezing starts at nuclei in the larger pores; ice crystals develop around these nuclei, drying the zones between them and compressing them (ice occupies a greater volume than the water forming it). Cultivation when the soil is moist causes local compression by shearing. Cultivating a wet soil causes smearing and puddling, while cultivating a dry one creates dust. Earthworms compress the soil passing through their gut.

The aggregates formed are stabilized by the physical binding of roots and fungal hyphae and by absorption on their surfaces of humus fractions produced by freshly decomposed residues. Much of the humus is ineffective. Aggregates are destroyed by the shattering action of rain drops and by cultivating or trampling the soil when it is too wet.

Soil Water

Water is held in the soil by surface tension. The suction force in a circular pore is proportional to the pore radius and to the surface tension between water and air. A wet soil drains from large pores until the retaining suction of the smaller pores balances the gravitational force exerted by the column of water in the wet profile below; the soil is then at "field capacity". The effective suction in British soils is found to be 0.5–1.0m (1.5–3ft) head of water, corresponding to a pore radius of 0.03–0.015mm. A freely draining wet soil takes from one to three days to reach field capacity. Thereafter, drainage is very slow and further water is lost from the soil by direct evaporation or by transpiration through the vegetation. If the ground is bare, the rate of evaporation depends largely on the rate at which water can diffuse from the moist subsurface to the surface.

Diffusion through dry soil occurs only slowly and when water from the top few centimetres has evaporated, the rate becomes very slow. Consequently, a bare surface will conserve soil moisture. With a cover of vegetation, the soil may be dried to considerable depth by transpiration. The driving force for transpiration is the evaporation of water from the leaves. The suction in the leaf tissues is transmitted via the xylem canals to the roots and from them to the soil.

If the soil is moist and roots are densely distributed, water moves to them easily and the amount of water lost from a given area of vegetation depends upon the energy it receives to evaporate the water. The energy is provided directly from the sun's radiation, but may also include energy carried by warm winds from adjoining areas.

If the soil is somewhat dry and the roots are sparse, movement of water to them may not keep pace with the loss by evaporation from the leaves. The stomata of the leaves then close. Evaporation may continue through the leaf-cuticle until a point is reached when the leaf cells lose turgor and, consequently, wilt. When the energy supply for evaporation is reduced, eg at night, the leaves can retain turgor. When, however, the soil water has been reduced to a suction of 100–500m (330–1,640ft) head of water, corresponding to a pore radius of 0.15–0.03μm, water movement becomes very slow indeed and plants wilt permanently. The amount of water held between field capacity and the permanent wilting point is the plant-available water. It amounts to about 2–2.5mm of water per 10mm depth in a loamy soil. A clay soil has many more fine pores and consequently holds more water than a sandy soil at a given suction, although it does not necessarily have more available water.

Depth of soil and depth of rooting are very important in determining the ability of vegetation to withstand drought.

Soil Air

To ensure adequate aeration it is important that the soil should drain freely. The ease with which a soil drains depends on the size and continuity of the large pores. A clay soil drains very slowly indeed if it is not aggregated, so the creation of a good structure providing large inter-aggregate pore spaces is most important for agriculture. Channels created by roots, by worms and other animals and by deep cultivation also assist drainage in heavy soils.

In a soil with aggregates larger than about 3mm (0.1in) the pore space between the aggregates may be adequately aerated, but within the aggregates the fine pores may be filled with water and the oxygen concentration too low for aerobic respiration of microorganisms. In these conditions nitrate may be reduced to gaseous nitrous oxide and nitrogen, and so lost; ethylene and hydrogen sulfide gases, which check root growth, may also be generated. In badly drained, water-logged soils such as gley soil (see Hydromorphic Soils, p. 137) these conditions are accentuated.

Pan
Pan loosely describes a soil horizon difficult for roots or implements to penetrate.

STRUCTURAL PANS are very compact horizons of low porosity. Important types are *planosols*, *periglacial pan* and *plow pan*. Planosols are gray soils with a close packed horizon of sand, silt and clay at a depth of 0.5–1.0m (1.5–3.3ft). They tend to be found on level terrain in regions where the subsoil dries regularly to the wilting point in summer. Periglacial pan is a compact subsoil layer often rich in aluminum oxide, developed in areas that have experienced alternating intense freezing and thawing. They are widespread in formerly glaciated regions of the Northern Hemisphere. Plow pans are formed by pressure of the plow sole under repeated plowing at a constant depth.

CHEMICAL PANS are not necessarily dense but are hardened by deposition of a cementing agent. Important examples are iron oxide pans, aluminum oxide or bauxite pans, calcium carbonate pans and silica pans. Iron oxide pans are formed particularly in tropical

A pan of iron oxide in Kano, northern Nigeria, exposed by excavation. The pan is the layer above the man's head, resting on a substratum of granite and with topsoil above it.

regions with alternating wet and dry seasons. In the wet period, ferrous ions in groundwater are absorbed on nuclei, often ferric or manganese oxides. In a subsequent dry period the ferric ions are oxidized to form ferric oxide which, if thoroughly dried, is deposited irreversibly to develop concretions which may eventually be linked to form a pan known as *laterite*. Aluminum oxide or bauxite pans are formed in the tropics by intense weathering of basic rocks, resulting in the removal of nearly all the other constituents. Calcium carbonate pans, or *calcrete*, develop in semi-arid regions from groundwater that is concentrated by evaporation. They are widespread in North Africa and are associated with calcareous rocks. Silica pans, or *silcrete*, form similarly in arid regions of internal drainage, notably in central Australia.

Pans at a shallow depth seriously impede plant growth, since roots cannot penetrate to lower layers and the plants are liable to suffer in droughts. Structural pans may be broken up temporarily by subsoiling, ie plowing to below the depth of the pan. Chemical pans are usually too hard for such remedies. Over very long periods of time they are undercut by erosion and so destroyed. Laterite pan under humid savanna will gradually dissolve if a forest cover can be established.

P.N.

Soil Acidity

Many plants require soils of a certain degree of acidity and some are completely intolerant of soil of certain kinds. To understand why this is so we need to understand something of the physics and chemistry of water.

The water molecule is composed of two hydrogen atoms and one oxygen atom, ie H_2O; however, a few of the water molecules split into hydrogen ions (H^+) and hydroxyl ions (OH^-), so water can be regarded as consisting of three types of particles: H^+, OH^- and H_2O. Pure water is neutral because the number of positive hydrogen ions equals the number of negative hydroxyl ions; however, when substances dissolve in water to form solutions, this equilibrium is altered. Acid conditions are created when circumstances cause the concentration of hydrogen ions to increase and the concentration of hydroxyl ions to decrease; conversely, alkaline conditions occur when the concentration of hydroxyl ions is greater than the concentration of hydrogen ions.

Acidity is conventionally expressed as soil reaction or pH, which can vary between 1 and 14. When the concentration of hydrogen ions exactly equals the concentration of hydroxyl ions conditions are neutral and the pH is 7. pH values in the range of 1–7 indicate acid conditions; values in the range of 7–14 indicate alkaline conditions. pH is related inversely to the logarithm of the concentration of hydrogen ions in a solution, so that a ten-fold increase in the concentration of hydrogen ions is required in order

A deep-plowed meadow soil. The organic matter content is high because the soil is rather wet, as shown by the bluish horizon in the subsoil (gley). If the soil were permanently wet, peat would form.

to lower the pH value by 1 unit. Thus soils with a pH of less than about 4 are uncommon, and those with a pH of less than 3 are rare.

Acidity in soils comes from three main sources: firstly, carbon dioxide in the atmosphere, which dissolves in water to give a weak acid (carbonic acid); secondly, some types of rock (eg sandstones) which, on weathering, give rise to soils of an acid nature; thirdly, carbon dioxide and organic acids produced by the metabolism of both soil and microorganisms and plant roots.

The tendency of soils to become acid is counteracted by the presence of compounds of alkali elements (also known as basic elements or bases) such as calcium, magnesium or potassium. Carbonic acid reacts with these elements to form predominantly bicarbonate salts. These salts dissociate, and the bicarbonate ions cause an increase in the quantity of hydroxyl ions, thus reducing the acidity.

In summary, acidity, which tends to increase due to the metabolic activity of living organisms, is counteracted by the presence of bases in the parent rock from which the soil is derived. Thus, rocks such as limestone, chalk, serpentine, and dolomite are rich in bases, and therefore usually give rise to soils which are neutral, or even slightly alkaline. At the other extreme, sandstone and some types of granite contain only small quantities of basic compounds, and form acid soils.

Another factor which increases the acidity of a soil is drainage. In wet climates, both temperate and tropical, water which falls onto the soil moves downwards into the groundwater system, and eventually drains into rivers; whilst the water is moving through the soil it dissolves the bases, which are thus removed. Soils which are depleted of bases in this manner are said to be leached. Because of the leaching process, soils in wet climates have a tendency to become acid, unless the concentration of bases in the parent rock is very high. In hot, dry climates moisture moves upwards through the soil because of evaporation at the surface, where dissolved bases are precipitated; in arid conditions, therefore, soils tend to be alkaline.

Calcicole (or chalk-loving) plants are those which are adapted to grow in base-rich soils, that is, soils with a pH greater than 7. *Calcifuge* (or chalk-hating) plants are those which cannot tolerate alkaline soil conditions; in fact they are usually found growing in soils with a pH lower than about 6.

The manner in which soil acidity affects vegetation is very complex, and by no means fully understood. However, there are some known effects. Acidity has a direct effect on the plant roots, all living tissues being sensitive to pH. It also has an indirect effect on plants by controlling the availability of chemicals in the soil. Certain elements such as copper, iron and zinc become readily available to plants in very acid soils, and although these elements are required by plants in small quantities, large amounts can be toxic. Another aspect of availability is that in very acid soils phosphorus becomes *less available*, so that plants growing on such soils frequently show symptoms of phosphorus deficiency even in the presence of considerable quantities of soil phosphorus.

Lastly, elements such as calcium, magnesium and potassium are important nutrients for plants, therefore acid soils are frequently infertile for the same reason that they are acid—the lack of basic elements.

J.WN.

The Soil and Plant Nutrition

Plant roots absorb nutrient ions from the soil solution at a rate related to their concentration. Most of the ions in the soil solution are in rapid dynamic equilibrium with much larger amounts adsorbed on the surface of the soil colloids—the clay and humus. The adsorbed ions buffer the soil solution concentration and serve to maintain the level of nutrients as these are taken up by the plant roots. Ions have to move to reach the absorbing surfaces of the roots, since only a very small proportion of the nutrients entering a plant comes into physical contact with root surfaces. Roots do not take up the nutrients which they disturb through growth, because the root cap is not an adsorbing organ. Ions move to the root or root hairs by diffusion and mass flow. They diffuse down the concentration gradient created when the root lowers their concentration by uptake at its surface. Ions are very much more mobile in the soil solution than when they are adsorbed on surfaces; hence they move very slowly through dry soil, which is one reason why uptake from dry soil is much reduced. Mass flow is caused by plant

transpiration, the ions being carried to the root in the water moving towards it. Ions whose concentration in solution is very much less than their concentration in the plant, eg phosphate and potassium, are largely supplied by diffusion. On the other hand, ions relatively more concentrated in solution, eg calcium, can be supplied by mass flow alone.

Seven elements, nitrogen, phosphorus, sulfur, potassium, calcium, magnesium and iron, are required in relatively large amounts and are usually referred to as *macronutrients*. The most important of these, nitrogen, phosphate and potassium, vary greatly in their condition and behavior in soils. Most nutrients are ultimately derived from rock weathering, but nitrogen is produced from the air by reduction of N_2 to organic NH_2 by such organisms as the nitrogen-fixing bac-

An important role of plants in the environment relates to the nitrogen cycle. Green plants contain large amounts of nitrogen compounds. These are released into the soil either directly by death and decay or indirectly after consumption by herbivores. Organic remains and products of excretion decay to ammonium nitrogen, which can either be reabsorbed by plant roots or be "nitrified" by bacteria to form nitrites (which cannot be taken up by plant roots), then nitrates (which can be taken up by plant roots). Some of the nitrates are lost from the soil either by leaching or by the action of bacteria to form atmospheric nitrogen. Some of the atmospheric nitrogen is either fixed by nodule-inhabiting bacteria and utilized directly by plants or fixed by free-living bacteria and blue-green algae, the breakdown products of which are released as ammonium nitrogen.

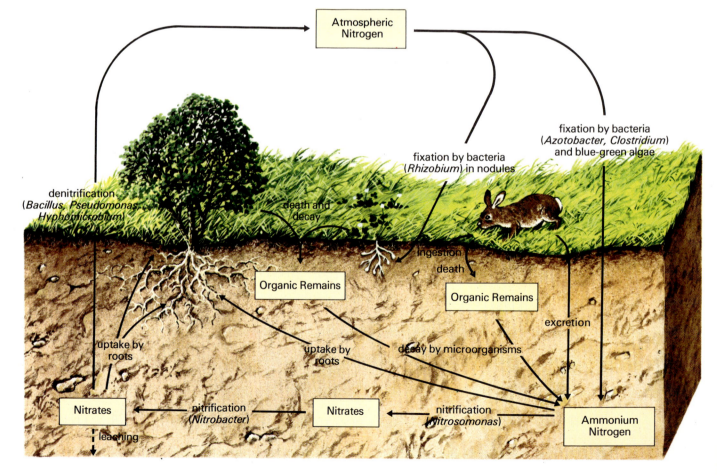

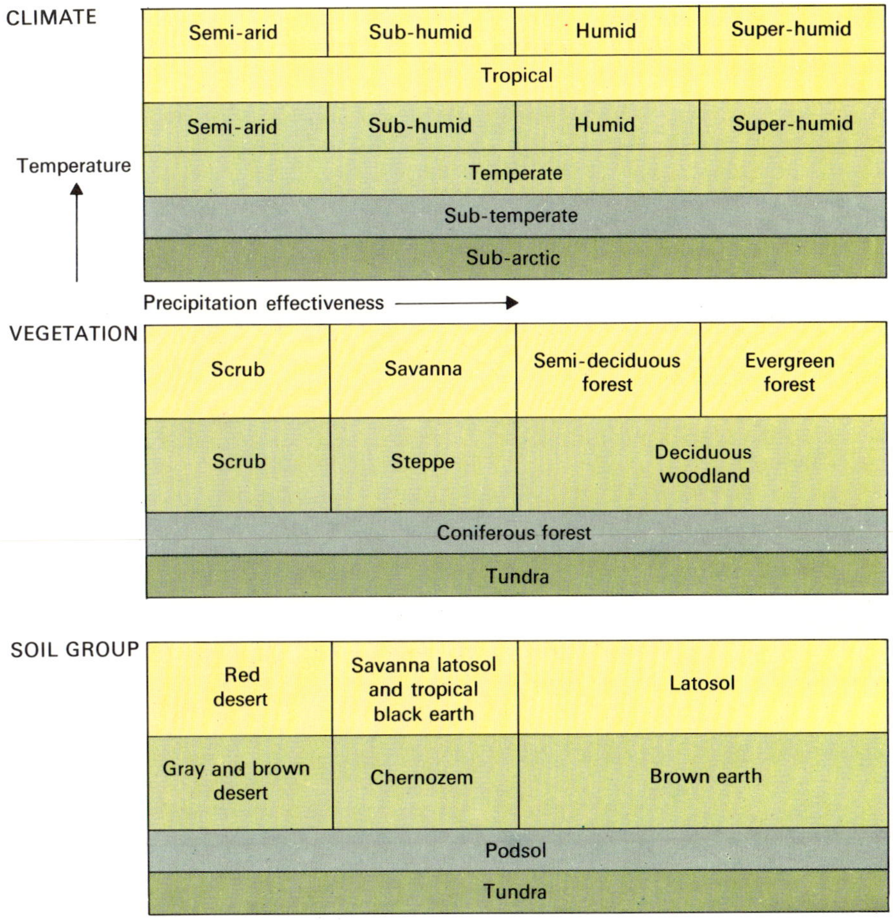

terium *Rhizobium* associated with the roots of legumes. Organic nitrogen is released to the soil when organisms die and it is incorporated in the soil humus. It is slowly mineralized by a variety of microorganisms to release the ammonium ion (NH_4^+) which is adsorbed as an exchangeable cation. In aerated soil that is not very acid, ammonium ion is converted by the bacterium *Nitrosomonas* to nitrite ion (NO_2^-) which is rapidly converted by another bacterium, *Nitrobacter*, to nitrate ion (NO_3^-). In these conditions plants take up nitrogen as nitrate ions rather than as ammonium ions because they move more easily to their roots. Under acid conditions the ammonium ion is the source of plant nitrogen. Nitrogen is returned to the atmosphere by reduction of nitrate to nitrogen and nitrous oxide gases by denitrifying bacteria.

Phosphorus is found in rocks as the mineral fluorapatite. On weathering, it occurs in soils as relatively insoluble hydroxyapatite and complex aluminum and iron hydroxyphosphates. In tropical soils it is often occluded in laterite concretions. Formation of stable soil phosphates is very slow and the concentration of phosphate in the soil solution is maintained by phosphate in amorphous complexes or adsorbed on the surfaces of calcium carbonate or iron and aluminum hydrous oxides. In the surface horizon of mineral soils up to half the phosphate may be in organic forms synthesized by plants and soil organisms. This phosphate is gradually released by mineralization and is an important source of phosphate in some humid tropical soils.

Potassium occurs in many rock minerals such as orthoclase, mica and hornblende. On weathering it is adsorbed by the soil exchange complex and behaves as an exchangeable cation.

In addition to the macronutrients it has been found that plants require at least seven other elements. These are boron, chlorine, copper, manganese, molybdenum, sodium and zinc. Because they are required in relatively small amounts they are called *micronutrients* or trace elements. Micronutrients are just as essential as macronutrients because they have specific functions which are indispensable and if any one is in insufficient supply, symptoms of deficiency become visible and the plants eventually die. Some, if not all, of the micronutrients are required to maintain the activity of enzymes upon which growth depends. Copper, for example, is a component of several enzymes concerned in biological oxidation, while manganese helps to catalyze a number of reactions involving phosphate. The precise functions of some of the micronutrients are still in doubt. Chlorine appears to be essential for photosynthesis, but it is not clear why, and the role of boron in plant nutrition is still a mystery.

As techniques improve, it is likely that the list of micronutrients will increase. Some of the essential elements may not be required by all plants or under all conditions. Cobalt has

CLIMATE	Semi-arid	Sub-humid	Humid	Super-humid
	Tropical			
	Semi-arid	Sub-humid	Humid	Super-humid
	Temperate			
	Sub-temperate			
	Sub-arctic			

Temperature ↑

Precipitation effectiveness ——→

VEGETATION	Scrub	Savanna	Semi-deciduous forest	Evergreen forest
	Scrub	Steppe	Deciduous woodland	
	Coniferous forest			
	Tundra			

SOIL GROUP	Red desert	Savanna latosol and tropical black earth	Latosol
	Gray and brown desert	Chernozem	Brown earth
	Podsol		
	Tundra		

A table illustrating the relationship between climate, vegetation and soil group.

been found to be an essential requirement for legumes when they use gaseous nitrogen as a nitrogen source, but not when they are supplied with nitrate. Vanadium has been reported to be an essential element for certain algae, but not for other plants.

Soil Types

Soils may be classified in various ways, depending upon the use to be made of the classification. While, for agriculture, the important features would include soil texture, water-holding capacity and nutrient reserves, for other purposes the mechanical properties and formative processes of the soil, for example, are more relevant. One of the more generally accepted systems recognizes three main orders of soils: zonal, intrazonal and azonal. *Zonal soils* are mature soils occupying large areas with generally similar climates; *intrazonal soils* are local variants of the prevailing climate type resulting from, for example, topography and parent materials; *azonal soils* lack a clear profile, either because of immaturity or special features of the parent material. Of great importance in the classification of soils is the vertical profile when the soil is mature; a fully developed profile in a humid soil can show the following horizons:

L   Litter on the soil surface
A_0 Fibrous layer containing roots
A_1 Layer of amorphous humus
A_2 Main eluvial layer, from which humus, sesquioxides and exchangeable bases are removed by leaching.
B_1 Illuvial layer in which humus is precipitated.
B_2 Illuvial layer in which aluminum and iron sesquioxides accumulate, the iron giving it a distinct brown color.
C   Partly weathered rock or substrate.
D   Unaltered base rock or substrate.

Zonal Soils

The principal zonal soils, constituting the great soil groups of the world, are:

TUNDRA SOILS. Dominated by cold and moisture, these soils show a minimal development and all overlie a layer of permafrost at about 0.5m (1.5ft). There is little leaching and no clear horizons, though they have a surface layer of partly decomposed peat overlying a viscous, blue-gray, reduced, gley horizon.

PODSOLS. These are typically cold-temperate and develop under a coniferous or heath litter that decomposes slowly to form a surface mat of raw humus (mor) and liberates polyphenols that cause iron and aluminum oxides and clay to disperse. These are leached, to leave a gray eluvial horizon overlying dark humus and orange sesquioxide illuvial horizons.

BROWN EARTHS. These are typical of areas which support, or have supported, temperate deciduous forests, and show considerable variation over the world. Brown earths contain 2:1 clays and a weakly developed illuvial clay horizon. They range from acid to alkaline according to the parent material. The profile may be leached free from carbonates, which are found only in the C horizon. Free ferric sesquioxide gives the characteristic color to these soils.

CHERNOZEMS. This group of soils is developed under subhumid grassland on calcareous parent material, often loess. A typical profile is a deep, free-draining, dark brown to black humic silty montmorillonitic clay with stable granular structure and calcium carbonate concretions merging at 1–2m (3.3–6.5ft) into gray loess. B horizons are absent. The term chernozem comes from the Russian, meaning black soil. With decreasing rainfall, chernozems grade into so-called *chestnut soils*, which differ principally in the lower organic content of the A_1 horizon, in which calcium carbonate may also accumulate.

DESERT SOILS. Developed under arid conditions, desert soils support a scrub vegetation at most. They are generally sandy, with the coarsest grains in the upper layer, and the surface may be stony because of wind erosion. They are shallow, with nodules and concretions of calcium carbonate and gypsum marking the limit of leaching during rains. In Australia and parts of North America desert soils are red, but the most widespread types are grey (sierozem) or brown.

LATOSOLS (LATERITIC SOILS, TROPICAL RED EARTHS, FERALLITIC SOILS). These are deep and strongly weathered and hence contain little silt. Their clay is kaolin, gobbsite, haematite and goethite, and the iron oxides give the soil a reddish hue. They are through-leached, and, in the superhumid tropics, acid (pH 4–5) throughout the profile. In the humid tropics recycling of cations by the forest brings the pH of the top soil nearer neutrality, some 2:1 clays are formed, and an illuvial clay horizon may develop. Laterite concretions are common. *Savanna latosols*, which develop over acid rocks, are similar. Over basic rocks, *black earths* (grumusol, vertisol) form. They contain montmorillonitic clays with deposits of calcium carbonate at depth. They crack severely in the dry season, the top edges fall in and the subsequent expansion causes the whole solum to turn over gradually—hence the term "self-mulching" soils. Despite their color their content of organic matter is low.

Intrazonal Soils

These are more or less local soils which owe their special features to topographic factors or to the parent material, with climate and vegetation being of secondary importance.

1 This soil is known as a ground-water podsol. It is a very acid (pH about 4). Humus and iron oxides are washed down from the subsurface leaving a pallid layer. They are deposited at the level of the ground water table.

2 Red lady (ultisol) in Australia. Such brown soils are formed over basic rock, eg basalt, under free draining conditions, and contain montmorillonite and kaolinite clays. Blocky aggregates are shown.

3 In sub-humid temperate regions a characteristic dark brown crumbly soil known as chernozem is formed over unconsolidated calcareous deposits. Here the deposits are gravelly.

4 Gray desert soil formed under short grass of the Great Plains, United States, in semi-arid temperate conditions. The subsoil is full of calcium carbonate. The organic matter content of the topsoil is low (less than 2%).

5 Brown earth (ultisol) over basalt in a mid-slope position where drainage is fair (South Australia). This yellowish-brown soil is formed on unconsolidated clay. Such soils need to be artificially drained to make them productive.

6 Kalahari red desert soil, showing the yellowish-brown blown sand overlying a thick layer of white calcium carbonate. Deposits of calcium carbonate at a shallow depth are typical of desert soils.

7 Savanna latosol near Esmeralda in the Upper Orinoco region of southern Venezuela. The vegetation is savanna grass and the soil is developed on quartzite. The subsoil is impermeable and water lying on the surface causes organic matter to reduce the iron oxides to ferrous iron salts which give a blue tint to the soil.

8 A tropical black earth (vertisol) in South Australia. Such soils are formed over basic rock, eg basalt, under poor drainage conditions, and contain the clay mineral montmorillonite, which characteristically forms a cracking black earth with blocky aggregates. Despite their black color, these tropical soils are low in humus.

Diagrammatic representations of grassland (**left**) and forest soils (**right**), showing the layers and horizons. A_0 is the layer of litter, which is thin in grassland. A_1 is the humus-rich layer and A_2 the leached top soil. Note that leaching is much less in grassland than in forests, where there is a very rapid transition from the A_1 to A_2 layers. The B horizon represents the subsoil in which leached materials accumulate. In forest soils there are two layers in the B horizon, deep tree roots often tapping the humus-rich B_1. The C horizon is composed of degraded parent rock.

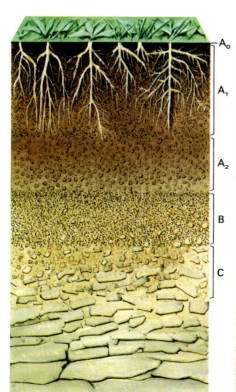

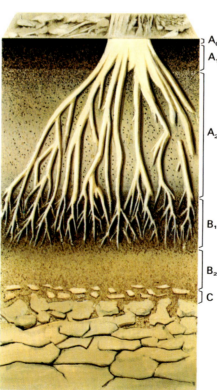

2   3   4

SALINE OR ALKALI (HALOMORPHIC) SOILS.
Where the relief is such that groundwater
coming into a depression is rich in sodium
salts, and the water table is near the surface,
saline or *solonchak* (white alkaline) soils form.
These occur in arid regions where evaporation concentrates the salts which form the
characteristic white crust on the soil, as
around the Caspian sea and the Great Salt
Lake in the United States. The soils are
neutral or weakly alkaline. If, as a result of
leaching, the sodium chloride content falls
sufficiently to permit hydrolysis of the clay-
colloids the released sodium can form sodium
carbonate, to give a strongly alkaline *solonetz*
(black alkaline) soil. This has a maximum
compression and the depressed surface often
supports black alkaline solutions of humus.

HYDROMORPHIC SOILS. In cool, humid regions
a high water table gives rise to *gley* (meadow)
soils. These show a permanently reduced
grayish or bluish "gley" horizon below an

5

6   7   8

orange, red and gray-brown mottled horizon formed under alternating oxidizing and reducing conditions. The superficial humus layer is of variable thickness. Where drainage is badly impeded the extent of organic accumulation reaches a point at which peat soils are formed.

PEATS. There are two broad classes of peat: bog peats and fen peats. *Bog peats* (sometimes called terrestrial peats) develop without direct association with flowing water. The water comes from precipitation, which does not contain a high concentration of mineral salts. Therefore, the acid products of microbial metabolic activity (eg humic and carbonic acids) are not neutralized, which causes the acidity to increase and metabolic activity to slow down.

Bog peats are composed primarily of the bog mosses (*Sphagnum* spp.), sedges (*Carex* spp.), and cotton grasses (*Eriophorum* spp.). Because of the slow microbial activity the plant remains are easily identifiable and the peat types are often named after the major plant group of which they are composed, eg sphagnum peat, sedge peat.

Fen peats develop under the influence of water flowing in from the catchment. The concentration of minerals in this water is usually far higher than in rain or snow, so that the acid products of microbial metabolism are neutralized, and fen peats do not become very acid, and may be slightly alkaline. Fen peats, which depend entirely upon waterlogging to slow bacterial activity, are always more highly decomposed than bog peats, because their less acid conditions

Peat exposed by erosion on a high part of the Brecon Beacons in South Wales, United Kingdom. The peat consists of partly decomposed plant remains that have accumulated in acidic and waterlogged ground over a long period.

The margin of a peat bank formed by *Chorisodontium aciphyllum* in the maritime Antarctic. Melt water has flowed out of the active layer of the bank over the permafrost surface to emerge and form icicles.

allow microbial decomposition to continue, and because the water flow to them fluctuates, thus periodically the peat surface is exposed to the atmosphere, which stimulates microbial activity. Fen peats are formed in the margins of lakes and rivers, so that reeds (eg *Cladium*, *Phragmites*, *Typha*) are amongst the more important species contributing to these types of peat.

In temperate situations, herbaceous species are the main peat-forming types. However, in the tropics, most peat is formed from woody species, with the herbs making

only a very small contribution to the peat mass.

Peat requires a certain minimum water supply for its development, and extensive peat deposits are consequently found only in the cool temperate and tropical regions of the world, principally in Canada, northern Europe, Russia, China and the Malay Archipelago. Smaller deposits occur in tropical Africa and tropical America, as well as in Peru, New Zealand, Tasmania and southern South America.

CALCINOMORPHIC SOILS. The calcareous nature of the parent material is the overriding factor in the properties of these soils. Over limestone or chalk the presence of free calcium carbonate leads to the development of *rendzinas*. They may be gray to whitish, but are usually darker, with up to 12% humus. The A horizon is not more than about 30cm (12in) thick and a B horizon is absent; the soils are usually slightly alkaline. Over hard limestones, especially in the Mediterranean region, *terra rossa* soils form. These are red, base-unsaturated, may be leached free of calcium carbonate and have a high content of clay; they may be alkaline or acid and are usually shallow.

Azonal Soils

Immature soils without a developed profile and governed by the parent material are said to be azonal. Coarsely grained rocky and gravelly soils on slopes where erosion is active are termed *lithosols*. They retain little water or organic matter. Over sand or loess a thin grey soil known as a *regosol* can form. It has a highly organic A horizon but no B horizon.

ALLUVIAL SOILS are often azonal since the constant influx of parent material, as in a river delta, prevents the formation of a true profile.

P.H.N.

Hard limestones often weather to form reddish-brown soils that form deep tongues in the subsoil. When the soil is bright red it is known as *terra rossa*.

Fire

Periodic or occasional destruction of above-ground plant parts by fire is a feature of many habitats. Communities in which the canopy or sward is continuous, particularly those containing high proportions of wood or fibrous material, may burn readily, especially if their water content is relatively low. Plant litter or organic material accumulated at the soil surface adds to the available fuel. Hence, among factors controlling the flora and fauna, influencing nutrient cycling, and affecting the physical and chemical properties of the substratum, fire is of fundamental importance in certain types of forest, scrub, savanna, grassland and heath, and often plays some part in other vegetation types.

Temperatures of over 1,000°C (1,832°F) may be reached in crown fires in forests, but in heath and grassland the range is normally from 200°C to 900°C (392°F to 1,652°F). The temperature reached depends on the amount and nature of the fuel, and on weather conditions, particularly wind. In a

Fire was once considered to be an "unnatural" environmental factor initiated by Man. However, it is now known to be a "natural" factor that is of major importance in forming a particular community. This dry-season bush-fire in Nigeria checks the development of scrub vegetation and many nutrients bound up in the plants are released into the environment. Communities such as this will contain fire-resistant species and fire is therefore essential if they are to retain their position in the community.

normal grass or heath fire, maximum temperatures are generally maintained for little more than one minute at any spot.

Vegetation fires may be started by natural causes, notably lightning. Forests and grasslands in areas subject to a high incidence of thunderstorms have been influenced by periodic lightning-induced fires throughout their history, as for example in the southeastern and central United States, eastern and southern Africa, India, Australia and Finland. However, the role of fire as an ecological factor has been greatly increased by Man. In addition to accidental wildfires, burning has been used for forest clearances, driving game and as a means of management to promote the regeneration of desired tree species or the regrowth of grazing plants including grasses and heather (*Calluna vulgaris*). While much effort is devoted to the prevention of destructive wildfire, "prescribed burning" (ie the use of controlled ground-fires) has a place in forest management (eg in the United States). In Africa, the Guinea-zone savanna is burned annually to stimulate the growth of forage grasses, while in western Europe, especially Britain, heathlands are burned at intervals of 10–12 years to maintain a dense uniform stand of heather and to encourage the production of edible shoots required by sheep or the game-bird, grouse (*Lagopus l. scoticus*).

Regular burning affects the composition of vegetation by eliminating many fire-sensitive species and permitting survival of those which are to some extent fire-adapted. Adaptations include the possession of thick, resistant bark in some tree species and the ability to regenerate rapidly after the passage of fire, either vegetatively from underground parts or by seed germination. Germination in some fire-adapted species may be stimulated by exposure of the seed to heat for short periods, while in certain conifers (eg *Pinus banksiana, P. serotina*) the cones are "serotinous", that is they require the heat of a fire to bring about separation of the cone scales and release of seed. Fire has undoubtedly contributed to the dominance of some forest trees, including the *Pinus* species mentioned above, *Pseudotsuga menziesii* and the *Eucalyptus* species of Australia.

Redistribution of the nutrient elements contained in the aboveground parts of the vegetation occurs on burning. The smoke contains a high proportion of those elements which form volatile compounds (carbon, nitrogen, sulfur). Fire temperatures of about 600°C (1,112°F) may release over 60% of the nitrogen into smoke, and at higher temperatures may release significant quantities of other nutrients. The remainder are contributed to the soil surface in the form of ash. Some, such as phosphorus, are largely

bound in insoluble complexes, but others such as potassium are readily leached from the ash by rain. Hence, the system may be depleted of certain nutrients as a result of fire, but this must be balanced against inputs during the periods between fires.

Fire may reduce the capacity of some forest and grassland soils for water infiltration, leading at times to increased runoff. However, the reverse seems to apply to heathland soils (except where a "skin" develops on most peat surfaces). Prior to regeneration of the vegetation, erosion losses may be considerable from the exposed surface of soil or organic material. This has been demonstrated in various types of forest, scrub, grassland and heath, but not generally in subtropical savannas.

C.H.G.

Biotic Factors and the Environment

As a result of their evolutionary history, plants have an inherited ability to live within certain limits set by the physical factors considered above. Where they actually occur, however, depends not only upon this

The tamarack larch (*Larix laricina*) is intolerant of shade. In its natural habitat, therefore (**left**), it is restricted to harsh, cool or cold climates where there is no competition from other shade trees, seen here in a typical tree island in the string-bog landscape of northern Canada. However, when planted for ornament or forestry (**right**), it will flourish in a wide range of soils and climates so long as it is kept away from competing shade-producing trees.

tolerance range of each species, but also upon its ability to reach particular areas and to interact with other living organisms once it arrives, which also depends on the species' genetic capability.

Tolerance and Plant Distribution

There are several hundred thousand species of flowering plants, and few, if any, have exactly the same geographical distribution. Furthermore, an examination of any specific locality where a variety of ecological conditions exists, such as a shore line, a mountainside, or a sand dune, will show that no two species have exactly the same distribution with regard to the habitat variables—pH, moisture, soil fertility, and so on. The differences in local distributions are due to the fact that each population of each species has a certain range of tolerance with respect to every environmental variable, which differs from that of other species. This also partly accounts for differences in geographical distribution.

It is interesting to note that living material has evolved to the extent that most environmental conditions encountered on the Earth's surface are tolerable to some form of life. For example, bacteria have been reported to survive and metabolize at temperatures as low as −20°C (−4°F) and as high as 104°C (219°F) and in pressures from zero to 1,400 atmospheres. Many organisms can survive in distilled water, and some fungi are capable of growing in a solution of 50% magnesium chloride. In fact, most of these and other records do not even represent the environmental limits of life, because in most instances the experiments required in order to gain this knowledge have not been carried out.

Extreme tolerance limits such as those

Colonizing plants have a high tolerance of extreme soil conditions. Here *Armeria* species are growing on zinc-rich mine spoil by the River Innerste, Harz, Germany.

These old spoil tips in the Harz Mountains, Germany, are rich in zinc and other heavy metals. After 50 years the dry tips themselves are still barren and on the nearby moist soil into which minerals have been washed the vegetation consists of a few metal-resistant flowering plants, lichens and blue-green algae.

mentioned above are only achieved by the less complex organisms. The limits which can be tolerated by higher plants are generally narrower, although many species of conifer are capable of *dormant* survival at temperatures as low as $-60°C$ ($-76°F$) in arctic and alpine situations. At the upper extreme of temperature, the stonecrops (*Sedum* spp.), as a group, are very tolerant, being able to survive in temperatures of up to $50°C$ ($122°F$), while the prickly pear (*Opuntia* spp.) can survive for short periods in temperatures of almost $70°C$ ($158°F$), provided that the atmosphere is dry.

Soil reaction (pH) is rarely too low for flowering plants of one sort or another, and where extreme conditions of alkalinity are encountered, such as in certain deserts, a vegetation cover is in any case precluded by the lack of moisture. In fact, just as almost every environment on the Earth's surface has been colonized by some form of life, however simple, so most of the land has been colonized by higher vegetation. Only about 6% of the land surface has no vegetation cover, mainly in the warm deserts, such as the Sahara, and very cold areas, such as the tops of high mountains and the polar regions. In the warm deserts there is simply a lack of water; in the cold regions the frequent low temperatures are directly hostile to higher plant growth because they cause a reduction of metabolic activity; they also act indirectly by freezing water, thereby making it unavailable to plants.

For any given plant species there is a range for each environmental factor within which it can survive; however, the range over which it can survive and prosper is considerably more limited. Many plant populations of restricted geographical distribution could survive in other areas, but fail to do so because of competition (see p. 146).

The colonizing plants in many successions are capable of tolerating the rather extreme environmental conditions which often pertain to the early part of succession, but these colonizers modify the environment to create conditions that are within the tolerance range of other species; these species, being better competitors, eliminate the colonizing plants, and so successional change in the vegetation occurs.

If we examine the local distribution of various species populations which contribute to vegetation, what we observe are the results of succession having graded the populations in terms of their tolerance ranges and the local environmental factors: the tolerance range of a population is, in turn, the result of evolution. The easiest way to examine the tolerance of a local plant population to various environmental factors is to consider a simple *environmental gradient* where the change in many factors is roughly parallel to some line—from the top to the bottom of a mountainside for instance, or from the upper to the lower part of a sea shore. When the distribution of forests on a mountainside or of seaweeds on a beach is examined, it is usually found that different species dominate the vegetation for different portions of the gradient. The proportion of the gradient where a particular species is found approximates to the tolerance range of that species; the true tolerance range is usually wider than is indicated by such observations because, as mentioned above, a population does not usually compete well near to the limits of its tolerance, and the full potential range could only be discovered by the elimination of competition.

The fact that different populations of plants have been observed to occupy different sections of gradients has been presented by some ecologists as evidence in support of the contention that species populations evolve in such a way as to avoid competition. This may be true, but competition operates on two time-scales: in the long term, where it is responsible for the evolution of populations, and in the short term, where it is a factor in succession.

The distributions that we observe today are certainly the results of competition and succession, but we cannot be sure to what extent these distributions represent evolution of population tolerances in such a way as to avoid competition, until we find the same population relationships to environmental factors in the absence of competition.

J.WN.

Indicator Plants

Indicator plants are those whose presence tells us something about the immediate environment. The types of plants which can grow in any place are determined by the complex of environmental factors operating there. Only those immigrant propagules which are able to tolerate that particular environment will survive. It follows that plants, especially those with narrow tolerance ranges, are good guides to the type of environment. Such indicator plants have often been used as a measure to predict environmental conditions, the predictions being based on what is known of their requirements elsewhere. The behavior of whole plant communities can be used in this way—indeed the fundamental basis of phytosociology (see p. 59) is that vegetation is a true reflection of its environment, and is its best indicator. Since plants with narrow ecological tolerances are obviously the most accurate indicators and are also among the more rare members of most communities, phytosociological classification tends to emphasize their importance.

Not all plants can be used as reliable indicators. Short-lived species, for example, occupy habitats sometimes for rather brief periods and their presence may not be so indicative of particular habitat conditions as that of long-lived plants. Another crucial point is that many species have a range of local races or ecotypes, in which local adaptation may mask habitat differences. In some instances, plants may be adapted to a very wide range of habitats; the occurrence of grasses like *Festuca rubra* or *Agrostis stolonifera* both on "normal" soils and on natural and artificial soils containing large amounts of usually toxic heavy metals such as lead or nickel, to which they have become adapted, is evidence of this. Such behavior is further support for the view that plants of narrow tolerance are the best indicators.

Groups of species or whole communities are better indicators than single species, since in a whole range of unrelated species all are unlikely to show similar local adaptations. Also groups of species can be used as a kind of scale to measure the intensity of a particular controlling factor. The use of a graded series of lichens radiating out from a city center has been especially instructive as a means of measuring atmospheric pollution.

Apart from their value to the ecologist in identifying environmental complexes, indicator plants have economic importance. They have been used as clues to the presence of underground water in arid areas and they also have considerable economic value as guides to the presence of exploitable deposits of metals. There is an impressive array of metals that can be detected by this means, as well as by noting toxicity symptoms in plants growing on or near the metal ores. Further, chemical analysis of plants can show whether the metals are present in exploitable quantities. There are drawbacks: some plants, for instance, are capable of absorbing enormous quantities of toxic ions and thus give a false impression of the amounts available in the soil, but for primary survey, plant indicators can be very reliable.

s.w.

Dispersal and Migration

Terrestrial plants, and many aquatic ones also, are essentially static organisms. On germination and establishment they develop

Lychnis alpina is one of the plants that serves as an indicator of metals present in the soil. This specimen, growing in western Greenland, signals the presence of copper deposits.

root systems or rhizoids which are essential for the supply of water and nutrients to the plant, and this precludes the possibility of the adult plant moving from one place to another. A few exceptions exist, but most plants remain in one place from germination until death. The term migration, when applied to plants, is therefore quite different in its connotations from the idea of migration in animals. Plant migration involves the transmission of propagules from an established adult, their germination and growth in a new site and, on attaining maturity, the production of new propagules which can continue the movement of the species. Plant migration is therefore a slow process, its rate depending upon the generation time and dispersal capacity of the species concerned. Many different methods of dispersal have evolved, some very simple, others very elaborate.

There are many terms, such as propagule, germinule or disseminule, which refer to the wide range of dispersal phases in plants. For our purposes, the most useful term is *diaspore*, since it refers to any dispersal phase rather than to specialized organs. Diaspores can be either vegetative (parts of the parent plant) or sexually produced spores, seeds or fruits.

By far the simplest method of dispersal in plants is the fragmentation and dissemination of existing parts or the formation of specialized vegetative organs. Rhizomes and stolons are commonplace for many vascular plant families, being the principal means of population expansion for herbaceous, short-lived perennials living in pioneering habitats with strong competition from other plant species. The breaking off or floating away of vegetative parts can frequently become the main mode of dispersal in some plant species. In sterile populations such as the Canadian pondweed (*Elodea canadensis*), stem and stolon fragmentation can adequately replace the seed habit.

Specialized, easily detachable dispersal organs are common, including the axillary, aerial bulbils of *Polygonum*, *Poa* and the common celandine (*Ranunculus ficaria*). In *Poa* the bulbils are wind-dispersed whereas in *Polygonum* they are mistakenly eaten as seeds by birds and in the celandine they are rain-dispersed. Some water plants, such as *Utricularia* and *Potamogeton*, develop turions, which are submerged fragmenting dormant buds that sink into the muddy bottoms of rivers and lakes for survival during the adverse conditions of winter.

For most of the plants that cover the Earth, sexually produced diaspores, ie the seeds and spores, are undoubtedly the normal organs of dispersal. However, there are many different structures for their method of actual dispersal, particularly for those which are combined into fruits. Since successive generations of all plants have to move from one location to another, dispersal mechanisms have evolved in response to those environmental factors which can transport diaspores. These structures fall into three major syndromes of animal, wind and water

dispersal. Additionally, there are those dispersal mechanisms for self discharge evolved by the plants themselves.

Animal Dispersal

Diaspores, particularly seeds, can either be carried on the outside of an animal, a phenomenon known as *epizoochory*, or pass right through their digestive systems in a process called *endozoochory*. Additionally, diaspores can sometimes be intentionally carried in the mouths or by the limbs of animals such as birds, mammals and ants, a process known as *synzoochory*.

Epizoochory is by far the most frequent method of dispersal in animals and the most unspecialized examples are those diaspores of waterside plants sticking to the feet of animals, particularly waterfowl. Epizoochorous transport becomes a very important process when the plants in question form part, or all, of the animals' main diet. Rodents, such as squirrels, rats and hamsters, and birds, such as jays and woodpeckers, generally destroy the fruits and seeds of oaks, pines and cereals on which they feed. They do become valuable dispersers, however, when they cache their food.

Epizoochorous systems abound and generally include adhesive projections such as spines, hooks and viscid exudates. They are often associated with fruits and seeds which detach easily from their typically low-growing mother plants. In some species of the bean family (Leguminosae), for example, viscid seed exudates are commonly secreted to aid dispersal.

Fruits with spiny appendages are very common. Trample burrs are usually found in plants of dry open plains such as steppes and prairies. Again, these are typical of plants growing close to the ground, the puncture vine (*Tribulus terrestris*) being an important example from the Mediterranean region. The spines on these fruits are so incredibly hard that they are capable of penetrating the feet of sheep or the hooves of horses, by which they are then transported. Hooking burrs attach to the fur or skin of passing animals; examples are the American cockleburs (*Xanthium*), the European goosegrass (*Galium aparine*), and *Acaena* of southern temperate regions.

The majority of endozoochorous diaspores are modified for dispersal by fruit-eating birds. Birds generally eat fleshy pulp from the soft fruits and then excrete the hard seeds undamaged; examples are cherries, olives and dates, which all possess features attractive to fruit-eating birds. Amongst the important features are visually attractive, usually non-odorous edible parts, protection by unpleasant tastes against premature eating etc, signaling colors in the mature fruit and seed, strong attachment to the mother plants to allow *in situ* consumption, soft skins and dangling, exposed seeds in harder fruits.

Adaptations to mammal endozoochory

differ markedly from those for birds because mammals generally possess a keen sense of smell, have strong biting teeth, have much larger bodies and rarely live an arboreal life. The diaspores, which are usually fruits, possess strong-smelling, often rancid but not necessarily visibly attractive edible parts, have stronger protections in the seed against premature eating (eg stony walls or toxic components), have various signaling devices when mature, usually fall off trees when mature, and frequently possess hard skins.

Wind Dispersal

Modifications for wind dispersal tend to predominate in regions of pioneer vegetation, shifting habitats and areas inhospitable to plant life. Characteristic modifications include dust diaspores of orchid and *Orobanche* seeds, balloon fruits of some of the Leguminosae, plumed diaspores of various grasses and Compositae, tumbleweeds, as in *Astragalus*, and winged fruits from trees.

Dust diaspores are tiny (it is estimated that some orchid seeds weigh only 0.001mg each). Plants with very small seeds invariably belong to those families of saprophytes, epiphytes and parasites occupying specialized habitats. These require a blanket cover of diaspores to be spread over widely dispersed suitable environments for successful colonization to take place.

Balloon diaspores are variously derived but always involve additional portions of the flower or inflorescence to give a larger

surface area to the ripe propagule and an effective overall decrease in density. The Mediterranean leguminous shrub *Colutea arborescens*, for example, has quite small seeds but these are enclosed in huge inflated pods which at maturity become dispersed by wind to new sites.

Plumed seeds and fruits are also typical of plants from open habitats. The origin of "sailing" mechanisms, by which the seeds float through the air, are diverse; they may be appendages of the corollas, calyxes, ovaries or styles. Notable examples range from the single long seed-hairs of the epiphyte *Aeschynanthus* to the multi-plumed seeds of *Epilobium* and the incredible parachutes of such Compositae as the dandelion (*Taraxacum*) and thistle (*Cirsium*).

Undoubtedly the most diverse wind-dispersed diaspores are those heavy seeds and fruits with wings and flat appendages modified for gliding or, when asymmetrical, for dynamic propulsion. They need to be launched from great heights, so it is not surprising that many forest trees have winged flying diaspores. The largest known winged seeds are those of the Malaysian climbing genus *Alsomitra* (Cucurbitaceae) which can grow up to 15cm (6in) in diameter at maturity. By contrast, the seeds of alder (*Alnus*) and birch (*Betula*) rank amongst the smallest winged diaspores. Perhaps the most spectacular of all gliding mechanisms are the rotating diaspores of *Cavanillesia*, where there are several wings up to 9cm (3.5in) in diameter, surrounding the fruit.

One-winged seeds, known as samaras, actually gain propulsion by aerodynamic principles of whirling flight akin to that of helicopters. The best-known flowering plant examples are the maples (*Acer* spp.), where single trees can produce thousands of whirling seeds in the autumn. Samara dispersal is a widespread method in the flowering-plant

Dandelion (*Taraxacum* spp.) fruits have a "parachute" of fine hairs (the pappus) to aid disperal by wind. This dandelion fruit is caught in a spider's web.

kingdom, reaching its greatest development in *Centrobium robustum*, the zebra-wood tree of the Leguminosae which produces wind-dispersed, winged pods up to 17cm (6–7in) long.

Diaspores dispersed by transmission of the whole or bits of the parent plant from one place to another are known as "tumbleweeds". Dispersal by tumbleweeds is very spectacular in windswept arid regions, such as in deserts and steppes where whole areas are sometimes filled with moving masses of vegetation. In the Mediterranean region, the fruiting inflorescences of *Fedia cornucopiae* are easily detached at maturity when they then simply roll away from the rest of the plant. The thistle *Grundelia tournefortii* behaves similarly, as do many steppe species of the Amaranthaceae and Chenopodiaceae.

Water Dispersal

For the most part, water dispersal mechanisms have evolved from other terrestrial methods, more often than not from wind systems. The two modes have much in common, in fact, since both are associated with unselective agents of dispersal coupled with problems of diaspore weight and surface area. The diversity of water dispersal mechanisms parallels that of wind dispersal systems, the submerged ones comparable to flyers and the floating ones to tumbleweeds.

In ecological terms, by far the most widespread class of water dispersers comprises the floating diaspores; amongst the obvious features associated with this mode of transport are small unwettable seeds which float by surface tension, and low specific weight achieved by air spaces or light corky materials. Among the most famous freshwater examples are the floating fruits of the lotus, *Nelumbo lutea*. In the harsher environments of fast-flowing rivers and the sea the

Seeds can be dispersed by many animals, including ants. Some seeds or fruits contain small bodies rich in fat or carbohydrate called elaiosomes, which are attractive to ants. The ants carry the seeds to their nests, eat the elaiosomes and discard the rest. In this photograph harvester ants are transporting grass stalks to their nest.

Dispersal by sea: coconuts (*Cocos nucifera*) sprouting on a beach on Cape York Peninsula, northern Queensland, Australia.

transport are rain-splash dispersal and rain-ballist mechanisms which play an important dispersal role in some of the lower plant groups (algae, fungi, mosses, liverworts). Rain wash is an unspecialized phenomenon, in that there seem to be few particular adaptations for it, but diaspores moved by dripping water and flooding have been recorded in rain forests and deserts. Rain-ballist mechanisms are on the other hand quite the opposite, occurring only in the highly specialized groups. The fungal family Nidulariaceae, for example, embraces a number of rain-splash genera which utilize the energy of rain drops to make diaspores jump out of dispersal cups (eg *Cyathus*).

Active Plant Dispersal

The final category of "autochory" or self-dispersal represents an assorted assemblage of differently derived systems with examples from many different plant families. These can be grouped into two major categories of generative diaspores, those spread by

diaspore will remain immersed longer than in gentle slow-moving streams, and much more sophisticated modifications are needed. Important adaptations include greater protective coverings, the provision of larger nutrient supplies for developing embryos, and additional floating tissue. Without doubt the most spectacular of the sea-dispersed fruits include the coco de mer (*Lodoicea maldivica*) and the coconut (*Cocos nucifera*). In the giant drupes of these coastal palms there is a hard, protective, impermeable inner layer, the endocarp (the "nut" of the coconut), a fibrous middle layer, the meso-carp, and a leathery exocarp. The milky endosperm is an important source of water on arid shores. Plants alongside rain-forest rivers and tropical seas provide the most conspicuous adaptations for water dispersal. Notable river genera include *Combretum*,

Eugenia and *Physostigma*, and a familiar oceanic example is *Crambe maritima*, whose articulate fruits are a common sight on temperate shorelines throughout the world.

Many aquatic plants possess diaspores which are not at all adapted for floating, but are nevertheless still carried by water currents. Predominant modifications here include the ability to survive submerged under water for lengthy periods, usually by the development of protective devices such as sticky slime coverings, increase in surface area by means of hairs or wings for current transport and the ability to become fixed on the river bottom by anchoring hooks or burrs. Typical examples here include the water chestnut (*Trapa natans*), the water primroses of *Hottonia* and the loosestrife, *Lythrum salicaria*.

Much less widespread modes of water

Below One of the fungi showing "rain-splash" dispersal of its spores is the bird's nest fungus (*Cyathus striatus*), illustrated here. The general structure of the fruit-body, with the peridioles in which the spores are contained, is shown in vertical section (a), together with details of a single peridiole before and after the funicular cord is released from the sheath (tunica). Dispersal is achieved by rain splash (b): when a raindrop falls into the fruit-body (1) it releases a ripe peridiole (2) which frees its funicular cord as the tunica ruptures, so that the basal hapteron catches onto any nearby plant (3) which the cord then twists around (4) before the spores are ultimately released.

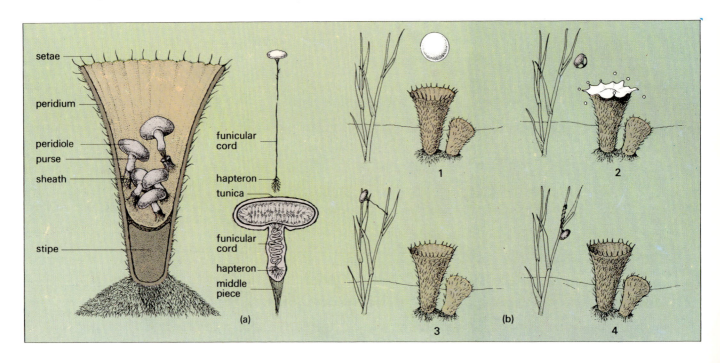

growth, and those dispersed by active or passive ballist systems. In most examples, diaspores spread by growth movement of the plant itself, typically when creeping stems drop seeds at different intervals as the plant moves along.

C.J.H.

Colonization and Establishment

Although, as has been seen, all plants disperse, to some extent, by the release from the parent of seeds, spores, vegetative propagules, etc, the effort is only worthwhile if this dispersal results in the growth of a new plant. Usually, the majority of such offspring survive within a relatively short distance of the parent, where the habitat conditions are likely to be suitable. However, some progeny will reach new areas, where they will grow, thus extending the species' range, and it is this that is usually taken to constitute colonization. A species may move into a new area because it is better adapted (more "aggressive") than the plants already there, or the area may be partially or completely virgin, as in the case of new volcanic islands or regions left exposed by retreating glaciers.

Their small numbers make primary colonizers very vulnerable to random losses which can make the initial stages of colonization a very uncertain business indeed. There is thus a strong advantage for species which can rapidly increase their numbers and this has led to a "colonizing syndrome" of such features as a short life cycle (annuals and biennials predominate among colonizers) and a high reproductive output. Since the latter would be difficult for cross-pollinating species, given the initial wide dispersion of individuals, the syndrome also includes at least the capability of self-pollination, or some form of nonsexual reproduction (see Apomixis, p. 107). One problem, however, is that continued self-pollination results in increased genetic uniformity within the populations, so that the necessary increase in numbers may be achieved at the expense of the genetic variability on which future adaptation and evolution can depend. For this reason many colonizing populations have chromosomal mechanisms that can maintain genetic variability in the face of inbreeding. Two of the commonest of these mechanisms are an increase in the number of chromosomes or a change in structure by inverting or interchanging segments following spontaneous breakage and reunion of one or more chromosomes. Since these mechanisms produce distinctive, visible chromosome patterns during pollen-formation, they provide a valuable means for identifying the "colonizing edge" of a species and thus a clear indication of its migration routes (see Centers of Origin, Diversity and Variability, p. 220).

Although, ideally, a colonist initially increases its numbers at an increasingly rapid rate, this exponential growth is gradually constrained by a shortage of resources or some other consequence of its own increasing density, such as a greater liability to damage by pests and diseases. Such density-dependent factors can lead to a stabilization of the size of populations and the gradual establishment of newer, more permanent communities. The plants of these communities tend to have late reproduction, long lives, larger but fewer seeds, and they often accumulate an ageing structure in the form of, for example, woody stems, rhizomes and rootstocks; the increasing prominence of perennial species after the earlier stages of colonization is a frequent indicator of establishment being achieved.

Although the process of colonization is most spectacular when a species arrives in a new area and spreads rapidly, it is fair to consider most species as having some colonizing ability. Certainly, most successional species are doomed in their present environment and their continued existence depends on colonization of new successions elsewhere (see p. 149). Even the plants of apparently stable communities (old forest, established grassland) are in continual flux as individuals die and are replaced by others in different positions within the community. Sometimes such a sequence of local death and new establishment elsewhere results in an internal cycle within vegetation so that each species in the community is endlessly colonizing the territory of another.

It is relatively rare for the early stages of natural colonization to be monitored, and much of what we know has to be inferred from observations on developed plant communities. However, the deliberate or accidental introduction by Man of species into areas where they are not native provides abundant opportunities to observe the initial patterns of colonization (see Weeds and Aliens, p. 259). For example, Darwin's observation that "several of the plants now most numerous over the wide plains of La Plata, covering square leagues of surface almost to the exclusion of all other plants, have been introduced from Europe" illustrates the role of exponential increase during early colonization. Thus, *Opuntia stricta*, introduced from the United States to Australia in 1839, spread rapidly until it occupied 4 million hectares (10 million acres) in 1920; by 1925, it was still colonizing Australia at the rate of 400,000 hectares (one million acres) a year, while half of the infested area was so dense as to be almost impenetrable to Man and large animals. Such aliens apparently owe their success to being introduced without their associated predators and parasites, and the introduction of biological control to reduce such weeds illustrates both this and the effect of the density-dependent factors referred to above. Interestingly, many common weeds show the syndrome of features mentioned earlier, stressing the parallel between the initial colonizers of naturally open habitats and those maintained in this condition by

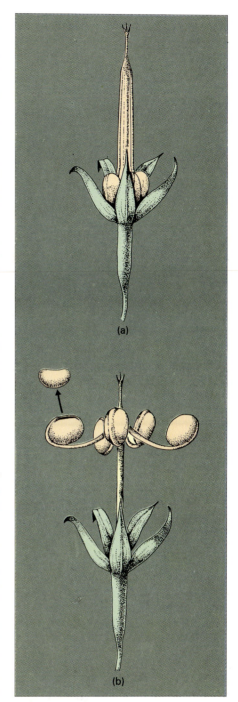

The best-known autochorous systems of seed-dispersal in flowering plants are shown by the active ballists, such as the cranesbill *Geranium dissectum* with its catapult mechanism, shown here. Before release, the fruit has the seed-chambers attached to each other at their bases (a). When the seed-chambers detach, often as a result of being knocked, their release from the central core of the style is accompanied by the sudden coiling of the catapult spring so that the seeds are violently ejected in the direction indicated (b).

human activities. Clearly, much of our future knowledge about colonization could come from studies of plants owing their distribution to the increasing influence of Man on this planet (see Chapter Seven: Man and the Green Planet).

J.L.H.

Some of the best colonizers are the worst weeds. Here the water hyacinth (*Eichhornia crassipes*), the world's most notorious aquatic weed, has completely filled a canal in Florida, United States.

Competition

As has been noted earlier, evolution by natural selection depends upon the differential ability of populations and species to contribute their characteristics to succeeding generations. At its starkest, this process has been emotively termed "survival of the fittest", with the protagonists competing in a trial of strength, speed or stamina to determine the winner. However, continuing in an anthropomorphic vein, it may be noted that competition ranges from the fight to the finish in gladiatorial combat to the fine tuning and compromise which permits individuals, often of different species, to exist together in a society.

The term competition, like many other everyday words used in science, thus has a variety of meanings. Some authors use it in a formal and restricted sense to describe the process by which organisms consume or deplete the resources needed for growth with the result that the growth rate of other members of a population is slowed down. Others use the term for all the processes by which an organism damages the environment of another, while some authors treat competition as the result of these processes rather than the processes themselves. The parts of a single plant are even sometimes said to compete with each other when a limited resource is shared between leaves or between developing seeds and fruits.

For plants and animals, the resources required for growth are limited and populations may reach densities at which neighbors deplete the resources available to each other. For green plants there is a limited number of

critical resources: light, water, nutrients (eg nitrogen, phosphorus, potassium, calcium, magnesium), carbon dioxide and oxygen. It is misleading to regard competition between plants as an active struggle like that between two dogs for a bone. Rather, each plant lives in a physical environment that may be modified by the presence of other plants. Thus if the leaf canopy of plant A overtops the canopy of plant B, A will grow in full light and B in shade. Similarly, if the root system of A depletes the level of nitrate ions near its roots, plant B rooted in the same zone will grow in a less nutrient-rich soil.

Some of the ways in which a plant influences its local environment are so narrowly restricted that a neighbor will rarely be affected; for example, phosphate ions diffuse for such short distances in soil that it must be extremely rare for phosphate depletion by one root to affect what is available to another. In contrast, nitrate and potassium ions diffuse for relatively long distances in the soil and depletion zones created by one plant can fairly easily extend into the rooting zone of a neighbor. Gaseous resources are so quickly diffused through the environment that the immediate neighbors are unlikely to be affected by, for example, depletion of atmospheric carbon dioxide or oxygen (except on an extremely still day). Insofar as the resources of the environment are depleted, carbon dioxide deficiency is likely to be diffused through a mass of vegetation and will not be sensed as a neighbor-to-neighbor reaction. It is resources that lie between the two limits of high and low diffusibility that are primarily involved in competition: nutrients like nitrate, and to a lesser extent potassium, water (at tensions at which it moves slowly but significantly through the soil) and light (because a shading effect is discrete yet is transmitted from one plant to another over a distance).

Within a population of plants at the same

age and stage of development, crowding produces certain characteristic responses. Phenotypes are developed with reduced branching and fewer vegetative parts. Competition generally reduces the number rather than the size of the plant parts. The strong effect of density on branching is reflected in a reduced number of flowers and fruits and so a reduced reproductive capacity. Competition that affects a plant during the phases of seed set and ripening causes an abortion of ovules before and after fertilization or whole ovaries may be lost. The mean weight per seed is, however, rarely much affected by competition from neighbors.

The dry weight of plants responds to density so as to compensate almost perfectly for such differences. Thus, dry weight of plants per unit area of land (see Biomass, p. 41) tends to be independent of plant density over a wide range, although at very high densities plants may suffer from disease or tissues be so weak that stems collapse (the phenomenon of lodging that is well known in cereal crops). When a population of plants is sown at a range of densities the yield of dry matter per unit area approaches a plateau of yield (this is the "law of constant final yield"), which implies that the individuals are limited in their growth by the availability of some resource.

The effect of inter-plant competition on seed production per unit of land (and some other components of yield such as latex production by rubber trees) is more complex. As plant density is increased the number or weight of specific components of yield rises to a peak and then falls. One effect of competition, therefore, is that the individual plant adjusts the allocation of resources to different structures; most plants when in competition with neighbors devote more of the limited resources to support structures and less to seed production. One consequence of this is that the optimal density for seed production by a crop is often quite different from the optimal density for dry matter production.

Within populations of single species sown at high density, individual plants vary in their success at preempting environmental resources. An individual germinating a few hours ahead of its neighbors will tend to place its leaves higher (and so shade the neighbor) and extend its root system further and faster and so tap limited resources more quickly. Such an initial advantage accumulates as the "winner", gaining a greater share of resources, continues to grow faster and thus extends its monopoly. Slight differences in the spacing between individuals also have a profound influence on the time at which competitive interactions occur. One result is that a population of plants (even of a single genetically uniform stock) comes to be composed of a hierarchy of dominant and suppressed individuals. When such a population includes genetic variants they also contribute to determining which individuals are successful in dominating the hierarchy.

The suppressed plants have a higher than normal risk of death, and a population of competing plants continually eliminates the smaller weaker members. The speed of this self-thinning process is directly related to the speed with which the survivors grow.

This appears to be true for a wide range of plant species including forest trees, grasses and annual herbs. Competition is not a haphazard process but conforms to rigid rules.

In mixed populations of two or more species the same general effects of neighbor on neighbor occur but the hierarchy that develops within the population tends to be a species hierarchy. It is difficult to imagine two species with such similar biologies that when they meet in a struggle for existence the outcome is evenly balanced. In general, it is found that slight differences in germination time, hypocotyl length or leaf area determine that one species emerges as dominant in the hierarchy and the other suffers the greater share of the mortality that occurs. Occasionally it is found that when two species are sown at high densities in a mixture and in pure populations, each suffers less from the presence of the other than it would from an equal number of neighbors of its own sort. These are ecologically compatible species that differ, for example, in the demands that they make on environmental resources. Mixtures of legumes (eg clover) and grasses are peculiarly compatible with each other when resources of nitrogen are limited: the legumes escape from this particular struggle by virtue of their ability to fix nitrogen from the atmosphere.

Competition for available nutrients is of great economic importance, especially in agricultural crops. This field of young wheat has been invaded by poppies (*Papaver rhoeas*) and cornflowers (*Centaurea cyanus*) which may seriously reduce the yield of the crop unless they can be controlled by selective herbicides.

Species may also escape some of the force of competition by rooting at different depths in the soil and so tapping partly independent sources of limited nutrients or by having different cycles of leaf production so that each species avoids shading by the other at critical phases of growth. Species that form stable mixtures in nature often appear to avoid the full force of the struggle for existence by such differentiation.

Competitive interactions can often be detected in the field by deliberately removing one species from a mixture and determining the effect on those that remain. This is the principle underlying weed control in crops but it can easily be extended to study interactions in natural vegetation. The most striking effects can be seen when a tree is removed from a canopy and neighboring trees grow faster and put on larger increments. This effect is so strong that it is sometimes possible to date the year at which a competing tree in a forest died or was removed by examining the growth rings made in the wood of neighboring trees.

There is much argument as to whether the effects of plants on the growth of neighbors is wholly accounted for by mutual demands made on limiting resources or whether some species may inhibit others by the production of toxins. As will be seen in the next section (Allelopathy), it can be shown under laboratory conditions that extracts from one species may inhibit germination and slow the growth of another, although it is extremely difficult to design experiments that prove toxic interactions in the field. Desert shrubs which have been shown to contain toxins (eg terpenes) that can damage other plants also harbor populations of seed-eating animals which have been shown to restrict the occurrence of various plant species in the area. Clearly, both toxins and herbivores, as well as other factors, are all involved in

competition within plant communities—one of the tasks of the ecologist is to determine their relative importance in different situtations.

It is dangerous to interpret competitive interactions in the field without extremely careful experimentation and equally, in cases of competition tested outside a natural community, the mechanism probably cannot be identified with certainty. Plants weakened by deprivation of resources may die from quite different causes; for example, densely growing plants are particularly sensitive to pathogenic attack. Similarly, plants damaged by some factor irrelevant to competition (eg by being trodden on or bitten) may as a result be placed at a disadvantage in obtaining limited resources in the presence of vigorous undamaged neighbors. Plants growing sufficiently densely to make mutual demands on limited resources are usually close enough to change the physical environment for pathogens and pests so that one species may force the elimination of another by indirect action, although it may appear to be direct competition.

The study of plant competition in simple model populations has developed into a sophisticated science. The analysis of its role in natural vegetation, however, remains technically very difficult.

J.L.H.

Allelopathy

One aspect of competition between plants which has aroused controversy and interest in recent years is allelopathy. This is the negative effect of one plant on another by means of chemical products loosed into the environment. As part of their struggle for existence, plants have to compete with each other for moisture, light and soil nutrients in the natural ecosystem. In the course of this struggle, they have developed various means of defense against their neighbors: whenever the defense is chemical in nature it is called allelopathy. The term is usually restricted to vascular plants, although lower plants are indirectly involved since soil microorganisms may metabolize the chemical substances produced. The phenomenon was recognized, even before the time of Darwin, by de Candolle in 1832, but it is only rather recently that allelopathy has been confirmed by careful experimentation and the chemicals involved properly identified.

Chemical defense mechanisms may be invoked both when plants of the same species compete with each other and when one type of plant, eg a shrub or tree, competes with another, eg a herb or grass. Such competition is most acute in extreme climates and some of the best examples of allelopathy have come from studies of desert plants. However, the effect has also been demonstrated in plants growing in a range of habitats, from humid rain forest to open grasslands. The chemicals involved are called *toxins* or allelopathic

substances and are usually fairly simple, low molecular weight compounds. Most of the compounds which have been positively identified are either volatile terpenoids or else phenolic substances.

One of the first allelopathic substances to be identified was 3-acetyl-6-methoxy-benzaldehyde, isolated by R. Gray and J. Bonner in 1948 from the leaves of the shrub *Encelia farinosa* growing in the Californian desert. The substance is washed from the leaves into the soil around the shrub and this prevents the germination and growth of other plants in the vicinity. Investigation of another desert plant, the rubber-producing composite *Parthenium argentatum*, showed that other modes of transport of toxin are possible. In this case, the toxin, *trans*-cinnamic acid, is excreted from the roots directly into the soil. Its main effect is on other plants of the same species, preventing individuals in the population from growing too near each other. This is obviously advantageous in a desert in an area of low rainfall where the amount of moisture in the soil is severely limiting on growth.

Allelopathy has been most extensively studied in plants of the Californian chaparral, an area of low foot-hills along the coast north

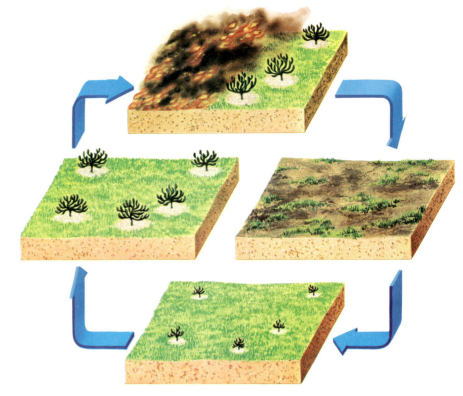

Above Chaparral fires destroy not only the shrubs but also the allelopathic substances (volatile terpenes) that they produce, so that immediately afterwards herbs and grasses dominate the vegetation. Eventually, new shrubs grow, producing more toxic terpenes which inhibit growth of nearby plants.

Below The allelopathic shrub, sagebrush (*Artemisia californica*) in the chaparral of California, United States.

of the city of Los Angeles. Cornelius Muller and his associates, working at the University of California at Santa Barbara, were first drawn to the problem by noticing that bare patches of vegetation characterically developed around certain shrubs, notably *Salvia leucophylla* and the sagebrush, *Artemisia californica*. After eliminating other ecological variables, they came to the conclusion that plant growth was prevented around these shrubs by the exudation of volatile terpenes which were constantly being produced in the leaves. Analysis of the soil around the plants showed that the plant terpenes were in fact saturating the soil beneath the shrubs. Two of the major terpenes implicated were cineole and camphor. Indeed, laboratory experiments, both with soil and plant extracts and with pure terpenes, confirmed that the seed of the annual grasses (wild oats, *Avena fatua*), normally part of the vegetation of the chaparral, were inhibited from germinating and growing by these toxins.

The chemical ecology of the Californian chaparral is complicated by the fact that the vegetation undergoes cyclical change as a result of natural fires, which occur on average about every 25 years. The destruction of the shrubs is accompanied by the burning of the volatile terpenes in the soil so that in the first few years after the fire the annual herbs and grasses dominate the landscape. However, inexorably there is new growth of the shrubs and slowly the toxic terpenes again accumulate in the soil. The characteristic circular patches of bare soil around the shrubs develop after some six or seven years and may continue to be observed until the next fire.

Also growing in the chaparral are other shrubs which practice a different form of allelopathy on competing herbs: two such plants are the rosaceous *Adenostoma fascicu-*

latum and the ericaceous *Arctostaphylos glandulosum*. The visible signs of chemical warfare are precisely the same as in the case of the sagebrush but the toxins here are water-borne phenolics. These are leached from the leaves by rain, fog and dew and so accumulate in the soil, thus preventing the growth of annuals. Among the phenolics present in the plants and in the soil around them are arbutin and ferulic acid but other simple phenolic acids are also involved. There is evidence that this is a more stable form of allelopathy, since phenolics, unlike terpenes, are only slowly metabolized by soil bacteria. In periods of good rainfall, allelopathic effects of the terpenes differentially disappear and are only apparent again in the chaparral during very dry spells.

Allelopathy has also been detected in other natural habitats, and under conditions of normal to heavy rainfall water-borne toxins seem to be the rule rather than the exception. This is true for example of the walnut tree, *Juglans regia* and the oak, *Quercus falcata*. With the walnut, the toxin in the soil under the tree has been identified as the quinone pigment juglone. This yellow substance, although it occurs in walnut shells, is not present in the leaves, which have instead a nontoxic water-soluble precursor called hydrojuglone glucoside. This compound is washed into the soil by rain and only then does it undergo hydrolysis and oxidation to the active toxin. Juglone is an example of a differential toxin, being effective in arresting the growth of some plants (eg ericaceous shrubs) but not of others (eg Kentucky bluegrass). In the deciduous rain forest of South Carolina, a similar water-borne toxin, this time salicylic acid, provides allelopathic protection to the dominant oak species, *Quercus falcata*.

Another aspect of allelopathy not so far mentioned is *autotoxicity*, whereby individual plants of a number of species which take part in ecological succession inhibit their own growth due to the excretion of toxins from their own roots. The autotoxicity of sunflowers can appear to cause a kind of fairy ring: on spreading outwards, these plants show maximal growth only when invading new soil around the periphery.

Although the toxins discussed here are either terpenes or phenolics, there is circumstantial evidence for believing that other types of secondary constituent provide chemical defense. Examples are the cyanogenic glycoside prunasin, exuded from the roots of peach trees, the mustard oil sinigrin excreted by the cabbage plant, and the alkaloid nicotine, produced in the roots of the tobacco plant. Although the study of allelopathy is still at a relatively early stage, it is probable that it will eventually be recognized as a universal phenomenon in plants and plant communities. It seems likely that a whole range of structurally different toxins is involved in these interactions, whose further study must be important in understanding the relationships between the plants comprising a community.　　　　J.B.H.

Succession

Succession is the sequence of changes in vegetation which takes place in any environment until the climax (see p. 43) is reached. If a field or garden is abandoned it will soon become overgrown by weeds. The first plants to appear will be those that can reproduce quickly, which for the most part will be annual or ephemeral herbs. These herbaceous weeds are then slowly eliminated by the woody perennials which eventually become established. Whenever we see such an area of abandoned land overgrown by herbaceous and shrubby weeds, we are observing the results of natural succession.

In many situations, such as on sand dunes or at lake margins, the sequence of plant communities is well understood, in which case the succession is called a *sere*. This term appears in many words which are used to describe different types of succession; thus any succession proceeding from a habitat devoid of vegetation is known as a primary succession or *prisere*. More specific terms can be employed, such as a *lithosere*, which starts on a bare rock surface or a *hydrosere*, which is the sequence of plant communities that accompanies the slow silting up of any body of fresh water. However, in many cases the

Important colonizers of harsh environments include lower plants like this fern, *Polypodium pellucidum*, which has gained a foothold on a lava bed in Hawaii.

process of succession starts on land which has been cleared or abandoned by Man, and where a soil already exists; in this case a secondary succession or *subsere* ensues. Human activity, such as felling, burning or grazing, gives rise to a sequence of communities which differs from the communities that would occur under completely natural circumstances; this type of succession is known as a *plagiosere*.

Succession starts with colonization by species which are adapted to the prevailing environmental conditions; these may be dry, wet, saline, sandy, etc. Such populations constitute the pioneer community. In a prisere the colonizing species may be algae, mosses or lichens; in many situations, however, such as sand dunes, higher plants may colonize directly. After colonization, the vegetation gradually closes until the plant cover is complete; there follows a sequence of population and structural changes leading to a final community, after which major changes no longer occur. This climax is often, although not always, woodland.

Details of each type of succession vary in different locations, the major influencing factors being the available species and the prevailing environmental conditions. However, the general nature of each type of succession is always the same, which enables us to understand vegetation changes without necessarily having a detailed knowledge

of the species populations involved.

The descriptions that follow are concerned mostly with higher plants. However, it should be borne in mind that a succession of living organisms occurs during the decaying processes of dead animals and plants, as well as on feces; even the sequence of bacteria involved in tooth decay is a succession.

Succession on Dry Mineral Substrata

This type of succession, called a *xerosere*, occurs on bare rock surfaces, cliff faces, lava flows, scree, blown sand, etc. Where the exposed mineral substratum consists of large blocks, the first colonizers will be bacteria, algae, and lichens; these organisms have the advantage of being able to adhere to surfaces, and therefore do not need a rooting medium. They are also capable of existing on nutrients dissolved out of the rock by rain water. Nitrogen presents a particular problem for these colonizing organisms because it is virtually absent from the Earth's crust. The only nitrogen available to these first species is

that dissolved by the rain or snow falling through the atmosphere (nitrogen itself is almost insoluble, but lightning converts nitrogen gas into the more soluble oxides, and it is these which are dissolved by precipitation). As a result of this nitrogen shortage, organisms capable of fixing atmospheric nitrogen usually play an important role in the colonization stages of xeroseres: such organisms will include certain types of pigmented bacteria, blue-green algae, and lichens with a blue-green alga as one of the symbionts, ie partners in symbiosis.

When the first arrivals die, their organic detritus may collect in cracks and crevices, giving rise to humus, which forms a growing medium for higher plants such as mosses (although some mosses, eg *Tortula* spp., *Eurhynchium* spp., appear capable of growing directly on rock surfaces). Among the breakdown products of dead organic matter are various acids, which in turn increase the rate of rock weathering by chemical action, and in due course a soil is formed which will support the growth of various herbs, shrubs and trees.

In situations where the mineral particles

Colonization of a sandy shore on a tropical island in the South China Sea. In the foreground the colonizing species are *Ipomoea pes-caprae* and *Scaevola taccada*. The feathery shrub in the middle-distance is a *Casuarina* species.

are sufficiently small to provide some form of water retention and anchorage for plants, it is possible for the mineral substratum to be colonized directly by ferns, flowering herbs, shrubs, or even trees. Direct colonization by higher plants can be seen on material such as mountain scree, or rock waste from mining operations. In such cases, both the chemical and mechanical activities of roots help to promote further breakdown of the rock.

Volcanic eruption can create sterile islands which offer the microbiologist an excellent opportunity to investigate details of the early xerosere. In recent years microbiological tests on several islands of this nature have confirmed that bacteria may be found before any visible signs of vegetation.

One volcanic eruption which has been of considerable interest with regard to the study of plant succession occurred in the Indian Ocean in 1883 when part of the island of Krakatau sank. The remaining 25km² (10mi²) were completely covered with ash to depths of up to 30m (98ft), obliterating all the vegetation. Since that event, regular visits by botanical expeditions have been made, providing a fine chronicle of this tropical xerosere. The first species were recorded about three years after the eruption, when it was noted that much of the ash was covered with a crust of blue-green algae. The vegetation was dominated by ferns, including species of *Pteris*, *Dryopteris* and *Pteridium*. Altogether 26 species of vascular plants were recorded, and the vegetation was open.

Eleven years later when the plants were next examined many changes had occurred, including the development of completely closed stands of vegetation. The seashore vegetation was dominated by *Ipomoea pes-caprae*, a member of the bindweed family (Convolvulaceae) which is abundant on tropical seashores, including those of the neighboring islands of Java and Sumatra. Several specimens of *Casuarina equisetifolia* were also noted. The ferns had by now almost vanished from the interior of the island; in their place was a savanna dominated by the tall grass *Saccharum spontaneum*. By 1906 shrubs and trees were scattered amongst the interior grassland, including species of *Ficus* and *Macaranga*, and after a further 20 years most of the savanna had been replaced by woodland dominated by these genera. Throughout the half century following the explosion on Krakatau the number of established vascular plant species had increased steadily until more than 250 were present in the closed woodland vegetation. Whether or not the vegetation has yet reached the climax remains to be seen; in any event it appears that the diversity (number of species per unit area) will have continued to increase until climax is reached: this is a characteristic of tropical successions, which contrasts with temperate situations where the diversity frequently declines at climax.

Thus the succession on Krakatau has developed along lines that are now well established as normal for a xerosere, which is a sequence of communities dominated in turn by microorganisms, cryptogams, flowering plants and trees.

Succession on Rock Surfaces

This is a special type of xerosere usually referred to as a *lithosere*. The sequence is as described above for a xerosere, except that microorganism and cryptogam stages are usually an essential prerequisite of the succession. In many parts of the world, plants in the early stages of a lithosere face a serious problem of water shortage, and therefore the colonizing species usually include lichens and certain mosses, most of which can survive dehydration. In addition, the first flowering plants may also be those with provision for water shortage, such as the stonecrops (*Sedum* spp.).

Succession on Sand

This particular type of xerosere, which is found in many coastal areas, is sometimes called a *psammosere*. The psammosere has been very widely studied because the nature of sand dunes is such that they often form in a chronological sequence, with the youngest dunes nearest to the sea, and the oldest dunes farthest from the sea. In the most straightforward situations the dunes will have developed as a series of ridges running parallel to the seashore, and a botanical walk from the most landward dune to the shore is much the same as walking backwards through time. Therefore the sequence of dunes enables us to study different communities, together with the soil changes that are associated with this type of succession.

Embryonic dunes develop on the uppermost part of the shore where they cannot be destroyed by high tides, and plants that colonize these embryo dunes must be able to withstand a high concentration of salt. In temperate climates the vegetation growing on embryonic dunes might include such species as the sea couch grass (*Agropyron junceiforme*), sea sandwort (*Honckenia peploides*) and the saltwort (*Salsola kali*); this vegetation reduces the wind speed just above the surface of the dune, causing blown sand to accumulate. As the embryonic dune becomes large so inundation by salt water becomes less frequent, and eventually leaching by rain lowers the concentration of salt to a level which can be tolerated by marram grass (*Ammophila* spp.). Marram grass can grow in the dry conditions of sand dunes because its leaves are modified to prevent water loss, and it possesses an extensive root system capable of reaching a low water table; furthermore, this grass appears to thrive in spite of being frequently covered by sand. These properties, together with the fact that the tall leaves of marram encourage the deposition of blown sand, mean that marram grass communities are often associated with rapidly growing dunes. Because much sand is visible between open patches of vegetation, these dunes are often called yellow dunes. Other species found on yellow dunes are sea couch, sea sandwort, lyme grass (*Elymus arenarius*), sand sedge (*Carex arenaria*) and sea bindweed (*Calystegia soldanella*).

With the passing of time, certain changes will take place on the dune: the vegetation will close, the sand will stabilize, and continual leaching of salts will cause the pH to become lower. As plants die, the organic content of the sand increases, helping to bind it as well as to improve its powers of water retention. Eventually conditions may become more suitable for lichens (eg *Cladonia* spp., *Peltigera* spp.) and mosses (eg *Tortula ruraliformis*, *Bryum pendulum*). Red fescue (*Festuca rubra*) and sand sedge become more abundant on these dunes, together with such species as restharrow (*Ononis repens*), bird's-foot trefoil (*Lotus corniculatus*), ragworts (*Senecio* spp.) and *Galium* spp., which are all rare or absent in earlier stages. Leaching and humus accumulation, together with the lichens, give a gray appearance to sand dunes at this stage of succession, and consequently they are often referred to as gray dunes.

Later stages of the psammosere, which can sometimes be seen on the most landward dune ridges, consist of shrubs followed by woodland. The exact nature of these communities is very dependent upon soil pH and the species available. In northern Europe, where the pH is high, vegetation on the gray dunes is succeeded by shrubs such as sea buckthorn (*Hippophae rhamnoides*), privet (*Ligustrum vulgare*) and barberry (*Berberis vulgaris*). Finally, woodland of birch (*Betula*

Succession on sand in which dunes are formed is termed a psammosere. The first colonizer of this embryonic sand dune in eastern Scotland is sea couch grass (*Agropyron junceiforme*).

spp.) and oak (*Quercus robur*) may develop, although this is comparatively rare. On acid dunes, heath plants such as ling (*Calluna vulgaris*) and heaths (*Erica* spp.) follow the gray dune stage; these in turn may be succeeded by such shrubs as gorse (*Ulex europaeus*) and blackberry (*Rubus* spp.).

In North America the basic sequence of communities is the same as in northern Europe, except that the species and genera are different to those mentioned above, especially in the shrub and woodland stages. Plants such as cottonwood (*Populus deltoides*), willows (*Salix* spp.) and pines (*Pinus* spp.) may precede the climax woodland, which itself may be composed of many different trees, eg oaks (*Quercus* spp.), maples (*Acer* spp.) and hemlock (*Tsuga canadensis*).

The succession of vegetation on sandy shores in the tropics was mentioned briefly with reference to Krakatau. In the Malay Archipelago, mobile sand is bound by vegetation dominated by *Ipomoea pes-caprae*; other common plants in this zone include *Spinifex littoreus* (a tufted grass) and *Canavalia* spp. (family Leguminosae). The shrub phase that follows includes species such as *Tournefortia* spp., and screw pines (*Pandanus* spp.); this in turn may be succeeded by the barringtonia woodland which is dominated by *Barringtonia* spp., and ironwood (*Casuarina equisetifolia*). It is interesting to note that *Casuarina*, like the north temperate dune shrub *Hippophae*, possesses nitrogen-fixing root-nodules, and tends to form pure stands.

Succession on Glacial Moraines

As a glacier melts and retreats, the entrapped rock detritus is deposited in a sequence, with the most recent deposits occurring adjacent to the snout of the glacier. Because it is possible to trace the path of a retreating glacier and to date the various deposits, these, like sand dune ridges, offer a convenient way of studying succession. The best documented account of succession on moraines comes from Glacier Bay, Alaska, where the glacier has retreated more than 70km (44mi) in the past 200 years. The heterogeneous glacial deposit is suitable for a variety of life-forms to contribute to the colonizing community, including mosses, herbs (eg *Dryas* spp., and *Epilobium* spp.) and trees (*Salix* spp., and *Populus deltoides*). The next stage of succession is occupied by alder (*Alnus* spp.) which, together with *Dryas* spp., of the earlier stage, increases the soil nitrogen content through nitrogen-fixing activities. This increase in soil nitrogen, together with a reduction of pH brought about by decomposition of the alder leaves, creates conditions suitable for spruce (*Picea* spp.) and hemlock, which comprise the final woodland stage.

Succession from Water

Any body of fresh water, such as a pond, lake or river, is liable to be converted into land as a result of shallowing and colonization by vegetation. This *hydrosere* may start with planktonic vegetation, including microbial forms such as diatoms and other algae, whose dead remains fall to the bottom of the water mass and form an organic sediment called *gyttja*. This organic material, together with deposited inorganic sediment, causes a gradual shallowing of the water, which continues until sufficient light reaches the bed to allow a community of rooted submerged plants to become established. In north temperate regions this community will include the Canadian water weeds (*Elodea* spp.), water milfoils (*Myriophyllum* spp.), hornworts (*Ceratophyllum* spp.) and pondweeds (*Potamogeton* spp.).

With the establishment of these submerged communities of higher plants, the rate of sedimentation due to silt and organic detritus becomes even faster. Eventually the water becomes sufficiently shallow to permit the growth of rooted plants with floating leaves, such as the broad-leaved pondweed (*Potamogeton natans*), water lilies (*Nymphaea* spp.), amphibious bistort (*Polygonum amphibium*) and the dropworts (*Oenanthe* spp.). Also abundant in this zone are floating plants which are not rooted to the substratum, such as the duckweeds (*Lemna* spp.), water buttercups (*Ranunculus* spp.), frog bit (*Hydrocharis morsus-ranae*) and, in warm climates, the water hyacinth (*Eichhornia crassipes*).

When the water is less than about 1m (3.3ft) deep it becomes possible for rooted plants with emergent leaves to grow successfully; this is known as the reedswamp stage. Plants found in this zone include the common reed (*Phragmites australis*), sedges (*Carex* spp.), the saw sedge (*Cladium mariscus*), reedmaces (*Typha* spp.), bur-reeds (*Sparganium* spp.), and bulrushes (*Schoenoplectus* spp.). In warm climates, papyrus (*Cyperus papyrus*) may also be an important constituent of reedswamp.

The above account of the early development of the hydrosere describes the main stages that might be found when succession commences in fairly deep water. However, if the water is sufficiently shallow initially, the early stages may be absent or insignificant, and colonization will give rise directly to the reedswamp stage.

Continued silting, together with accumulation of organic detritus from the reedbeds, may eventually cause the water table to lower sufficiently for the reedswamp stage to be replaced by other communities. However, the nature of succession beyond the reedswamp stage depends upon mineral supply and hydrology. In nutrient-rich conditions, lowering of the water table causes the reedswamp to be replaced by moisture-tolerant shrubs (eg alder and willows), and finally by climax woodland. In north temperate latitudes this climax may be composed of deciduous trees such as oak

(*Quercus*), maple (*Acer*) or birch (*Betula*).

Where conditions are nutrient-poor, lowering of the water table will lead to the reedswamp being replaced by sedges and finally by bog moss (*Sphagnum* spp.). In due course the water table may be lowered sufficiently for climax woodland to develop; in these nutrient-poor, acid conditions this woodland will consist of conifers such as pine (*Pinus* spp.) and spruce (*Picea* spp.).

Under certain conditions the various stages of the succession described above are manifested as a series of zones on the margins of ponds and lakes, with the submerged plants occupying the zone nearest to open water, and the shrub or woodland stages the landward zones. In classical situations terrestrialization proceeds by a slow movement of all zones towards the center of the lake, until most of the water is finally replaced by land.

Succession on Salt Marshes

Salt marshes offer yet another opportunity to study a succession, the *halosere*, which develops in response to environmental conditions that are quite different from any mentioned above. Plants occupying the early

Accumulated organic remains

Woodland

stages of this succession must be tolerant of shifting mud, high salinity and frequent inundation by tidal water. In the later stages, although the frequency of flooding is less, there is the problem of considerable fluctuations in salt concentration; the occasional high tides create saline conditions, but the subsequent degree of salinity depends upon either dry weather increasing the concentration of salt through evaporation, or wet weather decreasing it through leaching. (See also Salt Marsh, p. 204.)

Some Theoretical Aspects of Succession

Successional change in plant communities is essentially unidirectional towards climax, although in climax communities themselves the biological conditions may be such that the final stages of the succession are repeated, giving rise to cycles. For example, in many forests of beech (*Fagus* spp.), when a tree falls the opening is soon occupied by a rapidly growing scrub of birch (*Betula* spp.); only in the course of time will this clearing again be covered by beech forest.

Successions proceed as they do because with passing time environmental changes occur which create conditions more suitable for populations of species other than those already present. These environmental changes result from nonbiological processes such as leaching and silt accumulation, also from biological activities which give rise to humus formation, as well as to the creation of various microclimates. In the early stages of primary succession, nonbiological processes are generally more important than biological ones; as succession proceeds, however, environmental change occurs largely as a result of biological activities.

In the early stages of a succession, while the vegetation is still open, plant competition is of little consequence, and colonization is by populations of species that are able to tolerate the prevailing environmental conditions. Many of the species involved in colonization are capable of growing in less difficult conditions, but for various reasons are unable to compete; thus, by tolerating conditions prevailing during the colonization stages of succession they occupy a niche for which there is little competition.

In addition to their tolerance of extreme environmental conditions, colonizing species are usually fast-growing, shallow-rooted plants of low stature, which produce large numbers of small seeds, a high percentage of which never germinate; these features are typical of "opportunistic species". As succession proceeds there is always a change in the vegetation towards taller, deeper-rooted plants.

For convenience of presentation, and indeed of investigation, succession is usually conceived as a series of plant communities arranged in adjacent zones, or as a series of communities occurring at specific times in a chronological sequence. However, it is important to appreciate that these are only abstractions from what is, in reality, a continuum of change. Succession occurs because altered environmental conditions favor certain species, which therefore compete more successfully than others for light, space, nutrients, etc; the populations of well-adapted species replace earlier ones less well equipped to compete in the existing conditions. Thus, during the course of succession, individual species populations come and go, giving rise to a gradual progressive change in the community.

The ecological complexity of the later stages of succession is greater than during the early stages. This results from the fact that the later communities are more structured, and the food chains more elaborate.

A diagram illustrating an idealized succession of vegetation on the margin of a lake (a hydrosere). Note that the water table is lower at the woodland stage.

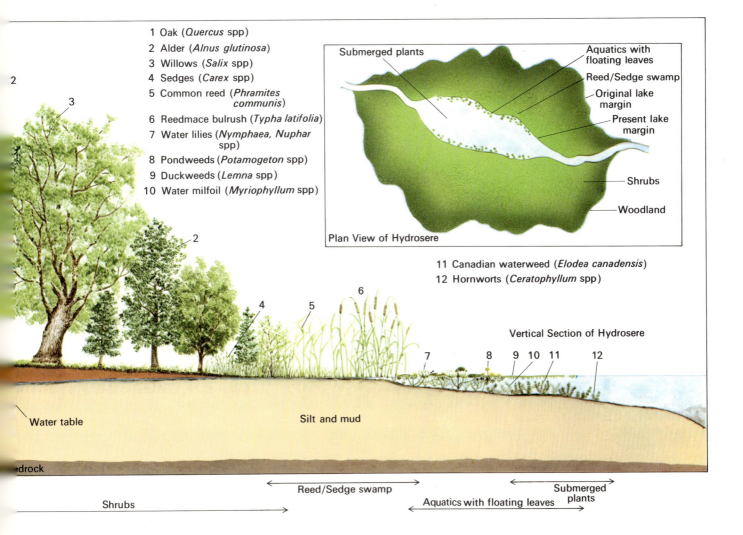

1 Oak (*Quercus* spp)
2 Alder (*Alnus glutinosa*)
3 Willows (*Salix* spp)
4 Sedges (*Carex* spp)
5 Common reed (*Phramites communis*)
6 Reedmace bulrush (*Typha latifolia*)
7 Water lilies (*Nymphaea, Nuphar* spp)
8 Pondweeds (*Potamogeton* spp)
9 Duckweeds (*Lemna* spp)
10 Water milfoil (*Myriophyllum* spp)
11 Canadian waterweed (*Elodea canadensis*)
12 Hornworts (*Ceratophyllum* spp)

Submerged plants

Aquatics with floating leaves
Reed/Sedge swamp
Original lake margin
Present lake margin
Shrubs
Woodland

Plan View of Hydrosere

Vertical Section of Hydrosere

Water table

Silt and mud

Bedrock

Shrubs

Reed/Sedge swamp

Aquatics with floating leaves

Submerged plants

The stability of climax communities can result from their greater diversity and complexity, although this diversity does not seem to be a feature of climax communities in nontropical regions. Such apparent climax stability may, in part, be produced by large, slow-growing woody perennials.

During the early part of succession, the amount of root and stem tissue is relatively low compared with the amount of photosynthesizing tissue in leaves; therefore the quantity of organic material needed for maintenance (via respiration, etc) is not great. In consequence, there is a surplus of material produced by photosynthesis over that which is required for growth and respiration. This surplus is the net production and it ensures accumulation of organic material in the ecosystem, and the biomass of communities involved in succession gradually increases until climax is reached. At climax the large amounts of supporting tissue and extensive root systems require considerable quantities of respirable substrate for maintenance, therefore net production is low, or zero, and further increase in the biomass does not occur.

J.WN.

Zonation

The early part of this chapter treated the environment as a set of interacting physical and biological factors, some, like light intensity and relative humidity, being closely linked and often varying in unison. Where such factors change in a regular, linear fashion in space, one can talk of an "ecological gradient". Some ecological gradients are steep, that is they demonstrate very considerable changes in certain factors over a short spatial distance; other gradients are more gentle or shallow in nature and have only gradual changes within them. Different organisms occupy various spatial positions along the gradient depending upon each one's optimal requirements, tolerance limits and competitive interaction with other species. The outcome is that there will be a change in vegetation along such a gradient and the rate of change will depend upon the steepness of that gradient.

In some situations the change in vegetation along a gradient may be so gradual as to be almost imperceptible, whereas in other situations it may fall into a distinct series of bands or zones. In the latter situation the vegetation is said to exhibit zonation. Zonation most typically occurs on steep ecological gradients where certain plants assume dominance over a section of the gradient and replace one another in a regular sequence. The replacement of one dominant species by another along the gradient has the effect of producing strictly demarcated zones and since the dominant plant effectively determines the microclimatic conditions within the zone, many other plants and animals may be typically associated with a particular zone.

When describing successions in which the sequence of plant communities at one point in space will change during the course of time, it was noted that this can be reflected in the spatial zonation of vegetation. Thus, in traversing a salt marsh, from the *Salicornia* communities at the lowest, wettest and most saline part of the gradient, to the stable mesic grassland at the furthest point from the ocean, the zonation of communities indicates what will happen through time, as the *Salicornia* initiates the succession by trapping silt, the first step in changing the environment. However, the zone sequence often does not represent a successional series, since the ecological gradient can be permanent enough to give a stable zonation. Two examples of this will be given.

Zonation on a Rocky Seashore

The steep ecological gradients which permit the development of zonation are most frequent in regions intermediate between two contrasting habitats, such as dry land and aquatic habitats, or marine and terrestrial ones. The description of a specific situation in which zonation is strongly developed may serve to illustrate some of the underlying features which are common to many such systems. As an example, the moderately sheltered rocky shore system of western Britain can be taken as one of the most conspicuous situations in which plant zonation is found.

The marine and the terrestrial environment could hardly be more contrasting, and the region intermediate between the two is often one of steep ecological gradients largely supporting brown seaweeds. The lowest zone of plants is dominated by the kelps (*Laminaria* spp.) which extend a little way higher up the shore than the extreme low water-mark of spring tides, yet which are generally intolerant of long periods of emersion (ie emergence from the water). The ocean is a very stable, unvarying environment in comparison with the land, and organisms that live within it are not subjected to the same fluctuation in water supply, temperature and chemical environment which are the lot of the land plant. As the tide retreats down the shore, the emersion of plants causes them to occupy this intermediate zone and be subjected to these stresses. Various seaweed species show differing degrees of tolerance to the stress. *Laminaria* species are intolerant of desiccation and this fact restricts them to the lower part of the ecological gradient. Their robust morphology results in their upper limit being a very distinctive boundary on the shore, which is usually employed to distinguish the eulittoral from the sublittoral zones (the eulittoral zone extends from the shoreline at extreme high tide down to extreme low tide level; the sublittoral zone begins at this depth and extends to the margin of the continental shelf).

Moving up the shore, the next zone is occupied by the serrated wrack (*Fucus serratus*). This species is still rather intolerant of the desiccation associated with emersion, but is capable of survival for longer periods out of water than are the laminarians. The lower limit is probably determined by competition with *Laminaria*, but within its zone it is able to outcompete other brown seaweeds from higher up the shore. It is eventually replaced by the knotted wrack (*Ascophyllum nodosum*), which is an extremely robust brown alga, forming large tussocks of fronds. Often this is interspersed with the bladder wrack (*Fucus vesiculosus*), which becomes the most abundant brown seaweed of this midshore zone under conditions of increased wave exposure. Within this midshore region, the algae may have to spend 50–60% of their time emersed, which renders the plants liable to drying in the sun and also to varying temperatures, from direct sunlight down to possible night frost. Rainwater can result in considerable salinity changes in the plants' immediate environment. These factors combine to make this region of the shore inhospitable and quite unsuitable for the growth of the lowshore algae.

Higher up the shore, the rocks are covered for shorter periods still and even *Ascophyllum* is eliminated. As growth of this robust alga becomes depressed by the increasing stress of emersion, so other species, better adapted to these difficult circumstances, are able to compete and maintain themselves. The spiral wrack (*Fucus spiralis*) is the alga which dominates under these circumstances, forming a distinct band along the upper part of the shore. Even *F. spiralis* is unable to survive in the upper regions where submersion may occur only during the higher tides. Here most terrestrial plants still fail to colonize because of the occasional saline conditions, coupled with the unfavorable substrate for rooting. Only some maritime lichens are able to occupy this region, and the resulting lack of competition leaves a vacant niche for one of the most desiccation-tolerant of the brown algae, the channeled wrack (*Pelvetia canaliculata*). This species is capable of growth where it receives a covering of seawater only every two weeks or, if the site is subject to salt spray, even less often than this.

Above the *Pelvetia* zone there may be further zones of orange and gray lichens before the few flowering plants which can tolerate spray and shallow soil begin to assume dominance. The overall pattern produced as a result of this interaction of varying physical stress and biological competition is a series of zones dominated by conspicuous and, for the most part, robust plant species. The precise pattern and the species involved, however, vary if a further physical factor, that of wave exposure, increases. Under such circumstances several of the brown algae forming zones on sheltered shores find it impossible to establish themselves. On many such shores, barnacles and limpets occupy the midshire region instead of *Ascophyllum nodosum* and *Fucus vesiculosus*. It can be demonstrated by experiment that this bare (that is, devoid of plants) midshore area is maintained by the intensive grazing pressure of the limpets (*Patella* spp.). If an area is experimentally cleared of these grazers, then sporelings of *Fucus vesiculosus*, although not of *Ascophyllum* (which is truly intolerant of wave action), are able to establish themselves and produce adult plants. In this situation, therefore, the overall zonation pattern observed on the shore is the result of a physical gradient and intensive depredation along a part of that gradient. The upper part and the lower part of the

Seashores, with their rapid transition from one environment to another, create striking examples of well-defined zones, with different species adapted to survive the varying degrees of immersion in salt water. This diagram shows the zonation and characteristic plants found on a temperate rocky shore.

shore do not suffer in the same way, although the precise algal species involved in the zonation change with increasing exposure to waves.

Zonation of a Valley

Zonation patterns also exist in terrestrial habitats which are not related to a successional series. For example, on heathlands in southern England the wet valley floors grade into drier ridges and a vegetation zonation can be observed which is related to the availability of water. Bog habitats in the valleys, dominated by *Sphagnum* mosses and cotton sedge (*Eriophorum angustifolium*), pass into a better drained, but still peaty zone in which tussocks of deer sedge (*Scirpus caespitosus*) are a conspicuous feature. The grass

Molinia caerulea and the cross-leaved heath (*Erica tetralix*) are also abundant in this zone. In drier areas, *E. tetralix* becomes less frequent, gradually being replaced by the bell heather (*Erica cinerea*), which may form a codominant with ling or heather (*Calluna vulgaris*) on the drier areas.

From these examples a number of general principles emerge relating to zonation of vegetation. There is always a fairly steep gradient of some critical physical factor or complex of factors involved (such as period of emersion or water table). This results in a spatial arrangement of plant species, in which factors such as competition and predation may also play a part in enhancing the frequently strict zone demarcation which is produced. The zone sequence may also represent a successional series, but this is not invariably the case.

P.D.M.

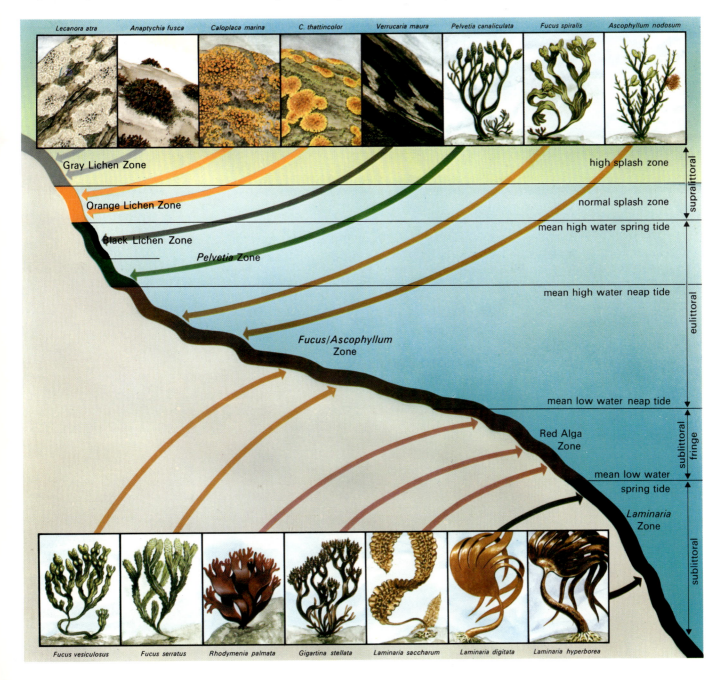

CHAPTER FIVE

Vegetation Today

The World's Main Vegetation Types

A whole range of features has been used to describe the different kinds of vegetation that have been observed in different parts of the world, the most common being physiognomy, dominant species and floristic composition. Criteria not derived from vegetation, such as the habitat or the presumed climax of the community, have also been invoked. Such features have also been employed in erecting hierarchical classifications of vegetation. However, there has been considerable divergence of views as to the validity of any system of classification because of the variety of concepts concerning the nature of the plant community. Despite these difficulties, classifications of vegetation can be produced, usually most effectively on a regional or world basis.

The world's vegetation is most readily described in terms of its major physiognomic classes, differing in the overall appearance of the plant cover. Most simply, these can be divided into forest and non-forest communities, but it is more usual to consider that, broadly speaking, there are five main physiognomic types: forest, scrub, grassland, desert and tundra. Each of these can be subdivided according to differences in the life-form of its dominant plants. For example, within the forest category the dominants in western Europe and the eastern United States are broad-leaved, deciduous trees which shed their leaves in the winter (deciduous summer forest), while those of many tropical areas are broad-leaved, evergreen trees with no seasonal change (tropical rain forest).

Within the formation-types, the use of the floristic composition as a tool for describing vegetation becomes increasingly important. At its simplest, this approach uses the dominant species in the communities to characterize the vegetation, but ideally it is based on an analysis of the full floristic complement. The floristic approach, which has reached its greatest refinement and formalization in phytosociology, is especially suitable for regional studies within the broad physiognomic framework of world vegetation. On these criteria, the formation-types can be subdivided into formations, which show regional differences in their floristic composition. Thus, for example, the deciduous summer forest (D.S.F.) formation-type, mentioned above, comprises at least four formations: North American D.S.F., West European D.S.F., East Asian D.S.F. and Andean-Fuegian D.S.F. Within each formation, a number of more local associations (eg beech forest, oak forest, ash forest) can be recognized. Intensive studies of small areas of vegetation require precise measurement of such features as abundance, height, age, structure and biomass of the component plants, individually and collectively.

In an account such as this it is impossible to describe the enormous diversity of plant cover of our planet. This chapter, therefore, outlines the major formation-types of world vegetation, generally corresponding to the main climatic zones of the Earth, with some mention of the principal regional variants (formations).

The natural vegetation occupying well-drained soils in a particular climatic zone constitutes the zonal vegetation. It need not, however, be confined to one climatic regime but may also occur as extrazonal vegetation in suitable environments under other climatic conditions. Certain types of vegetation are found in a range of climatic zones, perhaps because edaphic factors are more important than the climate in determining their features; such azonal vegetation includes that found in marshes, dunes etc.

When the disturbance caused by Man is essentially minimal, we have natural vegetation. Subnatural vegetation is somewhat modified by Man, while seminatural vegetation is much more so and grades into artificial vegetation, such as cultivated fields and lawns, which has its nature and composition essentially controlled by human activities. In the many parts of the world where such conditions exist now it has to be inferred what the potential natural vegetation would be in the absence of human influence. The types of vegetation can merge to give zones of intergradation, ecotones, of varying widths; indeed, they may be so wide and conspicuous as to be recognized as distinct units, and always make the boundaries between one vegetation type and another to some extent arbitrary. Bearing in mind these provisos, this chapter aims to outline the major zonal and azonal formation types that can be recognized in the world's vegetation cover.

Opposite Daffodils (*Narcissus pseudo-narcissus* subspecies *abscissus*) blooming in mass in an alpine meadow of the central Pyrenees early in the year. The alpine zone on mountains occurs between the tree line and the zone of permanent snow, and although equivalent to the tundra region of the Arctic, floristically it is much richer.

Zonal Vegetation

Structurally, the most complex types of vegetation are *forests*, which occupy a variety of climatic zones ranging from areas with more than 4,000mm (160in) of rain per year and no marked season, to those with an annual rainfall of about 500mm (20in) and sharply contrasting seasons. As the rainfall falls below 400mm (16in) per year and the hot dry season extends beyond about eight months, forests give way to *scrub*, communities dominated by shrubs rather than trees; many other areas dominated by scrub are, however, subnatural or seminatural, the vegetation resulting from human activities. With yet greater aridity the plant cover becomes more and more sparse until hot *desert* conditions are reached, although the virtually plantless deserts of popular fiction are encountered in relatively few regions, such as southern Arabia, western Australia and northern Chile. Seasonality, of course, can also involve an increasingly long cold season, and at higher latitudes the period when temperatures are high enough for growth is too short to support trees. Beyond the northern and southern limits of the forests in both hemispheres, therefore, the only woody plants are deciduous dwarf shrubs associated with herbs, mosses and lichens to form a *tundra* vegetation. In the Northern Hemisphere, such vegetation endures at least 177 days during the year in which the temperature does not reach 0°C (32°F) and, as the conditions become more severe, the vegetation becomes so sparse as to constitute a cold desert.

Grasslands are found in both tropical and temperate regions. There has long been speculation about the role of human activity in their formation and there is, indeed, no doubt that the grasslands of western Europe, for example, are due to Man's destruction of the forests which formerly covered the area. Nevertheless, the steppes of Asia, and the prairies of North America seem to be zonal grasslands in continental areas of less than 400mm (16in) annual precipitation, having cold wet winters and hot dry summers in which at least two months are so dry that trees cannot survive. Insofar as edaphic factors may be involved in this lack of available water, such grasslands are probably partly azonal, as may be the case in the South American pampas, which can occur in areas of up to 1,000mm (40in) rainfall, although with high summer temperatures causing high water loss. In the tropics, grassland is well represented by savanna, and it appears that an annual rainfall of 400mm (16in) again signals the point at which trees are unable to compete with grasses under the prevailing temperatures, although local soil factors may also be important.

It has long been known that with increasing altitude in mountainous regions the vegetation changes, apparently as a result of the decreasing temperatures, in a manner analogous to that which accompanies a move from lower to higher latitudes at lower levels. There are, however, a number of factors which cause *mountain vegetation* to behave in this way, and the popular idea that walking uphill is like walking to the poles is far from reality.

Finally, it must be noted that there is no general agreement about the names used for the various formation-types, or even their circumscription. In the account that follows, an attempt has been made to use the most familiar name, together with some of the commoner alternatives, but no attempt at completeness is intended, or even possible.

D.M.M

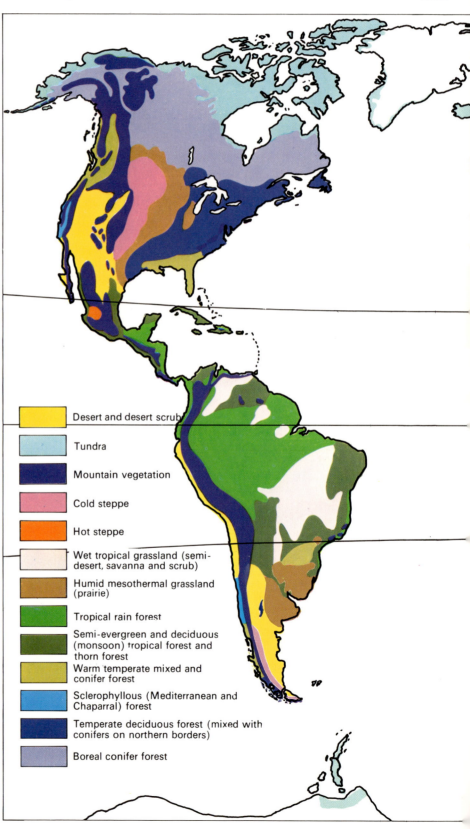

Desert and desert scrub

Tundra

Mountain vegetation

Cold steppe

Hot steppe

Wet tropical grassland (semi-desert, savanna and scrub)

Humid mesothermal grassland (prairie)

Tropical rain forest

Semi-evergreen and deciduous (monsoon) tropical forest and thorn forest

Warm temperate mixed and conifer forest

Sclerophyllous (Mediterranean and Chaparral) forest

Temperate deciduous forest (mixed with conifers on northern borders)

Boreal conifer forest

Forest

Forests are plant communities which have a closed canopy of trees, beneath which may be other smaller trees, shrubs and a ground layer of herbs. Forests are the climax vegetation over about 40% (4,500 million hectares or 11,115 million acres) of the world's land surface, although Man has substantially altered them or entirely removed them from vast areas—from much of western Europe, for example. Different types of forest are found in different climates, and there are striking convergences in forest structure and general appearance between different continents where the same type of forest occurs although composed of different species. The explanation of this similarity of appearance, or *epharmony* as it is called, is still largely unexplained, although physiological causes such as response to water stress or mineral deficiency are involved.

A map of the world's natural vegetation zones, which indicates the importance of climate in their delimitation.

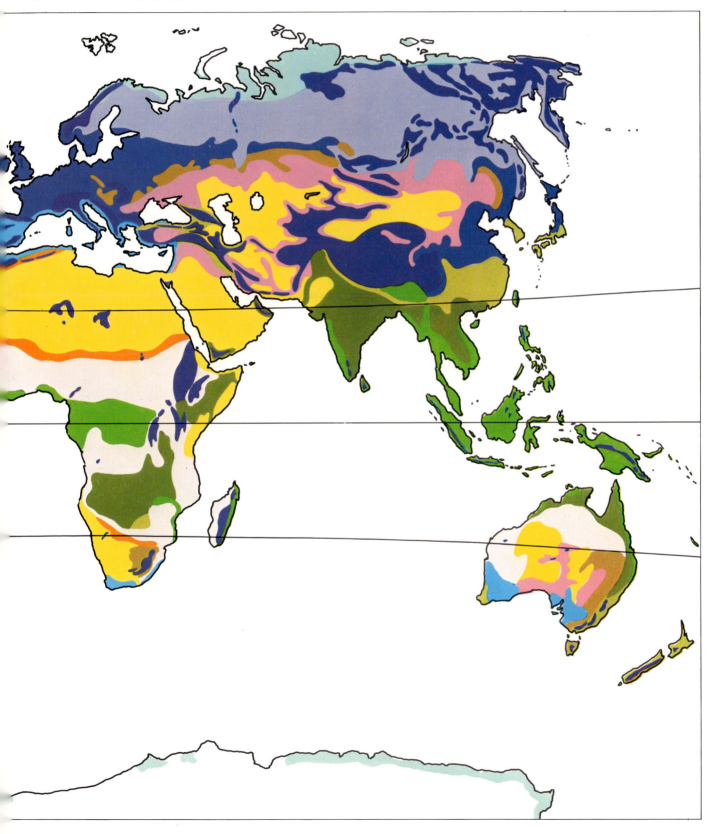

Forest Types

Tropical Rain Forests (Formation-type Pluviilignosa)

In the tropics and subtropics, evergreen tropical rain forests occupy about 1,000 million hectares (2,500 million acres) in the wettest climates. They occur in three great blocks centered on Amazonia, the Guinea-Congo region of Africa and the Malay Archipelago, the latter extending from the Western Ghats of India to the wet, high islands of the Pacific. There is also an isolated small zone in eastern Madagascar and the Mascarenes. Tropical rain forests are the most complex and species-rich communities in the world. They occur in the best conditions for plant life, with no dry or cold season to interrupt growth. The biggest trees average 30–45m (100–150ft) in height, although some reach 60m (200ft). Beneath them grows a dense profusion of smaller trees. The tallest trees commonly occur as isolated emergents standing head and shoulders above a continuous canopy. Shrubs and herbs are rare, the undergrowth plants consisting mostly of small trees. Different tree species reach different heights at maturity, and characterize different layers of the canopy; certain families are important components at the top, others characterize the lower levels. The emergent and tallest canopy trees usually have broad, umbrella-like (sympodial) crowns, composed of numerous small, rather dense subcrowns. Smaller trees commonly have conical crowns taller than they are broad, which frequently have a single main axis (monopodial). Buttresses, which may reach 10m (33ft) or more up the trunk, are an important feature of many kinds of rain forest, and in some types stilt-roots are common.

The bark is very diverse, varying from black through orange to white, and on stems larger than about 0.3m (1ft) in girth it is commonly sculptured with fissures, scrolls or scales. Leaves are principally of medium size (mesophyll), although larger leaves are often conspicuous. Leaves may have a prolonged,

Above Tropical rain forests form the richest and most complex vegetation type. High annual rainfall with no marked dry season, intense sunlight, warmth and high humidity create ideal conditions for growth. This photograph is of tropical rain forest in the mountains of Bolivia, at an altitude of 900m (3,000ft).

acuminate "drip-tip". Compound, pinnate and palmate leaves are also present. Climbers (lianas) and epiphytes are common in a great diversity of forms and species. Stranglers, which start life as epiphytes but then send down roots and ultimately engulf and kill the host tree, are prominent. Saprophytes and parasites occur, including, in the eastern rain forests, *Rafflesia*, which has the largest flowers in the world. The trees provide a complex framework for these other, dependent synusiae (see p. 41), and an intricate set of niches for animals. Both flora and fauna are exceedingly rich. For example, the rain forests of the Malay Peninsula have a total flora of about 8,000 species of vascular plants, including about 2,500 trees species, which is some six to seven times greater than the flora of England and Wales, of equal area.

The absence of any dry or cold season unfavorable for plant growth is believed to be an important contributory factor to floristic richness. Important also is the structural

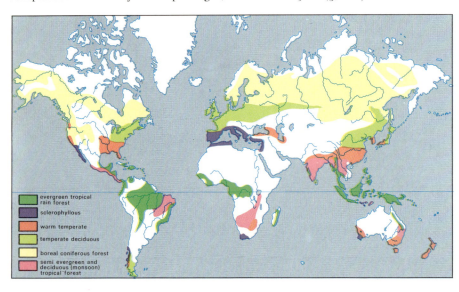

Left World map showing the regions and main types of forest.

evergreen tropical rain forest

sclerophyllous

warm temperate

temperate deciduous

boreal coniferous forest

semi evergreen and deciduous (monsoon) tropical forest

complexity of the rain forest, which provides numerous niches. Many botanists believe that it was in tropical rain forests that flowering plants evolved. Today, they certainly contain the greatest concentration of "primitive" groups. The richest and most complex forests occur on moderately moist sites. Southeast Asia is richer than Latin America. Africa is much poorer; for example, as many species of palm are found on Singapore island as on the entire continent of Africa. Away from the optimum, the vegetation structure becomes simpler and there are fewer different species, for example in swamps (see Ombrotrophic Mires, p. 193) and on mountains (see p. 183). One of the most strikingly distinctive lowland rain forest types is so-called *heath forest*, such as the *caatinga* forest in South America and *kerangas* of Borneo. This occurs on soils that, even by tropical standards, are impoverished: namely the coarse, freely draining, siliceous sands that become podsolized (see p. 135).

Below Within sight of the snow-capped Mt. Kilimanjaro giraffes graze in open parts of the seasonally dry tropical forest, which has been cleared in much of East and Central Africa.

Heath forest is of low stature, small-leaved (microphyllous) and very dense, with a uniform canopy top of high albedo (reflectivity). Its physiognomy is believed to be an adaptation to water stress, and/or mineral deficiency. In the eastern tropics this structure and physiognomy is also found in upper montane rain forest (see Cloud Forest, p. 184) and several similar species occur in both.

Dry (Seasonal) Tropical Forests
As the rainfall drops below 2,000mm (80in) and a distinct dry season becomes increasingly prominent, tropical rain forests change to various kinds of seasonal forest. At first, although the species composition alters, the genera and families present remain much the same as in the tropical rain forest, and only the taller trees are deciduous. Leaf shedding and flowering become synchronized and correlated with climatic seasonality. These forests are of much simpler structure and with fewer woody species than in the rain forests, often only a few being dominant. *Semi-evergreen seasonal forest* grades into *deciduous seasonal forest (formation-type Hiemilianosa)* in the strongly seasonal dry

tropics, where the dry season is approximately five months. In the so-called "monsoon forests" of India and Burma, for example, the sal (*Shorea robusta*) and teak (*Tectona grandis*) form major stands.

With lower annual rainfall (down to about 500mm; 20in) and a dry season extending to about seven months (four to five virtually without rain), the tree cover becomes more open as the forests are gradually replaced by woodland, scrub or grassland. The *savanna forest* in these areas is characterized by the often deep rooting and the flattened crowns of numerous trees and shrubs. Many of the dominants also have small leaves and, except in Australia, are thorny.

In all climates with a marked dry season, fire (see p. 139) is an important factor controlling the structure and species composition of the vegetation. In progressively drier climates, the total amount of moisture becomes more important than length of dry season. Soil factors interact, especially physical ones. Consequently, these monsoon and savanna forests can form a mosaic pattern with more closed, moisture-loving (hygrophilous), prominently evergreen, so-called gallery forests, which occur along water-

courses. Seasonally dry tropical forests are extensive in all three continents. The flora is rich, with many fire-resistant herbs (including bulb-forming geophytes) and grasses in the drier and more open types. Such forests support a rich fauna, including the spectacular big game of East Africa.

Sclerophyllous Forests (Formation-type Durilignosa)

By no means all climatic zones with an unfavorable growing season support deciduous forest. Evergreen forest may occur as well, and this is seen, for example, in the zonal woodlands of Mediterranean climates, in which the summers are hot and dry, and the winters warm and wet, due to cyclonic rain. Annual rainfall is 500–1,000mm (20–40in) but irregular, and there are

Below In many parts of the world bamboos can be prominent in warm-temperate rain forest; here, at an altitude of 1,200m (4,000ft) on the slopes of the Himalayas in Central Nepal, this stand of bamboos reaches 15m (50ft).

prolonged periods of low relative humidity. There is no really cold season and spring is the main growing and flowering season. The Mediterranean basin has been the center of civilization from ancient times, and deforestation, cultivation, grazing and soil erosion have destroyed most of the original forest. The original zonal vegetation was a comparatively open, evergreen forest dominated by a canopy of the holm oak (*Quercus ilex*), 15–18m (50–60ft) tall, often interspersed with the Aleppo pine (*Pinus halepensis*), for example, with shrubs and herbs beneath. Where the trees are cut about every 20 years forest is replaced by a dense shrub vegetation called *maquis*, which is very rich in species, including many geophytes. Maquis is subject to fire and becomes degraded by excessive grazing and burning to open *garigue* (see Chaparral, p. 167). Many species are adapted to Mediterranean climatic conditions by the possession of small, leathery, sclerophyllous, evergreen leaves which minimize water loss in dry periods (see Sclerophyll Trees, p. 51). Other species are deciduous in dry periods. Other common drought-resistant plants (xerophytes) in the Mediterranean area are the *malacophyllous* plants. These differ from sclerophylls by the absence of thick cell walls (so the leaves are not hard), and usually do not have such deep roots or a high heat resistance of the protoplasm; like sclerophylls, they lose water (transpire) slowly when the leaf pores (stomata) are shut and hence are highly resistant to dessication. *Cistus*, *Rosmarinus*, *Lavandula* and *Thymus* are examples.

Similar vegetation occurs in central and southern California, where the dominants again are evergreen oaks (eg *Quercus engelmannii*), often intermingled with *Pinus sabiniana*, for example. In the Southern Hemisphere, sclerophyll forests occur in Chile, where only remnants persist, the Cape of South Africa (where the flora is extremely

The extensive sclerophyllous woodlands of Australia are virtually exclusively dominated by gum trees (*Eucalyptus* spp.). This *E. fastigata* woodland in western New South Wales has an understory of *Acacia dealbata*, a sure indicator that fire has swept through recently.

rich, especially with members of the family Ericaceae) and southern and western Australia, where the genus *Eucalyptus* and members of the families Epacridaceae and Proteaceae predominate to give a distinctive appearance to the vegetation. The area occupied by dry seasonal sclerophyllous forests of all kinds is 1,400 million hectares (3,500 million acres).

Warm Temperate Evergreen Forests

In many parts of the world, a progression from the tropics to higher latitudes is accompanied, not by the development of a hot, dry season, but by a cool winter, which results in a distinctive series of forest-types. Warm temperate forests, which total 100 million hectares (247 million acres) in area, are found mainly on the eastern seaboards of the continents, exposed to monsoon or trade winds. Rainfall is plentiful, 1,500–3,000mm (60–120in), and well distributed throughout the year. In Southeast Asia, China, Korea and southern Japan, eastern Australia and southern Brazil there is a continuous gradation with increasing latitude from wet tropical to subtropical to warm temperate conditions. It is very difficult to distinguish zones in these evergreen forests. They are rich in tree species, especially broad-leaved evergreens, which characterize the most widely recognized formation-type, *warm temperate evergreen rain forest*. Some conifers are often present and where they become important an *evergreen mixed forest* formation-type can be recognized. Epiphytes and climbers are prominent, but less so than in the tropical rain forest. Plank buttresses are absent. Some trees are deciduous, giving marked seasonal differences in appearance. Leaf size decreases with latitude. Plants reproducing by spores (cryptogams) are abundant on the ground and on tree-trunks. Bamboos are common in some types, as are tree-ferns. There are marked similarities with montane tropical rain forests in structure, physiognomy and families represented. The climate, however, has a marked annual rhythm, whereas the montane tropics (see Montane Zone, p. 184) have a greater diurnal than annual climatic range. The general appearance varies with different regions and there is a complete change in flora from low to high latitudes. In Australia, it has been shown that tropical communities extend farthest south on the best soils and moistest sites.

The temperate rain forests are mainly dominated by species of *Nothofagus, Ceratopetalum* (coachwood) or *Daryphora*. In northeast Australia the conifers *Agathis*, bunya bunya (*Araucaria bidwillii*) and the hoop pine (*A. Cunninghamii*) become prominent. Similar forests, dominated by a variety of broad-leaved evergreens, including species of

Nothofagus, cover much of New Zealand, the conifers, predominantly *Dacrydium, Podocarpus* and the famous kauri pine (*Agathis australis*), entering the canopy in the northern half of the country. In Africa, only parts of the Drakensburg mountains and a narrow strip along the south coast have suitably moist sites for this type of forest; here species of *Podocarpus* provide the coniferous component. In South America, the "Valdivian Rain Forest", largely found west of the Andes in central southern Chile, is rich in broadleaved evergreens including *Laurelia, Drimys* etc, with which coniferous communities dominated by *Araucaria* and *Fitzroya*, for example, are intermingled.

In North America, the warm-temperate forests are relatively poorly developed because cold air masses move south as far as

No member of warm-temperate rain forests is more famous than the Sierra redwood (*Sequoiadendron giganteum*), seen here in the Yosemite National Park, California, United States.

the Gulf of Mexico, but they are found near the coast from Louisiana, Florida and Georgia to North Carolina. The tree flora is rich, including evergreen oaks (*Quercus*), a few palms, the evergreen magnolia (*M. grandiflora*) and some climbers. Bald cypress (*Taxodium distichum*) swamps occur in wet areas and fire-climax pine (*Pinus*) forests are found on dry sands. In California, north of 36°, the coastal strip is moist from summer fogs which result from cool onshore ocean currents. Forests of the giant redwood (*Sequoia sempervirens*) occur. Further north on the same coast there are magnificent forests of western hemlock (*Tsuga heterophylla*), western red cedar (*Thuja plicata*) and Douglas fir (*Pseudotsuga menziesii*). Although broad-leaved trees are not prominent in these forests, they are usually considered to be temperate evergreen rain forests rather than boreal coniferous forests (see p. 164), which occur further north.

In Eurasia, temperate rain forest only occupies extensive areas in southern Japan

and southern China, where numerous species of evergreen oaks (*Quercus*) and magnolias abound, together with some admixture of pines. This type of vegetation occurs around the Black Sea and includes the species-rich Tertiary-relict Colchic forest (with *Rhododendron*, *Laurus*, *Ilex*, *Quercus*, etc) of Transcaucasia, which lies in a region where the summers are mild and wet enough for tea cultivation to have replaced most of the zonal forest. It extends eastward as far as the shores of the Caspian Sea.

Summer Deciduous Forests (Formation-type Aestilignosa)

These are perhaps the most intensively studied forests of all. Their total area is 800 million hectares (2,000 million acres). They formerly covered most of western Europe and are still extensive in North America. They are virtually restricted to the Northern Hemisphere (apart from an area along the southern Andes and in Tierra del Fuego). Leaf fall is an adaptation to the marked but not very prolonged cold season when water is unavailable or restricted (by contrast with the tropics where it is an adaptation to drought). Annual rainfall is 700–1,500mm (30–60in). Evergreen broad-leaved trees cannot resist cold or winter drought and in western Europe, ivy (*Hedera helix*) and holly (*Ilex aquifolium*) are both Atlantic species absent farther east where winters are more severe. *Rhododendron* and *Vaccinium*, also evergreen, by contrast, survive winter cold below a snow covering. These forests are found on the eastern coasts of North America and Asia between the warm temperate forests and cold or arid temperate regions. In North America, they extend north to the Great Lakes and upper reaches of the Gulf of St. Lawrence, and west of the Mississippi. In

Asia, they occur in northern Japan and on the adjacent part of the continent. They are also found on the western edge of Eurasia in Europe, north of the Mediterranean zone, and where the Gulf Stream causes winter rains to be replaced by evenly distributed rainfall or rain with a summer maximum, and where the cold season is relatively short. Here they range east to the Urals as a wedge between the steppes and the boreal coniferous forests. Deciduous trees occur where there are four to six months with adequate rain and this forest is absent from extreme maritime climates of the western seaboard as well as extremely continental climates.

The temperate deciduous forest zone of western Europe is one of the most populous regions of the world because the climatic conditions also favor prosperous agriculture and grazing, and only tiny fragments of the forest remain, with virtually none in virgin condition. In England, for example, many woods are situated on the site of primeval forest (as is shown by the genetic diversity of their tree species), and there has been continuous forest cover since the last Ice Age. However, their structure and relative abundance of different species are strongly influenced by many centuries of human interference. Floristically, the western European forests are poorer than the others due to extinction in the Pleistocene ice ages. Beech (*Fagus*), oak (*Quercus*), lime (*Tilia*) and ash (*Fraxinus*) are locally dominant in the single tree layer. In wet places, alder (*Alnus*) and willow (*Salix*) become common. There is a single shrub layer in which hazel (*Corylus*), field maple (*Acer*) and hawthorn (*Crataegus*) are common, and a herb layer. There are few climbers and only cryptogamic epiphytes. The trees flower early, commonly before the leaves open, and most are wind-pollinated;

this allows a long period for fruits to form and ripen before the onset of winter. In early spring, before the canopy becomes leafy, the forest floor herbs flower, creating carpets of blossom (especially of bluebell, primrose or oxlip) which are one of the glories of these forests, equalled only by the spectacular yellow, orange and red tints of the dying foliage in autumn. The early spring temporal niche is succeeded by a spring one, occupied by other herbs, eg wood sorrel (*Oxalis*), which flower at the time of leaf flush. In Asia and North America, there are more genera and species in both tree and shrub layers, including magnolias, numerous maples (*Acer*), tulip tree (*Liriodendron*), buckeye (*Aesculus*) and hickory (*Carya*), as well as temperate outliers of such tropical families as the Staphyleaceae. The summer deciduous forests of southernmost South America, on the other hand, are dominated by only two *Nothofagus* species, N. *pumilio* and N. *antarctica*, which shed their leaves in winter.

Boreal Forest (Formation-type Aciculilignosa)

The boreal coniferous forest occurs to the north of the temperate deciduous forest in regions with colder and longer winters, with which it is linked by a belt of mixed deciduous-coniferous forest. It lies mainly between 45°N and 70°N and is very extensive—1,500 million hectares (3,700 million acres). The main dominants are conifers with xeromorphic needle leaves, which are more resistant to winter cold and drought than broad-leaved trees. Such trees can commence photosynthesis immediately conditions permit in spring, so are better adapted to exploit regions where the growing season is short. To maintain growth, deciduous trees need about 120 days per year with a mean temperature over 10°C (50°F), while conifers can manage with 30 days, although there are differences between species. The narrow, conical, monopodial tree-form with drooping branches is adaptive to regions of high snow fall. These forests have only a poorly-developed shrub and herb layer: shade is dense, decay of falling leaves is slow, so that undecomposed litter covers much of the surface, and the climate is worse. Spruces (*Picea*), firs (*Abies*), pines (*Pinus*) and larches (*Larix*) dominate in different places. The northernmost forest in the world is in eastern Siberia at 70°50′N, 105°E. It is dominated by a larch, *Larix gmelinii* (L. *dahurica*), which is highly productive in the very short summer, and one of the few deciduous conifers, losing its needles each winter.

At its northern limit the boreal forest

merges into *taiga*, which is open parkland having scattered groves of trees and a ground cover dominated by lichens and heath plants. Here, in addition to the conifers, various species of *Betula* (birch) may also be present. Proceeding northward, the trees become smaller and the woodland becomes more and more open, until finally taiga is replaced by open tundra (see p. 173). In the most northerly areas of taiga the only trees which can survive are larch and birch. Although heath species and lichens are the most abundant plants in the ground cover, mosses, sedges, dwarf willow (*Salix*), and dwarf birch are also fairly common. Lichens are an important constituent of the taiga, grazed upon by herds of reindeer; in fact, certain species of lichen are commonly known as reindeer mosses.

There is no comparable belt of coniferous forest in the Southern Hemisphere, where indeed there is no land mass at the appropriate latitudes. On north temperate mountains, a coniferous forest zone commonly occurs above the deciduous broad-leaved forest, reaching up to the tree line (see Timberline, p. 186).

Forest Ecology and Man's Intervention

It is apparent that the species composition of a forest is dependent at the grossest scale on plant geography, and within any region there is variation due to habitat, for example between swamp and dry land, and with different soil types. A further important variation arises from the complex structure of a forest community. At maturity a closed forest canopy casts dense shade. Plants can only grow under the canopy if they are able to succeed in conditions of low light and also of high root competition. In temperate deciduous forest many herbs avoid these limitations, to some extent, by making much or all of their growth (including flowering) before the trees come into leaf. Only certain tree species have seedlings which can grow up under a closed canopy. These are often called shade bearers (or tolerants).

Sometimes gaps form in a closed forest canopy. A storm may blow down isolated trees or fire may sweep through. These gaps are then colonized by tree species that have

The enormous numbers of pollen grains produced by pine (*Pinus*) and other conifers of the boreal zone is demonstrated by this pollen tide on a Canadian lake.

efficient means of seed dispersal. The seedlings are adapted to grow and succeed in the brightly lit, sometimes desiccating conditions of gaps. These species are often known as light demanders (or intolerants). The seedlings cannot grow up in shade, so the trees cannot replace themselves *in situ*. They always colonize gaps and are therefore sometimes referred to as pioneer species. Familiar English pioneer species are Scots pine (*Pinus sylvestris*) and birch (*Betula*). In fact, in many pioneer forests, there is a spectrum of types from obligate shade bearers (ie species which *must* have shade) to strict pioneers, especially in regions with a rich flora, as in the humid tropics.

In the summer deciduous forest of temperate North America, the pioneers *Pinus strobus*, *Quercus* and *Castanea* tend to be replaced in the absence of catastrophic forest destruction by the shade bearers *Acer*, *Tsuga* and *Fagus*. The two oaks native to England are also both light-demanding species. In West Africa, present-day extensive, tall species-rich forest containing much valuable timber in the form of several species of the family Meliaceae (African mahogany) is being replaced by a lower forest with fewer species of less commercial value. The principal timber species of Malaysia, the Philippines and Indonesia, light wooded *Shorea* species (meranti), are near-pioneers, favored by mild but not total forest disturbance. This dynamic aspect of forest composition, therefore, with different species adapted to different temporal niches in the canopy growth cycle, has important ecological and commercial implications. The science of silviculture is based on understanding and manipulating it (see Chapter Seven: Man and the Green Planet).

The syndrome of characters that pioneer species possess, which includes soft, pale timber and rapid growth, makes them especially valuable to foresters for artificial plantations. In the tropics today there is an increasing trend to replace natural forests with plantations. One of the most widely used genera for planting is *Pinus*, all of whose species are natural pioneers, most of them adapted to poor, often drought-prone sites. This trend has gone even further in temperate forests. Trees have long lives, and rare catastrophes leave their mark on forest composition for a century or more. Pioneers come up in large gaps as even-age stands and a rather coarse mosaic of large patches of different ages develop. Shade bearers succeed the pioneers, replacing each other in smaller areas so that ultimately a mixed-age stand develops with a fine-scale mosaic pattern of gap, building-phase and high, mature forest. But it is doubtful if there is ever an equilibrium state, or constant species composition; a catastrophe, or indeed secular (ie extremely slow) climatic change, sooner or later intervenes to cause gross alteration. In western Europe, Man's interference with the forests has been prolonged and profound and influences present-day structure and species composition. The dominance of light-demanding oak over much of England reflects its conscious selection by silviculture, not its ecology.

Various estimates have been made of the primary production of different types of forest, but there are formidable difficulties especially in structurally complex and species-rich types, and the often quoted annual figures of 6 tonnes/hectare (2.4 tons/acre) for boreal conifer forest, 10 tonnes/hectare (4 tons/acre) for temperate deciduous forest and 15 tonnes/hectare (6 tons/acre) for tropical rain forest need to be taken with a grain of salt. Because of high respiration losses, the net production of tropical rain forests probably does not greatly exceed that of temperate forests. The grand stature of tropical rain forest, dwarfing the human form, has sometimes been mistaken by Europeans to imply inherent fertility and high potential of rain-forest sites for agriculture, but scientists have long known that this

is a myth. Tropical rain forest differs from temperate forests in having nearly all of the plant material nutrients locked up in the plants. Very little is in the soil. There is a closed cycle. Fallen litter decays very rapidly and the minerals are taken up again by tree roots, so much so that runoff water from virgin rain forest contains a very low concentration of minerals. It follows that destruction of the forest cover entails removing much of the nutrient store, which is exacerbated by the increased erosion and runoff once the tree canopy has gone. On a global scale, the growth of tree crops as plantations for timber or other products should be concentrated on land converted from tropical rain forest, with annual agricultural crops concentrated in the zone of temperate deciduous forests.

T.C.W.

Scrub

Scrub (shrubland) is vegetation dominated by shrubs—woody plants usually with more than one stem. Many parts of the world support such vegetation as a result of Man's modification of forests but in other areas scrub is a zonal formation-type developed in response to climatic conditions. Restricted shrubby vegetation is found wherever conditions become unsuitable for tree-growth, as at the timberline (see p. 186), but the principal scrub formations occur on the arid side of forest or grassland formation-types as desert conditions are approached. Very broadly, they can be divided into two principal groups: those of semi-arid areas in continental interiors where drought persists throughout much of the year and the winters are often very cold, and those developed in areas of "Mediterranean" climate (see p. 121), with hot dry summers and relatively mild winters with some rainfall, which tend to occur on the west sides of the continents and support scrub that is widely termed *chaparral*.

Scrub of Continental Interiors

Probably the greatest expanse of tropical semidesert scrub occurs in a belt along the south side of the Sahara and other North African deserts, through the Arabian peninsula and into western Asia, from Mauritania in the west to the Thar region on the borders of India and Pakistan in the east. Rainfall is low, down to 120mm (5in) in some places, but it tends to be somewhat seasonal, occurring either in winter or summer, while average temperatures are generally high. Shrubs of many groups, which are either thorny (eg *Acacia* spp.) or succulent (eg *Euphorbia* spp.), predominate to give more or less open communities up to 2m (6.5ft) high, often with thorny herbs at ground level. Structurally similar formations occur in southwest Africa, from Angola southward along the east side of the Namib desert, although they are not very closely related floristically. In Australia, many parts of the arid interior support a semidesert scrub which differs from that in most other subtropical areas in lacking conspicuous succulents among the dominants, which include several species of *Eucalyptus* (mallee) and *Acacia*, together with the strongly drought-resistant grass, spinifex (*Triodia* spp.). Elsewhere, in a warmer part of the Old World, the succulent and thorny *Euphorbia* scrub of parts of Burma seems to be an azonal response to alkaline soils with impeded drainage.

In the other parts of Eurasia, deciduous scrub communities occupy an area from about the Caspian Sea eastward to Mongolia. Rainfall generally does not exceed more than about 300mm (12in) and the winters are

Semidesert scrub, principally dominated by candelabra-like cacti and xerophytic shrubs, on the Isthmus of Tehuantepec, Mexico. The canopy is open so that a conspicuous layer of drought-resistant grasses and herbs develops.

cold. Wormwoods (*Artemisia* spp.) are frequent dominants, along with shrubby members of many genera, including the buckthorns (*Hippophae*), *Astragalus* etc. Interestingly, species of *Artemisia*, especially the sagebrush (*A. tridentata*), are also dominant over large areas in the Great Basin region of North America between the Rockies and the Sierra Nevada. These scrub communities seem to have originally occurred in areas of 125–250mm (5–10in) annual rainfall, but they have extended into moister areas as a result of overgrazing by cattle and other domestic animals. Southward through southeastern California, Arizona and New Mexico a semidesert scrub develops as the rainfall decreases towards 80mm (3in). The creosote brush (*Larrea divaricata*) dominates many of the communities, while various cacti, such as the chollas (*Opuntia* spp.) and saguaro (*Carnegiea gigantea*), become conspicuous. Similar vegetation, with a great variety of cacti, other succulents belonging to genera such as *Agave* and *Sedum*, and thorny shrubs of, for example, *Mimosa* and *Cassia*, extends through tropical America in arid parts of Mexico, Venezuela, Ecuador and Peru to northwest Argentina and the southern fringes of the Atacama desert in Chile. Further south, a temperate scrub dominated by deciduous shrubs such as *Larrea*, together with numerous cushion-forming plants, many of them woody, occupies much of Patagonia east of the Andes and extends almost to the Straits of Magellan.

D.M.M.

Chaparral

The other major zonal scrub formations are usually grouped together under the name chaparral. This is dominated by evergreen shrubs and occurs in areas of the world with a Mediterranean climate. In this climate rainfall—300–900mm (12–36in) per annum—is concentrated (over 65%) in the winter months: November to April in the Northern Hemisphere, May to October in the Southern Hemisphere. The winters have at least one month in which the mean temperature is below 15°C (59°F), but the temperature rarely goes below freezing and there are usually no severe frosts.

This climatic type is found in narrow regions between desert and areas with true temperate climates. It provides more than adequate water through the active growing season, which is cool, but subjects the vegetation to considerable drought stress during the summer. These conditions largely occur in rather restricted areas on the western coasts of the six inhabited continents, mainly between latitudes 30° and 40°. About half the total area is found surrounding the Mediterranean Sea, but there are substantial regions in Australia and California and much smaller areas in Chile and South Africa. The climatic similarities between the areas have resulted in their possessing vegetation types with a very

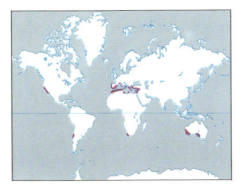

A world map showing the zones where chaparral vegetation occurs.

similar physiognomy. Known collectively as chaparral, they include *maquis*, *garigue* or *macchia* (Mediterranean and South Africa), *matorral* (Chile) and *fynbos* (South Africa).

The dominant plants of chaparral are broad-leaved sclerophyllous shrubs and small trees, but a special feature of this vegetation type is the importance of annual plants, which may account for up to half the total species; they are specially favored by mild, moist winters and summer drought. There are few deciduous shrubs; it would seem that the temperature is not cold enough, nor is the drought severe enough, to make the deciduous habit advantageous.

The Mediterranean climates owe their existence to the presence of offshore cold currents along the western sides of continents, where westerly winds prevail. These currents are associated with the presence of melting ice caps and thus cannot be more than a million years old. The plants that occupy these climatic regions may have evolved since the Pleistocene (probably most of the annuals did this). They may also have come (as did most of the shrubs) from tropical or near tropical ancestors, partly by selection of those which possess characteristics favoring survival in the Mediterranean climate and partly by evolution of new, tolerant forms. Whatever happens to the world's climate in future, whether it becomes cooler or warmer, it will probably eventually result in the elimination of these cold currents and with them the Mediterranean climate and the chaparral as it is today.

Although chaparral has been studied in all the areas where it occurs, that of southern California is best known. Most of the southern Californian chaparral is of two main types: *chamise chaparral*, consisting of needle-leaved evergreen scrub dominated by *Adenostoma fascicularis*; and *mixed chaparral*, dominated by a variety of broad-leaved sclerophyllous shrubs, including several *Ceanothus* spp., *Arctostaphylos* spp., *Cercocarpus betularis*, with several other genera including *Quercus*, *Arbutus* and *Umbellularia*. There is a third type, called *soft chaparral* or *coastal sage*, dominated by *Salvia* spp. and *Eriogonum* spp. This occurs on very dry sites, the mixed chaparral being on the most moist sites. Similar gradients from the wetter to drier sites, but of course with different

species, can be seen in chaparral elsewhere.

Although the environment is broadly suitable for deciduous shrubs, the soil moisture often falls below the wilting point, the point at which it has insufficient moisture for deciduous plants to replace the water they lose through their leaves (transpiration). Much of the precipitation is lost to the vegetation by runoff and by penetration of underlying strata. A great deal more is lost by evapotranspiration. Depending on the amount of rainfall during winter, the soils will fall from field capacity, when it is at its most moist, in winter to wilting point at any time from early summer to autumn. The availability of water naturally affects the amount of carbon fixation that the plants are able to achieve during the summer. Growth closely follows soil moisture content and thus the bulk of growth takes place in the wet season, at which time carbon fixation is at its highest.

Apart from the overriding effect of climate in shaping chaparral vegetation, another factor, fire (see p. 139), plays an important role. Fire is a natural part of the environment in Mediterranean climates and chaparral exhibits various adaptations to it. Fire occurs most frequently when the vegetation is at its driest, ie in the drought period. In the past it was usually caused by lightning but now it is often started by people. The inflammable state of the plants often leads to large areas being burned very rapidly and most of the vegetation and the surface litter are destroyed, releasing large quantities of minerals hitherto locked up in the plants. As well as these effects, fire has a dramatic effect on some characteristics of the soil.

Cassia closiana flowering in the Chilean chaparral at Maipu, Santiago.

A general view of chaparral near Tucson in Arizona, United States. Cacti, drought-resistant shrubs and annual herbs are typical of chaparral vegetation, which has to be adapted to hot, dry summers and cool, moist winters.

Chaparral plants produce quantities of such chemical compounds as essential oils, waxes, phenols and terpenes, some of which are deposited on the soil and some of which remain in the litter. These substances exert some control on microbial activity in the litter and upper layers of the soil, and some also are phytotoxic, affecting the growth of plants (see Allelopathy, p. 147), as well as inhibiting seed germination. In addition, hydrophobic substances leached from the plants or their litter are deposited in the upper layers of the soil and make them highly nonwettable. In unburned chaparral these substances accumulate with time.

The temperatures in chaparral fires can be very high, often over 200°C (392°F) below the soil surface, 650°C (1,202°F) at the surface and 1,100°C (2,044°F) above. Several phenomena result from this. The hydrophobic substances are distilled out and condense in lower soil layers, making them nonwettable and often resulting in surface erosion from movement of the now wettable surface layers. The phytotoxic substances are destroyed, while many of the organisms above and in the surface layers of the soil, and some below the surface, are killed.

Many of the shrubs and trees are capable of sprouting new growth within a few days of a fire, although often it is much longer before this occurs. Since fires usually occur in the dry season there is little growth of other plants and the advent of the rains often results in erosion and increased runoff of water. However, after the rains, immense numbers of annuals germinate. These annuals may have been absent from the vegetation for many years. The ground under the shrubs in chaparral is often quite bare, concealing the fact that the soil contains millions of dormant seeds. For a few years after a fire, large quantities of annuals grow, only to decline again as the shrubs regenerate. Some regeneration of the shrubs by seed also takes place after fires. After about 20 years, depending upon the local circumstances, the shrub canopy becomes more or less complete. After this, if no further fire occurs, the canopy goes on thickening. Normally fires are frequent, so unburned chaparral over 50 years old is rare. When a fire happens in very old unburned chaparral it is extremely damaging.

Chaparral has a special interest for Man because of the many early civilizations which developed in the Mediterranean region. Not surprisingly, here and elsewhere, human activity has profoundly modified the vegetation, and many areas formerly supporting sclerophyllous forest have been converted to chaparral-like vegetation as a result of burning, felling and overgrazing. It is probable that the Mediterranean vegetation known as *garigue* (*phrygana* in Greece, *tomillares* in Spain) has been largely formed in this manner. Furthermore, shallow and locally dry soils support chaparral communities in areas climatically suited to forest. For this reason there is a gradation between true, zonal chaparral and azonal communities usually grouped together as heath (see p. 188).

s.w.

Desert

Substantial parts of the Earth's surface have been termed desert; they include most areas which have low precipitation because of very cold or very warm dry air moving over them, so that plant cover is discontinuous and often virtually absent. Much of the vegetation in desert areas has been described in other sections of this chapter, since in all but the most extreme conditions open communities of, for example, dry scrub or grassland communities may be encountered. Consequently, we shall here concentrate more on the devices by which desert and semidesert plants cope with the extreme environmental conditions.

Desert Types

Subtropical or Warm Winter Deserts
These are the product of permanent high-pressure cells which dominate their climates. Descending air warms adiabatically (without gain or loss of heat energy) and arrives at the surface almost devoid of moisture, hence incapable of producing precipitation. These deserts, such as the Sonoran and Sahara, grade toward the equator into grasslands, where moist equatorial air penetrates during the high sun period, producing convectional showers. Toward the poles they grade into Mediterranean sclerophyll vegetation supported by winter rains.

These deserts have a very great summer heat, very high exposure to sun, great diurnal range, low to moderate humidity and winter temperatures normally above freezing.

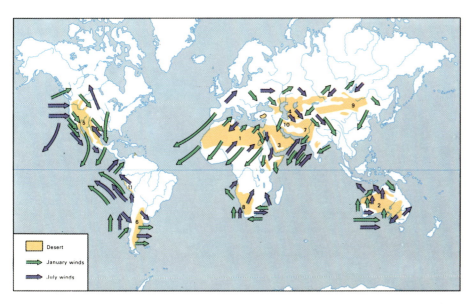

Above The major deserts (and desert scrubs) of the world, excluding polar deserts: (1) Sahara, (2) Australian, (3) Arabian, (4) Turkestan, (5) Great American, including the Great Basin, Mojave, Sonoran and Chihuahua deserts, (6) Patagonian, (7) Thar (India), (8) Kalahari and Namib, (9) Takla Makan including the Gobi, (10) Iranian, (11) Atacama. The green arrows indicate prevailing January winds and the purple arrows prevailing July winds. In deserts such as the central Sahara, Arabian, Gobi and Australian the main cause of aridity is their great distance from the sea and the lack of rain-bearing winds. The Namib and Atacama lie in coastal areas, but the cold oceanic winds bring only fog and mist, not rain. Monsoon-type summer rain, brought by the northward-moving July winds, occurs to a very limited extent in deserts such as the southern Sahara, Arabian and Thar. Subtropical deserts, such as the Kalahari and southern Sahara, are more or less permanently deprived of rain owing to semipermanent areas of high pressure, while the American and parts of the Australian desert are deprived of rain by nearby mountains.

Cool Winter Deserts

These occur in similar climatic regions but at higher altitudes; the Mojave desert of California is an example. They too have extreme summer heat and a large diurnal temperature range, but minimum winter temperatures are often below freezing and some precipitation occurs as snow.

Cool Coastal Deserts

These are found where warm winter deserts border on a cool sea on the western sides of continents. Here a layer of cool, moist sea air is drawn inland below the dry descending air, producing fog but no precipitation and very equable, mild conditions. They are virtually rainless but are drenched in sea fog much of the time. Daily and seasonal temperature ranges are low. This climate favors epiphytes and lichens.

Rain-shadow or Cold Winter Deserts

Such deserts as the Great Basin or Gobi are the result of the passage of air over high mountains, where it loses water and descends, becoming hot and dry on the lee side. Deserts like the Gobi in interior Asia, remote from the Atlantic and sheltered from the Pacific and Indian Oceans by mountains, are virtually free of any precipitation.

Polar Deserts

These have very low precipitation, although water is available to some plants in summer as the permafrost melts. Extremely cold air holds little moisture and the temperature all year round is near, below or far below freezing. Humidity is very low in winter and quite low in summer. The plants of these deserts face different problems from those of the remainder. (See also The Arctic Region, p. 214: The Antarctic Dominion, p. 218.)

Desert Climate

The distribution of deserts closely coincides with areas of high rainfall variation; in other words, precipitation in desert areas is unpredictable as well as low. Rainfall, when it comes, is often in the form of storms in which much water is lost (especially to shallow-rooted plants) because runoff is rapid.

Other factors common to deserts (except polar deserts) are high evapotranspiration (loss of water from the soil by evaporation and plant transpiration), usually low relative humidity and, in many cases, nights cold enough to allow dew deposition. Temperatures may reach over 40°C (104°F) and soil temperatures may go as high as 87°C (189°F).

Wind is a constant feature during the day, the result of convectional warming of the air and its replacement from elsewhere; this has a dual effect in that it moves dry air over the plants and carries suspended particles which can abrade plant surfaces. The net movement of water in the soil is upward, so that dissolved salts in the soil water are brought to the surface and redeposited there, often in great quantities, resulting in extreme salinity of some desert soils. In all it is an inhospitable environment for living organisms.

Because of the severe environmental conditions of deserts the plants are widely spaced and/or sporadic in their appearance. Consequently, physical features of the environment are prominent, and deserts thus provide an important natural laboratory for studying the ways in which plants respond to stress. This topic is clearly highly relevant to understanding the factors responsible for the increasingly open and less complex vegetation as forests give way to scrub in various parts of the world.

Water deficit and temperature stress are the two major problems of deserts. Salinity is a problem shared with salt marshes (see p. 204), but in deserts the salt is usually sulfate rather than chloride.

All vital activities require moisture; most cells and tissues fail to survive if their water content falls below 10%. Water normally enters plants through the roots and what is not metabolized is lost in transpiration through the leaves. When loss exceeds supply, drought ensues. This may result from a low soil moisture content which prevents the plant obtaining enough water, or from such a high rate of loss that even in a moist soil the plant cannot transport water fast enough to maintain the rigidity of the cells (turgor).

Tissue dehydration causes two kinds of injury to plants, mechanical and metabolic. Mechanical injury is probably more common, since when dehydration ensues mechanical injury has often taken effect before metabolic injury has come into play. However, when very slow drying out occurs metabolic injury follows.

In the 1930s the Russian plant physiologist W. S. Iljin put forward a theory of drought injury that has not been seriously challenged. He suggested that during both dehydration and remoistening there are mechanical stresses in cells. The vacuole shrinks during drying and this results in an inward stress on the protoplast. If the cell walls are firm, the protoplast may tear away from the wall as the outward stress of the mechanically strong cell wall resists the inward pull of the protoplast. Soft cell walls will collapse and many bryophytes (mosses and liverworts) resist drought injury in this way.

Metabolic injury is not very well understood. There are general trends in metabolism that follow drought: photosynthesis (see Light, p. 122) may actually increase for a while, although eventually it falls, and respiration also increases. Often this results in a net loss of carbohydrate in the plant. Normally there is an inhibition of starch-sugar conversion in plants, but under drought conditions sugars are often transferred from the leaves to the stem as a result of relaxation of this inhibition increasing the osmotic potential of the cell sap. Whether this results in better water uptake is not clear. Reallocation of nitrogen compounds within the plant may occur, protein breakdown going on in leaves and nitrogen compounds being transported to leaves nearer the growing point.

A desert landscape of sand dunes and scattered, stunted shrubs in Death Valley, California, United States. Only the most arid deserts are completely devoid of plants.

Methods of Plant Survival in Deserts

In order to survive in the arid conditions of deserts, plants must have some kind of resistance to drought. Drought resistance has been divided into three arbitrary categories: tolerance, avoidance and evasion.

Tolerance

This is perhaps the least common means of dealing with drought, at least in flowering plants. There are desert bryophytes and ferns which are extremely tolerant of drought, being capable of drying up until apparently lifeless and, on rewetting, resuming growth very quickly. One flowering plant well known for its ability to tolerate drought is the creosote brush (*Larrea divaricata*), which occurs around desert margins over a wide area of the southwestern United States, and also in parts of South America. In extreme drought, the lower leaves dry up and die, but the small leaves in the bud, although they dry out, turn brown and become quiescent, rapidly resume growth after rain, and develop into actively growing shoots. Total lack of rain for over a year is often not sufficient to kill this species. Probably partly as a result of its extreme drought resistance, *Larrea* occupies some sites with very low rainfall and in such sites individuals are spaced very widely.

Presumably in a species like *Larrea* the metabolism of the plant differs from that of more normal plants in ways not fully

understood. It seems likely that nitrogen metabolism and also the ability of the tissues to retain water more effectively are involved, but again the evidence is scanty. Knowledge of how such a plant survives in the desert could well be of considerable value.

Another factor which desert plants have to face is high temperature. Plant organs may reach temperatures 20°C (68°F) higher than that of the air. Yet there is little evidence of high temperatures causing injury to desert plants.

Avoidance

Used by most desert plants other than annuals, this covers a variety of means of dealing with drought. In fact, desert plants illustrate very clearly what has been called "community-wide adaptation", in that the problems provided by the habitat are not solved by one or even a few adaptations, but by many, in a variety of combinations.

One of the major sources of water loss in plants is through the leaf-pores (stomata), which must be open for at least some of the time for effective gaseous exchange, and hence photosynthesis, to take place. For plants with access to reserves of water deep in the soil, this is no problem, indeed it has advantages, since high transpiration rates keep the leaf temperature down because of the latent heat of vaporization of water. Plants that do not have access to unlimited water face the problem that, as the leaf temperature rises above that of the environment, the vapor pressure of the leaf cells also rises above that of the surrounding air and the drying potential of the environment increases. Stomatal closure, reduction of cuticular transpiration and reduction of leaf surface-areas are the main ways of dealing with this.

Stomatal closure is very important. Some

of the most successful plants open their stomata in the early morning, thus allowing free gaseous exchange, and close them as the temperature rises. Sometimes the closure of stomata is very rapid: *Spondias tuberosa*, a plant of arid areas in Brazil, can completely close its stomata within five minutes of having its water supply cut off. Furthermore, in some succulent plants, dark carbon dioxide fixation, with the fixation of carbon as organic acid at night, allows photosynthesis to proceed during the day without excess water loss (crassulacean acid metabolism).

Two families of flowering plants, the Cactaceae in the New World and the Euphorbiaceae in the Old World, have many succulent members, which reduce surface area by having leaves greatly reduced or absent, and cut down cuticular transpiration with the aid of waxy surfaces. Leaf shedding is common in desert species; as available water decreases, leaves are gradually discarded. One species conspicuous for this is the ocotillo or devil's coachwhip (*Fouquieria splendens*). The leaves have stout petioles about 15mm (0.6in) long, and near the base of the leaf-blade the mechanical tissue of the petiole ends as a spine. When the leaves fall this spine remains. After moderately heavy rains, leaves appear and last as long as the quantity of water allows; the plant can be induced to produce successive crops of leaves by watering.

Opposite Plants living in deserts are variously adapted to surviving long periods of hot drought and then making the most of short periods of rain. The two illustrations depict a desert scene from Arizona, United States, before (**top**) and after (**bottom**) rain. Typical plants are shown, although it is unlikely that all would be found together in this density in a natural habitat. Many of the plants are succulent cacti, such as the barrel cacti (*Echinocactus* sp.) (**left and middle foreground**), the giant, columnar saguaro (*Carnegiea gigantea*) (**left foreground**) and the prickly pear (*Opuntia* sp.) (**right foreground**). These plants have no true leaves but large, green succulent stems from which they draw reserves of water in the dry season. Other succulents, such as the yucca (*Agave* spp.), have basal rosettes of leaves from the center of which large candelabra-like flowering stems will emerge. The cholla (*Opuntia echinocarpa*) (**center foreground**), has succulent stems clothed with masses of spines, which reduces water loss and prevents grazing by animals. Deciduous shrubs are also represented, the creosote bush (*Larrea divaricata*) (**middle distance**) being particularly well represented and adapted to this environment. In the ocotillo (*Fouquieria splendens*) (**seen behind the prickly pear**) the leaves are lost in the dry season but the spiny stalks persist. When the rain comes the desert bursts into flower and the water reserves of the succulents are replenished. The annuals appear, for example the blue- and yellow-flowered sand verbenas, flowering and fruiting during the brief wet season. As well as possessing deep tap roots that draw up the water from great depths in the dry season, many desert plants also have numerous surface roots that absorb the rain as soon as it falls on the soil surface and penetrates the first few centimetres, for example the saguaro. Others have bulbous roots that store the water for the dry season.

The utilization of infrequent rain has to be rapid, as otherwise much of it is lost by evaporation. Some desert plants have exceptionally well developed root systems. Being often very shallow rooted, they can rapidly absorb water from even light showers that moisten the top centimetre of soil. Many cacti produce new rootlets and root hairs within hours of a shower, and species like the giant saguaro (*Carnegiea gigantea*) can absorb huge quantities of water, which they store in their tissues.

The possession of extensive root systems by desert perennials allows full exploitation of the whole soil volume; this in turn must result in intense competition for water and it is not surprising that in areas of very low rainfall the individuals are widely spaced; in fact in many populations there is a close correlation between density and rainfall. The regular spacing out of *Larrea divaricata* individuals shows this particularly well.

Evasion

This is the third method by which plants survive dry conditions, either by having a very short life cycle or by passing the unfavorable part of the year as an underground organ, as in bulbous plants (geophytes). Seeds are highly tolerant of environmental extremes and can often survive many years of drought in desert soils; their metabolism is slowed almost to zero and they offer a relatively safe way of passing through unfavorable times. In deserts that have rain very rarely, plants with short life cycles (ephemerals) may be almost the only plant that can survive. Along with the use of seed for survival comes, often, a shortening of the life cycle to a few weeks, effective seed dormancy, and a variety of dispersal mechanisms that enable the species to occupy areas which are suited to it and at the same time give it some ability to disperse to other areas. Further, there have evolved a series of germination responses that adapt the seeds most precisely to germinating them only when conditions are such that the life cycle is likely to be completed.

Detailed work has been carried out on seeds of a number of species collected in the Mojave desert, which has two rainy seasons, the main one being in winter between December and March and the secondary (more erratic) one in summer. Two types of ephemerals occur: winter annuals that only germinate in early or late winter, and summer annuals that germinate after adequate summer rain and remain dormant through the winter rains.

In experiments, when the day temperature was kept at 18°C (64°F) and the night at 13°C (55°F), only the winter annuals germinated and the summer annuals did not appear. If the day temperature was 27°C (81°F) and the night temperature 26°C (79°F), only the summer annuals germinated. This precise control of germination by temperature effectively determines the time of appearance of these plants in the desert. They are, as might be expected, related to the

temperatures experienced in summer and winter in the desert from which the seeds came.

Germinating only in response to temperature is not enough in such an extreme climate. Water, as we have seen, is the primary limiting factor in deserts and the behavior of the seeds reflects this. Many of them will not germinate unless a minimum of 15mm (0.6in) of rain—preferably more—has fallen within a short time, this despite the fact that 3mm (0.1in) is sufficient to make the surface layers of the soil quite wet. Seeds of desert annuals (and indeed of many annual and other plants in less arid habitats) contain germination inhibitors in their seed coats that require a certain amount of rain to leach them out and enable germination to occur. Normally this is adequate rain to allow completion of the life cycle. Some seeds require bacterial action on their seed coats; this needs prolonged soaking of the soil. In either case the main requirement is a heavy rainfall.

Many of these desert annuals are mesophytes which are related to or even of the same species as plants of moister habitats. They are distinguished merely by using the brief period of favorable climate, the one interval in a long period during which they are totally unable to grow to the reproductive stage.

Seed dispersal of desert annuals also shows some precise adaptations. Many if not most annual plants characteristically have small, light seeds that are capable of being transported easily. At the same time, many desert plants have very restricted habitats even within the already restricted possibilities of the desert as a whole. Clearly if a parent plant has successfully completed its life cycle in an area it is likely that its offspring can also. At the same time, the potential for dispersal to other suitable habitats must be available.

Hygrochastic plants (ie plants the fruits of which open by absorption) are found only in dry regions. They tend to retain their seeds during the dry season, but soon after the rains commence the seeds are dispersed. One such plant is *Plantago cretica*, from the Mediterranean region of Israel. As the plants dry out, the inflorescence-stalks (scapes), formerly curving upward, curve down and are appressed to the ground; the flowers are also held close to the inflorescence scape. After the rain the scapes curve upwards again, the bracts and sepals open, and seed capsules are exposed, and the rain washes out the seeds. The cell walls of the epidermal layers of the seed coat swell and become mucilaginous when wetted and this helps to make them sticky and prevents them being blown away.

Perhaps the most striking adaptation of this kind is that shown by *Gymnarrhena micrantha*, a member of the Compositae found in the Negev Desert. This has flowering heads that are borne in clusters, on very short branches at the soil surface. The fruits have a pappus of hairs, characteristic of the family, and are dispersed by wind. In

addition, there are subterranean inflorescences with one or two flowers whose corolla tubes open to the soil surface. The fruits of the buried flowers have a vestigial pappus and germinate *in situ*. Thus this species ensures continuing occupation of a favorable site, along with dispersal to other areas.

The complex of plant types that go to make up a desert flora combine to exploit the environment in a variety of ways. They occupy different strata above ground, from just on the soil surface to 10m (33ft) or more, and their roots are similarly stratified. They exploit different temporal niches, some being evergreen, some semideciduous, some deciduous, some succulent and some ephemeral. Some use only leaves for photosynthesis, some leaves and stems and some only stems. In any mature and stable plant community the members occupy the available niches in a variety of ways. The desert, because its extreme nature limits the number of niches available, is a habitat in which the diversity of means of exploitation of the available resources is much more clearly discerned than in more complex vegetation.

Deserts also exhibit clearly a phenomenon that is apparent in all kinds of vegetation, but perhaps much more so in extreme environments: convergent evolution. Similar climates produce vegetation of similar physiognomy, and—as much recent work has shown—plants of similar physiology. Some of the convergences are very striking. The North American *Fouquieria splendens* in the Fouquieriaceae resembles surprisingly closely a plant of Madagascar, *Euphorbia splendens*, in its possession of deciduous leaves, spiny stems and red flowers. The Cactaceae and Euphorbiaceae show numerous similar characteristics. Many of the adaptations described earlier can be found in desert plants from widely differing parts of the world. The plants of extreme environments sometimes yield information on their adaptations more readily than those of relatively favorable areas. Desert plants offer opportunities for much future research.

s.w.

Tundra

A Lapp term meaning barren land, tundra originally applied to treeless areas of northern Finland that were dominated by low shrubs and herbaceous perennials, together with mosses and lichens. It has now been extended to describe most of the vegetation beyond the forests and associated communities in polar regions, as well as much of the alpine vegetation above the timberline in mountainous areas. Thus, tundra covers some 700 million hectares (1730 million acres) of the far north of Eurasia and North America, together with zones in most mountainous regions of the world, these accounting for a further 32 million hectares (79 million acres). In the Southern Hemisphere, tundra communities occur in

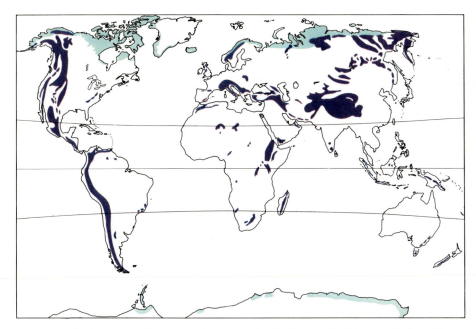

A map of tundra distribution. Blue areas indicate true tundra zones of the Arctic and Antarctic, while the purple areas indicate mountainous areas on which tundra-like vegetation occurs (see Mountain Vegetation, p. 183).

many parts of the maritime Antarctic and on certain of the sub-Antarctic islands, covering at most about 50 million hectares (124 million acres). Although this latter region represents only a small fraction of the tundra regions of the world it is of great interest because the vegetation has developed under very different conditions to that in the Northern Hemisphere.

Polar and alpine tundra extend over such tremendous ranges of latitude and altitude that it is difficult to generalize about the plants found there. The following account draws initially upon the similarities but points out areas of greatest contrast in both tundra environments and the plants found there, and then reviews the different assemblages of plants found in the various tundra regions of the world.

The Tundra Environment

A fundamental characteristic of the tundra environment is its inherent instability and frequent youthfulness. Land surface in tundra regions may be viewed as being available to plants on three time-scales. Firstly, there has been the long-term progression from one glacial maximum to the next during the Pleistocene glaciations. Secondly, there are short-term oscillations in climate such as occurred during the Little Ice Age and are seen in the present-day retreat of glaciers from the maximum extension in the late 19th century. Finally, there is an ever-present instability arising from the continuous frost-shattering of rocks and boulders, the movement of scree, the slow movement of soil down slopes (solifluction), and all the minor year-to-year variations in

climate that can lead to snow beds melting a critical few weeks too late for plants to make much growth.

In general, tundra regions are characterized by long cold winters and short cool summers, with a chance of frost at any time. The mean air temperature of the warmest month usually falls between 0°C and 10°C (32°–50°F), although there is a marked contrast between continental and oceanic sites, with a much restricted amplitude in the annual course of mean monthly temperature at the latter. The greatest contrast between polar and alpine tundra lies in the diurnal radiation regime. Progressing toward the poles from the Arctic and Antarctic Circles, the sun fails to set for increasingly lengthy

periods: 60% of the annual radiation received may fall in only three months. At the other extreme, in the alpine tundra of equatorial mountains (see Montane Zone, p. 184), day-length hardly varies throughout the year and plants may pass through a temperature range equivalent to that of a full year in polar tundra on each and every day of the year. The subpolar and subalpine tundra environment is therefore often colder than that of the polar and alpine tundra, this apparent paradox being caused by the sun dropping below the horizon each night in subpolar regions, while cold-air drainage occurs during the night from alpine down to subalpine tundra.

Permafrost is generally prevalent in arctic tundra, but is much less common in oceanic and alpine areas. When the active layer starts to thaw in spring the upper parts of the ground become saturated, with the surface of the permafrost beneath representing as much of a barrier to free drainage as solid rock. Impeded drainage means that bogs and swamps are common, despite the fact that annual precipitation is often equivalent to less than 200mm (8in) of rain. Since there is often little surface flow of water carrying in nutrients and low input from rain, it is hardly surprising that major nutrients (especially nitrogen) are commonly limiting factors for plant growth in continental arctic tundra. Very local amelioration is often evident around bird perches, such as the low mounds

The east coast of Signy Island in the South Orkneys. Ice-free ground in the maritime Antarctic is mostly restricted to narrow lowlands by the coast. The exposed areas of ground are known as nunataks (see p. 225).

used by snowy owls, and the entrances to the burrows of small rodents, where the grasses present are much more lush in their growth than their counterparts in the unmanured surroundings. It is with respect to nutrient availability that a great contrast is seen between oceanic and continental tundra sites. On the oceanic islands of the sub-Antarctic and the maritime Antarctic, much of the ice-free ground is in the coastal areas and there is a large input of nutrients from breeding seabirds (penguins and burrowing petrel) and moulting seals, as well as from sea spray.

In all tundra regions, both species composition and the growth-rate of plants are controlled by the length of the growing

Left Torne Trask, a mountain lake in arctic Sweden. In the foreground sparse vegetation including mosses covers an exposed rocky shoulder, while tundra grasses cover the wetter slopes below.

Below A tundra scene at Point Barrow, on the north slope of Alaska. In this photograph the surface layer has melted, but below it is a frozen layer of permanent permafrost which does not allow the melted water to soak away, thus creating boggy conditions.

season, which is in large part tied to the depth of snow cover to be melted in the spring. Snow provides effective insulation against the extremes of air temperatures that occur during the winter. Diurnal fluctuation in temperature are more or less restricted to the top 30–50cm (12–20in) of snow; once depths greater than this have accumulated, the vegetation beneath may remain at or slightly below freezing for much of the winter. Many vegetated areas are covered by much less than 30cm (12in) of snow and so are poorly insulated; in extreme cases plants are exposed to a risk of winter drought, when they may become freeze-dried. However, in such situations the snow melts more quickly at the onset of summer, so that the potential growing season is longer than in the better insulated sites, although this potential may not be realized because the supply of water is often limited to the time of the main spring melt. Lack of available water is a problem to be faced by many plants of the tundra.

At sites with the greatest depth of snow it is not uncommon for an air and/or liquid water layer to develop at the ground surface as the melt starts, so that root growth and the development of leaves may begin. As the snow cover becomes thinner, increasing amounts of light can penetrate to sufficient levels for photosynthesis to proceed slowly; indeed, when snow banks melt late, some mosses may make a large part of the season's growth under snow. When, as in tundra, wind can blow uninterruptedly across large areas, shearing effects are pronounced. In consequence, the smallest microtopographic feature can be associated with a great range of microclimatic conditions, especially with differences in radiant heating, as, for example, between the north and south faces of the same boulder or even across the hemispherical surfaces of cushion- and hummock-forming plants (see under Exposure, p. 128). Certain plants of arctic tundra, among them species of *Pedicularis*, are known as compass flowers, because the development of flowers is more rapid on the warmer south side of the flowering shoot.

Growth and Reproduction of Tundra Plants

Although most plants of the tundra fall into a number of diverse growth-forms (mosses, lichens, grass-like plants (graminoids), dwarf shrubs, herbaceous perennials, cushion- and mat-forming plants) and each of these is characteristically more likely to be found in one habitat than another, certain factors in the tundra environment will affect them all,

Tundra vegetation becomes increasingly sparse as the arctic ice is neared. In the extreme conditions at the edge of the ice-cap only lichens can survive to form a thin crust on the rocks.

to a greater or lesser extent. Because the growing season is generally short, with freezing temperatures possible at any time, plants which are able to grow and reproduce are likely to show some convergent adaptations. Not only is the season short but the melt may commonly occur after the summer solstice, so that the day-length will already be decreasing. All tundra plants must become active very quickly in order to make maximum use of this period for both growth and the production of flowers and seeds.

Growth

Two main methods of deployment of leaves are evident. In many grasses and sedges, as well as in deciduous dwarf shrubs and other herbaceous perennials, the full complement of leaves is produced very quickly with the aid of extensive reserves translocated from prostrate stems or underground storage organs such as rhizomes, often within less than one month of snow melt. This is the more remarkable since the soil may well be frozen at the time. An alternative strategy is shown by evergreen plants, in which leaves are produced much more slowly but may

Right Antarctic tundra. Beyond the colony of penguins on the coast of a maritime Antarctic island, green moss carpets and paler peat banks being formed by the moss (*Chorisodontium aciphyllum*) have developed on a gentle slope fed with melt water from the snow on the ridge beyond.

maintain photosynthesis over a number of seasons, as many as seven to 15 in some evergreen dwarf shrubs such as *Cassiope tetragona*. Mosses and lichens are also evergreen plants and in some mosses such as *Polytrichum* leaves may be photosynthetically active for up to three seasons. There is a strong inverse correlation between absolute rates of photosynthesis and the length of life of leaves. Evergreens generally show much lower rates than deciduous graminoids. Most plants found in tundra regions show an ability to sustain relatively high rates of net photosynthesis at 0°C (32°F), while the optimum temperature for net photosynthesis is about 15°C (59°F) under saturating light intensities.

In species that occur over wide ranges of latitude in both arctic and alpine tundra, such as *Oxyria digyna*, it has been found that respiratory processes can adjust to changing growth temperatures in much the same way in plants taken from anywhere in the range and grown under controlled conditions. There was marked ecotypic differentiation, however, between arctic and alpine populations in the potential for adaptation of net photosynthesis, the latter populations being much more flexible. This probably reflects differing selection pressures, since alpine environments show much greater variability than arctic environments.

Decreasing day-length, lower temperatures and drought all harden plants and induce dormancy toward the end of the season. As mentioned above, the greatest contrast between arctic and alpine tundra lies in the radiation regime and this difference is reflected in the control exerted on initiation and formation of overwintering buds. In the Arctic, *Oxyria digyna* forms buds as day-length decreases from 18 to 15 hours. By contrast, alpine populations do not form buds until day-length is about 12 hours and then they form quickly. Many mosses appear to grow right until the end of the season and the more delicate apices are then killed, new growth in the following season occurring from subapical buds.

Reproduction

In many perennial plants of the tundra, flowers are formed during one season and overwinter in the bud; flowering can then occur right at the start of the next season. This is seen at its extreme in plants such as *Ranunculus nivalis* and *Saxifraga oppositifolia*, which flower within three or four days of snow melt, often long before new growth is evident in most plants. Since the flower bud may have been initiated as much as two seasons before, it is evident that a season in which flowering is copious is not necessarily a season favorable to flowering but rather a reflection of the nature of past seasons. In many higher plants of the tundra, other than grasses, flowers are large and brightly colored and pollination is carried out by insects. In some instances the whole flower head follows the sun and the corolla may also act as a parabolic reflector, so that the

temperatures around the reproductive organs may be 5–10°C (about 9–18°F) above that of the surrounding air, encouraging insects to bask and at the same time transfer pollen. Flies (Diptera) and bumble bees are the most common pollinators. Some plants, including *Saxifraga oppositifolia*, *Dryas integrifolia* and *Salix arctica*, are almost completely dependent on flies, while bees are the chief pollinators of species of *Pedicularis*.

Even if a seed is set, the chances of successful establishment of seedlings are slight. For the seed to germinate, temperatures have to reach a threshold, and this must occur early enough in the season to allow growth to a viable size. Growth must also be uninterrupted by drought. A further hazard to seedlings, especially when growing on bare ground, is the small-scale disturbance caused by the development of needle ice crystals forcing up through and out of the soil. It is hardly surprising that annual plants are uncommon.

Most plants in tundra maintain their presence by vegetative proliferation. Underground branching occurs from rhizomes and stems and, as the older parts decay, fragmentation of the system occurs. By the time of this fragmentation, the new unit may already have attained considerable age and flowering may have taken place. It is therefore often difficult to recognize an individual plant, as is true of many perennial plants in other parts of the world. For example, there may be a number of vegetative shoots and flowering shoots of a grass showing above ground but below ground they are interconnected in a well defined

---

Below Most tundra plants have preformed flower buds, so that flowering takes place right at the beginning of the season. Here *Ranunculus nivalis* flowers within one week of snow melt, among grasses that have hardly started to produce new green leaves.

Bottom right *Salix arctica*, one of the many dwarf willows that grow on tundra. Small plants such as this one, some 15cm (6in) across, may be as much as 100 years old.

Opposite The club moss *Lycopodium selago* growing amongst prostrate dwarf shrubs in Arctic Norway.

pattern by a rhizome system, being of various ages, each having very different metabolic capacities and often being interdependent in many respects. Apart from vegetative fragmentation below ground, a few species, such as *Polygonum viviparum* and *Saxifraga cernua*, produce bulbils which drop to the ground and may take root.

Tundra Vegetation

The preceding sections have emphasized convergent aspects in the growth strategies of tundra plants. This convergence extends to another level, beyond that of the individual plant species, to the structure of the vegetation communities that they make up. For example, components of tundra vegetation such as wet meadows are superficially similar and consistently dominated by grasses and sedges, with mosses often forming a more or less continuous understory. Wet meadows throughout the arctic tundra, from Alaska, through Canada, Greenland, Spitzbergen and Iceland and on through the Scandinavian and Eurasian Arctic, differ mainly in the details of species composition. The mosaic of vegetation that develops in tundra areas not only reflects the varying availability of free water and length of growing season but also the relative stability of the habitat. Few plants are able to survive in habitats subject to frost action and abrasion from windblown dust and snow or ice particles. Despite the fact that plants have long been collected from tundra regions, it is only recently that detailed descriptions of the vegetation have become available. Descriptions of processes of change in arctic vegetation have been even rarer, despite the fact that the environment is unstable and change might be expected.

Plant Communities

In normally wet habitats such as stream sides and wet meadows, deciduous grasses (*Deschampsia caespitosa*, species of *Poa*, *Dupontia* and *Alopecurus*), sedges (species of *Carex* and *Eriophorum*) and herbaceous perennials

(forbs) are most common, all with extensive underground storage organs. An understory of mosses (species of *Cinclidium*, *Drepanocladus*, *Calliergon*) often covers most of the ground. The higher plants show high transpiration rates, associated with relatively high rates of net photosynthesis, and little ability to conserve water. Likewise, the mosses are also tied to wet habitats, needing high water contents to sustain maximal rates of net photosynthesis and having limited tolerance of periods of drought.

Because of low precipitation, ridges and areas of higher, free-draining ground are frequently liable to summer drought, often prolonged. In these more exposed habitats, often described as *fjeldmark* or *fell-field*, there are plants with dense growth-forms such as cushions or mats, including prostrate evergreen dwarf shrubs (*Cassiope* and *Loiseleuria*), deciduous dwarf shrubs (such as dwarf willows and birch) and forbs, together with many cushion-forming mosses and a number of lichens. Cushions and other growth-forms with dense, closely packed shoots are well adapted to such habitats. They trap rain and snow, the erosion of litter and soil is reduced and, indeed, cushions may represent small closed-circuit systems with recycling of mineral nutrients. These growth-forms may also generate within themselves a warmer, moister, calmer microclimate. Vascular plant cushions are of course able to draw upon water from below ground but moss cushions rely upon water at the ground surface or on precipitation. When this is not available they may dry out but, in common with lichens, they have an astonishing ability to resume normal metabolism when water becomes available, often within moments of being rehydrated. Hence, of all tundra plants, mosses, with lichens, exhibit the most opportunistic growth. Gradients in water availability may occur over very short distances, even as little as 1m (3.3ft), for example, between the drier rims of ice wedge polygons and the wet troughs between them.

Much of the tundra at the southern edge of the Arctic and extending into the sub-Arctic is covered with low shrubs of *Salix* and *Betula* with an understory of *Eriophorum* tussocks, other sedges, dwarf shrubs and mosses. Such vegetation covers much of northern Canada

and vast areas in Eurasia, and grades through scrub formations into boreal forest.

Tundra vegetation in the Southern Hemisphere differs greatly from that in the Northern Hemisphere. Because of their great isolation, subantarctic islands in the southern oceans and "islands" of snow-free ground at the margins of the main Antarctic continent support only a minute fraction of the variety seen in the vast, more or less continuous circumpolar tundras of the north, where free migration has been possible. The native flora of vascular plants on the subantarctic islands comprises only about 70 species. Many, however, are not circumpolar, most of the islands supporting between 21 and 36 native species. The maritime Antarctic, which includes all the islands of the Scotia Arc other than South Georgia, and also the western coast of the Antarctic Peninsula, is a unique region of the world in that its vegetation is almost exclusively cryptogamic. The topography is generally rugged and hence, in addition to outflow from the main ice caps, melt water from snow patches flows throughout the summer in many places. A great variety of moss growth-forms are present, ranging from continuous carpets in the wettest habitats to moss cushions in more exposed situations. In some localities, certain slowly decomposing mosses, particularly *Chorisodontium aciphyllum*, a species resembling the common Arctic moss, *Dicranum elongatum*, have built deep banks of moss peat, which may be more than 1m (3.3ft) thick. Insulated by the surface layers of moss, the underlying peat becomes permanently frozen.

Only two vascular plants occur on Antarctica, a grass, *Deschampsia antarctica*, and the cushion-forming plant *Colobanthus quitensis*, which in its vegetative state bears a close resemblance to a common plant of the Arctic, *Silene acaulis*, but differs in having colorless flowers. In the southernmost areas, on rock prominences (nunataks, see p. 225) in the main antarctic ice sheet and in other restricted ice-free areas on the coast, only a dozen or so mosses are found, but lichens are still common, often making their best growth at or even under the margins of snow beds.

Processes of Change in Tundra Vegetation

Changes in tundra vegetation are usually small scale and often cyclical; they are most likely to be occasioned by changes in the availability of water to a particular area as persistent snow beds change in size. For example, in both the Arctic and Antarctic, a sequence of vegetation development has been recognized in which mosses are replaced by peat-forming mosses with a lower water requirement. This is a process that is likely to occur even when water supply is constant, but will occur more rapidly when the supply is decreasing. During the drier stages of this succession in the Arctic the

willow *Salix arctica* is a common colonizer. Eventually, when the water supply is restricted to a brief period about the main melt, only prostrate, parched willows are left, since the moss that grew around and beneath the stems dries out and blows away. This process, from the initial growth of moss carpets to the erosion of all but the willows, can occur over relatively short periods of 30 or 40 years. The time-scale has in part been pieced together from counts of annual rings in the willows, some being more than 100 years old but covering much less than 1 square metre (10 square feet).

N.J.C.

Grassland

Grassland is a vegetation type in which the predominant species are perennial grasses with characteristic long, narrow, parallel-veined leaves and a fibrous root system composed mainly of adventitious roots formed at the nodes of the cylindrical jointed stems. Such vegetative features give rise to two main growth-forms—tufted or tussock and prostrate or creeping types—depending upon the pattern of development of the lateral shoots and the length of the stem internodes. Both growth-forms produce a mat of roots and shoots known as *turf* or *sward*, which is resistant to a variety of human-controlled influences from grazing and burning to recreation. Associated with the grasses are numerous perennial herbs, mainly plants which develop their buds just above or below the soil surface for protection (see Turf-dormant Herbs, p. 55), with an overwintering stage of a basal rosette of leaves or prostrate leafy shoots, several small annuals and also species of dwarf shrub. Woody shrubs over 0.5m (1.7ft) in height and trees are generally sparse or absent. As many as 75% of all species present fall into the hemicryptophyte life-form category. The herbs commonly have limited flowering periods and throughout the season of growth and reproduction confer characteristic aspects of color upon the basic green of the grasslands. Yellow is often the predominant color and this feature is thought to be an evolutionary adaptation to enhance chances of pollination, for many of the insect pollinators see yellow more clearly than any other color.

Climatic Distribution of Grassland Types

Extensive areas of grassland are to be found in all the continents of the Earth between latitudes 60°N and 50°S and they are primarily used for grazing by both domesticated and indigenous wild animals. Well over 1,300 million hectares (3,200 million acres), ie approximately 20% of the total land surfaces, are covered with grazing lands in the form of permanent grasslands; in

Oceania, these areas amount to 55% of the total land area; in Africa, 20%; and in all other continents, the coverage varies between 15% and 18%.

It has long been recognized that climate is the most important factor in determining the broad geographical distribution of grassland and there have been several attempts to define a "grassland climate", particularly one which enables the development of a "natural grassland climax vegetation". The existence of vast expanses of grass-dominated vegetation in the savanna lands of Africa, the steppes of Eurasia and the prairie region of North America led early ecologists to suggest that the climatic conditions in these regions were such that they favored grassland vegetation, while the development of a dominant woody vegetation was prevented or inhibited. But subsequent research has indicated that a variety of interacting environmental variables such as climate, soil, topography, fire and the influence of people have been and are the responsible factors. Nevertheless, climate in the form of annual seasonal variations in precipitation and temperature is a dominant factor restricting grassland distribution to four major zones: the wet and dry tropical, the dry steppe, the humid mesothermal (upper mid-latitude and humid subtropical) and humid microthermal (boreal and ice-cap) climatic zones.

This view in Tanzania, in front of the volcano Oldoinyo Lengai during its 1966 eruption, shows tropical savanna grassland. This is typically dominated by tall grasses with scattered trees and shrubs covering less than 20% of the ground. The flat-topped *Acacia* trees shown here are typical of such areas in Africa.

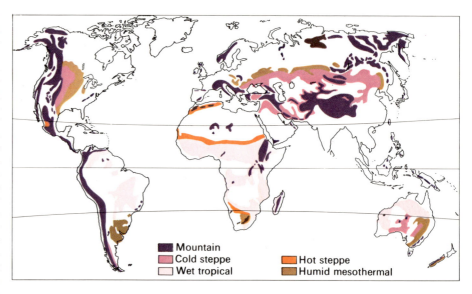

Mountain

Cold steppe

Wet tropical

Hot steppe

Humid mesothermal

A world map showing the main regions of grassland. Mountain areas are shown since grassland forms part of the zonation of the alpine belt (see Mountain Vegetation, p. 183).

Wet and Dry Tropical Zone—Savanna Grasslands

This zone, which covers large areas of Africa, South and Central America and northern Australia, has a winter dry season of three to five months followed by a rainy season and a minimum mean temperature of 18°C (64°F) in the coldest month. The range of precipitation falling in the rainy season varies between 200 and 1,000mm (8–39in) in areas where grasslands develop and surface flooding is a common phenomenon in wetter regions. These grasslands are known as

savanna, a word of Carib origin, first applied to those extensive, flat grasslands in the tropical regions of Central and South America which have a continuous cover of herbaceous vegetation over 80cm (32in) high and occasional scattered trees and shrubs. Since its adoption by the early Spanish explorers, the term has been extended to include virtually all types of tall, tropical grasslands in Africa and Australia and, also, indiscriminately, to a variety of tropical vegetation types from freshwater swamp to woodland. Three broad categories are recognized: *savanna woodland*, *savanna parkland* and *savanna grassland*, which represent a gradual transition from closed woodland with a canopy cover greater than 50% to an open grassland where tree and shrub cover is less than 20%. These major types can be identified in the true savannas of the West Indies, the *llanos* of northern Colombia, Argentina and Venezuela, the *cerrado* of central Brazil, the Guinea and Sudan vegetation zones of West Africa, and the rangelands of central and East Africa and northwest Australia.

Savannas occur on flat plains and plateaux and, because of the marked seasonal pattern of rainfall, there are pronounced differences in the seasonal appearance of the savannas. At the end of the rainy season, the vegetation changes from green to yellow as the tall grass dies and gradually falls over. As the dry season proceeds, the herbaceous vegetation is frequently fired, and a layer of ash remains (see Fire, p. 139). The ash is washed away by the early rains, the dominant grasses such as *Paspalum* spp., *Pennisetum mezianum* and *Themeda trianda* begin to shoot, and bulbous monocotyledons flower before the main period of grass growth and herb flowering in the five months of the wet season.

The concentration of rainfall causes extensive leaching, as a result of which the savanna soils are characteristically of low nutrient status, with an acidity below pH 5.0 (see Soil Acidity, p. 133), and usually with impeded drainage due to the formation of an iron or clay hardpan in the profile (see Pan, p. 133). The firing process is thus the main method of rapid replenishment of soil nutrients and also acts as the most important factor in the maintenance of open savanna grasslands for the large herds of grazing wild game and domesticated animals.

Dry Zone and Steppe Grasslands

The dry zone includes both desert vegetation (see p. 168) and steppe grassland, and is characterized by low and sporadic rainfall. There is a wide annual fluctuation in temperature, ranging from $-12°C$ ($10°F$) to over $30°C$ ($100°F$), and a wide diurnal temperature fluctuation; potential evaporation is always greater than precipitation. Two broad types of steppe are recognizable: *cold steppe* and *hot steppe*. In the former type, which includes the steppes of the central USSR, the mixed prairies of the foothills of the Rocky Mountains, the pampas of Argentina and the steppes to the east of the central Australian Desert, the mean winter temperatures fall well below freezing point and the rainfall is either concentrated in the summer months or evenly distributed throughout the

Savanna (llanos) on the lower slopes of the Venezuelan Andes near Trujillo. This type of grassland has a winter dry season, followed by a rainy season.

year. Alternatively, areas such as the tussock grasslands of central Mexico, western Australia, the central southern African grassveld and the Sahel steppe zone to the south of the Sahara desert have a concentration of rainfall in the winter months and winter temperatures remaining above freezing point. Such hot steppe grades into semidesert scrub and is of relatively limited distribution.

STEPPE. In its strictest sense, steppe describes the open, grassy plains of the temperate zone of Eurasia between latitudes 40°N and 50°N, from the Danube Basin in the west, eastward to the valley of the River Yenisei in central Siberia and the continental highlands of Mongolia and China. As we have seen above, this definition has been extended to include most of the temperate grasslands of the Earth, such as the prairies of North America and the pampas of South America while, on the other hand, plant sociologists describe steppe as dry grasslands of anthropogenic origin on lime-rich soils in eastern Europe. By referring to the original definition, however, as still used by Russian soil scientists, for example, the basic features of this formation can be understood. They describe steppe as a series of four climatic zones, running north-south, in the sequence: *forest-steppe, meadow-steppe, dry-steppe* and *desert-steppe*. There is a

parallel sequence of soil types.

The four steppe types have several features in common. First, the climate is typically continental: winter temperatures remain well below freezing point for at least four months when permanent snow cover exceeds 10cm (4in); high summer temperatures always cause evaporation from the steppes to exceed the annual precipitation and (along with an absence of summer rainfall) often cause late summer drought. Second, the steppe soils are all varieties of the chernozem (see p. 136) or chestnut soil-types derived from loess parent material. The former have a dominant, black surface-layer of humus often up to 150cm (60in) deep, are generally poor in carbonates and lack a horizon of clay accumulation. Alternatively, chestnut soils have a low humus content and are consequently brown in color with a paler, alkaline horizon of salt accumulation formed by the washing down by spring snow meltwater of salt particles deposited on the soil surface by summer winds from the deserts to the south. Third, species of feather-grass, notably *Stipa joannis*, predominate in a characteristic annual cycle of growth and development beginning with the melting of the snow cover in mid-April. In the meadow-steppe, a variety of colorful spring herbs flower until mid-May as the temperature rises and the soil remains damp. Towards the

The dry-steppes of Mongolia are dominated by low-growing grasses and are used for grazing.

end of May, the long awns of the feather-grasses appear and the swaying of the plumes against a profusion of color is a typical sight until July when rainfall ceases and the vegetation starts to wither. By the drought period of August, the steppes have assumed a dry and barren appearance, in which form they stay until the snow cover returns in November.

The dry-steppes are dominated by low-growing feather-grasses such as *Stipa capillata*, *S. pinnata* and sheep's fescue (*Festuca ovina*). They occur in the southern USSR in regions where the annual excess of evaporation over precipitation increases to between 300 and 500mm (12–20in), chestnut soils become more common and the chernozems are poorer in humus and richer in carbonates. The two remaining steppe types are transitional landscape types, desert-steppe falling between the dry-steppe and the central Asian desert, and forest-steppe falling between the meadow-steppe and the northern deciduous forests.

PRAIRIES. Extensive areas of grass-dominated vegetation in North America between latitudes 30°N and 55°N are known as prairies. They are essentially similar to the steppes of Eurasia in terms of climate, soil and floristic characteristics, but differ in that there is a marked east-west zonation instead of a north-south zonation as in the USSR. From the tall-grass prairies of such central states as

Wisconsin, Iowa and Oklahoma, there is a gradual transition through mixed prairie to short-grass types in the foothills of the Rocky Mountains in New Mexico, Nevada and Wyoming.

The tall-grass prairies were originally dominated by species of *Agropyron*, *Andropogon* and *Bouteloua*, which formed a more or less continuous sward, with "bunch grasses" such as *Sporobolus* and *Stipa* reaching 1m (3.3ft) or more and giving a "tussocky" appearance. The mixed prairie also has species of *Stipa*, *Sporobolus* and *Agropyron* with lower sward-forming grasses such as buffalo grass (*Buchloe dactyloides*) and gamma grass (*Bouteloua gracilis*). The short-grass prairie may be a response to more arid conditions but it is also thought to result from overgrazing, which has left the sward-forming grasses predominant. Indeed, most of the North American prairies have been substantially altered by grazing of domestic animals in the last two centuries or so.

Humid Mesothermal Zone

This climatic zone has at least one month with a mean temperature below 18°C (64°F), but above 10°C (50°F), and no month has a mean below −3°C (27°F). There are distinct winter and summer seasons; long periods of continuous heat or cold or continuous dry conditions are generally absent. The climate favors the development of deciduous broad-leaved and coniferous forests, but is also conducive to arable crop and grass growth, and in many regions the natural forests have been destroyed to provide lands for cultivation and grazing. At its extremes, this broad

climate type is capable of supporting grasslands similar to steppes and savannas but, unlike other climates, there are no extensive areas of grassland specifically associated with the climate. The most common form of grassland in the Southern Hemisphere is the tussockland of the eastern grassveld in South Africa, and of eastern Australia and New Zealand, while in the Northern Hemisphere large areas of the eastern United States, west and central Europe, China and southern Japan are covered with a variety of moisture-loving grasslands with a great diversity of grass and herb species.

MEADOW. Although grasslands now occur widely in such climates, they are largely the result of human activity and it is often impossible to determine the extent to which this is so. Consequently, a rather rule of thumb approach is usually adopted. Pastures are generally considered to be grasslands subjected to more or less continuous grazing while the more generally used term meadow is most commonly applied to a relatively flat area of herb-rich pasture which develops a flush of summer herbage that can be mown for winter silage, forage, thatch or mulch. Meadows occur on a variety of soil types throughout the subarctic, temperate and subtropical regions of the Earth and from sea-level shore meadows to the high alpine meadows or margins of the Himalayas at 3,500m (11,500ft). Seasonal temperature and availability of water are important factors in the productivity of the meadow vegetation and the summer growth period is characterized by stable, mesic soil and climatic conditions, maintained by gradual snow-melt in alpine regions, a period free from inundation by salt water in coastal regions and by the general lack of extremes of precipitation throughout the temperate zones. In general, meadow soils are richer in humus, nitrogen and phosphorus and have a greater water-retaining capacity than those supporting other types of herbaceous vegetation. Meadows are usually carefully managed by cropping and manuring to enhance the productivity, and play an important part in beef and milk production as well as the simpler peasant economies of alpine regions.

It is convenient to classify meadows using simple topographical and floristic criteria into four main types.

Temperate lowland meadows are found on mesic, lowland soils throughout the zones of temperate climate and several characteristic

genera of grasses are present: the meadow grasses (*Poa pratensis* and *P. trivialis*), oat and false oat grasses (*Trisetum, Helictotrichon* and *Arrhenatherum* spp.), meadow catstail (*Phleum pratense*) and meadow foxtail (*Alopecurus pratensis*).

Montane and *subalpine meadows* are found on the undulating foothills of the main mountain ranges from the subarctic to the subtropical regions; species of the lowland type are replaced by related species, for example *Phleum alpinum* and *Alopecurus alpinus* in northern montane regions and *Poa himalensis* in the Himalayan foothills.

Shore meadows are found in semi-saline, coastal areas, where sea meadow grasses (*Puccinellia* spp.) replace *Poa*.

Dwarf bamboo types contain tall (2–3m; 6.5–10ft) herbaceous vegetation which is dominated by genera of bamboo grasses such as *Sasa, Pleioblastus* and *Arundinaria* and are found mainly in the temperate and montane zones of northeast Asia from Japan to Nepal.

Humid Microthermal Zone

This is similar to the latter type, but differs in that the mean temperature of the coldest month is below −3°C (27°F), and below −38°C (−36°F) at winter extremes. This climatic type is mainly restricted to the Northern Hemisphere north of latitude 50°; most of Canada and the USSR fall within the

broad definition. The natural vegetation is mainly broad-leaved, mixed and coniferous forest; grasslands frequently occur in a mosaic of woodland clumps—for example, the poplar parklands of central Canada and the forest-steppe zone of the USSR. Alpine vegetation types above the tree limit also find a place in this category (see Mountain Vegetation, p. 183) and, in the Southern Hemisphere, microthermal grasslands are virtually restricted to the high altitudes of the main mountain ranges. In the tropics, the misty, wet and cold alpine belt between 3,500 and 4,000m (11,500–13,000ft), known as the *paramo*, has an important grassland component in the Andes, eastern Africa and the mountains of Indonesia.

The prominent tussock grasslands of the circum-Antarctic islands have proved difficult to place in any of the categories noted above. While they have been described as grass-tundra, it seems more appropriate to consider them as transitional between the humid microthermal grasslands and tundra. They occur in oceanic conditions in which the mean winter monthly temperatures vary from just above to somewhat below freezing point. Undoubtedly, a further factor in their distribution involves biotic influences. All of the large tussock grasses, such as *Poa flabellata* on the Falkland Islands and South Georgia, are intolerant of grazing by large

Below Tussock grassland or tussock land on the valley floor of the Hooker Glacier, New Zealand. This type of grassland is common in areas with a humid mesothermal climate, ie having a mean temperature of between 10°C (50°F) and 18°C (64°F) for at least one month of a year and in no month falling below −3°C (27°F). The tussocks shown here comprise the silver tussock grass (*Poa cespitosa*) and the thorny bushes are called matagouri (*Discaria toumatou*).

Southern oceanic tussock grassland is dominated by related species of *Poa* on the different circum-Antarctic islands. The species here is the tussac (*Poa flabellata*) at Port Stephens on West Falkland.

herbivores, none of which is native to the areas. For this reason, it seems likely that these very distinctive, though limited, southern oceanic tussock grasslands provide a striking example of the delicate balance between climatic conditions and the occurrence of the grazing animals that have evolved in continental areas (see Islands, p. 231).

Environmental Influences on Grassland Types

On to this basic pattern of the climatic delimitation of grassland types it is necessary to superimpose the diverse effects of the four environmental variables of soil, topography, fire and the influence of Man, if questions on grassland origin and ecology are to be answered.

Soil
The effects of soil are perhaps the most pronounced in producing variations in grassland structure and species diversity, from obvious differences between floristically rich swards developed over base-rich soils and species-poor pastures on acidic soils to subtle changes in humus and calcium carbonate levels within a single soil type over a wide geographical region. Caused by a climate-soil interaction, the latter situation applies particularly to the variation of steppe and prairie parallel to a variation in the chernozem and chestnut-arid brown soils of the USSR and the United States. The tall-grass prairies and their Eurasian equivalents, the meadow-steppes, both occur on thick chernozem soils, from the surface layers of

which calcium carbonate has been leached, whereas mixed- and short-grass prairies and feather-grass steppes are found mainly on the southern chernozems and chestnut-arid soils (see also Soils, p. 136).

Topography
It is possible to find situations where interactions between the effects of climate and topography and soil factors and topography produce distinct types of grassland. In many subalpine regions, several topographic features result in the formation of "frost pockets". Where there is a temperature inversion in a valley or basin area, normal forest development is prevented and is replaced by grassland. Accumulation of masses of cold air of a temperature less than $-7°C$ (20°F), at any time of the year, are common phenomena in many valleys of the Snowy Mountains of Australia. In the flat plains of the *llanos*, shallow depressions are subject to inundation by rainy season floodwater and their clayey soils are enriched with the inwash of nutrients. Here, the usual savanna grasses are replaced by flood-tolerant species—sedges and the occasional palm tree which gives such areas the name *palmares*. The East African rangeland equivalent is the *mbuga* depression in which the black, sodium-rich soil supports rich grassland and species of *Acacia* instead of palms.

Fire
The most ancient and perhaps the most important influence of Man on vegetation has been through the use of fire. Although fires are occasionally started by natural agents, it seems that many grasslands owe their origin and existence to deliberate firing. Evidence from pollen and macroscopic analyses of peat deposits in northwest Europe indicate that forest destruction and grassland establishment were well advanced by 2500 BC; radiocarbon dating and historical folklore suggest regular burning of forest and prairies by Amerindians and of the New Zealand vegetation by moa-hunting Maoris from 1000 AD onwards. Primitive peoples all over the world who rely on a simple pastoral livelihood still burn grasslands regularly to improve pasture and animal crop yields. Burning kills the old growth and releases nutrients more rapidly back to the soil which, in the case of a savanna, is often impoverished due to leaching and the formation of a hard pan or *arecife*. The morphology and annual growth cycle of perennial grasses are well adapted to burning. The buds for the next season's growth are normally close to the ground and are protected by a dense mass of dead leaf sheaths and much of the ripe seed is shed and covered by the surface soil before firing commences. Some species are more resistant and will regenerate more quickly than others. For example, in the hummock grasslands of northwest Australia, burning favors the development of dominance by the hard-leaved spinifex grasses (*Triodia* spp.) at the expense of the more palatable species (see also Fire, p. 139).

Influence of Man
In most of the countries of the humid mesothermal region, the influence of Man on grassland through mowing, grazing and controlled improvement by the use of fertilizer, selective weedkillers and more productive seed mixture is widespread. The terminology for the variety of grassland types is loose, but agriculturalists use a convenient subdivision into those areas, such as hill grazings, downland, fens and salt marshes which are uncultivated and those which are cultivated by regular mechanical and chemical treatment, although the turf is left unbroken. Hill grazings are common throughout the uplands of northwest Europe and are found mainly on either free-draining or waterlogged acidic soils. Species of fescue (*Festuca*), bent grass (*Agrostis*) and sweet vernal grass (*Anthoxanthum odoratum*) are the most palatable grasses, but the abundance of the tougher, wiry mat grass (*Nardus stricta*), the purple moor grass *Molinia caerulea*) and rushes (*Juncus* spp.) make the grazing less productive than their lowland counterparts on calcareous soils. In the latter type, sheep's fescue (*Festuca ovina*), upright brome (*Zerna erecta*) and several false oat grasses (*Arrhenatherum elatius*, *Helictotrichon pratense* and *Trisetum flavescens*) add further diversity to a herb-rich sward. Cultivated grassland may be subdivided into permanent types which include permanent pastures and meadows, and temporary types or leys.

Major Grasses Found in Grasslands

The variety of species of the grass family, Gramineae, can be classified, using leaf anatomy, into six subfamilies, some of which have interesting distribution patterns throughout the grasslands of the Earth (classification may also be based on spikelet characters, yielding a classification with some differences from that based on leaf-anatomy). Members of the bambusoid type are mainly restricted to tall, cane-grass meadows in the Far East and oryzoid species (rice-grasses) are dominant in marshy, lowland areas. A third group, the reed grasses of the *Phragmites*-type, are most common in the damper reedswamps, although species such as the tall and unpalatable *Danthonia pallida* often dominate moist temperate pastures in southern Australia similar to the ecological niche of *Molinia caerulea* in northwest Europe, and species of *Aristida* are important constituents of grassveld.

The majority of the more important pasture grasses fall within three types—chloridoid, panicoid and festucoid. The former have the most specialized leaf anatomy and are widely distributed in the humid mesothermal climatic regions, becoming dominant in such grasslands as the *Zoysia japonica* meadows of Japan, the sclerophyllous spinifex (*Triodia*) grasslands of Australia and the grama grass (*Bouteloua* spp.) prairies of North America.

The two remaining subfamilies tend to complement one another in general distribution pattern. The panicoid type has a high optimum temperature for growth, developing vigorously at 35°C (95°F) and extremely slowly below 15°C (59°F), responding mainly to short day-length. It occurs in its greatest diversity and with maximum dominance in the tropical and subtropical regions between 30°N and 30°S. Species such as *Pennisetum mezianum*, *Themeda triandra*, *Andropogon greenwayi* and *Panicum atrosanguineum* dominate the East African rangelands; the insinde or red grass (*Anthistiria imberbis*) is dominant on the grassveld, and the *llanos* are covered with species of *Paspalum*. One notable exception to this distribution rule is to be found in the prairie region where, although less than 5% of the total grass species are of the panicoid type, vast areas were dominated by three panicoid species before arable development—namely the big and little bluestem grasses (*Andropogon gerardii* and *A. scoparius*) and Indian grass (*Sorghastrum nutans*).

In the more westerly prairies, in Eurasia north of 30°, and most of South America and Australia south of 30°, the grasslands are dominated by festucoid grasses which have an optimum growth temperature below 27°C (81°F), relatively good growth below

Above A view of Annapurna South, in the Nepalese Himalayas, from the southeast, showing the zonation of vegetation from subtropical hills (**foreground**), where bananas and rice are cultivated, through rhododendron and coniferous trees (**distant slopes**), to alpine pastures at 4,500–5,500m (15,000–18,000ft), then permanent snow.

Below Mountains have several environments because the environmental factors differ with altitude. This also leads to a series of distinct vegetation zones. Shown here is the vegetation zonation on high tropical mountains from three different parts of the world.

10°C (50°F) and are adapted to long day-lengths. Species of fescue (*Festuca*) and feather-grass (*Stipa*) dominate the steppes, prairies, pampas and tussocklands, while the

pastures of the humid meso- and microthermal climates of Eurasia and North America usually contain representatives of the genera *Agrostis*, *Anthoxanthum*, *Avena*, *Bromus*, *Cynosurus*, *Dactylis*, *Festuca*, *Holcus*, *Lolium*, *Phleum* and *Poa*.

D.W.S

Mountain Vegetation

The zonation of vegetation evident as one progresses from the foot of a high mountain to its peak parallels in many ways that observed between low and high latitudes at lower elevations. As we shall see, however, several factors differ sufficiently to give mountain vegetation a distinctive stamp, the most important factor being temperature. On mountains it can vary greatly over a relatively short distance: air temperatures decrease with increasing altitude at the rate of about 0.5°C per 100m (0.3°F per 100ft). Further, the ground surface receives the maximum intensity of infrared radiation, which heats it when absorbed, only when it is at right angles to the radiation; therefore the steepness of slope also affects temperature. In the Northern Hemisphere, south-facing slopes are generally warmer than north-facing ones; in the Southern Hemisphere the reverse is true. Another feature associated with temperature on mountains is temperature inversion, both in the usual sense (air near to the ground cooling more rapidly at night than the air at higher levels), as well as cold-air drainage (cool air from the upper slopes slipping down into the valleys at night). When we try to interpret mountain vegetation in terms of decreasing temperature all these factors must be taken into account.

		Andes of Central Peru	Ruwenzori East Africa		Eastern New Guinea	
Continuous snow						6 000 m
Alpine zone	Steppe with tree-like Compositae		*Senecio-Lobelia* Scrub and grassland	Alpine grassland Alpine savanna		5 000 m / 4 000 m
Sub Alpine	Elfin woodland		Tree heath	High montane forest		
Montane Forest	Mossy forest		Bamboo zone	Mossy forest		3 000 m
Transitional Sub-Montane Forest			Mossy forest			2 000 m
Tropical Rain Forest						1 000 m / Sea level

In addition to infrared radiation, other wavelengths, including visible light and ultraviolet, increase in intensity with greater altitude, due to reduced filtration by the less dense atmosphere. Wind speeds and exposure to wind also increase with altitude. Wind exposure and radiation are, like temperature, also affected by slope and aspect.

Precipitation differs from these factors because it does not have a simple relationship to altitude from the bottom to the top of mountains. When moist air reaches a mountain mass it rises and cools. The amount of water vapor that air can carry decreases at lower temperatures, therefore as the moist air rises up the mountain slope, it cools until saturation point is reached, and the excess water is then released as precipitation. For this reason the lower slopes of a mountain side are generally more moist than the surrounding lowlands. However, the air carries only a limited amount of water vapor, and although it becomes even colder at higher altitudes, the chances are that little water vapor ·remains to be precipitated; furthermore, at higher altitudes the volume of the valley increases, whereas the volume of the mountain mass decreases, thus creating more opportunity for the air to pass through the mountain range with no additional increase in altitude. For these reasons, there is a decrease in moisture at the highest elevations of mountains in temperate regions, a feature which is exaggerated by increased exposure to drying winds. The water balance (precipitation/evaporation and transpiration) of mountains in temperate regions can therefore be summarized as follows: with the initial rise in altitude, the increase in precipitation causes the precipitation/evaporation (P.E.) ratio to increase: beyond the initial rise in altitude (ie at the higher altitudes) the lower precipitation and drying conditions cause the P.E. ratio to decrease.

Vegetation responds to the above environmental gradients by developing as a series of zones: *alpine*, *subalpine* and *montane*: below the montane zone an additional zone is sometimes recognized and given various names such as *basal forest* or *foothills vegetation*. These zones are not clearly demarcated; therefore the boundaries are to some extent arbitrary. Further, the altitude at which the zones occur varies according to local conditions and latitude: for example, subalpine vegetation will be found at higher altitudes on south-facing than on north-facing slopes in the Northern Hemisphere. With increasing latitude, there is a general decrease in temperature, thus causing all the vegetation zones to develop at lower altitudes.

Montane Zone

The effect of latitude on the altitude of zones is well illustrated by the boreal forests of North America. These forests occur at sea-level in eastern Canada, whereas the same forest forms the montane component further south on the Appalachians; on the Smoky Mountains they only occur at altitudes greater than 1,500m (4,900ft).

The upper part of the montane zone in the Rockies is occupied by Douglas fir (*Pseudotsuga menziesii*), which grows in dense closed stands, with subordinate white fir (*Abies concolor*) in the southern Rockies, and either grand fir (*Abies grandis*) or white spruce (*Picea glauca*) in the northern Rockies. In the lower part of the Douglas fir zone the ponderosa pine (*Pinus ponderosa*) becomes increasingly important until at lower altitudes the tree forms pure stands of open woodland.

Montane forests frequently contain upper and lower zones with trees of quite different character. Thus, in Europe, the lower zones are occupied by broad-leaved deciduous trees, eg oak (*Quercus* spp.), beech (*Fagus* spp.), chestnut (*Castanea* spp.), whereas the higher altitudes are dominated by conifers. In the eastern Himalayas, the zones exhibit even greater differences, with tropical rain forest below about 2,500m (8,200ft), and deciduous woodland above.

Owing to the high P.E. ratio, the montane zone in most temperate latitudes is occupied by woodland; however, if the air is fairly dry there may be insufficient precipitation, and the P.E. ratio will be too low to support trees. Thus, in the Mediterranean region, the lowest part of the montane zone is occupied by shrubs with hard, leathery leaves (sclerophylls); in Natal, the montane zone is occupied by grassland with occasional shrubs, and in parts of Asia conditions are so arid that the uplands are predominantly desert.

On tropical mountains the general distribution of environmental factors is the same as on mountains in temperate regions. However, the tropical atmosphere carries so much water vapor that the increase in humidity caused by air rising up mountainsides continues to much greater altitudes, so that the drier alpine conditions are found at the tops of only the very highest tropical mountains such as Mount Kilimanjaro in Tanzania, Kinabalu in Sabah, or Popocatepetl in Mexico. The warm moist conditions of tropical mountains raise the altitude at which trees are capable of growing. In Malaysia, for example, the tree line is at approximately 4,000m (13,100ft), whereas in Europe it is closer to 2,000m (6,550ft).

Forests in the tropical montane zone differ from lowland tropical rain forest in several respects. The trees are smaller than in the lowland (the height of trees decreases with increasing altitude), and so too are the leaves—a feature possibly associated with increased light intensity. Tree trunks are generally thicker than those of the lowland forests, and lack buttresses. There are also usually only one or two tree strata, a factor permitting greater light penetration than in lowland rain forests, with the consequent development of richer ground and herb layers; indeed the number of tree species in the montane zone is less than in the

Montane forest, dominated by magnolias, near Tragshindu, Nepal, at 3,000m (1,000ft) in the Himalayas. Low temperatures and high humidity are conducive to an abundance of lower plants in the ground flora.

lowlands. The lower temperatures of the mountains are more conducive to the growth of mosses and ferns, while owing to the greater humidity, the abundance of epiphytes is even greater than in lowland rain forests, especially in zones shrouded in perpetual mist and cloud.

Cloud Forest

Also called mossy forest, this is rain forest which occurs above the cloud level in the montane zone on tropical mountains. Its lower limit begins quite sharply at the cloud base, while the uppermost forest on the highest mountains is above the cloud and distinct from cloud forest, although this ecotone is usually broad. Cloud forest characteristically has a low, dense canopy of small trees with thick, gnarled crowns of tiny, leathery leaves (microphylls) and high reflective power (albedo). Trees and ground are thickly swathed in epiphytes, mainly filmy ferns and liverworts, but including mosses and flowering plants. *Sphagnum* moss often occurs in open places. In very humid climates peat (see Wetlands, p. 191) accumulates. The soil is waterlogged for all or much of the time, and is usually gleyed and podsolized (see Soils, p. 137). Much of the precipitation is derived by the fog (ground-level cloud) condensing on the vegetation. It is impossible to state whether the structure

Right Cloud forest on the upper slopes of the Nguru Mountains in Tanzania, East Africa. Such cloud is characteristic of these areas for most of the year.

and physiognomy of the forest and trees is adapted to mineral deficiency (oligotrophy) or to occasional drought (even in the wettest climates there may be a week or two each year when cloud is absent). The relative importance of each factor probably varies from place to place. Cloud forest occurs as low as 600m (1,970ft) in Malaysia and at 3,000–3,300m (9,250–10,830ft) in the main cordillera of New Guinea.

Cloud forest dominated by beech (*Fagus*) and fir (*Abies*) occurs in a belt of summer cloud forest around the Mediterranean. North California redwood (*Sequoia*) forest occurs in a coastal fog belt. (See also Forest, p. 159.)

Subalpine Zone

The characteristics of the subalpine zone derive from the fact that the environmental conditions are intermediate between the zones above (alpine) and below (montane). For example, temperatures are lower and exposure to wind greater than in the montane belt below; similarly the humidity is higher than in the alpine zone, but lower than in the montane. On most mountains of the Northern Hemisphere the upper boundary of the subalpine zone can be delimited by the timberline, but the lower boundary is not always so clear. In the Rocky Mountains, the lower boundary is taken to be where the dominant trees change from Douglas fir (*Pseudotsuga menziesii*) of the montane forests, to Engelmann spruce (*Picea engel-*

mannii) and alpine fir (*Abies lasiocarpa*); on other mountains, the closing of the woodland canopy may be arbitrarily taken as the lower boundary. As in all general accounts of mountain vegetation, the effects of such factors as latitude, slope and aspect make it impossible to ascribe the limits of this zone to particular altitudes.

In the lower part of the subalpine belt, trees may grow in closed stands, but with increasing altitude conditions become steadily more hostile to tree growth. Although the environment becomes drier, the cold climate prevents the rapid decay of litter, so that a thin organic soil develops which is capable of holding water both from the frequent summer showers and from the meltwater of winter snow. These thin moist soils favor shallow-rooted plants, especially herbs and grasses, at the expense of trees. Consequently meadows are a feature of many subalpine zones.

Elfin woodland in the cloud forest of central Tanzania. The tree branches are clothed with epiphytes, mainly mosses, liverworts and ferns. These plants rely on the saturated air for their moisture.

The windswept subalpine zone (*puna, puno*) of the Andes typically supports a treeless steppe dominated by grasses and other shallow-rooted herbs.

Wind is another disincentive to the establishment of trees. This is because wind desiccates seedlings, which therefore cannot become established. However, some trees do survive throughout the subalpine zone and, combined with the grassland, subalpine vegetation frequently has the character of parkland or taiga (see p. 164). With increasing latitude the trees become fewer and shorter. The wind also causes deformation of the stunted trees in this zone, giving rise to the term *elfin woodland*. Near the timberline, the desiccating effect of wind can kill buds on the windward side of trees, allowing growth to occur only on the leeward side. Wind may also give rise to a prostrate habit, with the crown leaves lying close to the ground in a leeward direction: this type of growth is known as *Krummholz*. Another feature of subalpine trees is that many of the species growing there do so, not for any special reasons of adaptation to environmental conditions of the zone, but simply because they cannot compete with species in the montane forests below. In the Rocky Mountains the bristlecone pine (*Pinus aristata*) is found in the subalpine zone, as well as on dry ridges in the montane zone below; in both situations it avoids competition from the montane forests. In Europe, *Pinus mugo* grows both in subalpine vegetation, and on lowland bogs.

For full development of the rich subalpine meadow flora, a certain level of moisture is required. Thus, on the eastern side of the Rocky Mountains, meadows are not characteristic of the subalpine vegetation; instead, the ground flora is dominated by members of the heath family (Ericaceae). It is in wetter mountain regions, such as the European Alps, that the subalpine meadows are most luxuriant. The European alpine meadows of flowers and grasses are not only features of outstanding beauty in the late spring and early summer, but they are also of economic value, because the rich herb swards provide both summer pasture and winter hay for cattle.   J.Wn.

Timberline

Also called the treeline or forest limit, this is the uppermost boundary of tree growth on mountains, often marking the transition from the subalpine to alpine zones. In temperate regions, for example the Alps, this limit is set by the shortness of the growing season as well as the long winter. During summer there is insufficient time for the conifers of the uppermost subalpine forest to mature their needles, which then succumb to frost drought in winter. Elsewhere, where beech (*Fagus*) forms the limit, late spring frosts determine the timberline, although toward this limit annual production is so small that growth is weak and seed production scanty. In the central Alps the timberline is now at 2,000–2,150m (6,560–7,050ft). Grazing of the pastures above the tree limit, which has gone on for over 1,000 years, has depressed it by 100–200m (328–650ft). The Alps and other major temperate mountain masses exhibit the "Massenerhebung effect", whereby all vegetation zones, including the timberline, are 800m (2,625ft) or more higher on central ranges than on outlying peaks. The effect has been explained largely in terms of continentality or oceanity of the climate, particularly in terms of summer warmth.

Usually the timberline is abrupt, although grazing may leave isolated trees above its current limit. In the North American Rocky Mountains, where natural conditions still prevail, an area of stunted trees (*Krummholz*) extends 50m (164ft) above the tree limit at 3,750m (12,300ft). This is characteristic of humid climates. Elsewhere, where there is much snow, small groups of trees occur on knolls above the general limit. In arid conditions the timberline is diffuse. Aspect can be important; for example, in the Himalaya range there is a difference in the height of the timberline between north and south slopes. In the humid tropics and subtropics, forests decrease in stature upward and occur as a low, compact closed canopy only 1–2m (3.3–6.5ft) tall at the timberline. On the main cordillera of New Guinea this is at 3,720–3,840m (12,200–12,600ft), but it is commonly depressed by burning of the mountaintop grasslands and herb fields. It is further complicated by the presence of frost pockets in grassy hollows and gullies, which occur below the general tree line.

T.C.W.

Alpine Zone

The highest zone of vegetation on mountains is referred to as "alpine", or sometimes alpine tundra. In the Northern Hemisphere this belt is generally regarded as stretching from the timberline to the area of permanent snow. In the Southern Hemisphere, the lower boundary is less easily delimited because many mountains do not have woodland, or, as in New Zealand, because subalpine grass and moorland occur above the tree line.

Like all mountain vegetation, the alpine belt is found at lower altitudes with increasing latitude. In the Sierra Nevada Mountains, alpine vegetation starts at about 3,500m (nearly 11,500ft), whereas in the Canadian Rockies it begins at about 2,000m (over 6,500ft) and at even lower altitudes further north. Because of the cool summer temperatures and short growing season, there are many similarities between arctic tundra and alpine vegetation; however, these similarities decrease with increasing distance from the poles owing to differences in other aspects of climate and illumination. All alpine vegetation is floristically richer than that which occurs in the arctic tundra.

Plants growing in the alpine zone show numerous features which indicate adaptation to the rigorous alpine environment. The most immediately obvious problem is that of water shortage, which is brought about by such factors as low precipitation, very thin soils (if developed at all), high winds which

The timberline or treeline in Big Thompson Valley, Rocky Mountain National Park, near Denver, Colorado. At this altitude the trees give way to high alpine meadows. Tree growth is prevented on the valley floor by the accumulation of cold air in winter.

Alpine vegetation on tropical mountains appears at much higher altitudes than in temperate zones. Here at 5,000m (17,000ft) in the Dolpo Valley, Nepal, the only plants which can survive the thin (or absent) soils and high winds find footholds in the stable screes and in the shelter of rocks, which afford some protection.

enhance transpiration and evaporation, and frozen soils in winter which prevent water present from being absorbed by the roots. The prostrate habit of alpine shrubs such as willows (*Salix* spp.), as well as the creeping, cushion and rosette forms of many herbs (eg *Sedum* spp., *Dryas* spp., *Saxifraga* spp.), are clearly of value in such situations. Although most alpine vegetation is well suited to survive in exposed, dry situations, small peat bogs do occur in protected areas suitably located to trap snow meltwater; in such conditions sedges (*Carex* spp.) and cotton grasses (*Eriophorum* spp.) can be found.

The short growing season presents problems for annuals which need to complete their life cycle in one season; consequently, most species in alpine locations are perennial herbs, or low shrubs. These have food reserves which enable the plants to commence growth without delay when conditions are warm enough, thus making full use of the growing season; the food reserves are replaced towards the end of the summer growth cycle.

On clear days, plants in these high alpine situations are exposed to greater intensities of light, infrared and ultraviolet radiation than lowland vegetation; this is due to the thinner, clearer atmosphere. The brighter light is obviously an advantage to any photosynthesizing organism, provided it is fully utilized. It appears that alpine plants do in fact make use of this bright sunlight, because their photosynthesizing mechanism is not saturated at lowland light intensities. Exposure to infrared radiation is also advantageous because it allows the mat of low or prostrate vegetation to be warmed to temperatures higher than in the surrounding air (which, because of its clarity, does not absorb the infrared radiation to the same extent as the lowland atmosphere). Ultraviolet wavelengths can damage leaves, and it is possible

that the various nonphotosynthesizing pigments, which are common in the leaves of alpine plants, afford protection by filtering out some of this damaging radiation.

J.WN.

Azonal Vegetation

Although the major classes of zonal vegetation have been shown generally to be correlated with the world's principal climatic zones, it has been pointed out repeatedly that they can occur outside these zones as extrazonal vegetation in regions with particularly favorable microclimatic or edaphic conditions. This trend can continue to give azonal vegetation-types, which cover significant areas of the world as a result of major environmental factors not necessarily directly linked with the climate. The most important of these factors seem to be soil fertility and exposure which, together with Man, are apparently responsible for the development of *heath* vegetation; salinity, which has profound effects on *coastal* vegetation, as well as that of some inland areas; and nonsaline groundwater giving very wet soils which

support plant communities generally described as *bogs*, *fens*, *marshes* and *swamps*. It must be remembered, however, that the interplay of these and other factors with each other and with the more general climatic conditions can give rise to vegetation-types which, in the present state of knowledge, blur the distinction between azonal and zonal categories. Indeed, these lend support to the view that it is counterproductive to seek to impose too rigid and detailed a classification on vegetation when so much still needs to be learned about the factors responsible for differences in its physiognomy and the consequences of the interactions of the environmental variables.

D.M.M.

Heath

Heath is properly defined as a plant formation in which trees and tall shrubs are sparse or lacking and the dominant life-form is that of the ericoid dwarf shrub. The word has, however, also been applied to other types of vegetation on acid soils, such as grass-heath, moss-heath, lichen-heath and even the so-called "tree-heaths" of East African mountains. In its strict sense, however, heath vegetation is characterized by woody, much branched dwarf shrubs, mostly

evergreen, with small, usually sclerophyllous leaves, forming a canopy at a height not exceeding 1m (3.3ft) above the ground. Depending on the density of this canopy, other dwarf shrubs, herbs and mosses or lichens may form subordinate strata. The height of the dominants, however, varies considerably with habitat, so that it is necessary to introduce the term "tall heath" for the communities which attain 1–2m (3.3–6.5ft) in height and usually grade into scrub (see p. 166), while very low communities, up to 25cm (10in) high, are often called "dwarf heath".

Heath Types

Northern Hemisphere Heaths
Since ideas concerning heath vegetation developed in western Europe, we can take this area as a starting point. Heath is, or has been, widespread in the oceanic and suboceanic belt bordering the Atlantic Ocean and North Sea, which includes the British Isles and extends from southwest Norway, through southern Sweden, Denmark, the

Heathland on the Yorkshire moors of northern England dominated by heather or ling (*Calluna vulgaris*), the commonest heath species over much of western Europe.

The black crowberry (*Empetrum nigrum*) is an evergreen dwarf-shrub common in heathlands, especially on upland areas, in the Northern Hemisphere. A similar species (*E. rubrum*) occurs on Southern Hemisphere heaths.

Netherlands, Belgium, northern Germany, northern and western France to northern Spain. The annual rainfall in much of this region lies between 600 and 1,100mm (24–43in), well distributed throughout the year; the summers are seldom hot—the mean temperature of the warmest month is usually less than 17°C (63°F)—and winters are mild. Substrata include sand, gravel, freely drained podsolic soils, peaty podsols and peat. The heath region is completely included within the distributional area of heather (*Calluna vulgaris*), which is dominant on all but the more exposed sites and the wet or poorly drained substrata. There is, however, considerable variation in community composition, extending from the northern heaths, which may include *Vaccinium myrtillus*, *V. vitis-idaea* and *Empetrum nigrum*, to southern examples containing *Genista pilosa* or *Erica scoparia*. In the more oceanic districts (including the British Isles) *Erica cinerea* is abundant, with *Ulex gallii* and *Erica vagans* in parts of the extreme west.

A further series of communities, collectively described as wet heaths, occurs on poorly drained soils and wet peats. Many of these have *Erica tetralix* as the leading dwarf shrub instead of *Calluna vulgaris* and include various *Sphagnum* mosses, grasses and Cyperaceae.

The northern part of the western European heath zone falls within the limits of the coniferous boreal forest (see p. 164), while its southern part is contained within the area of the deciduous broad-leaved woodland. There is ample evidence from pollen analysis and other sources that the majority of heaths have been derived from one or other of these main regional types, and in the absence of a biotic factor, or management, would revert to scrub or woodland, except perhaps in the most exposed coastal sites or where severe

degradation of the soil has occurred. Heath may also develop as a seral stage, both in the succession following the drying out of bogs (see p. 191) and in that on noncalcareous sand dunes (see p. 202). Heath also occurs where adequate humidity prevails in the subalpine or even lower alpine zones on mountains. Characteristic species of mountain heaths include *Vaccinium myrtillus*, *Cassiope tetragona* and *Phyllodoce caerulea*.

In North America, heath communities occur as a thin, interrupted belt along the north Pacific and Atlantic coasts and across the northern parts of the continent. They tend to be somewhat localized and restricted to acid soils in extreme situations with humid, maritime climates. Structurally most similar to the west European heaths are lower alpine communities of the Cascade mountains, dominated by the ericoid shrubs *Phyllodoce empetriformis* and *Cassiope mertensiana*. Similar communities occur northward into the margins of arctic tundra and eastward across the continent. *Empetrum nigrum*, *Phyllodoce glanduliflora* and species of *Cassiope* and *Vaccinium* are important in this heath, which grades on the one hand into the

tundra communities and on the other into the boreal coniferous forest. On the Atlantic coast, in the maritime provinces of Canada, for example, lowland heaths occur in which species of *Vaccinium*, *Gaylussacia*, *Gaultheria* and *Corema* are particularly prominent. Fossil evidence suggests that climatic deterioration, followed by fire when Man occupied the area, induced the replacement by the heath of an earlier birch-alder (*Betula-Alnus*) woodland. The heath now seems to be a stable formation on the shallow, infertile soils it occupies. Further south, the summits of many of the Appalachian mountains support heath-like vegetation, the so-called "balds", as islands within the pine and oak-pine forests.

Southern Hemisphere Heaths

In southern South America and Tristan da Cunha, oceanic heaths dominated by *Empetrum rubrum* are often structurally reminiscent of their European counterparts. The dominant can form a dwarf heath on rocky, shallow-soiled sites among the cold, dry grasslands east of the Andes, while in the higher rainfall further west it gives rise to a tall heath reaching 2m (6.5ft) in height and forming a series of communities which grade into the deciduous and evergreen forest formations. The wet heath communities of southern Chile contain other ericoid dwarf shrubs, such as *Myrteola*, *Pernettya* and *Gaultheria*, along with *Empetrum*. These, together with other associates, are all present in the very wet, exposed, tundra-like "Magellanic moorland" of the western archipelagos, which is thus floristically and structurally closely related and often included with the heath formations.

In Australia, heath vegetation has been described from temperate to tropical climatic zones, although it seems to be absent from the more arid regions. In all cases, the soils

The cool-temperate heathlands of southernmost South America and adjacent areas are dominated by the red crowberry or diddle-dee (*Empetrum rubrum*), as shown in this photograph near Port Stanley, Falkland Islands.

supporting heath have been shown to be of low nutrient status, with lack of phosphorus and nitrogen being particularly important. Where seasonal waterlogging of the soils (podsols and peats) is prolonged or, on the other hand, where the soils are extremely well-drained (eg deep sands), the most obviously characteristic heath communities are found, both in the lowlands and under subalpine conditions. Many genera considered to be ericoid or semi-ericoid are present in these heaths, and include species of *Banksia, Leptospermum, Epacris, Kunzea, Prostanthera, Callistemon, Hakea,* etc. Similar vegetation occurs at high elevations in New Guinea. Such communities merge into grasslands or bogs in many areas, especially at higher elevations, or into forest of various kinds, particularly those dominated by sclerophyll trees (see p. 51). Indeed, extensive areas of southern Australia support communities in which a lower heath stratum is increasingly overshadowed by trees; when the canopy becomes closed the heath species largely disappear. There is a good deal of evidence that sclerophylly is a response to soils of poor nutrient status and low water availability and it is not surprising, therefore,

that some workers have enlarged the category of heath vegetation to include open sclerophyllous forest with a heathy understory. Whether this is correct, or whether such vegetation is better considered part of an ecotone between heath and sclerophyllous forest, must remain in some doubt.

In southern Africa a situation exists which is very similar to that just described for parts of Australia. The so-called *fynbos*, which occurs widely in the southern part of the Cape Province of South Africa, for example, usually has an upper story of variable development dominated by broadleaved sclerophyllous shrubs (see Chaparral, p. 167). However, in the "mountain fynbos" a series of low open communities, which must be recognized as heath, even in the limited sense, is widely developed. Here, numerous *Erica* species are important in a vegetation-type usually of about 1m (3.3ft) in height, which also includes many other ericoid shrubs, such as *Sympieza, Phylica,* and *Roella.* As in the Northern Hemisphere heaths, those in Australia, South Africa and other parts of the Southern Hemisphere have been greatly extended as a result of fire, grazing and other factors resulting from human activities.

Heath Exploitation and Management

It is clear from the above account that heath vegetation in virtually every part of the world has been greatly extended in area as a result of human activity. There is considerable evidence that heath is a response to naturally infertile soils but, by grazing and fire, Man has reduced the soil fertility of other areas and prevented the growth of trees which would overshadow the heath plants. Since the effect of human activity is best documented in western Europe, it seems appropriate to consider this here as a guide to the sorts of anthropogenic factors which have been involved in dramatically extending the world's heath vegetation.

In certain northern districts of western

Periodic burning is the best method of heathland management for sheep and grouse, but heathlands may degenerate to podsol, so that afforestation has replaced heaths in many areas of northern Europe, and is even encroaching on this grouse moor dominated by *Calluna vulgaris* in northeast Scotland.

Europe (parts of Norway, northern Scotland) a major factor promoting replacement of forest by heath was climatic change, particularly the onset of more oceanic conditions (cooler and wetter) about 2,500 years ago. However, heaths have developed from forest at many different times, and throughout much of the region their origin is associated with signs of human intervention. Human impact on forest vegetation extends at least as far back as Neolithic times. At first, as indicated by sequences in the pollen record, clearances were temporary only. Felling, accompanied by burning, opened up limited areas to cultivation, followed (on acid soils) by invasion of heath plants. These were probably grazed for a time and then abandoned to recolonization by forest. Several such sequences may have preceded more permanent forest clearance. While a few heathlands have probably been in existence continuously since Neolithic times, the origin of others can be traced to the Bronze or Iron Ages or later. Pollen diagrams in the heath region frequently show a marked decline in the proportion of tree pollen coincident with a rise in ericaceous pollen, dated at about 500 BC, the beginning of the Iron Age in western Europe. At about this time, the role of Man in reducing the extent of forest began to be reinforced by climatic change as conditions became cooler and wetter. Thereafter the process of conversion of forest to heath was continued, and locally intensified throughout historical times. Increased demands for timber, for constructional purposes and for use as charcoal, contributed to this trend. Forest destruction made way for the continued extension of heathland up to the 19th century.

Once heathland was established, effort was devoted to its maintenance because it proved useful for domestic herbivores—both cattle and sheep. With the advent of hardy breeds which could survive on the heaths throughout the year (especially the blackface sheep in northern Britain), heaths dominated by heather (*Calluna vulgaris*) became valuable grazing ground on which large flocks were maintained. A woody dwarf shrub, however, is not an ideal grazing plant. During the course of its life-history, which lasts normally between 30 and 40 years, heather passes through relatively distinct phases of growth—a pioneer phase (establishment), a building phase (development of dense canopy and maximum productivity of young shoots), a mature phase (formation of a gap in the center of the bush) and a degenerate phase (death of the central frame-branches). Throughout this sequence there is a progressive decline in the food-value of successive generations of young shoots and an increase in the proportion of the total biomass consisting of wood. Very heavy grazing, particularly in summer, amounting to removal of about 60% of the current year's production of leaf-bearing shoots, can maintain heather plants in a juvenile phase and prevent passage into the mature or degen-

erate condition. However, owing to the shortage of herbage in winter it is never practicable to maintain sufficient stock on heaths for this purpose and recourse must be made to other methods of managing heather stands to keep them as far as possible in the building phase.

Hence, from the earliest times heaths have been managed by burning (see Fire, p. 139). At first this was spasmodic, mainly to prevent colonization by trees and shrubs. In sheep-farming areas, however, it became the accepted method of management, and burning at regular intervals of about 10 years has been the recommended practice. Particularly in northern Britain this system has operated on many heaths since before 1800. It has been maintained despite a decline in the profitability of sheep-farming on heathland, because of an increase in the numbers of red grouse (*Lagopus l. scoticus*), an indigenous game bird, following the expansion of heathland. Heather constitutes a very high proportion of the diet of grouse which, like sheep, requires the young green shoots. For grouse, as for sheep, burning is the appropriate method of heathland management. The development of interest in the sport of grouse shooting is therefore an additional factor contributing to the continued existence of heaths in Britain. Burning of heather is carried out at about the same frequency as for sheep, but usually in smaller patches to ensure that each grouse territory contains stands of several different ages (including some older stands for shelter and nesting).

Much interest has centered on the ecological effects of repeated burning. A normal, well-controlled fire produces temperatures in the canopy of the bushes not much exceeding 500°C (932°F), while those at ground level should not rise beyond 400°C (752°F). Under these conditions, while the bulk of the aboveground parts of the heather is removed, the base of the stem, partly buried in the surface litter and humus, remains undamaged and the humus horizon of the soil does not ignite. Sprouting from the stem base then brings about rapid regeneration of the stand and complete cover may be re-established after two to three years. The capacity for vegetative regeneration declines with increasing age of the stand, especially beyond about 15 years. If old stands are burned or if, for other reasons, an excessively hot fire occurs, vegetative regeneration may fail. The return of heather then depends on seedling establishment and as this is considerably slower than vegetative regeneration a secondary succession involving lichens, mosses and other heath plants such as *Erica cinerea*, *Deschampsia flexuosa*, *Empetrum nigrum* or *Vaccinium myrtillus* may precede the return of *Calluna vulgaris*.

Burning management is designed to maintain dense, pure stands of heather, most of which are in the building phase when the competitive vigor of *Calluna* is maximal. This, together with the elimination of fire-sensitive species, reduces the floristic diversity of these communities. It is also likely that the

ecosystem is progressively depleted of certain nutrients. Considerable quantities of nitrogen can be lost in smoke and other elements may be lost by solution in rainwater from the ash and subsequent drainage or runoff. Although it has been shown that in oceanic climates incoming rain contributes more than enough in the period between fires to compensate for the loss of most nutrients, this apparently does not always apply to available nitrogen, and the concentration of phosphorus in rainfall is low. This element is not readily dissolved from ash, but other pathways of loss include surface erosion and the removal of animal produce. Attempts to establish tree plantations or sown grassland, in both hemispheres, have shown heathland soils to be notably deficient in phosphorus and available nitrogen.   C.H.G./D.M.M.

Wetlands

Wherever conditions are such that the water table is at or near the soil surface for much of the year the consequent waterlogging results in plant communities which are collectively referred to as wetlands. However, they can be subdivided into a number of categories which, although there is much popular misuse of the terms, will be followed here. For example, although the word *swamp* is widely used for many kinds of wetland, it is correctly applied to any state of the hydrosere (a succession that starts in water) in the tropics. It is therefore a rather imprecise term since it includes coastal vegetation (see p. 201), in which water salinity is of prime importance, as well as marshes, nonsaline wetlands in which peat is not important, or in which it is and the water is nutrient-rich, to give *fens*, or nutrient-poor, to give *bogs*. Consequently, the tropical wetlands, swamps, will be dealt with under the appropriate headings in this section and in that which follows, on coastal vegetation. Therefore, in this section, wetlands will be dealt with in two general categories, those developed over peats (bogs and fens) and in the absence of peats (marshes).   D.M.M.

Bog

Any area of wet peatland which includes both vegetation and a peat substratum may be described as a *bog*. Peatlands can be conveniently subdivided according to the nature of the water-supply. Where the water-supply is eutrophic (nutrient-rich), peatland is called a *fen*. When the water-supply is oligotrophic (nutrient-poor) the peatland formed is described as a bog. The most important source of nutrient-poor water is precipitation, and this fact has given rise to the term *ombrotrophic mire*, which is a synonym for bog. The subject is complicated further by the fact that many intermediate types of peatland exist between those of extreme fen and extreme bog; those which

A typical northwest European nutrient-poor valley bog with a dense carpet of the bog mosses *Sphagnum pulchrum* and *S. papillosum*, and pools containing *S. auriculatum* and *S. cuspidatum*. Under such waterlogged, anaerobic conditions sphagum bog peat forms.

are neither eutrophic nor oligotrophic are called *transition mires* (for rheotrophic mires, see Fen, p. 196).

General Types of Bog and their Characteristics

By far the most important plants in ombrotrophic mires are the numerous species of *Sphagnum*, commonly called bog mosses. The significant feature of sphagna is that they have many perforated, empty cells both on the surface of their stems and intermixed with narrow photosynthetic cells in their leaves. These cells enable the plant to absorb water which permits peat bogs to grow very rapidly. Different *Sphagnum* species have different tolerances of water level, for

example *Sphagnum cuspidatum* grows submerged in pools or continuously wet hollows in the mire surface, whereas *S. fuscum*, *S. rubellum* and *S. magellanicum* can tolerate fairly dry conditions, and are found on hummocks. A hollow-and-hummock structure is characteristic of certain bog types, and is believed to play an important part in their growth. Other *Sphagnum* species appear to survive best in an intermediate position where they may be subjected alternately to periods of dryness and wetness; this group includes such species as *S. papillosum*.

On the wetter parts of ombrotrophic mires, the sedges (family Cyperaceae) are the most important flowering plants (angiosperms). The most common species are the few-flowered sedge (*Carex pauciflora*), mud sedge (*C. Limosa*), cotton grass (*Eriophorum angustifolium*) and the beak-sedges (*Rhynchospora* spp.). On drier parts of the mire, heath plants (family Ericaceae) are abundant; these may include heather (*Calluna vulgaris*), marsh

andromeda (*Andromeda polifolia*), and the various types of whortleberry (*Vaccinium* spp.). In North America, greenland tea (*Ledum groenlandicum*), *Kalmia* spp., and *Chamaedaphne* spp. are also important components of the drier parts of bogs, in addition to the genera already mentioned.

If the hydrological situation of a mire is such that some of its water is received from the surrounding mineral soil, then the nutrient supply from rainfall will be augmented, and the bog is not strictly ombrotrophic; it is then described as a transition mire. This decrease in oligotrophy manifests itself in the fact that certain species are present which are not found in the extreme ombrotrophic situations. Among the most common of these are the slender sedge (*Carex lasiocarpa*), tufted sedge (*C. elata*), bottle sedge (*C. rostrata*), hare's-tail (*Eriophorum vaginatum*), bogbean (*Menyanthes trifoliata*), fennels (*Peucedanum* spp.) and the marsh bedstraw (*Galium palustre*). Most of the sphagna mentioned in connection with purely ombrotrophic situations may be found on transition mires, together with *Sphagnum palustre*. With further increase in nutrients, there appear such plants as broad-leaved cotton grass (*Eriophorum latifolium*), water mint (*Mentha aquatica*), reed (*Phragmites australis*) and the lesser spearwort (*Ranunculus flammula*). The vegetation of transition mires, therefore, includes some species generally associated with oligotrophic situations, some associated with eutrophic situations, and some which grow most successfully in the mesotrophic environment of transition mires. This serves to emphasize the fact that mire vegetation shows a continuum of change, and that the distribution of any species over the different types of bogs is a function of its ability to compete, which depends, at least in part, upon its nutrient and hydrological tolerances.

Tree roots do not generally survive in waterlogged conditions, but if for any reason there is a lowering of the water table on bogs, then it is possible for woodland to become established, usually as a transition to boreal or temperate forest. Woodlands on ombrotrophic mires in Europe are dominated by pine (*Pinus* spp.), which gives way to birch (*Betula* spp.) on less oligotrophic sites. The hollows and hummocks are found in these woodlands just as in the open, but lowering of the water table, which is a prerequisite of woodland development, will tend to eliminate pools from the hollows, and with them such water-demanding *Sphagnum* species as *S. cuspidatum*. Light-demanding species will also be replaced by shade-tolerant ones, for example *S. recurvum* will fill the niche occupied by *S. rubellum* in the open. Mire woodlands on North American bogs are dominated by black spruce (*Picea mariana*) and tamarack (*Larix laricina*).

Formation and Growth of Bogs

The two main requirements for the initiation of bog development are firstly that the input of water by precipitation should exceed losses by evapotranspiration, and secondly that drainage is impeded to create waterlogged conditions. Under these circumstances, plant remains do not decompose fully and peat is formed. Bogs develop only in the presence of an oligotrophic water supply, when the acid products of decomposition are not neutralized by bases; this causes the pH to drop and the process of decomposition to slow down even more. This fact, together with the great water-holding capacity of *Sphagnum*, makes it possible for bogs to grow much faster than fens.

Situations that can give rise to the hydrological conditions described above are numerous; however, it is possible to reduce these to two types, in which the circumstances creating the criteria are basically different. The first occurs in upland areas where the water-supply may come directly from precipitation or be drained from the surrounding land. In the latter case the water will not have traveled any distance through mineral soil, and will not, therefore, have dissolved nutrients to any great extent. This

A blanket bog in Sutherland, Scotland. The peat forms numerous small pockets within the water catchment area.

water will accumulate in depressions, and these pools will then be colonized by *Sphagnum* mosses, especially *S. cuspidatum*, which absorb water and cause the water table to rise by capillary action. In consequence of this rise in water level, the bog can continue to grow upwards and even spread sideways over the shallow ridges. In this way, the separate patches of peat link up to form *blanket bog*, which is very characteristic of upland areas in many parts of the world. Blanket bog is also found at lower elevations in northern latitudes, where it may cover vast areas of land; in Canada, this type of peatland is called *thermal blanket mire*.

The second situation giving rise to oligotrophic conditions is one in which the growth of peat in a river valley or lake simply raises the vegetation above the influence of the mineral-rich ground water, thus creating ombrotrophic conditions. These bogs usually have a base of fen peat beneath the ombrotrophic peat. The exact nature of bogs formed in this manner is very dependent upon climate. In northern Scandinavia and, more rarely, in Britain, such bogs may consist of small hummocks, capped with *Sphagnum fuscum*, and surrounded by hollows of fen or marsh. Such skeletal mires, in which the peat never grows to any great thickness, are

called mixed mires. Further south in Europe, bogs formed from lakes or rivers may grow to several metres above the original water table, and extend well beyond the confines of the original basin, by means of water-absorbing sphagna. Such peat mosses are described as raised bogs, and can be several kilometres across. The peat moss of a raised bog may become convex, provided that the original basin is of sufficient size and the precipitation not so great as to create a flow which would erode the bog surface. These forms are called *domed raised bogs*, or Baltic raised bogs, because they are well developed in countries adjacent to the Baltic Sea, where climatic conditions are suitable. The oligotrophic water that drains from a raised bog, with or without a dome (*cupola*), eventually combines with eutrophic water flowing from the surrounding mineral soil. Because of this, raised bogs are usually surrounded by a drainage system (*lagg*), in which the vegetation resembles that of a transitional mire, or even a fen.

Ombrotrophic Mires in the Tropics
The main factor that restricts peat accumulation in tropical countries is the high temperature which stimulates decay processes wherever plant litter is exposed to the

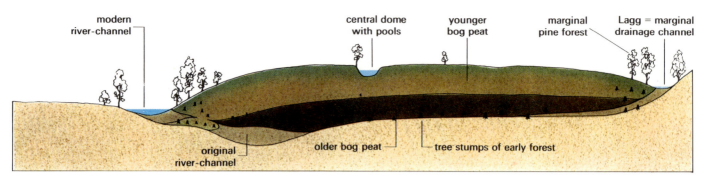

modern
river-channel

central dome
with pools

younger
bog peat

marginal
pine forest

Lagg = marginal
drainage channel

original
river-channel

older bog peat

tree stumps of early forest

A cross-section of a domed raised bog that has formed in a river valley.

air. This is why fens (rheotrophic mires) are so rare in tropical countries; in fact, until the thick peat deposits in Sumatra were discovered at the beginning of this century, peat was unknown in tropical regions. These tropical ombrotrophic mires, which require more than 2,000mm (79in) of rain per year, have now been recorded in Indonesia, Malaysia, Borneo, New Guinea, tropical Africa, and tropical America, but only those of Malaysia and Indonesia have been studied in any detail.

In Sarawak, as in other parts of Malaysia, ombrotrophic mires develop as raised bogs, several kilometres in diameter, on the coastal plains, usually adjacent to the wide rivers. Furthermore, they are often found behind mangrove swamps or alluvial mud. In these marginal areas there are many forms of soil ranging from pure alluvium, through muck soils (soils containing between 35 and 65% organic matter) to peat soils (more than 65% organic matter), depending upon the amount of flooding. These marginal soils are locally very important for agriculture.

The peat masses, which may be up to 20m (66ft) thick in places, are lens-shaped with cupolas up to 10m (33ft) high. The peat matrix is a semiliquid mass of dark brown material, in which the partly decomposed remains of plants can be found. Unlike their temperate counterparts, the peat in these tropical bogs is composed almost entirely of the remains of woody plants. Tree trunks can be found at most depths below the surface, and present a considerable problem to investigators trying to obtain cores for pollen analysis, etc. In general, peat from the lower levels is much more decomposed than peat from the surface layers of the bog. The surface slope of these domed masses decreases towards the middle, so that the central areas of most bogs are completely flat (this also applies to the domed raised bogs of temperate regions). There appears to be a definite correlation of cupola height with distance from the coast.

These mires are completely wooded by tropical rain forest vegetation very similar to that growing on well-drained mineral soils, except that the flora is somewhat less rich. Over 200 species of trees have been recorded from the Sarawak peat forests, but less than 20 species of herbs. All the usual rain-forest species are present, and epiphytes are especi-

ally abundant. The vegetation is zoned in a concentric manner, and the variation is such that the inner and outer zones differ markedly in species composition, only one or two species being common to all zones (see Zonation, p. 154). Just as in wooded Baltic raised bogs, the trees are tallest on the outside, where they can be 30m (98ft) high, and decrease in height towards the center where most trees are less than 10m (33ft) tall. Many of the important rain-forest families are present, including the Myrtaceae, Dipterocarpaceae, Euphorbiaceae, Rubiaceae and Anacardiaceae.

The central zone of these bogs is particularly interesting because the vegetation of stunted trees takes on the form of an open parkland, which is in marked contrast to the closed canopy of the outer zone forests. In this central zone the vegetation is morphologically adapted to dry conditions (xeromorphic), and insectivorous pitcher plants (*Nepenthes* spp.), which are present in all zones, become even more common. Xeromorphy and the presence of insectivorous pitcher plants are both characteristic features of temperate raised bogs (there is no satisfactory explanation as to why some bog plants exhibit xeromorphic features).

Cores to the bottom of the peat have shown

the base to consist of alluvial silt overlain by mangrove (see p. 206) clay; this is followed by a succession of communities which is a parallel of the vegetation zonation existing today. The question then arises as to how these ombrotrophic peats developed directly on the nutrient-rich mangrove clay; this is particularly relevant considering that areas of ombrotrophic peat in the tropics are very small in relation to the extensive developments of non-peaty swamp forests. This problem has not been solved; however, it is likely that impediment of the drainage by the mangrove and alluvium, together with a supply of nutrient-poor drainage water, are necessary prerequisites for the early stages. Once the first layers of peat have accumulated, the mire vegetation is soon isolated from flowing ground water, and the raised bog proceeds to grow utilizing nutrients supplied from rain water only.

The vegetation zones reflect differences in peat chemistry, water-table fluctuations, and water flow in different parts of the peat dome.

Carnivorous pitcher plants, such as these *Sarracenia* species growing on tropical ombrotrophic peat in Sarawak, make up for the lack of nutrients available in boggy habitats by catching small animals such as insects, which then decompose in the pitchers.

Some of these differences are known; for example, water flow is greater in the peripheral zone, where the water table does not fluctuate greatly, and the concentrations of certain elements in the surface peat, including zinc, aluminum, iron and phosphorus, decrease toward the center of the peat dome.

In Guinea, peat forests occur which are similar to those in Borneo. These forested mires, which are known locally as *pegass*, contain peat up to 2m (6.5ft) thick, and are found mostly in coastal areas where they are intermixed with the non-peaty swamp savannas. They also develop inland at the upper ends of rivers which drain water from adjacent oligotrophic soils.

At high altitudes in some tropical areas the cooler climate permits development of ombrotrophic mires closely resembling those in temperate lands. On the mountains of East Africa, hummocks dominated by *Carex monostachya* and *C. runssoroensis* are interspersed with *Sphagnum*-dominated hollows. The peat in these mires, which can be up to 2m (6.5ft) thick, is formed in valley bottoms receiving drainage water from nearby acid woodland. On the Ruwenzori range, these peats grow over steep slopes, and coalesce to form blanket mire.

World Distribution of Bogs

It was mentioned above that there must be an appropriate excess of precipitation over evaporation if ombrotrophic mires are to develop. Consequently they are found in abundance only in the cool-temperate, subarctic, subantarctic and tropical regions of the Earth; subtropical and warm temperate areas are too dry for anything other than local development on mountains.

Extensive ombrotrophic mires exist in Borneo and in many parts of the Malay Archipelago; they also occur in Africa and Central America. The complete distribution of ombrotrophic mires in tropical South America is still not fully known.

Climatic conditions suitable for ombrotrophic mire development in cool climates are generally found only at latitudes greater than about 40°. There is very little land south of the 40° parallel in the Southern Hemisphere, therefore cool-climate ombrotrophic mires are a landform primarily of the Northern Hemisphere. Russia and Canada alone have about two million square kilometres (750,000 square miles) of bog; large areas also exist in northern Europe and in northern parts of the United States. In the Southern Hemisphere, cool climate ombrotrophic mires are restricted to Tasmania, New Zealand, and the southern tip of South America. One interesting feature of the Southern Hemisphere mires is that in many places the *Sphagnum* mosses are replaced by cushion plants as the main peat-forming species.

The climatic requirements for the formation of domed ombrotrophic mires are more exacting than those for ombrotrophic mires in general, with the result that such mires have a very restricted distribution. The most

extensive development of domed raised bogs is found in northwest Europe in the region of the Baltic Sea; fairly large developments also occur in eastern Russia. Smaller areas of domed raised bog occur on the eastern seaboards of Canada and the northern United States, as well as in the vicinity of the Great Lakes. In the Southern Hemisphere, domed bogs occur only at the southern tip of South America. Like ombrotrophic mires in general, the extent of domed raised bogs in tropical countries remains to be explored; however, large areas of such bogs have been discovered in Borneo, and it appears likely that accurate leveling could eventually prove many tropical ombrotrophic mires to be of the domed form.

Forestry and Agriculture

The types of peatland discussed in this section have certain features in common: they are acid, waterlogged and oligotrophic. In the case of ombrotrophic bogs they may be

A thermal blanket mire in northern Ontario, Canada. The blanket of peat serves to insulate the frozen ground in winter and so reduces the depth of the summer thaw.

extremely acid and nutrient-poor. These are the problems that must be overcome if ombrotrophic peatlands are to be utilized for either agriculture or forestry.

The natural vegetation climax of oligotrophic peats in temperate lands is a forest of pine (*Pinus* spp.) or spruce (*Picea* spp.) and it is these species which have been used most widely and successfully in afforestation. The first step in the process of afforestation is to lower the water table by cutting ditches to improve drainage. In parts of northern Europe, especially Scandinavia, many of the bogs are tree-covered, and the water table is lowered simply to encourage natural regeneration of the woodland. On treeless or previously felled bogs, the improvement in drainage is accompanied by planting seedlings. The water table is lowered to about 50–80cm (20–32in) below the surface; however, no general rule can be given, because the amount of lowering required depends partly on the local climate. Various forms of fertilizer are frequently used in conjunction with drainage; phosphorus and potassium in particular have been shown to increase the volume of timber that can be produced.

Afforestation is carried out in many northern European countries, but it is especially important in Finland where about 32% of the land surface is covered by peat. This situation has stimulated Finland into becoming one of the world's leading pioneers of peatland exploitation. In addition to forestry, Finnish peatlands have also been very successfully utilized for agriculture. In the south of Finland, peat supports crops of barley, winter rye, oats, spring wheat, and various root crops; however, the north of Finland has a climate which is unsuitable for these arable crops, and there the peats are largely cultivated for pasture. The crops are produced only with the aid of yearly applications of fertilizers: up to 6,000kg/hectare (5,300lb/acre) of ground limestone may be used, in addition to potassium, phosphorus and nitrogen. However, in both agriculture and forestry, recovery of fertilizer in the crop grown on the peat is good, occasionally exceeding 80%. Peat soils are sometimes improved by mixing with mineral soil; if the peat is not too thick this can be achieved simply by deep ploughing.

In other parts of Europe, peatlands are also used for both forestry and agriculture, and these activities are particularly important in Ireland, Russia and Poland. In the United Kingdom, blanket bog, which covers most of the upland, is now treeless and drained, giving rise to heather moor which is used as rough grazing for sheep and the rearing of grouse for sport (see Heath Exploitation and Management, p. 190).

In New Zealand, upland sedge moors support good pastures provided they have been prepared; this involves draining, followed by removal of the natural vegetation by burning. Lime is spread to raise the pH to about 5; sulfur and phosphorus are also added. However, after the initial application of these fertilizers, only small dressings are required for maintenance. Some deficiencies of trace elements need to be corrected, including shortages of molybdenum, boron, selenium, copper and iron. Some cultivated peatlands in Finland also suffer from the deficiency of certain trace elements, especially boron and copper.

Oligotrophic peatlands in the United States have been used for growing cranberries (Vaccinium macrocarpon), and in Japan large areas of peatland are being reclaimed for rice production. Both of these crops are well suited to a high water table, therefore little or no drainage is required.

The most extensive utilization of tropical bogs to date has been in West Malaysia, where 6% of the country is covered by peat. Pineapple cultivation on peat started in Malaya in the late 1930s after attempts to use upland mineral soils for this crop had met with only limited success. Peat proved to be satisfactory, and today most pineapple cultivation in Malaysia takes place on coastal peatlands. All tropical bogs are forested, so after draining it is necessary to clear the trees. The pineapple crop must also be protected from weeds by manual hoeing once a month for the first 10 months, and subsequently every three months. Other crops are occasionally grown on the West Malaysian peats, including green pepper (Capsicum), chillies, ginger, papaya (Carica) and tapioca; coffee has been grown on muck. In Sarawak, bogs are still largely exploited for their timber, and the only crop widely grown is sago.

j.wn.

Fen

Fen is a wetland area where vegetation grows on peat, which has developed in the presence of moving, mineral-rich water; the presence of peat distinguishes fen from marsh (see p. 200). There are numerous synonyms for fen, including *minerotrophic mire*, *rheotrophic mire* and *valley bog*.

Fen Vegetation

Fens are associated with early phases of the succession that starts in water (hydrosere), the type of fen vegetation present depending upon the extent to which the succession has progressed towards firm ground, and on whether the water moves above or below the surface of the peat. If the water moves below the surface, the vegetation is in the nature of a turf of floating and weakly rooted plants, which, in Britain, may include certain mosses, sedges, kingcups (Caltha palustris), the bogbean (Menyanthes trifoliata), broad-leaved cotton grass (Eriophorum latifolium), marsh bedstraw (Galium palustre) and the lesser spearwort (Ranunculus flammula). However, when water flows on the surface for even part of the year, the only plants that will survive are those well rooted into the substratum; these include such species as the common reed (Phragmites australis), reed-maces (Typha spp.), saw sedge (Cladium

Cutting peat from infertile acid soil in Ireland. The cotton grass (Eriophorum sp.) in the foreground is an important component of acid peat bogs.

mariscus), and the reed grass (*Glyceria maxima*). This second type of fen is sometimes called a *fluvial fen*, a term which recognizes that the water is visibly flowing, unlike *turf fens* where the water movement is not obvious.

The foregoing division into fluvial and turf fens is an oversimplification; the most floristically rich fens are those that contain a mixture of turf and fluvial plants, so there is really a gradation between the two types of fen. The exact nature of fen vegetation will depend upon the water's chemical composition, frequency of flow and average level, as well as on the frequency and size of water-level fluctuations, and will alter with any change in the water regime. The fen may eventually become colonized by woody species if the water level drops sufficiently; these include sallow (*Salix atrocinerea*), buckthorn (*Frangula alnus*) and alder (*Alnus* spp.); when the woody cover is complete the vegetation is referred to as a *carr* (see p. 200).

Development of Fen Peat
The essential prerequisite of fen (or rheotrophic) peat formation is a continuous or intermittent supply of mineral-rich (eu-

trophic) water, which flows at or near to the peat surface. This means that the peat formation is always intimately associated with the drainage of an area, and occurs in such locations as river valleys, lakes and deltas. Peat develops from the partially decomposed remains of plants; therefore in lake basins the first deposits may consist of diatoms (unicellular green algae with siliceous walls), which fall to the floor of the lake after death; immediately above this will be the remains of floating plants (the megaplankton), which often comprise the first stage of the hydrosere. These diatom-rich sediments (*gyttja*) are frequently found at the base of lake peat deposits and represent the first stage of water shallowing which leads in due course to formation of land. When the water is sufficiently shallow, reed beds develop, as well as lawns of bryophytes and herbs. As this semiterrestrial vegetation dies the waterlogged conditions prevent complete breakdown by bacteria, so that the peat deposit becomes thicker.

Fen vegetation develops under the influence of drainage water, which is subject to seasonal fluctuations, ie will be low in summer and autumn, but high in winter and

Fringed by reeds (*Phragmites australis*) and backed by fen woodland (carr) the quiet water of a fenland drainage dike reflects the sky. A fen with surface water, as here, is called a fluvial or reed fen, as opposed to a turf fen, where the water table remains below the surface.

spring; during certain times of the year, therefore, the upper layers of peat may not be waterlogged. Taking into consideration the fact that fens are continually provided with a supply of eutrophic water, these periods of aeration stimulate intense bacterial activity, and for this reason fen peats are generally more decomposed that their ombrotrophic counterparts, bogs.

The development of firm ground can only partly explain the formation of fen peat, because a consistent shallowing of the water would give rise to corresponding consistent changes in vegetation. In addition, the surface would soon grow above the influence of the flowing water, and terrestrial, non-peat-forming conditions would be created. In contrast to consistently changing assemblages of plants, sections through peat frequently show thicknesses of 2–3m (6.5–9.8ft) which have clearly been derived

The common reed (*Phragmites australis*) makes an ideal, but expensive, thatching material that lasts much longer than the wheat straw generally used. This thatcher in Norfolk, eastern England, is stacking the bundles of reeds that he himself has cut from the reed beds.

from the same vegetation type. This indicates that the water level must have slowly risen as the peat developed, so that the relationship between the peat surface and the water level remained unchanged over a long period. This raising of the water level occurs because the developing peat behaves like a sponge and absorbs water as it forms.

If the water table drops for any reason, the peat decomposes very rapidly, and becomes colonized by woodland. Thus, the presence of amorphous peat in a section (highly decomposed and structureless), with or without woodland, indicates a period when the water table was lowered relative to the surface. The nature of the peat, as seen in cuttings, can, therefore, provide valuable clues regarding conditions prevailing during the time of deposition: it is possible to conclude whether the water flow was fast or slow; whether the water level was changing or steady in relation to the surface; and whether the vegetation was turf, reed or woodland. These types of observations, when combined with analysis of the fossil pollen grains preserved in the peat, constitute the science (or art?) of peat stratigraphy (see Palynology, p. 66).

Fens of Eastern England

Since these fens are much the best studied, they are described here as a case history of this type of vegetation. They occupy a part of East Anglia which includes the counties of Norfolk, Suffolk and Cambridgeshire. This region, together with its counterparts in Denmark and Holland, is today one of the most important agricultural areas of Europe. This emphasizes the fact that fen, or rheotrophic peat, develops under mineral-rich conditions and, when drained, gives rise to fertile agricultural land.

The fens of East Anglia date back to the pre-Boreal period (about 8000 BC) (see The Quaternary Period, p. 80). During this time eastern England was connected to northwest Europe by land that today is covered by the southern part of the North Sea; trawlers fishing in the region occasionally remove chunks of freshwater peat from the seabed. This peat contains the remains of pine, birch and hazel, all characteristic of the Boreal woodlands. During postglacial times the glaciers slowly melted, causing a worldwide raising of the sea level, and slow flooding of the land connection between Britain and Europe. But the process was not a straightforward continuous sinking. Many soil profiles in East Anglia show layers of freshwater peat alternating with layers of estuarine silt. Therefore, at times, the land of East Anglia must have stretched far into the now southern North Sea as an area of swamps and marshes, drained by the rivers of eastern England which laid down deposits of freshwater silt and peat. At other times this land was inundated by the sea with the deposition of estuarine silt. Today, this delicate balance manifests itself in the fact that areas near the coast are generally occupied by marshes (see p. 200), whereas fen proper is found at the more inland locations.

All stages of the hydrosere are found in the fens of East Anglia: floating plants, reedswamp, alder and sallow woodland are widely developed, but they form as a patchwork rather than as clear zones of the classical hydrosere.

HUMAN INFLUENCE. Man has made use of the East Anglian fens since at least Roman times. Since the dikes frequently have a lower water level than the rivers into which they drain, a pumping system was developed, necessitating the windmills so characteristic of the East Anglian landscape. In recent years, wind has been replaced by steam, oil or electrical power. Fenland, drained and cleared of woodland, provides rich agricultural land which is used for grazing and numerous arable crops as well as for fruit and bulbs. This drained land may be peat (true fen) or silt (marsh), both of which can give rise to very fertile soil.

Draining the fens has been the major method by which Man has changed the fenland landscape; however, the removal of fen peat for fuel has also been significant. It is now generally believed that the inland waterways (broads) themselves are not natural features, but merely flooded peat excavations. Yet another of Man's activities affected the broadland fens in a more indirect way. In 1929 the coypu was introduced from South America so that it could be bred for its valuable pelt; eventually some escaped and became naturalized in the region. The simplest effect of these animals was to reduce the area of the reed beds by using them for food; however, they have also affected the later stages of succession in a more indirect way. In this region, fen woodland or carr (see p. 200) usually starts with sallow (*Salix*) and alder (*Alnus*) seedlings growing on tussocks of the panicled sedge (*Carex paniculata*), which grows on the still fairly thin peat mat.

By this means, the seedlings grow in an environment where the water table is sufficiently low. Unfortunately, the coypu is rather partial to the rhizomes of plants which constitute the peat mat, and their feeding reduces the strength of the mat, thus creating an unstable base for tussocks with carr trees growing in them. These tussocks may then topple over and succession towards climax woodland is arrested, giving rise to extensive areas of *swamp carr*, which under normal circumstances is a short phase in the succession. Another problem created by the coypu is not as dramatic as the two described above, but equally problematic for the biologist; this is the fact that selective feeding has created unnatural species assemblages, making investigation of the natural ecological processes very difficult.

Man has directly affected the fenland vegetation simply by cropping it for kindling, basket-making, thatch, hay etc. The beds of saw sedge in particular have been reduced because of the value of this species for thatching and kindling. Where cutting has occurred once every four years, purple moor grass (*Molinia caerulea*) becomes co-dominant with the saw sedge. With more frequent cutting, purple moor grass may become the single dominant of a herbaceous community, which in the past was cropped for cattle bedding. The anthropogenic plant communities prevent the hydrosere continuing to the carr stage. However, this develops very rapidly when cutting of the beds is stopped.

The Pleistocene changes in sea level which gave rise to the East Anglian fenland also affected Holland, with the consequent development of marshes and fens. However, such areas in Holland have, if anything, been more intensively utilized than in England, even to the extent of reclaiming large areas of salt marsh from the sea by means of dikes and drainage systems (dike means *embankment* when used with reference to low country in Holland, in contrast to a *ditch* in the English fenland). The Zuyder is the most famous of these reclaimed areas which provide very fertile agricultural land. Although large areas of fenland proper once existed in Holland, very few natural fens remain. Most of the 70,000 hectares (173,000 acres) of fen peat have been drained, and now support high quality grassland.

Tropical Fens

In fens, the decaying organic material may not always be completely saturated with water, which means that aerobic conditions will prevail for at least some of the time. The high temperatures of tropical climates greatly stimulate the activity of aerobic decomposing bacteria, so that fen peats are comparatively rare, with most rheotrophic mires taking the form of reed or forest swamp with little organic material in the soil. However, limited developments of fen peat have been discovered in Africa, America, and in the Malay Archipelago, where they are usually referred to as *valley bogs* or *topogenous moors*.

East Africa contains extensive areas of reedswamp dominated by the common reed, saw sedge and papyrus (*Cyperus latifolia*). Fen peats rarely develop to any thickness in the lowlands. However, with increasing altitude, the temperature drops, slowing down bacterial decay and allowing thick fen peats to develop. Above about 2,000m (6,500ft)

During the early stages of the development of a carr (**below**), reed beds are invaded by young trees, mainly alder (*Alnus glutinosa*); as the carr matures (**bottom**) the alder trees reach their full size and a rich ground layer of herbs dominated, as seen here in northern England, by large tussocks of the panicled sedge (*Carex paniculata*) develops.

papyrus does not survive, but the other reeds, as well as grasses and sedges, continue to form fen-peat deposits of considerable thickness. The royal fern (*Osmunda regalis*), which is occasionally found in temperate fens, is common in these high altitude East African fens. In Uganda many of the upland peats have been drained, and support crops of maize, sweet potatoes and sorghum; they are also occasionally used for grazing cattle. Where the water table has been sufficiently lowered, these peats can become infertile owing to the buildup of acidity: pH values lower than 2.5 have been recorded. However, the situation can be remedied by further lowering of the water table.

j.wn.

Carr

Carr is damp woodland developed on fen peat. Such woodland is distributed throughout most of the wet-temperate part of the Northern Hemisphere. Although a variety of tree species may be found in carrs, alder (*Alnus* spp.) is the most characteristic, and the term *alder-carr* is frequently used. Other trees that may be present are birch (*Betula pubescens*), ash (*Fraxinus excelsior*) and oak (*Quercus* spp.), with alder buckthorn (*Frangula alnus*) and sallow (*Salix atrocinerea*) being the two most common shrubs. The nutrient-rich fen peat combined with the relatively open canopy of these woodlands permits the development of a rich and varied ground cover. Among the more frequent species in this layer are the stinging nettle (*Urtica dioica*), meadow sweet (*Filipendula ulmaria*), the panicled sedge (*Carex paniculata*), king cup or marsh marigold (*Caltha palustris*) and the yellow flag (*Iris pseudacorus*). Also to be found in the herb layer are many ferns, such as the marsh fern (*Dryopteris thelypteris*) and horsetails (*Equisetum* spp.). Occasionally, the royal fern (*Osmunda regalis*) is found in alder-carrs. A synusia (see p. 41) of climbing plants is usually well developed, with hop (*Humulus lupulus*), bittersweet (*Solanum dulcamara*), and ivy (*Hedera helix*) being among the most common. The atmosphere inside carr woodland is very moist, and an abundance of lichens, mosses and liverworts can always be found covering the tree trunks and branches. The above list is only a very small selection of the more abundant species, most of which can be found in carrs throughout northern Europe, and some in North America.

Carr woodland develops on sedge or moss peat, which is always very dark, and highly decomposed near the surface because the nutrient-rich conditions and slight lowering of the water table (necessary for carr development) encourage microbial activity. In some areas, especially near to the base of tree trunks, mossy hummocks grow above the peat surface so that they are unaffected by the periodic flooding. In these situations, plants usually associated with drier, more acid woodlands can be found; these include the glittering feather moss (*Hylocomium splendens*), fork mosses (*Dicranium* spp.) and the red-stemmed feather moss (*Pleurozium schreberi*). Locally, conditions on these hummocks may become sufficiently acid to allow bog mosses (*Sphagnum* spp.) and sundew (*Drosera anglica*) to grow.

The ecology of carr woodland can be more fully appreciated if its position in the hydrosere is well understood. Carrs develop in open fen if the accumulation of fen peat stops; this usually happens when, for some reason, the water table is lowered relative to the peat surface. Fen vegetation then proceeds toward its climax. The first stage of this is the colonization of the fen surface by alder-buckthorn and sallow; from then on, even if no other hydrological factors come into play, the water table will be lowered owing to the

One of the most conspicuous dominants in western European reedswamps is the common reed, *Phragmites australis*, seen here in winter. The common reed is an abundant, dominant species in waterways throughout the world, often forming pure stands.

increased water loss (transpiration) caused by the shrub canopy. This further slight lowering of the water table creates conditions suitable for alder, which then becomes increasingly abundant. As the alder forest canopy closes, and the water table is lowered even further, sallow and alder-buckthorn become less common. The relative abundance of alder, sallow, and the alder-buckthorn gives an indication of the stage in succession between open fen and closed alder woodland. Another indication of this is the presence of various reeds which may persist in the ground flora through the initial shrub stages, but which are much less common in the later stages as the fen changes into alder-carr.

The alder-carr is considered by some to be the climax woodland of fen-peat. However, there is little doubt that in time alder-carrs do change to woodlands dominated by other trees, eg birch (*Betula*), oak (*Quercus*) or pines (*Pinus*). Therefore, it is probably more accurate to consider alder-carr as a sub-climax, which persists so long as the peat is subjected to periodical inundation by water.

j.wn.

Marsh

The term marsh is used when referring to an area of soil and vegetation in which the water table is close to the soil surface, creating waterlogged conditions for at least part of the year. The soil is composed predominantly of silt with very little organic matter, whereas in fens the substratum is peat. Marshes may develop wherever the situation is suitable for the deposition of sediment, eg river banks, flood plains, shores of lakes and deltas. Marshes which are subjected to periodic inundation by the sea are called salt marshes (see p. 204). The present section deals with freshwater marshes only.

Freshwater marshes can develop in any location where water movement is slowed down sufficiently for the silt load to be deposited. To this extent, marsh development can be associated with the hydrosere, where the early stages of submerged and floating aquatics serve to slow down water flow. Although such a situation may occur around lakes, it is fairly unusual, because if the water flow is slow enough for floating aquatics to develop, it is unlikely to carry much of a silt load. In addition, organic remains of the plants are likely to accumulate, thus giving rise to the formation of fen, in which case the marsh stage would not persist for long.

Therefore, although marshes may be an intermediate stage of the hydrosere in lakes, they are more frequent in situations where water movement is slowed by geomorphological features such as deltas or flood plains.

The plant species that grow on marshes are much the same as those growing on fens, with reeds (*Cladium*, *Phragmites*, *Typha*) and bulrushes (*Scirpus*) in the more swampy areas, followed by a rich herb flora as the marsh surface develops further from the water table. The commoner herbs in the later phase of European marshes include the rush (*Juncus conglomeratus*), marsh marigolds (*Caltha palustris*), pennywort (*Hydrocotyle vulgaris*), the lesser spearwort (*Ranunculus flammula*) and purple loosestrife (*Lythrum salicaria*). In addition to floristic differences associated with terrestrialization (formation of land), there are also differences which result from water movement through the marsh. This feature is well illustrated by marshes in the southwest of England, which develop on alluvium deposits. Some species, such as the unbranched bur-reed (*Sparganium simplex*), the great hairy willow herb (*Epilobium hirsutum*) and the water forget-me-not (*Myosotis palustris*), are common in all situations: others, such as reed grass (*Phalaris arundinacea*), the marsh foxtail (*Alopecurus geniculatus*) and the rush (*Juncus conglomeratus*), are found only on marshes with very slow-moving water; whereas the marsh bedstraw (*Galium palustre*), the hard rush (*Juncus inflexus*) and water mint (*Mentha aquatica*) are associated with moving water.

With continued lowering of the water table, marshes, like fens, can be invaded by shrubs, and subsequently by trees (see Carr, p. 200). In most lowland situations the silt gives rise to a nutrient-rich marsh, and in consequence the woodland is likely to be dominated by alder (*Alnus* spp.)

The marshes described above usually form in the lower part of drainage systems. However, the largest areas of marsh in the world have developed in low-lying regions where ground waters may seep to the surface almost anywhere, not only in valleys. Such an area serves as catchment for the Pripet river in the Polesie region of Poland, and gives rise to many thousand square kilometres of marsh. In these Pripet marshes the water table is at the surface in winter, but is so low in summer that peat cannot form and fen development is arrested, although some humus may develop in certain areas. Marshes that develop in the lower regions of drainage systems do so with water which has become rich in nutrients while moving through mineral substrata. However, ground water which gives rise to marshes such as the Pripet marshes may be nutrient-rich or nutrient-poor depending upon how far it has moved through the substratum before emerging to form the marsh. Therefore, marshes adjacent to fens will have been formed by much the same nutrient-rich water that gave rise to the fen, and have a

flora similar to that described above, followed by an alder-wood climax. However, adjacent to bogs the water is likely to be nutrient-poor and the marsh flora will contain such plants as bog mosses (*Sphagnum* spp.), cranberries (*Vaccinium oxycoccus*) and marsh andromeda (*Andromeda polifolia*), with pine (*Pinus* spp.) and purple moor grass (*Molinia caerulea*) in the drier areas.

Low-lying water catchments have given rise to vast areas of marsh in parts of the Mato Grosso in Brazil, water from these marshes draining into southern tributaries of the Amazon. A similar situation occurs in Uganda where over 10,000 square kilometres (3,900 square miles) of poorly drained swamps and marshes serve as catchment for the river systems. In addition to these catchment marshes, there are vast areas of valley marshes in both Uganda and other parts of East Africa; many of these valley marshes have developed as a result of the main drainage system being reversed during the Pleistocene period. Papyrus (*Cyperus papyrus*), wild rice (*Oryza bartii*) and the tufted grass (*Miscanthidium violaceum*) are locally important in these marshes, as are the common reed (*Phragmites australis*) and the saw sedge (*Cladium mariscus*), both of which are commonly found in temperate marshes.

There are many large wetland complexes in North America, the main ones being situated in the New England states, the Great Lakes area, and the Fraser River Delta. Within all of these complexes there occurs a mosaic of fens, bog, and marshland proper; exactly which wetland type develops at any location depends upon the local geomorphology and stage in the hydrosere.

Further south in North America, marshland becomes the most important component of wetland complexes because the drier

climate and higher summer temperatures are not conducive to peat formation. In Nebraska, for example, tributaries of the Missouri River give rise to over 40,000 hectares (99,000 acres) of wetland, most of which is either lake or marsh. These marshes contain all the usual marshland reeds and rushes, but in addition the Nebraska marshes carry an exceptionally rich and varied herb flora.

In the United States, just as in northern Europe, the area of marshland has diminished as a result of drainage and reclamation. This is especially true in California where the once extensive marshlands which occupied the central valley have been largely reclaimed to provide the fertile agricultural land so famed for its produce today.   j.wn.

Coastal Vegetation

Among the prominent environmental factors which plants have to face, one of the most severe is encountered at the point where the continents give way to the world's oceans. The high salt content (salinity) of sea water has been a major factor in preventing most higher plants from straying far from the land. Those plants which are salt-tolerant are called *halophytes*. The marine habitat, permanently aquatic and saline below the low-tide mark, is chiefly occupied by the larger algae, of which the kelps are vegetationally dominant. Among the vascular plants there are few to challenge the supremacy of these seaweeds, but the eel-grasses *Zostera* and *Posidonia* succeed in covering large areas of

A bed of papyrus reeds (*Cyperus papyrus*) growing by the shore of Lake Naivasha in Kenya. This sort of vegetation, dominated by species rooted in the submerged mud, is characteristic of the extensive marshy areas of East Africa.

Mature coastal sand dunes near Newquay, Cornwall, England. In the foreground are fruiting plants of marram grass (*Ammophila arenaria*), a deep-rooted species which helps to bind the sand together.

sandy seabed in shallow subtidal water. Here they experience conditions of nearly constant salinity (3.5–4.0%), which are less severe than situations where salt concentration is liable either to fluctuate widely within a few hours or to reach extreme concentrations during a prolonged period of evaporation.

Between high and low tide marks throughout the world there are habitats exposed alternately to inundation by saltwater and then to exposure to air with accompanying evaporation, the relative duration of these phases depending on position above or below mean tide level. On sandy, silty or muddy sites within this zone there are two types of halophytic vegetation: within the tropics tree-dominated *mangrove swamp*, and outside the tropics herb-dominated *salt marsh*. Plants of both formations are halophytes which exhibit succulence of their nonwoody parts, eg the leaves of mangrove trees and the shoots of glasswort (*Salicornia*). Coastal cliffs also have a proportion of halophytes in their vegetation, a response to the salt spray which they receive. *Sand dunes*, which are a prominent feature of coastal areas in many parts of the world, face the joint problems of salinity and a generally unstable substrate.

The problem for halophytes is to match the external salt concentration in their cell sap and so combat osmotic forces which would make intake of water impossible. The typical succulent leaves of halophytes contain a sufficient reservoir of sap to meet this demand and to store water during periods when further intake is prevented. In arid regions inland, drainage of rainfall carries salts into depressions where they become concentrated in the soil by evaporation, so halophytes provide the only vegetation here too.   D.M.M.

Dunes

Dunes are mounds and ridges of sand that have been piled up by the wind. Sand dunes may be coastal or inland, the former originating from windblown sand from the foreshore, the latter occurring in extremely arid deserts. Because of the aridity and mobility of the sand desert, dunes have very little vegetation, apart from a few specialized plants that can tap deep reserves of ground water. Coastal dunes, both on maritime and inland lake shores, can have well-developed vegetation in a succession from strandline (the shoreline above the high water mark) to scrub and occasionally woodland. The stability of the vegetation increases inland, and on seacoasts there is an associated decline in the effects of salinity.

The first detailed studies of plant succession were made on dunes along the shore of Lake Michigan. Most work on sand dunes, however, has been on seacoasts, where the large tidal range has important effects on the beginning of dune formation. After spring equinoctial tides, a belt of litter is left on the strandline and it contains the seeds of strandline plants. The litter retains moisture and reduces temperature fluctuations in the underlying sand, as well as providing a rich source of nutrients. A number of annual plants can exploit this habitat, but tend to be confined to it and are destroyed by autumn gales. Strandline plants play little part in dune succession but they do help accumulate sand and litter above the mean high water

mark of spring tides, and this is colonized by perennial grasses which form the first (embryo) dunes. Two such species are *Agropyron junceiforme* and sand lyme-grass (*Elymus arenarius*); they possess both horizontal and vertical rhizomes that may grow fast enough to allow 50–60cm (20–24in) of sand to accumulate in one year. The possession of groups of tillers at the ends of the near-surface rhizomes results in the development at these points of small tufts of rhizomes, at the ends of which more tillers may be produced. The foredunes may reach 2m (6.5ft) or so in height. What brings about the decline of these pioneer plants is not clear, but it may be partly the result of the reduction of water and nutrient supply in higher dunes and partly a response to the shading and more rapid sand accumulation caused by the even more vigorous grasses which follow, especially marram grass (*Ammophila* spp.).

The rhizome growth of *Ammophila* is almost unlimited, both vertically and horizontally, enabling it to withstand continual burial by windblown sand. Because of this it can form very high dunes, and the first main ridge of dunes inshore from the foredunes is often covered almost exclusively by *Ammophila*. The aerodynamics of sand movement are complicated; one of the factors involved is that the presence of plants slows down wind movement, as a result of which windborne sand particles drop. The characteristic shapes of dunes (best seen in deserts where there is little or no vegetation to change the windflow patterns) are the result of interactions of wind and topography, modified by plant cover.

Dune systems may either increase seaward, new sets of dunes forming in front of the old and creating a series of ridges and hollows or they may be more mobile; where onshore winds are strong, the seaward face may erode and whole dune ridges move inland. This phenomenon can create problems since agricultural land, forest and even settlements can be buried. These effects may be accentuated by the activities of Man, who can cause erosion by breaking through the thin stabilizing crust of vegetation.

The water relations and nutrient status of dune soils vary considerably with rainfall, size and composition of the particles of sand and the effects of the vegetation itself. Rain is almost the only source of water, although condensation of dew may contribute a little. In the hollows, or "slacks" between ridges, the water table may be near the surface, in winter even rising above it and forming pools. Nutrients come from seaspray, and systems with much calcareous sand (especially those replenished by shell fragments) are more species-rich than those deficient in calcium. But in all dune systems leaching through the well-drained sand removes nutrients rapidly. Accumulation of plant litter and animal remains gradually enriches the soil, and although little is known about them at present it is evident that bacteria and blue-green algae in the soil fix atmospheric

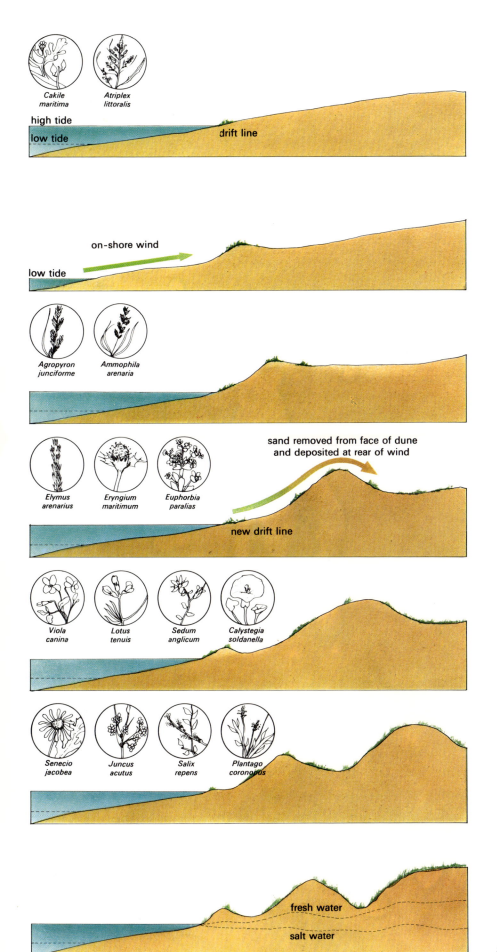

At the limit of the highest tides (spring tides of autumn and early winter) a band of debris, including seeds, collects. Seeds of plants able to tolerate these conditions, eg sea rocket (*Cakile maritima*) and sea orache (*Atriplex littoralis*) germinate to form a line of annual vegetation.

At low tide the sand dries and is blown up the beach by on-shore winds. The drift line vegetation reduces the wind speed near the ground and some of the sand is deposited amongst the plants.

Pioneer grasses colonize the embryo dune. Sea wheat grass (*Agropyron junciforme*) able to tolerate 6% salt will grow on areas still affected by the highest tides, while marram (*Ammophila arenaria*) (which is slightly less salt tolerant) grows above the tide limit. The grasses arrest more sand blown from the beach and the dune continues to grow.

The dune continues to grow and moves slowly inland as sand is removed from the face and deposited at the rear. Colonization by further species occurs: sea lyme grass (*Elymus arenarius*), sea holly (*Eryngium maritimum*) and varius spurges (*Euphorbia sp*). These species are all capable of growing back to the surface, usually with division and expansion, and producing new leaves after being deeply buried in sand. The growth and movement of the dune results in a new drift line being formed on the seaward side.

The number of species increases and the dune is almost completely covered in vegetation. The movement of sand is greatly reduced, and a new dune is beginning to form at the new drift line.

The growth of the dune has virtually stopped (late yellow dune phase) and biennials eg common ragwort (*Senecio jacobea*), and even some annuals join the plant community. The depression between adjacent dunes is called a low, and is much damper (if, as sometimes happens, it is very wet it is called a slack). The low is colonized by water loving plants: rushes (*Juncus*) and creeping willow (*Salix repens*). There is often a high salt content in the lows and salt-tolerant species such as stag's horn plantain (*Plantago coronopus*) are then common.

The dune has now reached the gray, or fixed stage. The surface of the sand is covered with mosses and lichens and there are many annual plants among the flora. From now on the normal processes of erosion cause the dune to spread and flatten, becoming lower with age.

Stages in the development of sand dunes.

nitrogen in dunes and thus contribute to the soil nutrient pool.

The flora of coastal dunes is rich and diverse, the large number of adventive species found on almost all dune systems enhancing this diversity. The diversity is in part the result of the dune habitat remaining "open" for a long time, allowing the establishment of many species characteristic of unstable situations. Special interest attaches to the weedy element in dune floras. There are several closely related pairs of species or subspecies with coastal and weedy habitats, such as the mayweeds *Tripleurospermum inodorum* and *T. maritimum*, dune and inland races of the silverweed (*Potentilla anserina*) and the coastal and weedy subspecies of the curled dock (*Rumex crispus*). The ecology of such pairs has great evolutionary as well as ecological interest.

Apart from the open vegetation with relaxed competition, the great topographical variation resulting from the presence of windward and leeward, seaward and inland,

Temperate salt marshes, which are dominated by herbs, usually start with the halophytic glassworts, such as the *Salicornia europaea* seen here at the lowest part of an English salt marsh.

sunny and shaded slopes, and slacks of various wetness, must make a great contribution to diversity. The microclimatic variations resulting from this topography are many and their cumulative effects on plants and animals throughout the year are profound.

As dunes become stable and the rate of sand accumulation slows, *Ammophila* becomes senescent. Other species can now invade and the dune-building species die out. At this stage, species richness is very high; if in time scrub develops, the diversity declines.

Dune slacks are relatively little studied. They may range from being almost totally waterlogged to the so-called "dry" slacks in which the water table is so far down that only deep-rooted species can reach it. Within a single dune system one may find continuous variation from one extreme to the other and the complexity of such a situation defies easy description.

Because of their coastal situations, dunes have come in for more than their share of human interference. Even if they are not destroyed by industry or housing, their use for recreation creates great problems. Litter accumulation can change their nutrient status and of course reduces their visual

attraction. Some introduced plants, like the sea buckthorn (*Hippophae rhamnoides*), rapidly shade out the native flora and, since it is a nitrogen fixer, it rapidly colonizes large areas and enriches the soil. Dune systems make good golf courses, and although the greens are well fertilized, a well-managed golf links should be able to combine a high recreational value with a diverse wildlife. The use of dunes by large numbers of people in an uncontrolled fashion can soon result in deterioration. Tramping kills the vegetation and along well-used paths the fragile surface protection it provides is broken through, allowing wind erosion to take place, which if not arrested can destroy dune systems. Clearly, it is important that sand dunes, intrinsically beautiful and diverse in plant and animal life, should be treated as a valuable resource of any country with a coastline.

s.w.

Salt Marsh

In temperate coastal areas, where water is shallow and coasts are protected by sand banks or spits, and in sheltered bays and estuaries, fine material carried in suspension in seawater is slowly deposited on the seafloor. This gradually builds up until the seafloor becomes exposed at low tide. Algal growth then commences which consolidates the mud. Other plants follow in time, and their presence slows down water movement and accelerates silt deposition. A succession of plants ensues, new species becoming established as the soil level rises. This vegetation is called salt marsh, and it can be found in suitable places along coasts from arctic to warm temperate regions, whenever there is sufficient tidal movement. A few inland salt marshes occur in areas where salt springs emerge, but they are relatively unimportant.

Salt-marsh development is not uniform, in that the deposition of mud is not even. The mud surface develops areas of slightly higher and lower topography, and water flow becomes concentrated in the lower areas. These areas become better defined as the overall level rises and they form drainage channels, or creeks, which are very characteristic of salt marshes. From the air, a salt marsh can be seen to have a complex ramifying network of creeks along which both inflow and outflow of water are channelled. Water scour deepens and extends these creeks, and on old marshes they may be very deep.

Salt-marsh surfaces also develop large numbers of hollows or "pans" which may be formed in several ways, including blockage of creeks. Initially they are more or less devoid of vegetation. Recolonization is slow, probably partly because of the drastic salinity changes occurring in pools which retain water for long periods.

Salt-marsh communities, superficially simple, are in fact subject to complex environmental changes and the distribution

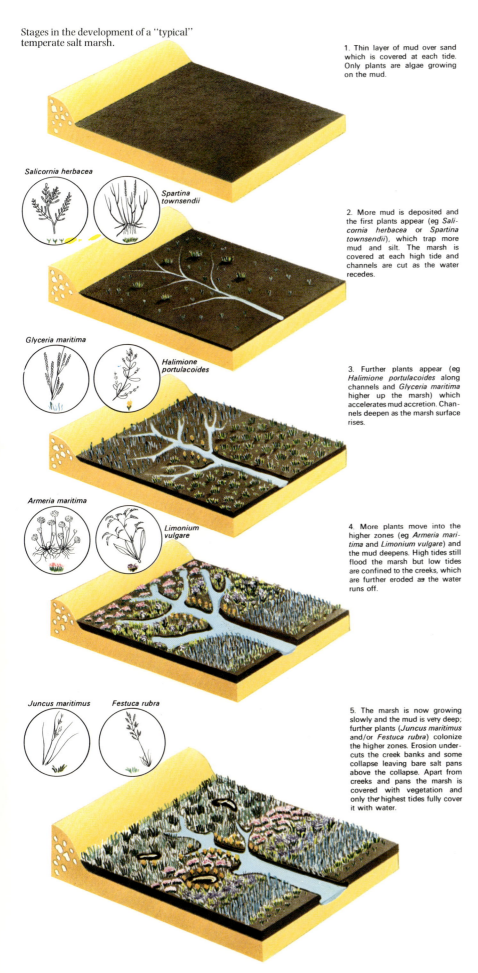

Stages in the development of a "typical" temperate salt marsh.

Salicornia herbacea

Spartina townsendii

1. Thin layer of mud over sand which is covered at each tide. Only plants are algae growing on the mud.

2. More mud is deposited and the first plants appear (eg *Salicornia herbacea* or *Spartina townsendii*), which trap more mud and silt. The marsh is covered at each high tide and channels are cut as the water recedes.

Glyceria maritima

Halimione portulacoides

3. Further plants appear (eg *Halimione portulacoides* along channels and *Glyceria maritima* higher up the marsh) which accelerates mud accretion. Channels deepen as the marsh surface rises.

Armeria maritima

Limonium vulgare

4. More plants move into the higher zones (eg *Armeria maritima* and *Limonium vulgare*) and the mud deepens. High tides still flood the marsh but low tides are confined to the creeks, which are further eroded as the water runs off.

Juncus maritimus

Festuca rubra

5. The marsh is now growing slowly and the mud is very deep; further plants (*Juncus maritimus* and/or *Festuca rubra*) colonize the higher zones. Erosion undercuts the creek banks and some collapse leaving bare salt pans above the collapse. Apart from creeks and pans the marsh is covered with vegetation and only the highest tides fully cover it with water.

of any species may result from a variety of factors. These factors can best be summarized under the headings: inundation, instability, aeration, salinity. Adaptations to all these have been identified but the true nature of adaptations to the salt-marsh environment is by no means fully understood.

Inundation and instability are related, both being the direct result of tidal action. Salt marshes are flooded twice daily during spring tides and remain largely uncovered at neap tides. They can be divided into lower and higher areas, the dividing line being at about mean high water level. The lower areas normally undergo more than 360 submergences every year, the average daily period of submergence in daylight being over 1.2 hours and the maximum period of continuous exposure never exceeding nine days. Upper areas experience less than 360 submergences annually; their submergence during daylight is less than one hour daily, and their minimum continuous exposure period exceeds 10 days. Thus lower (submergence) marshes are dominated by factors associated with submergence and upper (emergence) marshes are dominated by factors associated with emergence.

It follows that deposition of silt is more rapid in submergence marshes and the continuous rise in soil level exerts a constant pressure on plants, which must either be too short-lived to be affected, or must grow upward and keep pace with accretion. Both strategies exist, although salt marshes contain only a few annual species (eg *Salicornia* spp., *Suaeda maritima*) and short-lived perennial species (eg *Aster tripolium*). Most salt-marsh species are either herbaceous perennials (eg *Puccinellia maritima*) or woody perennials (eg *Limonium* spp., *Armeria maritima*, *Plantago maritima*). Annuals and short-lived perennials, such as *Aster*, are most vigorous in the lower, more open, rapidly growing marsh areas, while herbaceous perennials are good colonizers of waterlogged mud because they are capable of rapid vegetative reproduction. The woody perennials, in contrast, tend to do best in emergence marsh—they may be very long-lived and form dense compact growth which is sometimes an advantage in competition.

The inundation period may be critical for establishment, and seedlings tend to be limited by a minimum period of continuous exposure (eg three days for *Salicornia europaea*), in order for their roots to become long enough to withstand tidal scour. Other species may have upper limits determined by maximum tolerable exposure periods. Inundation is also critical for seed distribution, especially the autumn equinoctial tides, since salt-marsh plants rely mainly on the tide for dispersal.

Another consequence of inundation is poor aeration of the soil. While little research has been done on the responses of salt-marsh plants to waterlogging, the fact that soil-atmosphere oxygen levels may fall as low as 1–2% and carbon dioxide may rise to over 4%, must have a considerable influence on plant performance and indeed may limit the range of some species.

The single factor that probably has most effect in salt marshes is salinity; in fact the relative paucity of the salt-marsh flora is possibly a reflection of the difficulty encountered by species in adapting to a high and fluctuating salinity. It is perhaps significant that the salt-marsh flora includes only a few flowering plant families, notably Plumbaginaceae and Chenopodiaceae. Seawater not only poses problems of osmotic balance, but it also contains a complex structure of ions, including especially sodium, magnesium and chloride, which are toxic to most plants in the amounts present. The ways in which salt-marsh plants deal with salinity are not well understood. Like so many other ecological problems, there is more than one way of adapting to a particular set of environmental conditions. Indeed, the fact that different species solve the same problem in different ways enables them to exploit the environment more fully and share its resources. This phenomenon has been called "community-wide adaptation".

For some species at some stages in their life cycle, osmotic effects are undoubtedly important. Current work on gas exchange in relation to salinity changes has shown that when *Limonium* and *Aster* are transferred from nonsaline to saline conditions, all the responses can be explained in terms of osmotic effects. Seed germination in *Limonium* is apparently stimulated by exposure to seawater, a response that has been attributed to osmotic shock.

Ionic effects are probably crucial in determining which species are able to inhabit salt marshes. There is a variety of possible adaptations used by plants that tolerate salty soil (halophytes) and some species may use more than one. Ion exclusion, or at least selection, is used by some algae, which can take up ions such as potassium selectively from seawater, despite the high sodium content. Such exclusion and preferential uptake is used in other elevated ionic environments, eg on serpentine soils, and there is no reason to suppose that some salt-marsh plants cannot also use this mechanism. Ion accumulation may occur in many species. Long-lived perennials like *Limonium* shed their leaves every autumn and with them shed accumulated ions. Senescence of leaves of *Plantago maritima* in late summer may result from toxic amounts of ions that have accumulated. Ionic excretion is widespread and well known. *Limonium* and *Spartina*, for instance, have special salt glands from which almost solid salts are exuded in dry periods, so that the leaves may be coated with powdery salt on hot summer days. Within the plant, ions may be diluted and the occurrence of some succulent species may be a response to the necessity to reduce internal ionic concentrations. Yet another possible mechanism is the localization of ions in parts of tissues where they will not interfere with metabolism.

It can be seen that salt-marsh plants face a peculiar combination of environmental conditions and it is perhaps not surprising that few species have been fully successful in dealing with them.

Salt marshes have been utilized by Man in several ways. In some areas, for example, they provide valuable grazing. Grazing modifies the vegetation considerably, producing a close-cropped turf instead of a relatively tall, open cover. Some salt-marsh species are highly intolerant of grazing. In some places, the turf from the upper parts of marshes is used for lawns. The fact that salt-marsh extends the coastline seaward has been used in two ways: as a protection for the coast, and for reclamation. Large areas of marsh have been "reclaimed" by building sea walls and enclosing the marsh behind them; after a few years the salt is leached out and the land can be used for crops. Because they are level and easily filled with solid spoil, salt marshes have been extensively reclaimed for urban and industrial use. They are not regarded as attractive by most people, so until recently little attempt was made to conserve them. However, they are a valuable natural resource in several ways. Much of the production of salt marshes is removed by the tide and is available for the inshore marine ecosystems; this can be especially important for inshore fisheries. Salt marshes are important winter feeding areas and summer breeding sites for many seabirds and waders, especially in estuaries; their destruction reduces the populations of these birds, which is one of the reasons why various attempts are being made to conserve representative salt marshes in several parts of the world.

s.w.

Mangrove Swamp

Mangrove swamps are found on coasts of a geographical belt surrounding the equator. This belt reaches latitude 32°N, but in the Southern Hemisphere it extends as far as South Australia.

Mangroves develop on sheltered muddy shores of deltas and estuaries exposed to the tide. They vary in width, some reaching up to several kilometres. The species are evergreen, with the thick leathery leaves frequently associated with plants of saline soils. Many species have seeds which begin to germinate before the fruits are shed. Various types of aerial root are another characteristic of these plants. The vegetation is almost entirely woody, varying from low shrub to

Tropical salt marshes are typically dominated by woody mangroves, as exemplified by *Avicennia marina*, seen here colonizing the mud of a rocky shore on the coast of southeastern Australia.

rain forest 30m (98ft) high; a ground flora of herbaceous plants is uncommon, although it may be present in certain zones.

Mangrove swamps are one of the most unpleasant types of vegetation for the human visitor: the deformed tree trunks are frequently supported on prop-roots which, together with the breathing roots, make progress very difficult. Having negotiated the tangle of aerial roots the visitor is then likely to sink knee deep into the mud, which releases the unpleasant fetid gases that accumulate in the anaerobic conditions. The myriad biting insects are yet another disincentive. However, these swamps do contain one very fascinating animal, the mangrove fish (*Pariophthalmus*), which crawls out of the mud and up the trees.

Mangroves generally contain few species compared with other tropical formations. There are two quite distinct phytogeographical regions: the eastern mangrove formation, extending from Asia to East Africa, and the western, found on the coasts of America and western Africa. Although the genera in these two regions are similar, the species are quite different, the eastern mangroves being floristically richer.

There is a zonation of vegetation in mangrove swamps associated with the degree of immersion. Different species have different competitive power according to their tolerances of salinity, moving mud, and changing sea level. However, no general rule can predict the exact sequence of genera in this zonation because salinity conditions are a function of rainfall as well as immersion. For example, on the east coast of Africa, species of *Rhizophora*, which have long prop-roots and are well able to withstand the shifting tidal muds, generally occupy a more seaward zone than *Avicennia*, which has greater salt tolerance. In East Africa there are long dry periods during which evaporation from the inner zones, not regularly immersed by the tide, causes the salinity to increase toward the land. On the coasts of Malaya, however, the heavy rainfall causes leaching of salt from the higher zones so that the salinity of the mud increases to seaward, and here we find that colonizing species are usually *Avicennia*, with *Rhizophora* occupying a more landward zone.

Many unusual physiological and morphological features are found in mangrove species, especially concerning the root systems and the seeds. Root systems of mangrove plants are particularly interesting because of the aerial forms developed. Stilt- or prop-roots develop in many species. Prop-roots are slender outgrowths which grow downward from the main trunk at various distances up to 1m (3.3ft) or even 2m (6.5ft) above the ground. These roots, which develop from adventitious roots on the trunk, may branch or join before they enter the soil; under the soil surface they develop a secondary root system of their own. Prop-roots also occur on species growing in well-drained rain forests. However, some Malayan species, which grow both in swamps and in non-

swampy rain forest (eg *Myristica* spp., *Eugenia longiflora*), have prop-roots in the swampy environment but not in dry (seasonal) tropical forests (see p. 161). A similar plasticity has been observed in species growing in African and South American rain forests.

It is generally assumed that prop-roots give greater stability to the trunk when it is growing in shifting mud. Such a suggestion appears reasonable when one considers the shifting mud of mangrove swamps, but stilt-roots are also very abundant in many freshwater swamps where stability is not such a problem. Moreover, small trees tend to develop prop-roots while tall trees do not, but it is the tall trees which need supporting— indeed, tall trees are frequently blown over by wind. The exact function of prop-roots, therefore, is not understood.

The other type of aerial root found in mangrove and other tropical swamps is the so-called breathing root (pneumatophore) which may take on either finger-like or looped forms. The straight finger-like roots stick out of the mud for up to 15cm (6in). They are unusual in that they grow away from the force of gravity (ie they are negatively geotropic), unlike normal roots. Well-known examples of finger-like aerial roots are found in mangrove swamps on *Avicennia* spp. and *Soneratia* spp.; they also occur on various freshwater and peat-swamp forest species. Looped roots are found in numerous species of the mangrove *Bruguiera*, and in *Mitragyna ciliata*, which grows in the freshwater swamp forests of West Africa. Aerial roots of these species tend to bend into very tight loops, sometimes also called knee-roots, whereas the aerial roots of many swamp species develop as a tangle of loose loops up to 0.5m (1.6ft) high.

Pneumatophores of both types have pores in the bark (lenticels) which allow gas, but not water, to pass through. These lenticels are believed to be concerned with aeration of the root system. Measurements of carbon dioxide production indicate that pneumatophores have a high rate of respiration. When the root is submerged oxygen is used up and the carbon dioxide produced dissolves in the plant liquids. This creates a negative pressure so that, when the plant emerges from the water or mud, air rapidly enters the tissues in order to bring the pressures into equilibrium. The pneumatophores also bear fine rootlets which are concerned with nutrient absorption. These are of special value in mangrove swamps because they keep the absorbing region of the root system near to the surface of the swamp when silting causes the surface to rise. Controlled uptake of nutrients is an energy-demanding process; therefore roots concerned with this need to be well aerated so that adequate energy can be supplied by oxidative respiration (anaerobic respiration is a much less efficient source of energy). Thus the high respiration rate of pneumatophores is quite consistent with their possible dual function of aeration and absorption.

First stages of mangrove formation on the coast of South Australia. The growing tangle of finger-like pneumatophores (breathing roots), the fetid mud and trees with buttress roots are characteristic of this type of unpleasant vegetation which occurs on many tropical coasts.

The viviparous seeds of many species give rise to seedlings that hang from the trees until they fall and stick in the mud. Lateral roots grow very fast and in a matter of hours the young mangrove plant is secure against being washed away by the tide. Only if the seedlings fall at high tide will they float away and be lost.

Water and salt balances in the plants present a considerable problem. Some salt must be taken in, both for direct use by the plant and to create appropriate osmotic pressures; but if the salt uptake were not controlled, transpiration would soon cause lethal concentrations to be reached.

The cell sap of viviparous seedlings has an osmotic pressure which is much lower than that of seawater; considerable salt uptake must therefore take place soon after the seedlings stick in the saline mud, otherwise they would dehydrate. The cell sap of most mature mangrove species has an osmotic pressure corresponding approximately to that of the substratum, which may be much higher than seawater. *Avicennia*, which can tolerate very high salt concentration, has glands on the underside of its leaves which secrete salt to be washed away at high tide; however, for most mangroves the exact mechanism of salt control has yet to be discovered.

j.wn.

CHAPTER SIX

The Realms of Plants

Modern Distributions of Plant Groups

While the study of the vegetation types covering the world, examined in the previous chapter, has its roots in ecology, the subject of this chapter grew out of taxonomy. Here we deal with the actual areas of distribution of the various taxonomic categories from the species upward. As stressed in the first two chapters of the book, this subject, often called chorology, depends upon the highest levels of accuracy in recognizing and circumscribing the taxa and their distributions. Furthermore, the recent and more distant history of landmasses, and the evolutionary background of plant groups, are profoundly important, as we have seen in Chapter Three; so, too, is the environmental tolerance of the species and their component populations (see Chapter Four: Environmental Factors).

Taxa can be observed to have distributions of all kinds, from widespread to highly restricted. Not surprisingly, large and diverse families, such as the Gramineae, Cyperaceae, Ranunculaceae and Compositae, are cosmopolitan or almost so. Similarly, there are some large genera, such as *Senecio* and *Euphorbia*, which occur in most parts of the world. Such distributions testify to the ability of some plant groups to evolve in response to a wide variety of environments. The few species that occur naturally in many parts of the world, such as *Phragmites australis*, do so because they are adapted to relatively specialized habitats that are of wide occurrence and also because they comprise a series of genetically diverse populations. At the other extreme are those taxa of comparatively limited distribution. Species, genera and even families restricted for historical or adaptive reasons (endemic taxa) have proved important in understanding plant evolution.

The lessons to be learned from plant distributions differ according to the scale at which they are observed. On the global scale, the influence of geological history, together with the effects of climate, are paramount. When distributions are considered within small areas, however, other influences, such as soil types, microclimate and competition, become much more important. On the whole, this chapter will deal principally with large-scale distributions.

When the distributions of as many taxa as possible are plotted and compared it is found that, in spite of their number and diversity, they fall into a relatively restricted number of distribution patterns. These groupings of taxa have formed the basis for recognizing a number of floristic units (phytochoria) in the world. Within the major floristic divisions, increasingly smaller units can be recognized which, taken together, form the basis for charting the patterns of plant distribution on this planet. Since completely unrelated taxa are often involved in these groupings, the reasons for their occurrence must be sought outside the evolutionary processes which have produced the diversity within plant groups such as families and genera. For this reason, the floristic units observed can provide crucial information on the physical factors that have influenced the ways in which plants have moved over the face of this planet.

As far as can be determined, it has been traditional to describe the various floristic regions in terms of the conspicuous taxa present in them, endemic families, genera and species being used to define their boundaries. More recently, whole floras of different areas have been assessed statistically to determine the degree of floristic similarity between different parts of the world and to provide firm guidelines for delimiting phytochoria of various importance. Interestingly, however, in those instances in which comparisons have been made, the statistical approach has generally supported the delimitation of floristic regions previously based upon indicator taxa.

Taxa which occur in more than one of the floristic regions (liaison taxa) have been particularly important in illuminating the movements of plants over the world's surface. To take but one example, the links between the floras of widely separated continents provided strong evidence for continental drift long before geophysical data became available to provide the coherent pattern now accepted for plate tectonics and the changing panorama of the continental masses through geological time. Carefully mapping the distributions of groups of related taxa can also provide useful information on the history and past migrations of genera and species, but there has been considerable discussion on the value of using present centers of distribution and diversity in determining centers of origin and hence the evolutionary pathways of such groups.

Opposite Larches (*Larix* spp.), unusual among conifers in being deciduous and here showing their autumn coloring in the Black Forest, Germany, are among the trees reaching the furthest north in the Holarctic realm to which they are confined.

The World's Floras

The world's land surfaces, including continents and islands of all sizes, form the arena in which the vascular plants have evolved over various periods of geological time; in doing so, each great order and family has established and extended its occupation of territory according to the extent of land surface existing at the time. The acquisition of additional areas of territory by plants is a consequence of their reproductive success, aided by effective means of dispersal, resulting in the establishment of expanded populations and outlying daughter colonies. Concurrently with these continuous events, natural variation has constantly produced the capacity in some members of the population to tolerate more extreme conditions, and thus, as opportunity arose, to exploit situations hitherto unoccupied by that plant group. So the occupation of territory by related branches of a family is a product of evolutionary tendencies and geographical circumstances (see Chapter Three: Evolution of the Green Planet). The end result is that most parts of the Earth's surface can be characterized by the taxonomic composition of their plant populations, which is otherwise referred to as their floristic character or more simply their flora.

B.S.

Realms, Regions and Provinces

The floras of the various parts of the world, although distinct in a number of ways, can be grouped, according to the varying degrees of affinity between them, into floristic units of different extent. The floristic units (phytochoria) have customarily been arranged into a hierarchical system, although the names given to the different levels of the system have varied. English-speaking authors have termed the largest phytochoria *realms* or *kingdoms*; within these are *regions*, which are themselves divided into *provinces* (or *domains*), in their turn further subdivided into *districts*; intermediate, as well as lower, categories are sometimes also recognized. The criteria used for delimiting these phytochoria have varied considerably and few authors consistently provide information on those employed. In general, however, floristic realms tend to be characterized by particular families or substantial parts of them, while regions seem to be distinguished by the occurrence of 20–30% of genera which are endemic (not found elsewhere). Provinces are usually recognized on the basis of the endemic species they contain.

When species characteristic of one phytochorion are found in another they are said to constitute an *element* in the latter. Thus, for example, the boreal forest zone (see p. 164) has a characteristic group of species which extends in a broad band eastward across Europe to the Ural mountains; some of the species, such as *Linnaea borealis*, constitute the Boreal element in the flora of the British Isles, while species such as *Arbutus unedo* and *Viola lactea* exemplify the Lusitanian element in Britain, representing the distinctive flora of southwestern Europe. The occurrence of

A typical scene to be found under nutrient-poor conditions in western Europe—part of the Holarctic realm. Shown here is a successional sequence beginning in water—a hydrosere. The stages range from floating aquatic plants, eg *Stratiotes aloides*, which constitute the early part of the succession, to reeds and sedges dominated by *Carex appropinquata*, through to the climax vegetation of mature woodland in which Scots pine (*Pinus sylvestris*), birch (*Betula pubescens*) and alder (*Alnus incarna*) are the codominants on the stablized, relatively dry land.

elements in the flora of a particular region can indicate possible migration routes, as well as outlining the environmental tolerances of the species involved.

Liaison taxa (*polychores*) are those which occur in two or more phytochoria; the more of such taxa there are the more closely related the phytochoria and the greater the likelihood of their having had a common floristic history. A development of this long-known principle has provided the basis for recent objective approaches to circumscribing phytochoria on a defined and repeatable basis. For example, the Dutchman M. M. J. Van Balgooy investigated the phytogeographical subdivision of the Pacific basin by determining the affinities, at the generic level, between the floras of the various islands. Every island or archipelago was compared with every other to determine the degree of generic similarity (or dissimilarity) between them. Thus, if A genera occur only on one of two islands, B genera occur only on the other, and C genera are present on both, then the degree of floristic dissimilarity between the two islands is given as a percentage from the formula:

$$\frac{A+B-2C}{A+B-C} \times 100.$$

This "demarcation knot" (quotient of dissimilarity) varies from 0% when the islands are floristically identical to 100% when their floras are completely different. Although this approach is somewhat unsatisfactory for comparing floras of areas of very different sizes, in practice it provided an objective set of

criteria for delimiting phytochoria in the Pacific, with the realms, regions and provinces being distinguished at given values of floristic differentiation—demarcation knots. Interestingly, the divisions arrived at by these means accorded to a remarkable extent with earlier schemes based on more traditional, less clearly defined, procedures. Nevertheless, it is to be hoped that it will eventually be possible to define phytochoria throughout the world by methods similar to those used for the Pacific, although accurate data to permit this are lacking for many, particularly continental, areas; recent chorological studies on the ferns of Europe, however, based on detailed mapping of all species, show the potential of this approach.

Floristic Realms

The origin of our modern continents by continental drift (see Continental Drift, p. 72), occuring at a time of rapid evolutionary development in the plant kingdom, led to a degree of isolation within certain groups of plants. This is to some extent reflected in the global distribution patterns of many families and genera of flowering plants. Nevertheless, the major floral realms do not coincide with the geographical boundaries of the continents, which are themselves defined more by convention than by nature in some regions, as in the isthmus of Central America, in the Middle East (Asia and Africa) and between Europe and Asia for example, where events

The Cactaceae are one of the families which serve to distinguish floristically the Neotropics from tropical and subtropical parts of the Old World, where they are not native. This barrel cactus (*Echinocactus acanthodes*), growing in the desert region of the southwestern United States, reaches about 1m (3.3ft) in height.

other than continental drift have been important in shaping the areas available to plants.

The Holarctic Floral Realm

This includes North America, Greenland, Europe and Asia apart from the southwest, southeast and India, which became attached to the mainland during Tertiary times. The mountain chains extending from the Atlas and Taurus to the eastern Himalayas constitute the most important land boundary in the Northern Hemisphere, separating the Eurasian part of the Holarctic realm from the Paleotropic realm. As a major floristic boundary this is the southward limit of the Northern Hemisphere coniferous family of pines (*Pinus*), cedars (*Cedrus*) and firs (*Abies*) and at the same time it is the northernmost limit of palms (Palmae), with the minor exception of *Chamaerops* in the Mediterranean.

Correspondingly, in North America the firs and pines occur southward across the highlands of Mexico as far as the Isthmus of Tehuantepec, denoting the extent of the northern flora. The present division within the Holarctic realm, resulting from the development of the Bering Sea, is very recent in origin, being the product of the Pleistocene Epoch (2.5 million years ago to the present). North America and Eurasia therefore have many families and genera in common. One can often find ecologically equivalent species of a genus on both sides of the Atlantic; for example the deciduous woodlands of south-eastern Canada have *Quercus alba*, *Fagus grandifolia*, *Fraxinus nigra* and *Tilia americana* as important tree species. In Europe these are replaced by oak (*Quercus robur*), beech (*Fagus sylvatica*), ash (*Fraxinus excelsior*) and lime (*Tilia cordata*). Some common American tree genera, such as hemlock (*Tsuga*) and hickory (*Carya*), are no longer found in Europe, but they were evidently there before the Pleistocene glaciations (see The Tertiary Period, p. 77), which pushed them south until they were eliminated when further retreat was barred by such mountain ranges as the Alps and Pyrenees; the north-south orientated mountains of North America caused no such problem.

Tropical Realms

As noted above, the Isthmus of Tehuantepec in the Americas and the mountains extending from the Atlas to the Himalayas in the Old World, mark the southern limit of the Holarctic floristic realm. Apart from *Sabal*

An unusual scene from the Paleotropical realm. The distinctive "tree senecios" of the high mountains of East Africa, such as *Senecio johnstonii* ssp. *barbatipes* seen here in the alpine belt of Mount Elgon, Kenya, are a reminder that such "temperate" areas rising above the "tropical" lowlands, are "islands" no less than the pieces of land isolated in the world's oceans.

and *Washingtonia*, in Florida and southern California respectively, and *Chamaerops* in the Mediterranean region, all palms (Palmae) occur south of this line. Apart from the palms, families such as the Bombacaceae, Podostemaceae, Rhizophoraceae, Rafflesiaceae and Theaceae occur more or less throughout tropical zones but there is a marked taxonomic gulf between the Old and New Worlds. This has resulted in the designation of two tropical floristic realms: the Neotropical (Central and much of South America) and the Paleotropical (Africa, apart from the Mediterranean coast and the Cape, southern Asia Minor, India, Southeast Asia and Polynesia). This difference reflects the long period of separation—perhaps 70 million years—since these two continental masses split apart.

THE NEOTROPICAL REALM. This is characterized by such families as the Malpighiaceae, Cannaceae, Columelliaceae, Aextoxicaceae, Malesherbiaceae, Bromeliaceae and most Cactaceae, none of which occur in the Old World except by human introduction. The majority of families, however, occur throughout the tropical realms (pantropical) and often beyond, but numerous genera are characteristic of the neotropics. Within the palms, for example, *Chamaedorea*, *Mauritia* and *Euterpe* are restricted to the New World, as are such distinctive genera as *Yucca* and *Agave* in other families.

THE PALEOTROPICAL REALM. This is characterized by families such as the Pandanaceae (screw-pines). Nepenthaceae (pitcher plants), Melianthaceae and Lauraceae, and numerous genera. Referring to the palms again, Old World genera include *Borassus*, *Raphia*, *Phoenix*, *Areca* and *Nipa*, as well as, for example, *Ficus* and *Musa* from other families.

The Austral Realm

In the Southern Hemisphere the great separation of land-masses (whether island or continent) by huge expanses of ocean is reflected in a greater measure of distinction between their floras than is the case in the Northern Hemisphere. There is no universal agreement on which territories comprise separate floral realms, but it is accepted that the southern parts of Africa and South America are strongly distinguished floristically from the major parts of those continents and that to some degree they show affinity with the floras of Australia and New Zealand as, for example, in the possession of the families Myrtaceae, Proteaceae and Restionaceae. All these may be regarded as constituents of a great Austral (southern) realm, or variously accorded the status of separate realms, eg Australian, Cape (South African) and Antarctic.

Floristic Regions

As we have just seen, there has been a certain amount of disagreement as to which phytochoria should be recognized at the level of realms in the extra-tropical Southern Hemisphere. Similar differences of opinion between different workers have also been

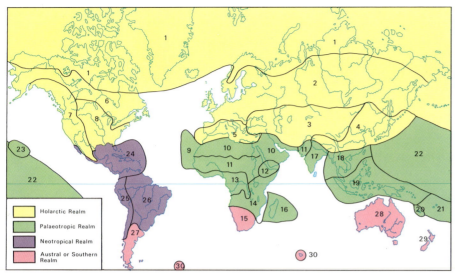

A map showing the floristic realms and floristic regions of the world. The floristic regions generally recognized are: 1 Arctic; 2 Euro-Siberian; 3 Irano-Turanian (Central Asiatic); 4 Sino-Japanese (including the Himalayas); 5 Mediterranean; 6 Hudsonian; 7 Pacific North American; 8 Atlantic North American; 9 Macaronesian (Islands); 10 Saharo-Arabian; 11 Sudanian-Sindian; 12 Ethiopian; 13 West African; 14 East African; 15 South African; 16 Madagascan; 17 Indian; 18 Southeast Asian; 19 Malaysian-Papuan; 20 New Caledonian (Islands); 21 Fijian Pacific (Islands); 22 Polynesian Pacific (Islands); 23 Hawaiian; 24 Central American (Caribbean); 25 Pacific South American (Andean); 26 Parano-Amazonian; 27 Argentinian; 28 Australian; 29 New Zealand; 30 South Oceanic (Islands).

evident over the years with regard to which floristic units constitute regions. For example, within the Paleotropical realm the most conspicuous phytochoria have often been termed regions, qualified by the geographical areas where they are developed, thus giving the African, Madagascan, Indo-Malaysian and Pacific (Polynesian) regions. In most modern treatments, however, these floristic units are placed at the rank of *dominion* or *sub-kingdom*, within each of which one or more regions are recognized.

The divisions do not depend only on the distribution of members of a single family, however importantly it may be regarded, but take account of the boundaries reached by other families, tribes and genera. For example, the Indo-Malaysian dominion is characterized by possessing the entire territorial extent of most genera of the family Dipterocarpaceae, which are among the dominant forest trees of Southeast Asia. The Madagascan dominion, comprising Madagascar, the Comoro Islands, Aldabra, the Seychelles and Mascarene Islands, has eight endemic families, including the Didiereaceae (a curious family of columnar cactus-like plants), about 300 endemic genera and with over 80% of all the species found nowhere else in the world. Although not far from Africa, the distance being only 400km from the closest point on Madagascar, their floras are very different and this is thought to be a result of their long isolation (probably since the Jurassic) and the relative environmental stability of Madagascar during periods when considerable changes took place on the continent.

The recognition of floristic regions depends more upon the limits of distribution of plant genera than of families and upon the quite separate concepts of centers of diversity (see p. 220) and endemism (see p. 227). Often it is found that a region does not have unique possession of a family or genus but it may have a greater number of separately evolved forms (ie either more genera within a family or more species within a genus) than any other region in which the group occurs. For example, the potato family, Solanaceae,

extends to temperate regions where its species, few in number, are members of a single genus (*Solanum*). In tropical South and Central America, however, a cornucopia of Solanaceae genera includes 36 found in no other region. Even more extensive as a family, the euphorbias (Euphorbiaceae) exhibit their greatest evolutionary expansion in the African dominion of the Paleotropic realm, within which centers of species-richness characterize the Ethiopian region and the South African or Cape region. The tamarisk family (Tamaricaceae) embraces the Mediterranean area and southwest to central Asia, but has its greatest wealth of species in the Irano-Turanian region of the Holarctic realm. The nutmeg family (Myristicaceae), although a widespread tropical family, and the vast genus of figs (*Ficus*), are both particularly rich in species in Southeast Asia, especially in the Malaysian-Papuan region where they contribute much to its floristic character. The family Pittosporaceae is represented throughout the Paleotropic realm by species of parchment-barks (*Pittosporum* spp.), but all other genera occur in the Austral realm, where they are exclusively confined to the Australian region, which is thus a center of diversity in this family, as it is also for the eucalyptus family (Myrtaceae). The rush-like family Restionaceae has two centers of generic and species diversity which help characterize the floras of the South African and Australian regions; the Proteaceae, similarly, has the same two centers.

The impression already given conveys something of the intrinsic quality of floristic regions as areas of the world which particular plant groups (of various levels in the systematic hierarchy) now occupy and evidently have frequently occupied for long periods of Earth history, during which they

have diversified by evolution. Thus, floristic regions are strongholds of kinship demarcated not only by their current climatic conditions but also by their geological history, including former land connection affording routes for migration, and former natural barriers conferring isolation (see Continental Drift, p. 72; Land Bridges, p. 222).

In most modern treatments, there is a total of about 30 floristic regions and each of them has such a high degree of integrity that its flora deserves treatment as an entity. These regions differ greatly in area and to a lesser extent in the abundance of their floras, depending upon their past histories and modern conditions.

The spiny stems of *Alluaudia humbertii* growing in the western highlands of Madagascar. This species belongs to the family Didiereaceae, and is one of the many endemic families and genera which make the Madagascan floristic region so distinctive.

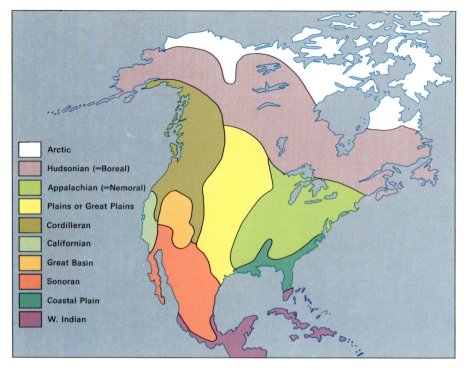

A map of the ten floristic provinces normally recognized in North America.

Floristic Provinces

By dividing each large region into provinces, the areas are further reduced to a size more immediately comprehensible to the traveler and more amenable to the production of catalogs and descriptions at species level. By way of illustration, the floristic provinces of Europe will be mentioned. The far north of Scandinavia, the Kola Peninsula etc, being part of the circumpolar Arctic region (see p. 214), comprises an Arctic province within Europe. Its equivalent in North America extends westward from Southern Labrador to the Brooks Range in Alaska. To the south of this in both continents there is a Boreal province which in Europe is included within the Euro-Siberian region and has a southern border drawn from the upper Ural River to the Gulf of Finland, thence at latitude 60°N (see Boreal Region, p. 216). To the west of this line lies a Middle European province, having its boundary with the Mediterranean province to the south at the Balkan Mountains. In the far west, from Northern Spain to Jutland and coastal Norway is an Atlantic province. In the southeast of Europe, broadly surrounding the Black Sea, is a Pontic province.

Each of these areas may be characterized by the distribution of certain genera but more especially by the occurrence of individual species from various genera whose climatic tolerances coincide with prevailing climatic conditions across the province. For example, most European species of *Erica* (Ericaceae) fairly represent the Atlantic province (with the exception of the tree heath (*Erica arborea*), which belongs to the Mediterranean). The Atlantic province is also well characterized by the distribution of ivy (*Hedera helix*) and to some extent by holly (*Ilex aquifolium*). The central European province is nearly co-terminous with the distribution of hornbean (*Carpinus betulus*) and eastward with mistletoe (*Viscum album*). The Mediterranean province is well indicated by the occurrence of oleander (*Nerium*) and, more approximately, by the tree heath, the strawberry tree (*Arbutus unedo*) and lentisc (*Pistacia lentiscus*).

The territorial unit here described as a province is typified therefore by the presence of species limited at their boundaries by climatic influences. Certain of these plants, for convenience, can be regarded as index species. As a consequence, provinces contain unique and equally characteristic types of vegetation, eg Atlantic bog (see p. 191) and heath (see p. 167), Mediterranean maquis, Central European deciduous forest (see p. 164), Pontic steppe (see p. 179), Boreal coniferous forest (see p. 216) and Arctic tundra (see p. 173). The region is conceived as a broader territory than the province, uniting related species which, although variously adapted to the wider range of conditions, share a common ancestry.

Floristic regions, therefore, contain tracts of the Earth's surface within which there has been relatively unhindered migration during the evolution of existing genera, allowing diversification of species within them. They are separated from one another by barriers to migration and to adaptation that have existed through significant episodes of Earth history and not merely by present-day conditions of climate or sea level.

B.S./D.M.M.

Some Floristic Divisions of the World

Even if the whole of this book were to be devoted to describing the major features of the world's phytochoria there would still not be nearly enough space. Consequently, it has only been possible to outline above the factors taken into account in delimiting the major geographical categories into which the world's floras can be placed. However, in order to get some "feel" for these floristic units this first part of the chapter concludes with descriptions of three phytochoria which have been afforded rather different places in the hierarchy and which cover rather different territories. Thus, the northernmost part of the Holarctic realm, recognized as the *Arctic* region, emphasizes the strongest floristic links known between the Old and New Worlds, which undoubtedly reflects the virtually continuous land around the North Pole, only broken significantly by the North Atlantic Ocean. South of the Arctic, where the Atlantic and North Pacific oceans have longer constituted wider breaks in the land surface, it has frequently been customary to describe a *Boreal* region traversing North America and Eurasia. However, most modern workers recognize much larger, different floristic regions in the two continental masses, with the Boreal zone constituting the northern parts of these regions, usually at the level of province in each. The wide separation of the continents at latitudes in the Southern Hemisphere comparable to those just mentioned results in conspicuous floristic differences within what is here considered as the *Antarctic* dominion of the Austral realm. In order to appreciate the extent to which the floristic divisions of the world do or do not accord with the vegetational divisions described in Chapter Five, the following accounts of the three phytochoria include some mention of the major vegetation-types found within their boundaries.

D.M.M.

The Arctic Region

In marked contrast to the Antarctic (see p. 218), where there is a large, ice-covered continent surrounded by frozen ocean, the greater part of the Arctic is made up of frozen sea encircled by the extensive northern margins of the Eurasian and North American continents. There are also a great many islands, ranging from the continent-sized island of Greenland to the numerous smaller islands of the Canadian Arctic Archipelago. There is no simple definition of the Arctic, since the Arctic Circle cannot sensibly be regarded as its southern boundary, at least not in biological terms. Rather a combined

Map legend

- Arctic
- Hudsonian (=Boreal)
- Appalachian (=Nemoral)
- Plains or Great Plains
- Cordilleran
- Californian
- Great Basin
- Sonoran
- Coastal Plain
- W. Indian

biological and climatological definition has been developed, invoking the rough correspondence between the timberline and the isotherm north of which the mean temperature of the warmest month is not greater than 10°C (50°F). Timberline itself is a rather loose term in this context, since there is gradation from forest to scrub and thickets of willow or birch in many areas; indeed tree-forming genera, such as *Betula* (birch) and *Salix* (willow), are represented as prostrate shrubs a few centimetres high as far as the northernmost coasts of the continents.

Despite the presence today of permafrost and in some instances ice caps and glaciers, the vegetation in many areas of the Arctic, including large parts of the North Slope of Alaska, eastern North America and Greenland, grows on top of extensive unglaciated deposits of beach or river sands and gravels. By contrast, other stretches of the coastline are made up of deeply indented fjordlands surrounded by high mountains and still have glaciers flowing directly into them. Since the inception of the Pleistocene, however, periglacial conditions must have existed throughout the area currently recognized as the Arctic, with periodic incursions of such conditions southward as, for example, prevailed across southern Britain during past glacial periods.

Many general classifications of the Arctic have been proposed, usually dividing it into three zones. Thus, a recent workable division has been into Low and High Arctic, with the further zone of polar desert, the last-named covering the largest area and being the least studied.

Map of the Arctic region showing the zones of permafrost—continuous, discontinuous and sporadic—which exerts such an influence on the plants able to survive the difficult conditions.

Vegetation

Although the vegetation of the Arctic is restricted in the sense that its canopy is usually simple, being only two layered, and species-diversity is much reduced when compared with temperate regions, there is still very great variety in the growth-forms of its plants and their habitats, often reflecting gradients in the length of the growing season, the duration of snow lie, water availability and exposure. Such gradients

In places where the snow melts late or not at all, vegetation is dominated by mosses and willows (*Salix* spp.). This glacier-edge scene is on Baffin Island in the Canadian Arctic.

may occur over very short distances, as for example, on the scale of patterned ground features such as tundra polygons, which develop where intense cold causes the ground to contract and crack. These cracks fill with water at the time of the melt, before the bulk of the active layer of the permafrost has thawed, and some of the water freezes. This process continues year by year until large ice wedges have formed which push up the ground on either side, thus giving two raised rims with a trough between, often with a height difference of no more than 1m (3.3ft) but with very great differences in the environmental conditions of the microenvironment.

In some areas of the Low Arctic, water is freely available, either from snow melting slowly throughout the summer or else because free drainage is impeded by the permafrost. Where there is little exposure to desiccation by wind, wet tundra meadows develop, dominated by grasses and sedges such as species of *Carex*, *Eriophorum*, *Dupontia*, *Alopecurus* and *Puccinellia*, together with an understory of mosses. Within these communities, populations of herbivorous rodents, the lemming in particular, are common.

Normally, vegetation is only to be expected on ground free of snow for much of the summer, yet in places where the snow melts very late in the season, snow-bed communities can form, dominated by lichens, liverworts and, particularly, mosses and species of *Salix*. Even within the snow of the permanent

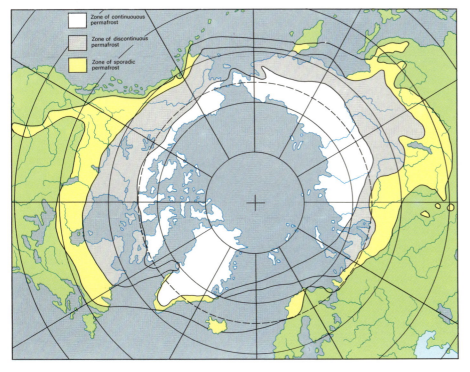

snow fields, plants may be found. Here minute, mostly unicellular algae, utilizing radiant energy directly in photosynthesis, with sufficient nutrients available from the snow, appear as red, green or yellow patches, in much the same way as marine phytoplankton, upon which the food chains of the Arctic ocean are based, grow beneath sea-ice.

Where the ground is stable and sufficiently drained, a tussocky tundra develops, dominated by bog cotton, commonly *Eriophorum vaginatum*, associated with dwarf shrubs such as species of *Betula* and *Salix*, various broad-leaved herbaceous perennials (forbs) and mosses with some measure of resistance to desiccation, together with lichens. Large animal populations graze in such areas during the summer, including reindeer, caribou and musk oxen, as well as lemmings. Large summer bird populations of waders and passerines also use these areas for breeding, feeding on the large populations of insects that appear. Low shrub tundra, in which species of *Salix* and *Betula* growing to a height of between 1 and 3m (3.3–10ft) are dominant, covers large areas of the southern Low Arctic and sub-Arctic, especially in the USSR.

Some areas of the Low Arctic are barren purely because there has been insufficient time for vegetation to develop as, for example, on the glacial outwash plains and moraines exposed by the retreat of a glacier or the melting of a permanent snow bed. The gravel flats of river sides may be too unstable for vegetation to develop, since they are usually disrupted by the force of water during the spring melt. By contrast, in the more exposed habitats the potential growing season is likely to be curtailed by drought. Throughout the Arctic, *fell fields* or *fjeldmark* and beach ridges tend to have a superficially similar flora of cushion-plants, dwarf shrubs, scattered forbs, mosses and lichens, the cushion being a life-form well suited to sheltering relatively delicate growth points.

Most of the High Arctic has a very sparse flora, often with as little as 5% of the ground covered by plants, rarely reaching 20%. In many places, mosses and lichens are the commonest plants and may make up to 50% cover locally. In many respects the vegetation resembles that of the more exposed habitats of the Low Arctic. In sheltered areas, often coastal lowlands, small areas of Low Arctic tundra may still occur, just as stunted spruce and other species of trees may spread along sheltered valleys from the sub-Arctic into the Low Arctic. Large herbivores may survive north of their main range by moving from oasis to oasis, both in the Low Arctic and the polar desert. The latter occurs in most northern areas, wherever summer temperatures are very low because the land is almost permanently ice-locked, and almost everywhere in the Arctic above 200–300m (about 650–1,000ft). In such areas the plant cover is almost exclusively of lichens and mosses, with vascular plants being few and far between.

History and Origins of the Flora

The Arctic flora comprises some 900 species of vascular plants, of which 200 can be described as circumpolar, occurring virtually throughout the region and giving it its conspicuous floristic unity. As might be expected, the Low Arctic is richer floristically than the High Arctic; as the environment becomes progressively more severe, north of 75°N, there is a progressive reduction to between 50 and 70 species as, for example, on the northernmost islands of the Canadian Arctic Archipelago.

The origin of the Arctic flora is evidently tied to the time of development of tundra and other Arctic habitats. On High Arctic islands, up to 80°N, which now support less than 20 species of vascular plants, spruce (*Picea*) forest grew at the beginning of the Pleistocene, approximately 2.5 million years ago, so that it appears that the present Arctic flora must have had its origins in the alpine parts of neighboring cool-temperate regions. Present-day alpine habitats in arctic regions are probably of late Tertiary origin, so that there would again have been no flora *in situ*. The hypothesis that the Arctic flora is of relatively recent origin is reinforced by the fact that there are few purely Arctic genera and only a moderate number of well-defined Arctic species (but many subspecies). *Saxifraga*, for example, contains more than 300 species but only 25 are found in the Arctic, belonging to 15 sections having widely differing centers of diversity (see p. 220). It has been concluded that the Arctic flora is an impoverished miscellany from various regions.

Some species common to arctic and alpine regions may have been largely pre-adapted by their evolution in the latter to spread and flourish in the former. Other species may have had a propensity for becoming conditioned on exposure to the arctic environment, as is seen in the acclimatization to changing growth temperature of, for example, optimal temperatures for net photosynthsis in some plants. Plasticity may also be expressed in morphological terms. However, apparent plasticity within a species may equally have a genetic basis, representing an early stage in speciation of plants adapted specifically to the arctic environment.

N.J.C.

Features of a patterned bog—a typical form of vegetation found in the Boreal region. (a) Plan view (derived from an aerial photograph) showing "islands" of peat supporting tamarack trees (individual trees are marked by dots), fringed by "strings" transverse to the elongated axis of the islands. The surrounding *Carex* fen has pools of deeper water separated by incipient "strings" having the same alignments as those of birch. (b) A transect, shown by the thick line in (a), across the boundary between a tree island and *Carex* fen, showing the plants characteristic of each zone.

The Boreal Zone

Geographically situated to the south of the Arctic and generally to the north of latitude 50°N, the Boreal zone spreads east and west to encircle the Earth and incorporates major portions of North America and Eurasia. Because of the dominance of coniferous forest throughout this area, criteria based on the structure of the vegetation have frequently been used in its delimitation, separating it from the tundra to the north and, usually, the grasslands or summer deciduous forest to the south. However, there are very evident floristic links throughout the zone which in the past have led to the recognition of a Boreal region within the Holarctic realm. Nevertheless, while most recent workers acknowledge these intercontinental links, they consider that, on a full evaluation of the floras, there are two floristic regions, the Euro-Siberian in Eurasia and the Hudsonian (Atlantic North American) in North America, in each of which the Boreal floras constitute a northern province.

Vegetation

The characteristics of the Boreal region are the presence of needle-leaved trees of spruce (*Picea*), fir (*Abies*), pine (*Pinus*) and larch (*Larix*) forming extensive forests, the occurrence of considerable areas of bog, swamp

Forested Muskeg
(Tamarack, Juniper)

and peatland known as *muskeg* and, in oceanic districts, the abundance of dwarf-shrub vegetation or heath. The forests normally have a one-layered canopy, larger shrubs are mostly absent, the field layer is dominated by dwarf shrubs such as bilberries and cranberries (*Vaccinium* spp.), leatherleaf (*Chamaedaphne calyculata*) and Labrador tea (*Ledum groenlandicum*) and there is a well-developed ground layer of mosses and lichens. Drier pine forests typically have herbs such as *Linnaea borealis*, *Trientalis europaea* and the wintergreens (*Pyrola* spp.), as well as saprophytes such as the orchids *Goodyera repens* and *Corallorhiza trifida*.

The peatlands occupy 1.3 million square kilometres (500,000 square miles) in Canada and a similar extent in Russia and Finland. Much of this is partly forested but the forest patches occur in a complex mosaic with bog and swamp vegetation, which develops several distinctive patterns. In southern parts of the Boreal zone raised bogs occur, mounds of peat several kilometres across, which rise gently towards the center, where the surface is pitted with pools and *Sphagnum* mosses are important in forming the peat. The steeper sloping perimeter of the bog may be forested with pine.

On gently sloping ground the bog surface is crossed by nearly parallel low ridges or "strings" built of peat and vegetated by peat-forming plants. The intervening areas are large shallow pools usually completely occupied by sedges (*Carex* spp.). Dispersed at intervals throughout this *muskeg*, rather like a fleet of ships, there are "islands" of trees, tamarack (*Larix laricina*) and black spruce (*Picea mariana*) in North America.

In the more northerly parts of the Boreal zone is found a bog which, when viewed from aircraft, consists of a mosaic of raised mounds of peat, irregular in outline, frequently forested and set among expanses of wet fen dominated by such plants as sedges (*Carex* spp.) and cat-tails (*Typha* spp.). This is typical of areas having discontinuous permafrost and it is the formation of convex masses of ice beneath the peat that causes peat mounds or *palsas*.

History and Origins of the Flora

The Boreal coniferous forests are considered to have developed in the Miocene era (about 7–26 million years ago), possibly in the mountains of the temperate zone. As the climate deteriorated toward the glacial period, the Boreal forests extended across the lowlands to form a circumpolar belt, within which there was relatively unhindered migration. The advance of the Pleistocene glaciations forced the plants southward to give the geographical separation in Eurasia and North America that still persists, although there has been a movement northward in the relatively short time (12,000–16,000 years) since the ice began to retreat. In addition to the physical gaps provided by the Atlantic and Pacific Oceans, the Boreal zone is crossed from north to south by three unbroken mountain ranges which create barriers to plant dispersal: these are the Rockies, the Scandinavian mountains and the Urals. They also obstruct the passage of oceanic winds and cause climatic differences between their east and west sides.

The occurrence of the physical barriers just described has demonstrably influenced the differentiation of the Boreal floras. Thus, although the genera occur widely, there is no circumboreal tree species. For example, while *Pinus sylvestris* is present from western Europe to the Urals and Siberia, the related *P. banksiana* is restricted to North America; *Abies sibirica* and *A. balsamea* are a comparable pair of species. Similarly, the white spruce (*Picea glauca*) is a dominant tree of forests from Newfoundland to Alaska, its equivalent in Eurasia being Norway spruce (*Picea abies*) west of the Ural Mountains and Siberian spruce (*Picea obovata*) to the east. Associated deciduous tree genera, such as *Alnus*, *Betula* and *Sorbus*, show a similar pattern of differentiation. While in North America the forests are evergreen across the continent, in the eastern Siberian part of Eurasia the evergreen species are replaced by a deciduous conifer, Dahurian larch (*Larix gmelinii*).

Interestingly, in distinction to the trees, a significant number of herbs, such as *Listera*

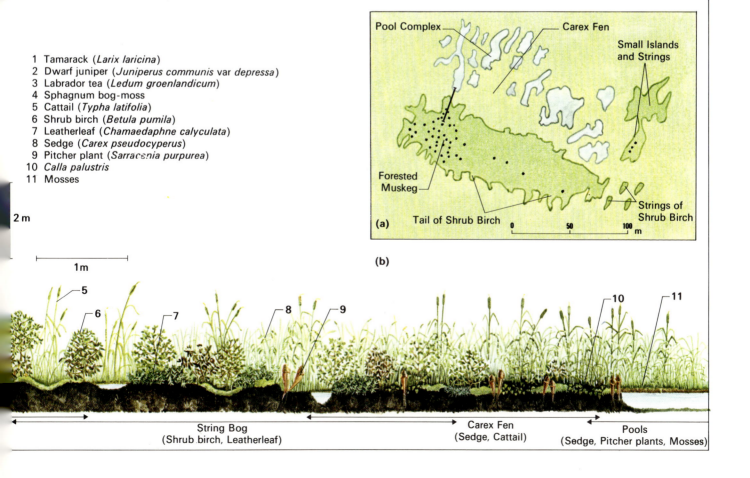

1 Tamarack (*Larix laricina*)
2 Dwarf juniper (*Juniperus communis* var *depressa*)
3 Labrador tea (*Ledum groenlandicum*)
4 Sphagnum bog-moss
5 Cattail (*Typha latifolia*)
6 Shrub birch (*Betula pumila*)
7 Leatherleaf (*Chamaedaphne calyculata*)
8 Sedge (*Carex pseudocyperus*)
9 Pitcher plant (*Sarracenia purpurea*)
10 *Calla palustris*
11 Mosses

cordata, Circaea alpina, Galium triflorum and *Empetrum nigrum* (crowberry), present in the lower layers of the coniferous forests, are circumboreal. The differences in the herbaceous communities of the Boreal zone in the Old and New Worlds, which contribute to their recognition as separate floristic regions, are considered to be due to relatively recent immigration from adjacent areas following the physical separation of the formerly continuous Boreal flora by the Pleistocene glacial advances.

On mountain ranges to the south of the Boreal zone there are comparable forests at higher elevations above the typical vegetation of their respective regions. Resemblance extends to the composition of

the mountain floras which include species of herbs shared with Boreal latitudes. However, the mountains often have tree species with a distinct identity although closely related to those of the Boreal forests. Thus, in the Appalachians we find red spruce (*Picea rubens*) and Fraser fir (*Abies fraseri*), in the Alps the silver fir (*Abies alba*) and European larch (*Larix decidua*) and in the Rocky Mountains there is Engelmann spruce (*Picea engelmannii*) and alpine fir (*Abies lasiocarpa*). Boreal bog vegetation is not well represented in mountain terrain and the climatic conditions even at mountain altitudes in regions south of latitude 50°N do not in all respects resemble those of the Boreal zone.

B.S./D.M.M.

The Antarctic Dominion

The Antarctic dominion or subkingdom is one of the important floristic divisions of the Austral realm. It is generally taken to include the major cool-temperate areas of the Southern Hemisphere, although it is sometimes restricted to areas south of the Antarctic Convergence between about latitudes 50–55°S. It is of particular interest because the lands, whose plants show strong links, are separated by the wide expanses of the southern oceans and there has been considerable discussion about the different roles of continental drift (see p. 72) and long-distance dispersal (see p. 142) in producing the present far-flung and disjointed distribution.

The Antarctic dominion is usually divided into three floristic regions, each with several provinces. The first, the Argentinian or Patagonian region, comprises the provinces of Patagonia, Fuegia, the southern Andes and the Falkland Islands. The New Zealand region contains the South Island but usually not the North Island of New Zealand, the Kermadec and Chatham Islands, Auckland Island and Campbell Island. Finally, the region of the Southern Oceanic Islands includes the islands encircling the Antarctic from Macquarie Island in the east to South Georgia in the west, northward to the Tristan de Cunha group and southward to include the Antarctic peninsula, the only part of the continent where vascular plants occur. Outside the Antarctic dominion, floristic elements having strong Antarctic affinities occur in Tasmania and through upland southeast Australia to the mountains of Malaysia, and in South America northward from Fuegia along the Andes, particularly on their western flanks.

Vegetation

The vegetation of the lands of the Antarctic dominion varies from luxuriant forest to bleak tundra, depending upon the local conditions. Where there is adequate rainfall (and, at higher latitudes, protection from wind) the antarctic forest, characterized by the southern beeches (*Nothofagus* spp.), is developed. *Drimys, Maytenus,* and *Libocedrus* are also important constituents of these forests in southern South America, as are *Dacrydium, Podocarpus* and *Weinmannia* in New Zealand.

The lower rainfall areas of Patagonia and Fuegia, which mainly lie north and east of the mountains, develop a grassland pampas dominated by *Stipa, Festuca* and *Poa* or, in the Falkland Islands, *Cortaderia*, while in the extensive grasslands of New Zealand

Nothofagus antarctica on Navarino Island, Tierra del Fuego. This southern beech grows closer to Antarctica than any other tree: the amount of dead wood indicates the severity of the conditions. The present-day and fossil distributions of *Nothofagus* are important indicators in the geological history of the Southern Hemisphere.

Danthonia, Poa, Festuca and *Chionochloa* figure prominently. One of the most characteristic features of coastal vegetation in southern cool-temperate lands is grassland dominated by various species of *Poa* which form large tussocks up to 3m (nearly 10ft) tall. The six species of *Poa* which dominate these grasslands are all closely similar and are strikingly different from species elsewhere in the world, but the different species replace one another geographically throughout the Antarctic zone.

In more exposed areas, particularly in western Fuegia, southern New Zealand and the islands, as well as at higher elevations, an open tundra vegetation is developed. These areas support a variety of dwarf shrubs, such as *Empetrum, Gaultheria* and *Luxuriaga*, but their most conspicuous feature is the large number of mat and cushion plants (see p. 54) belonging to such genera as *Abrotanella, Azorella, Astelia, Colobanthus, Donatia, Gaimardia, Oreobolus* and *Plantago*. These plants, which belong to a wide variety of families, apparently show close similarities in habit in response to the extreme conditions of the habitats in which they live. Not only is the low-growing compact habit better able to withstand the constant buffeting by the strong winds of the region but it has been shown that within and near the surface of the cushions the temperature is significantly higher than in the surrounding air. This is important not only for the effective metabolism of the plant itself but also because insects attracted by the greater warmth can be important for pollination.

With increasingly severe environmental conditions, the importance of mosses and lichens in the vegetation becomes greater. This reaches its clearest expression on the continent of Antarctica where, apart from two species of flowering plants, the only colonizers of the areas free from snow are lichens and, where there is sufficient moisture, mosses and some semi-terrestrial algae.

History and Origins of the Flora

The fact that the strong floristic links within the Antarctic dominion span major oceanic gaps, allied to the fact that most of the Antarctic continent is a cold desert, has meant that over the years there has been considerable discussion as to how the plants achieved their discontinuous distributions. Basically, the controversy has centered on whether the plants concerned migrated across continuous or almost continuous land by means of normal dispersal mechanisms or whether, by means of long-distance dispersal, they were periodically able to cross wide oceanic barriers of up to 2,000km (over 1,200mi). It is now generally accepted that, at different times, both methods have been involved.

The virtual absence of flowering plants from Antarctica is a relatively recent situa-

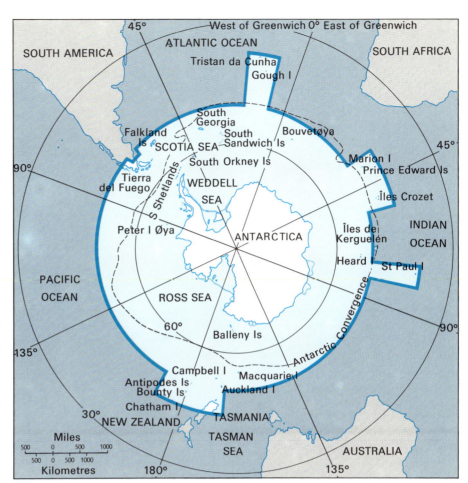

Map of the Antarctic dominion. Although sometimes restricted to areas within the Antarctic convergence, the dominion is usually taken to include all the land bounded by the blue line.

tion. Prior to the last glaciations, about 10,000–15,000 years ago, at least the coastal part of the Antarctic continent enjoyed a cool-temperate vegetation much like that now found in southern South America and New Zealand's South Island. Indeed, fossils of *Nothofagus* and some members of the Loranthaceae (mistletoes) and Myrtaceae (myrtles) are known from late Cretaceous or early Tertiary rocks in New Zealand, Antarctica and South America, at a time when these lands were united with Australia and South Africa in the ancient southern "supercontinent" of Gondwanaland (see Continental Drift, p. 72). Clearly, in such cases the present discontinuities date from the fragmentation of Gondwanaland into the modern continents and the plant groups concerned have undergone a considerable amount of evolution since that time. For example, New Zealand, Tasmania and southern South America have different species of *Nothofagus*, these ancient disjunctions being reflected by affinities at generic or even family level.

On the other hand some species of, for example, *Acaena, Apium, Colobanthus, Hebe, Ranunculus* and *Juncus* occur widely in the lands bordering Antarctica, many of them being found on islands which are known to

be completely covered with ice during the most recent glaciation. During the 16,000–10,000 years that have elapsed since the ice retreated there have been no land connections between these areas. Consequently, the occurrence of closely similar populations of these species in such widely separated regions can only be explained by long-distance dispersal of seeds or spores across the intervening oceanic gaps. Most of the species involved in these recent disjunctions are good colonizers, with the ability to form seed of high germination level by self-pollination. Furthermore, several of them have devices by which the seed can be dispersed over considerable distances. Thus, the far-traveling birds of the southern oceans are often considered to be involved in the transport of species of, for example, *Acaena*, which has conspicuously hooked fruits, while oceanic currents are likely vehicles for genera such as *Sophora*, which have seeds capable of floating in seawater for up to three years and still retaining their ability to germinate.

Whether or not Antarctica was ever the "Cradle of the Flowering Plants", as has been suggested by some botanists, it is certain that the affinities of these southern floras will continue to be the subject of research and speculation, just as they have since Sir Joseph Hooker drew attention to the links between them well over 100 years ago. (See also Tolerance and Plant Distribution, p. 140.)

D.M.M.

Distribution and Distribution Patterns

Once a taxon has evolved as a distinct entity it may be expected to increase its distribution by migrating into new regions (see Chapter Four: Environmental Factors) until it reaches areas in which the environmental factors are beyond the scope of its genetical tolerance (see Tolerance and Plant Distribution, p. 140). These limiting factors may be temperature, water availability, exposure, soil conditions etc (see Physical Features of the Environment, p. 117), and in the case of any taxon they may vary in importance from one area to another. Thus, for example, the thyme species *Thymus serpyllum*, which occurs in Europe northward from northeastern France and eastern England in the west and the northern Ukraine in the east, seems to be limited by very low winter temperatures in the northeast, cool summers in the northwest, high winter temperatures in the west and southwest, and by soil-types in the southeast. Clearly, of course, with sufficient genetic resources, adaptation (see p. 96) will gradually take place so that the tolerance limits can be extended, to some extent at least.

Age and Area

The above sequence of events led the British plant geographer, J. C. Willis, to propose his logically satisfying theory of "Age and Area", in which he postulated that, statistically, the area occupied by a species is directly proportional to its age as a species; in other words, the longer a species has existed the further it can spread. Although he contended that comparisons were only valid between groups of related species and that other things should be equal, the numerous exceptions have led to a general lack of support for the theory. However, perhaps it is possible to take a more positive position. Since Age and Area has a certain logic it should provide a basis for assessing the unequal opportunities which give rise to the many exceptions. For example, distribution maps show that some of the species occupying very wide areas occur in the Arctic region. About 16,000 years ago this was covered by ice, so that the range of the species is not related to their age but reflects their ability to rapidly colonize the virgin areas left by the retreating ice. At the other extreme, taxa of unusually limited distribution, endemics, should, according to Willis' theory, be newly evolved—the reasons why this is not always so help in understanding the histories of the plants present in the world today.

Continuous and Discontinuous Distributions

A study of distribution maps of species, genera or even families, shows that they can broadly be divided into two types—those in which the taxa seem to occupy a more or less continuous area of the world's surface and those in which there are one or more striking discontinuities in the range. While no species has a completely continuous distribution, forming a mosaic with other species, discontinuities are generally recognized when they exceed the normal dispersal distance of the species; since this is not usually known with any precision, attention has been focused on those cases where the discontinuity is great enough for there to be no doubt that "normal" dispersal cannot be involved. As the American plant geographer S. A. Cain noted, "the commonplace and continuous types of distribution are equally as important as the unusual and discontinuous in the solution of problems in historical geo-graphy". There can be no doubt that this is true, but plant geographers have been more interested in discontinuous than continuous distributions, perhaps because of the apparent lack of information and means for interpreting the latter, and this is reflected in the topics dealt with in this part of the chapter.

D.M.M.

Centers of Origin, Diversity and Variability

As was noted in Chapter Three, evolutionary and geographical studies of plants are inextricably intertwined, and nowhere is this more evident than in the interpretation of distribution maps. The particular instance of taxa having limited ranges will be considered later (see Endemism, p. 227) but widespread taxa, whether continuous or discontinuous, challenge our ability to unravel their history of migration and differentiation following their evolution. There has been much discussion as to whether it is possible to use

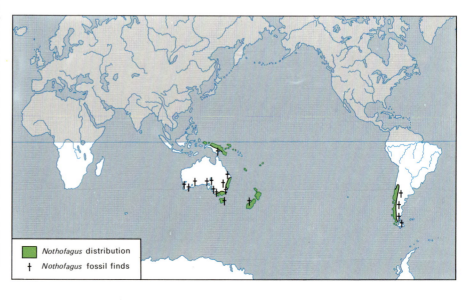

| | Nothofagus distribution |
| | Nothofagus fossil finds |

distributional data to determine areas where taxa evolved, about which it seems clear there can be no certainty.

Centers of diversity can be determined by using *isoflors* on distribution maps. Isoflors (isopleths) are lines bounding areas having equal numbers of related plants, usually members of the same genus or family. They resemble contour lines about a hill in having one or more limited areas of maximal numbers surrounded by irregularly concentric bands of diminishing numbers until, at the periphery, one or another species is the sole representative of the group. The area containing the greatest number, the "center of diversity", may be where the group originated and evolved. However, the frequent occurrence of two or more " peaks" serves as a reminder that conditions stimulating differentiation and evolution may be encountered in areas to which a group has migrated subsequent to its origin. Tongue-like extensions of isoflors mark routes of migration, especially when they are the same for different genera. Isoflors run close together where inhibiting environmental factors occur to which the group has not adapted.

Migrations of species can often be determined by mapping their patterns of variability. This can be done by plotting *isochars*. These are like isoflors except that, instead of linking areas with equal numbers of species or genera, they bound areas having the same numbers of contrasting characters. Areas of maximum numbers of characters are centers of variability and are generally considered to have been occupied longer by the taxa concerned than those regions where variability is low. This rests on the assumption, justified in many instances, that populations resulting from recent migration into peripheral areas will carry with them only a fraction of the genetic variation present in

Left Major discontinuous distributions which cannot be explained by recent dispersal from a central area. **Bottom** The principal areas of the southern beeches (*Nothofagus*) are considered to have been connected in one continental land mass (Gondwanaland) which sundered by continental drift. **Center** A comparable explanation apparently applies to the pepper bushes (*Clethra*), but the fossil occurrences (red) in Europe show that climatic changes subsequent to the major sundering of the landmasses have also been involved.

Right Disjunct distributions in three species. **Top** *Campanula uniflora* has a bicentric relict distribution in Norway. During glacial periods it survived in two zones (refugia) and, although suitable habitats now exist between the northern and southern refugia, they have not been colonized because of the species' poor dispersal. **Center** *Cherleria sedoides* once had a more widespread distribution but with changed climatic conditions it has become increasingly confined to restricted montane areas. **Bottom** *Erica arborea* is primarily a Mediterranean species but its occurrence on the mountains of Ethiopia, East Africa and Cameroun indicates a wider distribution when cooler, less arid conditions prevailed at lower elevations in northern Africa.

more central areas of the distribution where interbreeding between less isolated populations, and a longer history, maintain high levels of variation. As we saw earlier (see Colonization and Establishment, p. 145), populations at the colonizing edge of a species or genus may have distinctive num-

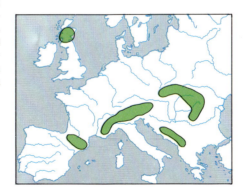

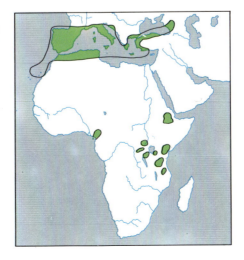

bers or arrangements of chromosomes in their cells, and/or they may be capable of self-pollination; such features as these can often permit some understanding of the routes taken by a species or genus in extending its distribution.   D.M.M.

Discontinuous Distributions

In many instances, as taxa have extended their ranges, major discontinuities have developed. Studies of these discontinuous distributions have proved important in understanding the factors involved in giving some of the major geographical patterns shown by plants today. In Chapter Three we have referred to the physical and evolutionary backgrounds to the modern distributions of plants, including those showing discontinuities. Here, therefore, reference will only be made to those aspects (*polytopy*, *vicariance*, *land bridges*) which arise directly from describing and interpreting the disjunct ranges of taxa, together with a consideration of polar distributions, which are much referred to in this context.

                        D.M.M.

Polytopy

When a species, or other taxonomic unit, occurs in two or more completely separate areas it is said to be *polytopic*. The word is normally used in the field of botany only and some disagreement exists as to its precise definition. One important problem presented by the concept is the degree of separation between the areas that it requires since, for example, two populations of oak trees in adjacent woods are not said to be polytopic. Rather the areas in question have to be disjunct and quite isolated from one another. The degree of separation is greater than the normal dispersal distance of the plant concerned and frequently it involves a specific physical barrier such as a mountain range or an ocean.

The interest of polytopy is centered in the origin of the polytopic distribution pattern. To explain this, several theories have been proposed, including special creation, though only a few are now considered likely. The most widely accepted theory suggests that the groups in each area are directly related to one another and that either they were connected in the past by a series of geographically intermediate populations or that the intervening gap was crossed by an unusual occurrence of long-distance dispersal (see p. 142). In the first case, the intervening populations are supposed to have died out—perhaps, for example, through climatic change—thus leaving two or more disjunct populations. In the second, chance is considered to have caused some diaspore to be dispersed considerably further than is usual. This has been followed by establishment at its point of landing.

Another theory invokes the occurrence of convergent evolution. In this, two unrelated groups are supposed to have evolved separately and independently to become similar to one another. Indeed, some authorities regard only this process as being authentic polytopy. A classical example of polytopy presumed to have occurred in this way is provided by the sand dune ecotype of *Hieracium umbellatum* agg. Various populations of this occur in separate sand-dune systems around the coasts of Sweden. They share certain specific adaptive characters, such as the potential for rapid shoot regeneration, but differ in minor leaf characteristics. In these respects they are more closely related to neighboring inland populations. The inference is thus that each sand dune ecotype has evolved independently but in the same direction in each area in response to the nature of the habitat. (See also Genecology, p. 36.)

B.WR.

Nassauvia serpens (Compositae), an endemic of the Falkland Islands, is a vicariant with the closely related *N. latissima* in the mountains of Tierra del Fuego, South America, and *N. magellanica* more widely distributed in the Southern Andes.

Vicariance

Groups of plants which are descended from a common ancestor and which are adapted to the same environmental conditions are said to be *vicariant*—ie they could be substituted for one another. The concept of vicariance must be distinguished from that of adaptive radiation (see p. 103), by which a group evolves from a common ancestor in different directions under a series of different environmental conditions. Vicariant groups, strictly speaking, are those whose divergence is largely the result of spatial isolation alone; in other words, vicariance is a "passive" condition, while adaptive radiation results

The genus *Donatia* contains two species which are vicariants in the bogs of the Southern Hemisphere, dominated by cushion-forming plants. *Donatia fascicularis*, pictured here in western Tierra del Fuego, is the southern South American species whose closest relative is *D. novae-zelandicae* in New Zealand.

from active selection by different environmental conditions. In this sense the differences between vicariant species or subspecies result from the lack of gene-exchange between them so that they evolve in isolation from one another (see Isolation, p. 108). The differences also perhaps reflect the results of genetic drift (see p. 104), the chance differences in genetic composition of the original, founder populations when they were separated.

The 30 or so species of the Hawaiian palm genus *Pritchardia* are probably vicariants; no species is found on more than one of the eight islands, while the species on any one of the islands are spatially isolated—each of the nine species on Oahu, for example, is found in a separate valley. Similar situations also occur among the plants of the Canary Islands and other isolated archipelagos but it is clear that detailed studies are often necessary to determine whether or not the differentiation

is primarily the result of environmental selection/adaptive radiation, or spatial isolation/vicariance, and it is doubtful whether the two are always mutually exclusive.

Since vicariance is the result of divergence from a common ancestor, it may indicate connections between the areas occupied by the vicariant groups. For example, the vicariant species of several genera now isolated on the mountains of Corsica, Spain, Italy and North Africa reflect the existence of an ancient Tyrian flora which split up during the Tertiary period. Similarly, the many vicariant species which span the Atlantic, such as *Anemone nemorosa* and *Hepatica triloba* in Europa and *A. quinquefolia* and *H. americana* in North America, are associated with a tree flora descended from ancestors which ranged widely round the Northern Hemisphere in late Tertiary times. The modern vicariant species are probably descendants of a common ancestry in the earlier circumboreal flora (when there were still land connections across the North Atlantic) which were driven southward and separated by the Pleistocene glaciations. Pairs of vicariant species on either side of the southern oceans, such as *Drosera uniflora* and *Phyllachne uliginosa* in southern South America and *D. arcturi* and *P. clavigera* in New Zealand, have frequently been cited as evidence of earlier land connections between the areas. Modern work on continental drift (see p. 72) supports such connections but they existed so long ago that the vicariants would be expected to have diverged very much more than they have. Consequently, these vicariants, like populations of a single species (eg *Ranunculus biternatus*) showing a comparable disjunction, are generally considered to result from relatively recent transoceanic long-distance dispersal.

It should be noted that, over the years, the term vicariance has been employed in a much wider sense than that used above. For example, it is sometimes used where two taxonomic groups are not closely related but occupy the same ecological niche in different habitats. In this sense the horseshoe vetch (*Hippocrepis comosa*) and the birds-foot trefoil (*Lotus corniculatus*) are vicariants on older limestones in Britain, while in deserts and semideserts the Aizoaceae in South Africa and the Cactaceae in the Americas are vicariant groups. Vicariance has also been used in an ecological context. Vicariant associations, for example, are said to be spatially separated but consist of numerous similar species, as in the case of some deciduous forest communities in North America and Europe.

D.M.M.

Land Bridges

When a plant species occurs on landmasses that are now separated by an extent of ocean over which the dispersal of propagules seems unlikely, such a disjunct distribution can sometimes be explained by the past existence of a land connection between them. However, before such a suggestion can be accepted it must be demonstrated that the proposed land bridge is possible in geological terms and also that the efficiency of seed or vegetative dispersal is inadequate as an explanation for the distribution.

The concept of land existing where now there is sea is supported by the extensive beds of submerged peat, often containing the remains of trees, which are abundant in temperate latitudes. Such changes in sea-level during the recent past can be explained by reference to the melting of the ice caps since the end of the last period of glaciation, which has caused sea level rises of over 100m (about 350ft) in some parts of the world. Sea-level changes caused in this way are termed *eustatic*. The post-glacial rise in ocean level has not been uniformly experienced over the whole globe; land areas which once bore a heavy burden of ice, such as Scotland and northern Canada, have risen since the melting of the ice, because of the reduced load on the Earth's crust. Changes in relative land/sea level resulting from crustal warping of this type are called *isostatic*. In some regions these two processes have compensated for one another in post-glacial times, rising oceans being matched by a recovering land surface.

In some areas, however, such as the region now submerged beneath the English Channel and the Irish Sea in Europe, and the Bering Straits in the North Pacific, extensive areas of land existed at the end of the last glaciation, some 10,000 years ago. The opportunities for plant migration were con-

Map of North America and northeastern Asia showing the Bering Strait region. The brown area indicates the probable extent of Beringia, an extensive land mass present during the glacial stages of the Pleistocene period, some two million years ago. The dots indicate the present-day trans-Beringian distribution of the woodrushes: a continuous distribution of *Luzula wahlenbergii* (blue) may have existed in the days of the land bridge, but *L. piperi* (red) has evidently been dispersed between the Aleutian Islands and has thus crossed the Bering Sea.

siderably better then than they have been for the past 7,000 years, when the eustatic sea-level rise reached its maximum. Climatic conditions in those early post-glacial times may locally have been quite favorable for warmth-loving species (which had survived the glaciation in lower latitudes) to extend their range. Among those in Europe were some Mediterranean species, including the strawberry tree (*Arbutus unedo*), which are thought to have spread quite rapidly along the oceanic, low-lying lands of western Europe as far as the southwest of Ireland. The coming of taller, more competitive trees probably eliminated *Arbutus* from many areas between the Iberian peninsula and Ireland, and the rising sea-level across this region has subsequently submerged all record of its precise migration route. *Arbutus* now has a disjunct distribution, isolated populations being separated by extensive areas of sea. Only the former existence of a land bridge along which migration could have taken place adequately explains this distribution pattern.

The great landmasses of Asia and North America are separated by the Bering Sea, which, despite being in places only 80km (50mi) wide and 50m (165ft) deep, now provides an effective barrier to the migration of many plant species. However, the floras of the two continents in this area are very similar, and many species, such as the mountain sorrel (*Oxyria digyna*) and the opposite-leaved saxifrage (*Saxifraga oppositifolia*), have distributions which exhibit no break corresponding with the Bering Sea. Most of these species, however, are of a boreal or an arctic type and some recent analyses of lake sediments containing pollen grains dating from the end of the glaciation, when a land bridge would have existed, show that the vegetation of the Bering land bridge was indeed then tundra. Thus, although a migration route existed across the Bering Sea, only species capable of tolerating cold conditions would have been able to pass across it. The same, of course, would have applied to the animals which crossed the land

bridge, including Man.

One must be cautious, however, in the application of the land bridge idea. It has sometimes been invoked to account for distributions which could have alternative explanations. For example, the occurrence of isolated populations of such small rodents as the wood mouse (*Apodemus sylvaticus*) on the outer isles of western Scotland, even on St. Kilda, has been explained in terms of late-glacial land bridges. However, the existence of such bridges is very doubtful for the reasons of isostatic depression and recovery already described. It is much more likely that such creatures have colonized the islands with the assistance of wandering human populations, such as the Vikings.

Changing sea-levels are not the only possible way of explaining the existence of land connections in the past; the role of continental drift in giving rise to disjunct distributions in many groups of plants, especially in the Southern Hemisphere, has been described earlier (see p. 72).

There are still some problems which are difficult to solve and for which the land bridge explanation is a possible one. For example, some plant species are found on both sides of the Atlantic Ocean, such as *Eriocaulon septangulare* and *Koenigia islandica*. Europe and North America parted between 50 and 70 million years ago and deep seas have separated them even during the lowered sea-levels of the Pleistocene glaciations of the last two million years. Either such species had already evolved and were widely distributed before the splitting of the continents or they have succeeded in crossing the ocean barriers. Perhaps the reduction in the extent of these barriers during the ice ages assisted such transmission. *Koenigia* is a remarkable plant in that it also occurs in the Southern Hemisphere, which suggests that this arctic species has dispersal powers that are sufficient for it to cross the unfavorable tropical zone, perhaps as seed on the feet of migrating birds. Thus the dispersal potential of some plant species should not be underestimated; such *long-distance dispersal* has certainly been important in causing some discontinuous distributions. Land bridges should only be invoked where there is support from geological evidence. P.D.M.

Polar Distributions

Certain plants are found growing mainly in the extreme environments of polar regions, and possess a number of special characters that enable them to do this (see Tundra, p. 173). Among these plants with predominantly polar distributions a number of more detailed distribution patterns have been recognized. Present-day distributions are the product of a history of great change in polar regions, of the advance of ice sheets and the subsequent revegetation of these regions after the retreat of ice. The present-day distributions of polar plants may provide some clues to this history.

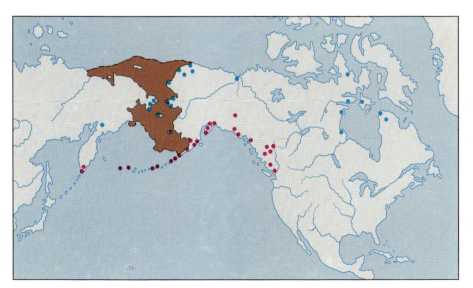

Distribution Patterns

Some polar plants are restricted to one hemisphere. Within that hemisphere they may have a completely circumpolar distribution, as for example *Oxyria digyna* in the Arctic, or it may be only partial. In the Arctic there is, for example, a predominantly west Arctic group of species (found in North America, Greenland and exceptionally Iceland or Norway) and an east Arctic group (Eurasia but sometimes Greenland and eastern North America). Circumpolar distributions have also been recognized in the Southern Hemisphere, although the history and origins of the flora of the Antarctic dominion (see p. 218) are markedly different from that of the Arctic flora. Not surprisingly, species found in polar regions also make up an element of the high altitude ("alpine") flora of lower latitudes. The disjunct southern circumpolar genus *Colobanthus*, for example, has one species, *C. quitensis*, which is present on the Antarctic peninsula and reaches the northern Andes and Mexico at increasing altitudes.

While some plants are restricted to the polar regions of only one hemisphere, certain species have bipolar distributions. In some cases, a circumpolar boreal species is represented by only a few outlying populations in austral regions. An example of such a species is *Carex lachenalii*, which is widespread throughout the Arctic and Boreal zones but also has outlying populations in the Southern Alps of New Zealand. One of the few annuals found in polar regions, *Koenigia islandica*, is a further example. Mosses and lichens are more common than vascular plants on the margins of the Antarctic continent and on many surrounding oceanic islands; many species common in the northern polar regions are present, having circumpolar distributions in the south.

Koenigia islandica has a bipolar distribution, occuring in cool-temperate areas of Eurasia and North America and in the southern part of Tierra del Fuego, where this picture was taken.

It is commonly found that, in addition to a predominantly bipolar distribution, there are often intervening locations, usually at increasing altitude the nearer the locality is to the Equator. For example, *Phleum alpinum* is found at 4,760m (15,600ft) in Mexico but at or near sea-level on both the sub-Antarctic island of South Georgia and in southernmost South America, as well as throughout northern polar regions. Distributions of this sort also occur at the generic level, as in the case of *Empetrum*, species of which are found throughout northern polar regions, with one species, *Empetrum rubrum*, found in southernmost South America, the Falkland Islands and Tristan da Cunha.

It should be recognized, especially in the case of cryptogams, that descriptions of distributions are often based on the examination of species lists from sources with a variety of basic criteria for the delimitation of species. This is a common problem arising out of much of the original work describing plants found during the exploration of high latitudes in the Southern Hemisphere. Morphological differentiation

The genus *Empetrum* (crowberries) has two species, *Empetrum nigrum* (including *E. hermaphroditum*) from northern Eurasia and North America and the red-fruited crowberry (*E. rubrum*) from Tristan da Cuna and cool-temperate South America. *E. rubrum* is here seen growing among *Sphagnum magellanicum* on Navarino island, south of the Beagle Channel in Chilean Tierra del Fuego.

between populations, phenotypic plasticity and the inherent belief of many early taxonomists that wide geographical separation was concomitant with speciation, led to the proliferation of names. Many of these names have become synonymous, often with northern species. For example, of 52 bryophyte specimens collected by the German International Polar Year expedition of 1882–3 on the subantarctic island of South Georgia, 51 were named as new to science. The vast majority have now been recognized as synonyms.

Origins of Polar Distributions

Present-day patterns in polar distributions evidently reflect a number of major changes on the Earth's surface in the past (and that still continue), such as continental drift, the existence of periodic land bridges and especially the waxing and waning of glaciations. In the Americas it was possible for elements in the periglacial flora to migrate towards the pole as the ice retreated to mingle with plants that survived in refugia (habitats that have escaped drastic environmental changes, thus enabling species and populations to survive) and on nunataks. Such migrations were more restricted in Eurasia because of the general east-west trend of mountain chains.

Two dominant patterns of evolution are likely in polar floras, either by depletion during migration or as a result of the depletion of a formerly widespread species. Divergence of populations by migration and depletion is shown, for example, in the rush *Juncus trifidus*. Alpine populations on Mount Washington in New Hampshire resemble

World map showing the polar distribution of *Phleum alpinum*.

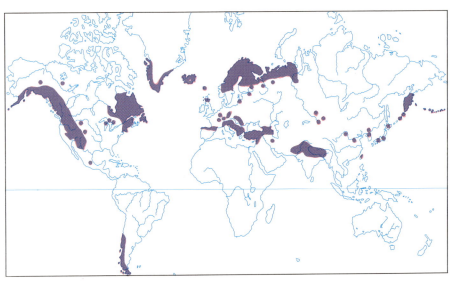

A scene in Antarctica. The exposed rock surfaces are nunataks, where mosses, lichens and algae may grow, either as survivors from earlier times or as new colonizers, depending on whether the rocks have been exposed for a long time or just recently.

arctic populations in Greenland and Scandinavia more closely than they resemble montane populations in the putative area of origin of this species in southern Europe. It has been suggested that many plants with circumpolar distributions in the Arctic may well have spread as propagules blown over the frozen Arctic Ocean or transported directly on the ice.

There has been much speculation about the origins of bipolar distributions but broadly there are two main lines of thought: migration may have occurred between Boreal and Austral regions, utilizing mountain ranges such as the Rockies and Andes, or else long-distance dispersal may have occurred. It is becoming increasingly apparent, as a result of three main lines of approach, that the relationships between different populations of species are extremely complex. Cultivation of plants from these populations under the same conditions and observations of morphological changes, examination of their physiology, not only in the field but also in parallel with cultivation experiments, and finally the use of new, biosystematic techniques such as chemotaxonomy and cytotaxonomy, all assist in the unraveling of the evolution, migration and basis for persistence of species over wide areas. As just one example, the circum-Antarctic species *Acaena magellanica* has a chromosome count of $2n = 42$ around the sub-Antarctic islands and in easternmost Tierra del Fuego but $2n = 84$ in the Falkland Islands, the forested parts of western Fuegia and in southern Chile. It appears that the $2n = 42$ race must have survived Pleistocene glaciations on islands that were only partially glaciated or else further north in South America, spreading after the retreat of the ice. The southward spreading forest zone, on the other hand, was colonized by the $2n = 84$ race.

Nunataks

In considering polar distributions it is important to remember that the ice-sheets are not always unbroken, and this has led to recognition of the role of *nunataks* in past and present plant distributions. A nunatak is any area of ground exposed within a snow- or ice-field that is contiguous with the sub-snow or sub-glacier surface. While nunataks may act as refugia for plants that were formerly more widely spread during pre-glacial periods, in some cases a nunatak may consist of ground that has been re-exposed as persistent snow and ice have retreated, thus providing a new surface for colonization.

In many ways, a new nunatak is similar to a new volcanic island rising from the sea and, indeed, the colonization of Surtsey, such an island off the coast of Iceland (see p. 234), has been studied in comparison with the colonization of some new nunataks on Vatnajökull, the main ice cap. Within two years of the exposure of one new nunatak, some 13km (8mi) from the ice cap margin, 1km (0.6mi) from a nunatak 30 years old, and 3.5km (2mi) from some older nunataks, three species of moss and one vascular plant had become established. Two years later this had increased to eight or nine mosses and fifteen species of vascular plants.

While these Icelandic nunataks are not remote from sources of immigrant organisms, within the Antarctic nunataks occur many hundreds of kilometres from the coast, which is itself inhospitable to plants. On such nunataks only a few mosses and lichens occur, together with certain algae. The vegetation is at its richest and shows greatest diversity where meltwater trickles around rocks during the summer and where there is some supply of nutrients in bird droppings from colonies of snow petrels.

Although a picture has commonly been painted of great ice-sheets covering vast areas during the ice ages, it is interesting to note that when an ice-sheet expands, the horizontal component is infinitely greater than the vertical component, so that present-day nunataks in Greenland and the Antarctic may well have persisted during past glacial maxima. When the persistent snow and ice eventually disappear the existence of nunatak refugia may be recognized long after by the distribution of particular species, as for example, in the bicentric distribution of *Campanula uniflora* and a number of other species in southern Norway and in northern Scandinavia. This probably reflects the existence of refugia that would have been similar in nature to parts of the coast of Greenland at the present time, with coastal lowlands and mountainous nunataks near to the edge of an inland ice-sheet, but still supporting some plant-life.

N.J.C.

Restricted Distributions

With reference to the theory of Age and Area (see p. 220), taxa having restricted distributions would be expected to be newly evolved. Although this can be the case, it is often not so, as we shall see next in considering relicts and the interrelated but much wider topic of endemism.

Relict

A relict is a remnant of an otherwise extinct population, species, flora or community of plants. The term relict may be applied in certain circumstances to the entire population of a plant species or just to those populations that are geographically isolated by great distance and natural barriers. The essential condition in the first case which qualifies a plant as a relict is its existence in an area geographically remote from the territory in which its most closely related species occur. For example, the European larch (*Larix decidua*) is found native at altitudes above 1,500m (4,800ft) in the European Alps and Tatra Mountains. Generally no larches occur in the adjoining foothills or lowlands and the nearest occurrence of a related species. Siberian larch (*L. russica*) is over 1,600km (1,000mi) away to the east of Lake Onega and the White Sea. These larches are separated by the vast lowland of the European Plain, where large trees are known only as rare exceptions from a few Polish forests, which are themselves widely scattered. These are relict stands surviving from an earlier period in forest history, as is illustrated by the discovery of fossil larch cones and leaves in many areas in Poland. Comparably, in the mountains of New Mexico there is a species of maple, *Acer grandidentatum*, which is isolated by 1,200km (750mi) of treeless plains from its nearest neighbors in the Ozark Mountains to the east where the westernmost outposts of sugar maple (*Acer saccharum*) are found. Several reasons suggest that species such as these, usually occupying comparatively small areas of physiographically distinct territory, eg a range of mountains or hills rising from extensive plains, cannot have originated in the isolated circumstances in which they now survive. If this were the case, seed of an ancestral plant must have been transported across the great distance of country in which conditions were unsuitable for its germination and reproduction and yet by chance have reached the remote district where it now grows, carried in a single flight by wind or birds. Whatever the means of its first arrival, natural genetic variation and selection have produced the differences by which the relict population is distinguished as a species during a long period of isolation.

The accepted explanation, supported by the discovery of fossil and buried plant remains, accounts for the origin of relict species by recognizing that climatic change has taken place both progressively and

repeatedly in the last two million years. Therefore, the territory that at present separates a relict species from its nearest relatives, although now experiencing conditions inhospitable to their requirements, formerly possessed the climate and moisture that enabled the plants to spread gradually until they reached the district occupied by the relict. Adverse changes in climate and other consequential environmental conditions severed the distant populations from their original territory leaving them as relicts to survive by chance where tolerable conditions remained. In many cases, such situations are found in mountains because the rainfall which they attract provides the moisture which the adjoining lowland no longer has. The retreat of forest and woodland which had earlier occupied the plains of North America and the first development of prairies, resulted from a climatic change of this kind occurring late in the Miocene period. Additionally, mountains preserve at their higher levels a cool environment, accentuated by sheltered ravines and shaded cliffs not reached directly by sunlight, even in periods when the general climate of a region changes to become warmer than before. This

has been the case in the mountains of Europe and North America in each of the interglacial phases of the Pleistocene period. Indeed the arctic-alpine species now found in isolated populations on individual high mountain peaks are truly glacial relicts in those places, eg *Salix herbacea* in the mountains of Wales and Scotland, *Dryas octopetala* in the Tatra Mountains, *Dryas drummondii* on Mount Evans, Colorado. In these examples, the relict populations consist of the same species that occur over large areas in arctic latitudes, having been continuous with them during the glacial episodes when arctic climate and tundra extended far to the south of their present limits. Post-glacial warming of climate led to their retreat northward and also upward to higher altitudes on mountains where consequently they become isolated.

The geologically recent date of these events has in most cases not allowed sufficient time for natural variation to produce distinct species in the relict populations. As the age of such populations increases, they become progressively distinct, so that, as we shall see, they provide a bridge to the important topic of endemism.   B.S.

Opposite top A large number of rare and unusual plants occur today in the general area of Upper Teesdale in Yorkshire, northeastern England. Some of these plants are characteristic of the flora that existed soon after the last Ice Age, while others have survived from subsequent climatic regimes, which in some cases were warmer than the present climate. Such surviors from earlier conditions are termed relicts and an area like this where they are found together is known as a refugium. The main reasons for the present assemblage of rare plants in Upper Teesdale seem to be: firstly, the unusual soil ("sugar limestone") which by a combination of its high base status, concentration of heavy metals and erodibility has meant that the area has never had a complete forest cover since the last Ice Age, and that many common dominating plants of upland areas have been excluded; secondly, the severe climate, little warmer than that of southern Iceland, which has discouraged the vigorous growth of other plants; and thirdly, a long history of grazing, either by wild ungulates or sheep, which has probably always maintained an open habitat, even when there was some forest cover.

Opposite center Two unusual species that occur in the Upper Teesdale refugia.

Right *Helianthemum canum* (hoary rock rose). which forms vigorous populations in Teesdale, has its main area of distribution in southern Europe, and it occurs here under cold and windy conditions as a relict of warmer times. The small Teesdale population has developed slightly different morphological characteristics as a result of the conditions and its long years of isolation.

Left In Britain *Viola rupestris* (Teesdale violet) is confined to one hill in Upper Teesdale, and two other similar refugia in northern England. It is only 1–2.5cm (0.5–1in) high and is shown here growing on the obviously granular and sugary "sugar limestone". Its seeds seem to need the open conditions provided by the eroding limestone to allow them to germinate and it is a relict of more widespread open conditions.

Endemism

Endemism is the condition of being limited to one natural region or habitat. The exact delimitation of the area is, however, by no means clear. It can be precise, as in the case of islands, isolated mountains or strongly marked habitats such as serpentine soils or a particular swamp, but it is often much more arbitrary. The important thing is that the endemic should be conspicuously more limited in its distribution than might be expected. Consequently, since endemism can refer to categories other than species, an endemic genus can have a wider range than an endemic species, while families are often considered to be endemic when they occupy a whole continent. Endemics have long attracted the attention of botanists intrigued by the reasons why a species or genus occurs in one small area and nowhere else in the world. The general rarity of endemics has also accorded a certain "prestige" to the areas where they occur, so that their numbers have often been swollen by students of particular territories or countries. Clearly, the recognition of endemic taxa depends on an adequate taxonomic study of the groups concerned. More recently endemic taxa have also attracted attention because, due to their limited range, and frequently the particular regions where they occur, they are susceptible to the destructive side of human activities; many have already become extinct and conservationists are very much concerned with those that remain.

Causes of Endemism

In considering the causes of endemism it is necessary to distinguish between apparent and real endemics. Different areas of the world vary considerably in the extent to which their floras are known. Botanical exploration of poorly known areas, therefore, is always likely to yield species previously thought to be endemic to other areas, as well as newly discovered species endemic to the area being explored. Similarly, the incidence of endemism in a group of plants can, as noted above, vary with the level of taxonomic study. Investigations on a continental scale have repeatedly shown that species considered endemic to particular territories are more correctly part of a wide ranging taxon, while, conversely, detailed studies often reveal the existence of hitherto unsuspected endemics.

While bearing in mind these causes of spurious endemism, they should not be unduly overemphasized—there are without doubt a very large number of clearly endemic taxa among plants. There are several reasons for their very restricted distribution. They may be limited to a particular isolated area, such as an oceanic island, from which dispersal is difficult; they may have limited powers of dispersal or be adapted to a highly specialized type of habitat, which occurs only

Below Present and past distribution of the European larch (*Larix decidua*), the Siberian larch (*L. russica*) and their ancestral species *L. eurasiatica*. The latter species had a continuous distribution during the late Pliocene period (about 5 million years ago) from northern Asia westward to the Rhine. (a) During glacial periods the larches retreated to the south and east beyond the reach of the advancing ice-sheets (shown by the cross-hatched line), surviving in two separate areas: southern Europe (purple) and Siberia (orange). (b) Early in temperate periods between the glaciations the ancestral larch moved north across Poland from the southern European refuge and west from the Siberian refuge, but broadleaved forest (green arrows) prevented their further expansion and relinking. (c) More recently separate species evolved, the European larch in the south (purple) and the Siberian larch in the east (orange), the former again retreating south to the mountains leaving the scattered relict populations shown here. Thus the present distribution of these larch species is a relict of the continuous distribution of the single ancestral species. The two present-day species are still separated by a band of broadleaved forest.

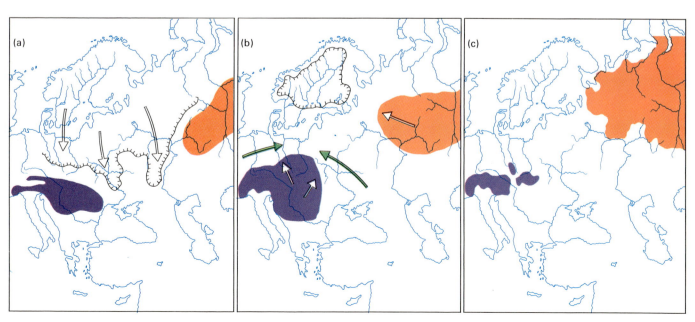

sporadically; they may have arisen relatively recently and not have had time to spread far from their area of origin or they may have formerly had a much wider distribution which, by extinction, has been reduced to the present very limited area. These last two reasons caused the great German plant geographer Adolf Engler to distinguish between newly arisen neoendemics and older paleoendemics, into which he considered all endemics could be divided. Although this may not be completely true, and in any case the evidence is not always available to decide firmly one way or the other, it does stress the importance of evolution, with the twin

processes of extinction and speciation (see p. 105), with which any consideration of endemics is inextricably intertwined and which also embraces the other causes of endemism mentioned.

Types of Endemics

Engler's recognition of only two categories of endemics oversimplifies the diversity of evolutionary situations, which they in part represent, resulting from the various ways in which extinction and speciation interact. Furthermore, when determining whether an endemic is new or old, there is a tendency to confuse absolute age in years and relative evolutionary age, which are not necessarily the same things.

There is a great need for further study of the categories into which endemics can be divided, but some of the currently most widely recognized types of endemics can usefully be mentioned here. These include Engler's categories, as modified by the Swiss botanist Claude Favarger, who has used data on chromosome numbers to assess relative evolutionary age. Further potentially valuable cytogenetical and other types of data could usefully be incorporated but so far have not.

Paleoendemics, which have also been called epibionts (epibiotics) and sometimes, relicts, represent taxa toward the end of their evolutionary life. As such they may be expected to be taxonomically isolated, all their close relatives having become extinct, with their present range reduced from a much wider area of distribution and containing a relatively small number of biotypes able, because of the low genetic variability, to grow in a very limited range of habitats. The maidenhair tree (*Ginkgo biloba*), now restricted to a small area in eastern China, the giant redwoods (*Sequoia sempervirens* and *Sequoiadendron giganteum*), limited to small areas in California and Oregon, and the dawn redwood (*Metasequoia glyptostroboides*), discovered in central China as recently as 1944, fulfill these requirements and demonstrate their age by fossil records showing them to be relicts of genera with many species widely distributed throughout the Northern Hemisphere in Tertiary, Cretaceous and even Jurassic times. Interestingly, however, the greatest geographical and genetical reduction of *Ginkgo*, at least, has taken place between the Upper Pliocene and modern times, during the last five million years of an evolutionary life extending back at least 30 times as long, so that absolute age is not necessarily linearly correlated with paleoendemism. For this reason, endemics which are taxonomically isolated but for which there is no fossil record known, as in the case of many herbaceous groups, may with some confidence be considered paleoendemics in a comparative evolutionary sense. Paleoendemics may be diploid or polyploid, but since the latter appears particularly common the distinction is sometimes made between paleopolyploids and paleodiploids.

Neoendemics represent plant groups near the beginning of their evolutionary life. Consequently, they are not taxonomically isolated, because their close relatives, which may or may not include their ancestors, are still present, often in the same or adjacent areas, and there is no evidence of a former wider distribution. Perhaps because the techniques of biosystematics, so prominent in taxonomic studies during the past 50 years, have mostly been applied to understanding the evolution and relationships of related groups of species and genera, where neoendemics are to be sought, the above definition of neoendemism has been shown to encompass several rather different situations.

Even by applying relatively crude cytological criteria it has been possible to distinguish several categories of endemics which can surely be refined and modified yet further by employing data now available. When an endemic species (or subspecies) has a diploid chromosome number and its close relative(s) is polyploid, it has been termed a *patroendemic*. Since, generally, polyploids are derived from diploids it follows that patroendemics are evolutionarily older and, like paleoendemics, represent a conservative or

Below Fossil and present distribution of two paleoendemics: the coast redwood (*Sequoia sempervirens*) and dawn redwood (*Metasequoia glyptostroboides*). Such paleoendemics are relicts of much wider distribution, representing taxa at the end of their evolutionary life.

Bottom Fossil and present distribution of the paleoendemic maidenhair tree (*Ginkgo biloba*). In proportion to the time interval, the reduction in distribution has been greatest in the past 5 million years (since the Pliocene) than in the previous 180 million years since the Jurassic.

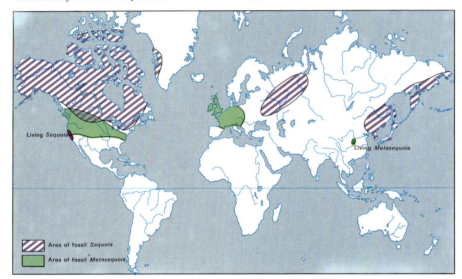

Living *Sequoia*

Living *Metasequoia*

Area of fossil *Sequoia*
Area of fossil *Metasequoia*

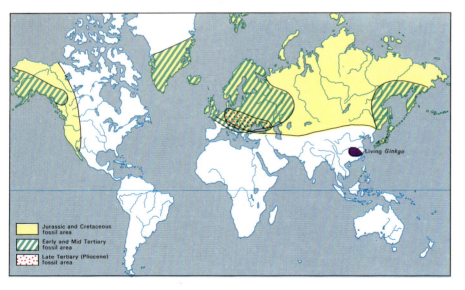

Living *Ginkgo*

Jurassic and Cretaceous fossil area
Early and Mid Tertiary fossil area
Late Tertiary (Pliocene) fossil area

passive facet of endemism, although obviously not to the same extent. *Apoendemics*, which are polyploid, are derived by auto- or allopolyploidy from one or several diploid relatives and constitute the converse situation, being more derived, more advanced and representing a more active facet of endemism. *Schizoendemics* are the result of gradual speciation, that is, more or less simultaneous divergence from a parent. Schizoendemics have the same chromosome number, whether diploid or polyploid, as their close relatives, which may also be endemic, and are considered to constitute an active component in endemism.

The definition and recognition of the various types of endemic just mentioned depends to a considerable extent upon the level of information available. The division into paleoendemics and neoendemics can be achieved to a significant extent by morphological and geographical studies, which are the only ones to have provided data on a sufficiently worldwide scale to allow useful comparisons to be made between the floras of the continents and major territories. Of the newer sorts of information, that on chromosome numbers, although still insufficient, is available for a greater diversity of taxa from more areas throughout the world than are other types of data. Consequently, in the cytologically best studied floras, such as those of Europe and North America, the recognition of apo-, schizo- and paleoendemics is both practicable and useful. Where, however, particular regions or particular groups, still relatively few, have been subjected to more intensive study it becomes clear that further refinements are desirable and possible. For example, the recognition of endemics having the same chromosome number as their close relatives as schizoendemics cannot be universally upheld. Although very many endemic taxa, such as those restricted to soils over serpentine or soils derived from guano of sea-birds, have clearly diverged by selection from their more widespread relatives, there are several well-documented instances of diploid neoen-

demics being produced by hybridization between diploid species which are still extant. Examples are *Penstemon* and *Delphinium* in California, in which, respectively, *P. spectabilis* (*P. centranthifolius* × *P. grinnellii*) and *D. gypsophilum* (*D. hesperium* × *D. recurvatum*) have arisen by this means. Clearly, schizoendemics cannot be recognized by chromosome number alone and some modification seems to be necessary. Again, what are clear patroendemics need not have polyploid relatives. In California the outcrossing diploid *Gayophytum eriospermum*, restricted to parts of the Sierra Nevada, is not only a patroendemic in relation to the widespread polyploid *G. diffusum* but also to the more widely distributed *G. heterozygum*, an inbreeding diploid species apparently derived by hybridization between *G. eriospermum* and another patroendemic, *G. oligospermum*. A final point which should be emphasized is that newly arisen endemics are often very difficult to distinguish. The more recent their origin the more they will resemble their relatives, so that in areas or groups of active evolution the frequency of neoendemism is likely to be underemphasized without extensive and detailed study. The term *cryptoendemism* has been coined for such situations.

Active Epibionty
In addition to the cases of endemism which show the variety of intermediates between paleoendemics and neoendemics already noted, several situations are known where the two are effectively combined to give a distinctive evolutionary and distributional pattern. Isolated areas, such as oceanic islands, often contain genera or distinctive parts of genera which, because of their relative taxonomic isolation and often on the evidence of their fossil record, are considered to be paleoendemics. However, these groups frequently show evidence of active speciation in response to the climatic, ecological and topographical diversity of the area, often by adaptive radiation (see p. 103), so that a phase of neoendemism, more precisely schizoendemism, is effectively superimposed upon a paleoendemic base. This situation, which demonstrates that paleoendemics need not necessarily be genetically impoverished and at the end of their evolutionary life, is termed "active epibionty".

This phenomenon is shown by several groups in, for example, the Canary Islands. The giant sowthistles (*Sonchus* subgenus

Dendrosonchus) provide a good example, as do the Canarian chrysanthemums now placed in the genus *Argyranthemum*. This woody genus is endemic to Macaronesia (Madeira, Salvage and Canary Islands) and is undoubtedly one of the many Tertiary relicts in the archipelago. In isolation it has diversified into 22 species and numerous subspecies which show strong evidence of adaptive radiation in response to environmental and topographical features. Furthermore, detailed studies have shown that this schizoendemism is still very active, with divergence continuing at the population level, and the paleoendemic group has a promising evolutionary future.

Right *Cheiranthus scoparius* growing in the subalpine zone of Los Cañadas del Teide, Tenerife. This particular species, the Canary Islands wallflower, is endemic to the Canaries.

Left The popular name Mount Cook lily is a clear hint as to the limited geographical distribution of *Ranunculus lyallii*. Also known as the great mountain buttercup, this flower is in fact endemic to the Southern Alps of New Zealand, of which Mount Cook is the highest peak. The plant grows at an altitude of 210–450m (700–1,500ft) and flowers from November to January.

All members of the Xanthorrhoeaceae, "grass trees", are endemic to Australia, New Zealand, New Caledonia and New Guinea. This rather bizarre view of *Xanthorrhoea* species growing in burned *Eucalyptus* woodland on Cape York Peninsula, Queensland, Australia, is a vivid reminder of the distinctive flora of that island continent.

Areas of Endemism

It has long been known that endemic taxa are not scattered evenly over the world but rather they are concentrated in particular regions. Because of the evolutionary situations represented by endemics, areas where they occur in significant numbers have attracted attention for the information they

Dicoma incana, one of the few trees in the family Compositae, is an endemic of relatively old origin occurring in Madagascar.

can give on the history of the flora and even of the region itself. Isolation (see p. 108) has undoubtedly been important in leading to a high level of endemism, because it reduces the rate of immigration and consequently of competition from plants which have evolved in other areas. This is attested by isolated oceanic islands, such as Hawaii, St. Helena and New Zealand, where respectively 90%, 85%, and 75% of the vascular plant species are endemic, when compared with continental islands like Ceylon, Jamaica, and Britain, with 34%, 20% and less than 1% endemism respectively.

Although oceanic islands have obviously figured prominently in studies of rich endemic floras, "habitat" islands can show the same situation. High levels of endemism are shown by the high-altitude floras of the African mountains, isolated from other temperate regions by the surrounding tropical and subtropical lowlands, while "habitat islands" on a much larger scale are suggested, for example, by tropical American ferns. These are centered in three principal areas—South Mexico, North Andes and south central Brazil, where over 60% of the species are endemic. Since ferns have excellent mechanisms for long-distance dispersal it is clear that ecological factors have been important in isolating these centers of endemism.

Isolation itself, however, is not sufficient to lead to significant levels of endemism, since the age of the area, or at least the time it has been effectively available to the plants now present, is also very relevant. Many of the islands in the southern oceans are isolated by very many kilometres of water from the nearest continents, as is Iceland, for example,

in the Northern Hemisphere. These islands have very low levels of endemism, much less than 1% and often nonexistent, because they have been available for colonization only in the few thousand years since the retreat of the Pleistocene ice-sheets. This, as well as their proximity to continental Europe, also accounts for the very low level of endemism in the British Isles, already alluded to above.

Isolation and age, then, would appear to be necessary prerequisites for endemism, and the greater the age the greater the likelihood that the endemics will be paleoendemics. The small island of New Caledonia, for example, situated in the southwest Pacific Ocean, is notable for the large number of endemic genera that have very ancient affinities, while having a very poor representation of more advanced groups of relatively recent origin. This flora appears to be a surviving, modified sample of the late Cretaceous flora of eastern Australia from which it separated by continental drift some 80 million years ago. Australia itself lost most of this ancient flora as it moved northward into subtropical latitudes after the breakup of Gondwanaland and the large number of endemics now characterizing that continent arose subsequently in response to the increasingly arid climates it encountered during the Tertiary period. Southern India, which also arose from the breakup of Gondwanaland, has a relatively impoverished flora which is not expecially rich in endemics and this would accord with the view that it rafted northward across the equator, losing most of its rich Cretaceous tropical to subtropical and austral temperate flora during its movement across the various climatic zones, a process hastened by the later increased aridity in the zone it now occupies.

These data suggest that the occurrence of old paleoendemics, relicts, depends not only upon the isolation and age of the area but also upon some degree of environmental stability. It is significant that New Caledonia, rich in paleoendemics, has not moved far from its position when the early flowering plants seem to have been evolving, nor have New Zealand and Fiji, for example, which are also rich in paleoendemics. Furthermore, all these areas have maintained mesic, equable conditions because of their oceanic climates. Perhaps for the same reason the island of Madagascar has a rather rich flora of paleoendemics, many of them showing links with Australasia, which have become extinct in the adjacent, more continental climate of Africa. These reasons would also explain why in much more recent times the oceanic Canary Islands have harbored the relicts of the mesic Tertiary floras of the adjacent, now arid, Mediterranean-North African region. Such findings have focused increasing attention upon the relationships between endemics and the environmental stability of the areas where they are especially prevalent. California, in which about one-third of the native species are endemic, contains obvious paleoendemics, such as the giant sequoias, referred to earlier, and genera such as

Clarkia, which are demonstrably in a phase of active evolution and producing neoendemics at the present day. A survey of the endemic flora in relation to the rather varied environments available showed that the paleoendemics are largely confined to mesic and xeric areas which seem to have persisted since at least the Tertiary period. Patroendemics, another relatively conservative type, are most abundant in areas which, although exhibiting a diversity of habitats, do not seem to have been subjected to great environmental instability during at least the Pleistocene climatic vicissitudes. Patroendemics seem to occupy areas of similar history in southern Europe. On the other hand, apoendemics, representing "active" neoendemics, are most abundant in borderline zones between mesic and xeric climates where even small climatic shifts will change local conditions to give unstable environments conducive to rapid extinction and evolution, resulting in the production of new forms. Detailed fossil studies in other areas have to some extent confirmed these findings. For example, paleoecological studies in Cuatro Cienegas, northern Mexico, have shown that an area renowned for its high frequency of paleoendemics has experienced no climatic fluctuations during the past 31,000 years.

In conclusion, it is instructive to refer to recent studies on the plants of the Galápagos Archipelago, the islands where the now famous finches demonstrated the endemism and adaptive radiation which were instrumental in leading Charles Darwin to his theory of evolution. These islands have three major ecological zones differing in their

These unusual, tree-like cacti (*Opuntia echios* var. *echios*), growing at Conway Bay on Santa Cruz, one of the Galápagos Islands in the Pacific Ocean, represent one of the several varieties of the same species occurring in the archipelago. Such diversity, linked to isolation and differing habitats, observed by Darwin during his visit to the Islands, was a major influence on his developing evolutionary theories.

immigration and extinction rates and in their ages. The littoral zone has a high immigration rate of widespread seaborne plants which, because of storm disturbance and the small area of this zone, suffer a high extinction rate. This zone has the lowest level of endemism. The arid lowlands, making up the major part of the islands in the archipelago, have an age of several million years and contain most of the Galápagos plant endemics. The moist uplands contain relatively few endemics, most having close relatives on the mainland of South America. This is unusual, because most tropical or subtropical islands have the highest proportion of endemics in the moist uplands, according with some of the earlier generalizations about stable mesic conditions. However, there is clear evidence that the moist upland habitats of the Galápagos Archipelago are relatively recent, being only 10,000 years old at most, thus explaining their relative paucity of endemic taxa. It is perhaps appropriate that this archipelago, which may be considered the cradle of current evolutionary concepts, should draw attention to the relationship between isolation, age, area and environmental stability, and endemism, which is both a product and a reflection of evolution. D.M.M.

Islands

Most of the topics discussed in this section of the chapter, devoted to distribution and distribution patterns, contribute to the great phytogeographical interest of islands which, as we shall see, play a central role in considering plant distribution and evolution. Strictly speaking, islands are areas with particular features different from those of the region in general. Thus, ponds, lakes, mountain summits or tracts of woodland in agricultural land, for example, may be considered as biological islands, sharing a number of features with the terrestrial islands scattered across the world's oceans and which usually spring to mind when considering this topic. These latter islands, however, possess additional characteristics which have long made them of special interest to biologists and they will be the principal subject of this account.

Inevitably, isolated pieces of land encountered after wearisome sea voyages have had a particular fascination for people, and biologists have been intrigued by the reasons for the often highly distinctive plants and animals present on islands, and the means by which those plants and animals reached them, after apparently crossing great expanses of open ocean. Very often the total number of species is more restricted and the ecosystems simpler than in comparable continental areas, which makes them ideal places for studying the interactions of plants and animals with each other and with the environment. Their very isolation has often preserved these "natural laboratories" from outside interference, to which their small size

makes them particularly vulnerable, so that the arrival of people on many islands has led to great modification of the habitats and the loss or endangering of their frequently unique plants and animals. Consequently, islands also figure prominently in conservation programs.

The diversity and affinities of plants we find on islands, as in other areas, are a reflection of their isolation, age, environmental stability, habitat diversity and size. In consequence, the study of island plants has proved important in increasing knowledge of the ways in which plants have evolved and migrated to give the distribution and diversity of the world's flora that we see today.

Isolation and Age

Continental and Oceanic Islands
In considering the effect of isolation on their floras it is customary to distinguish continental and oceanic islands. The former occur adjacent to a continent, with which there is evidence of former, sometimes relatively recent, land connections; the latter are far distant from the continental shelf and without obvious evidence of former direct connections. Island floras are derived, by some means, from the floras of continental areas—the source areas—and the flora generally becomes much less diverse with increasing distance between the island and its source area. An indication of the diversity is given by the ratio between the numbers of families and genera present in a flora, the higher the ratio the greater the floristic diversity. On this basis we see that continental islands such as Jamaica (154 families: 810 genera—5.3), 600km (370mi) from Central America, and Fernando Po (100 families: 470 genera—4.7), 35km (22mi) from West Africa, have larger and more diverse floras than those of oceanic islands such as Hawaii (83 families: 221 genera—2.6) and St. Helena (18 families: 34 genera—1.9), isolated in the middle of the Pacific and Atlantic Oceans respectively. Of course, this relationship between isolation and floristic richness is not always clear, because other factors are involved. Smaller areas tend to have poorer floras than large areas, while topographic diversity is important. The continental Bahamas, for example, have virtually the same area and degree of isolation as Jamaica, but a much lower floristic diversity (104 families: 290 genera—2.8), comparable to that of Hawaii, resulting from poor soils and uniform low altitude topography.

Continental islands are sufficiently close to the source areas for there to be a reasonable number of immigrants which can maintain their floristic diversity and their affinities with the continental flora, of which they often appear to be a somewhat impoverished fragment. The isolation of oceanic islands, on the other hand, results in greatly reduced numbers of immigrants, so that inevitable losses are not readily made good. However,

Opposite Kuta Bandos, one of the clusters of coral islands or atolls that make up the Maldives in the Indian Ocean. Islands have long fascinated biologists, who have been intrigued by the reasons for the highly distinctive plants and animals which occur on them. Coral atolls are particularly interesting since they frequently project so little above sea-level that their vegetation is on constant risk of destruction during storms.

those plants which are present are able to evolve in isolation, free from competitors, which is one of the reasons why so many oceanic islands are rich in plant species found nowhere else in the world. These endemics include such striking plants as the gumwood (*Commidendron*) and black cabbage tree (*Melanodendron*) of St. Helena, which are robust trees of 4.5m (15ft) or more related to *Aster*, the cushion-like "vegetable sheep" (*Raoulia eximia*) of New Zealand, also a member of the family Compositae, or the succulent *Dendrosicyos*, an endemic of Socotra Island in the Indian Ocean, which forms a giant elephantine stem up to 3.7m (12ft) high and which belongs, together with cucumbers and pumpkins, to the vine-like family Cucurbitaceae.

However, the distinctiveness of an island's flora, usually expressed as the percentage and level of endemism, is not solely, or even principally, dependent upon its degree of isolation. It is dependent also on the effective age of the flora, and this correlation is of considerable importance in considerations concerning the origins of that flora which, in the case of oceanic islands, bring into question the much-debated relative merits of long-distance dispersal and continental drift. The history of the flora of a continental island, on the other hand, is usually a reflection of the floristic history of the continent.

Long-distance Dispersal and Continental Drift

Over the years, discussions about the origins of the floras of oceanic islands have usually concentrated on three possibilities: that seeds or spores have been dispersed from the source area(s) across the intervening ocean by some means, such as wind, water currents or migrating birds; that the island was formerly part of a continental mass from which it sundered by plate tectonics (see Continental Drift and Plant Distribution, p. 75) or that it was connected to a continent by a now submerged land bridge. Although there is overwhelming evidence that continental islands such as Britain and the Falkland Islands were connected to the adjacent continents in the relatively recent past, this is much more debatable for truly oceanic islands, for which continental drift and long-distance dispersal seem the most likely mechanisms.

New Zealand, 1,600km (nearly 1,000mi) from Australia, and New Caledonia, 1,200km (750mi) from Australia and rather further north of New Zealand, have been islands since they separated from the rest of Gondwanaland about 80 million years ago. New Caledonia has 655 genera (in 142 families), 16% of which are endemic. Furthermore, a high proportion of these endemics belong to what are generally believed to be primitive families such as Coniferae, Myricaceae, Monimiaceae, Proteaceae, Winteraceae, etc and few to relatively recent families such as Campanulaceae, Compositae and Labiatae. Many are taxonomically isolated within their families and some have been proposed as distinct families. The flora of New Caledonia, rich in paleoendemics, thus has an ancient origin and appears to be a surviving, modified sample of the temperate late-Cretaceous flora of Gondwanaland. Likewise, much of the lowland flora of New Zealand is similar to that of temperate Gondwanaland 80 million years ago. At that time it was at approximately latitude 60°–70°S and has been moving northward ever since. During most of this period it was within temperate latitudes, subtropical climates only appearing in northern North Island in the Miocene (about 20 millions years ago). Its balanced flora, with 10% of genera and 81% of species endemic, contains such groups as Proteaceae, Winteraceae, various austral gymnosperms and *Nothofagus* which, as in New Caledonia, reflect its antiquity. The more impoverished flora of New Zealand, with fewer relict types than New Caledonia, reflect its more southerly position, the considerable reduction of its area during marine transgressions in the Oligocene (about 30 million years ago), when it comprised a chain of low islands, and the extensive Pleistocene glaciations. Certainly, many taxa of old groups such as Proteaceae, *Casuarina* and *Araucaria*, for example, became extinct in New Zealand during the period when these events took place.

In comparison with the ancient areas just mentioned, the Hawaiian Islands are young, having apparently arisen as volcanic cones in the middle of the Pacific Ocean less than 20 million years ago and never having been connected to a continent. The diversity of the flora, with 83 families and 221 genera, is significantly less than those of New Zealand and New Caledonia, although the proportion of endemic genera (19%) is comparable with that of New Caledonia. However, most of the endemics are not very isolated taxonomically, no separate family having ever been proposed for even the most distinctive of them. They belong to such relatively recent families as Compositae and Campanulaceae; families indicating antiquity, such as Winteraceae, are absent, in accordance with the geological youth of the islands. Hawaii, then, appears to have derived its flora by transoceanic dispersal from the New and Old Worlds bordering the Pacific, with the latter (especially Malesia) being the more important. It has been estimated that the present flora of about 1,700 species could be derived from about 350 original immigrants. The subsequent considerable speciation seems to result, as we shall see later, from the environmental diversity of the islands.

Comparable with the Hawaiian Islands, the Tristan da Cunha archipelago arose in the middle of the South Atlantic during the late Tertiary period (about 20 million years ago). It has a much harsher climate than Hawaii, and considerably less topographic diversity, which at least partly explain its much poorer vascular flora (70 species belonging to 31 families and 51 genera). Although 37% of flowering plants and 42% of ferns are endemic to the islands they are relatively new, having rather close relatives in other areas. The greatest affinities of the flora, including the endemics, are with South America, although South Africa is closer. This is a consequence of the clockwise motion of the strong winds and ocean currents around the world in these latitudes—South America is "upwind". The flora is unbalanced, comprising an unrelated group of plants which have in common only their propensity for long-distance dispersal. Thus, the high proportion of ferns (54%, compared to 25% in a continental flora) reflects the easy wind-dispersal of their light spores, which is reflected in many of the flowering plants, such as *Sophora*, which has been shown to be capable of dispersal over considerable distances by ocean currents, or *Acaena*, the fruits of which are carried attached to the bodies of migrating birds. Even greater youth is shown by the floras of many subantarctic islands, such as Macquarie Island, 960km (600mi) south of New Zealand, which were completely covered by the Pleistocene glaciations. Their present impoverished floras have resulted entirely from long-distance dispersal after the ice retreated, and the extremely low level of endemism (or lack of it) reflects their very recent origin.

Environmental Stability

As mentioned previously, the present floras of islands are a reflection not only of age and isolation, but also of their environmental stability. To some extent the conditions on all islands are buffered from climatic extremes by the surrounding water. Thus, the broad-leaved forests in the humid zones of the Canary Islands are a relict of the extensive forests of the Mediterranean region which were destroyed by the climatic changes of the late Tertiary period. Madagascar retains remnants of an earlier African flora decimated by the increased aridity of the continent, while New Caledonia and New Zealand, as we have seen, have retained floras similar to those of temperate Gondwanaland, although less so in New Zealand, which has suffered more vicissitudes. Among other fragments of Gondwanaland, Australia moved north into subtropical regions and lost most of its old, cool-temperate flora, while India rafted across the tropics into the Northern Hemisphere, losing all its original flora on the way. Conversely, it has been demonstrated that the evolution of new forms is stimulated by the new challenges offered by changing environments. The

The impoverished floras of southern cool-temperate islands are particularly susceptible to human interference. On Macquarie Island, 960km 597mi) south of New Zealand, the distinctive "herbfields" (**above left**), of which *Stilbocarpus polaris* and the silver-leaved *Pleurophyllum hookeri* are the most conspicuous members, are among the communities most sensitive to disturbance. In many areas rabbits introduced by Man have so affected the plant cover that only bare, eroding soil is left (**above right**).

distinctive flora of the island continent of Australia, for example, with the proliferation of such genera as *Eucalyptus* (gum trees), seems to accord with its drift into new climatic zones. New Zealand which, as we have seen, suffered impoverishment of its ancient lowland flora by the marine transgressions of the Oligocene period (about 30 million years ago), underwent considerable environmental changes during the late Pliocene period (about 10 million years ago), when its extensive mountain systems arose to provide new habitats, which were colonized by such genera as *Hebe*, *Epilobium* and *Celmisia*, arriving by long-distance dispersal from Australia. Similarly, environmental changes have brought about evolution of species in other islands. During the recent glaciations of the Pleistocene period the glaciers advanced, "pushing" the plants before them, and retreated, the plants following, several times. Montane plants were thus periodically forced together in ice-free lowlands—refugia—and isolated in mountain areas as the ice retreated. This "pump-action" of the glaciers has been shown to be a potent factor in the evolution of several plant groups and probably accounts for the endemic species of, for example, *Nassauvia* in the mountains of the "continental" Falkland Islands and Tierra del Fuego, which diverged from their closest relatives in the southern Andes following their retreat from the ice-free lowlands now occupied by the sea off the east coast of Patagonia.

We can thus see that environmental changes can act as a stimulus to the origin of

new plant groups, while at the same time being responsible for the extinction of many more ancient forms. In relatively recent times the greatest environmental upset for cool-temperate plants was caused by the Pleistocene glaciations. The sporadic occurrence of the cold-temperate family Hectorellaceae, with *Lyallia* in the subantarctic Kerguelen Islands and *Hectorella* on the mountains of the South Island of New Zealand, suggests massive extinction of a pre-Pleistocene family having affinities with the Portulacaceae and, perhaps, the Caryophyllaceae. The sporadic occurrence of peculiar broad-leaved herbs of such genera as *Stilbocarpus*, *Pleurophyllum*, *Pringlea* and *Myosotidium* in the islands south of New Zealand may also reflect such extinctions. The impoverished floras of the British Isles and Iceland, for example, with their low levels of endemism are a reflection of their recent return after the retreat of the Pleistocene glaciers as well as, in the former, proximity to a continental source area. Greenland and, particularly, Antarctica are very large islands with fossil records showing extensive former floras. Both are only just emerging from the Pleistocene "deep-freeze" and Antarctica, for example, which once had temperate forests dominated by *Nothofagus* and other genera, now has only two native species of flowering plants: *Colobanthus quitensis* and *Deschampsia antarctica*, both relatively recent arrivals from South America.

Colonization and Establishment

As we have just seen, the enormous island continent of Antarctica, emerging from its ice cover, is now beginning to be recolonized from adjacent South America. The process of colonization, so important in considering how all plants extend their distribution, is forcibly brought to the attention of those studying island floras, particularly where the colonists have to cross definite oceanic

barriers. Usually, the colonization of an island or other areas has to be inferred from the properties of the plants present today, sometimes aided by evidence from fossils, and direct observations are rarely available. However, two islands—one tropical, the other cool temperate—provide information on what is involved for plants to gain a foothold on an island once their propagules have reached it.

In 1963 the new island of Surtsey was formed by volcanic activity 40km (25mi) south of Iceland and it has been under constant surveillance ever since, the positions of all plants and propagules being carefully noted each year. The first colonizers were fungi, algae and flowering plants. Mosses arrived after three years, rapidly spreading on the lava, and lichens after eight years. The flowering plants are almost all coastal species, brought as viable fragments or seeds by wind, water, trapped in floating "mermaids purses" (the egg cases of various fishes), or in the gizzards of birds such as the snow bunting. Sea rocket (*Cakile maritima*), seeds of which arrived in 1964, had 30 plants in 1965 and one in 1966, none of which flowered; 14 of 38 plants flowered in 1967 but only seeds were found in 1968. Similar patterns are shown by the other early colonists, notably lyme grass (*Elymus arenarius*) and sandwort (*Honckenya peploides*). These tremendous fluctuations show the uncertainty of colonization once dispersal has been achieved. It is possible that initial establishment has now been effected, with small dunes formed by lyme grass, a rudimentary vegetation of chickweed (*Stellaria media*) and scurvy grass (*Cochlearia officinalis*) present on the steep cliffs, and the total number of plants is increasing, most dramatically between 1972 and 1973 when it rose from 119 to 1,273 individuals, mostly of scurvy grass and sandwort. It would be premature, however, to claim that they will not suffer further vicissitudes before establishment is fully achieved.

The volcanic island of Krakatau, situated 48km (30mi) from Java and 32km (20mi) from Sumatra, suffered a cataclysmic

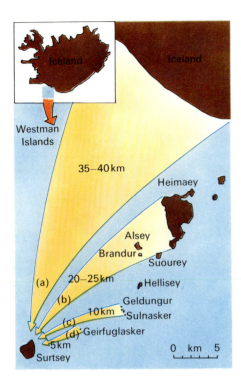

Areas from which colonizing plants have been brought by wind, ocean currents and birds to the new island of Surtsey since its origin off the south coast of Iceland in 1963. (a) Bryophytes from Iceland; (b) *Angelica archangelica, Cakile edentula, Carex maritima, Cystopteris fragilis, Elymus arenarius, Honckenya peploides, Mertensia maritima* and about half the bryophyte species from Heimaey and adjacent islands; (c) *Festuca rubra* and *Stellaria media* from Geldungur and Sulnasker; (d) *Cochlearia officinalis, Puccinellia maritima* and *Tripleurospermum maritimum* from Geifuglasker. Although this diagram is based on the nearest areas to Surtsey, where the species mentioned grew before the origin of the island, some of them may well have migrated over greater distances.

eruption in 1883 which apparently destroyed its rich tropical flora and fauna. After three years, 16 species of flowering plants were found on the shores and eight inland; there were also 11 species of ferns. Fifty years later the total was 271 species but in the meantime 115 species had become established and then disappeared. The diversity of the flora of Krakatau continues to increase, being still low relative to the rich forests of Java and Sumatra, but the fragmentary information shows the problems of colonization. The primary colonizers, arriving by sea, wind and birds, must be able to cope with open habitats—sand, rock, lava. They are often unable to survive in the denser vegetation produced by later arrivals and so the flora is in a perpetual state of flux until some level of equilibrium is reached as the appropriate climax vegetation is attained.

Both Surtsey and Krakatau, however, are rather close to their source areas. Isolated oceanic islands have a much lower number of immigrants, which result in even fewer becoming established following the uncertainties of colonization seen on Surtsey, for example. The consequences of the low level of establishment vary, depending upon the properties of the immigrants, the time which has been available since they arrived and the diversity of habitats available to them. The small number of immigrants often expands into a much greater number of habitats than are normally available to them on continents, where competition restricts them to conditions to which they are best adapted, and the species of many cool-temperate southern oceanic islands, for example, colonized relatively recently, are characterized by a much greater ecological amplitude than their mainland populations. Given time, and the genetic potential, these immigrants will evolve to give a series of related species adapted to exploiting the range of habitats available to them (see Adaptive Radiation, p. 103). The occasional immigrant can only bring a small portion of the genetic endowment available in the original source area so that island populations are often an unrepresentative sample of the variation shown by the species on the mainland and, evolving in isolation, can rapidly diverge to give a new island species (see Genetic Drift, p. 104).

Immigrants to oceanic islands require a good means of dispersal to arrive at all and the ability to propagate themselves once they arrive: the more isolated the island the more likely that seeds will arrive singly to give an isolated plant which must either reproduce vegetatively or produce seed by self-fertilization for establishment ever to be a possibility. Once on the island, good dispersal may be a liability, since seeds would generally disperse into the ocean! The occurrence on many islands of flightless insects and birds having winged relatives on continents suggests that there is selection against dispersal in island populations. There is some evidence for this in plants too. The bur-marigolds or tickseeds (*Bidens* spp.), for example, get their name from the two barbed prongs on the seed which readily attach to clothing, fur or feathers and are a very effective dispersal mechanism. The Hawaiian endemic *Bidens populifolia*, however, whose ancestors crossed large water gaps, has fruits with only two small ineffectual projections. Similarly, continued self-fertilization, especially in small island populations, can lead to loss of the genetic variability so important for evolving in response to the new habitats. The change from self-fertilization to outcrossing, with male and female flowers on different plants (dioecism), is relatively easy to achieve and the high proportion of dioecism in some oceanic island floras (27% in Hawaii, compared with 2% in Britain) suggests that this may be an evolutionary trend of importance in their diversification.

Man and Island Floras

The fascination that islands hold for people is probably matched by the havoc they have caused when they reached them. The flightless dodo of Mauritius is a universal symbol of extinction and serves as a constant reminder of the many island animals, such as the New Zealand moa hunted by the Maoris, which have been exterminated by Man. Island plants have fared no better. In some cases the species have become extinct before they could be described, as happened to three endemic ebonies (*Diospyros* spp.) in the Mascarene Islands. The most usual cause of destruction has been the introduction of animals, such as goats and rabbits, which have disrupted the whole plant ecosystem. The Atlantic island of St. Helena provides a sad example. When first discovered in 1501 it was densely covered with luxuriant forest, dominated by St. Helena ebony (*Trochetia melanoxylon*). Goats were introduced in 1513 and increased to many thousands by 1588, stopping natural regeneration by eating herbs and tree seedlings. Use of mature trees for fuel, in lime-kilns and in tanning, hastened the end so that by 1810 the forests had been completely destroyed and a barren landscape left as the heavy tropical rain washed away the soil. Nowadays, many endemics, including the ebony, are extinct, and fragments of the unique flora of this island are restricted to a few ridges and precipices; the remainder is dominated by European, American and Australian weeds.

As has been mentioned, much of the unique interest of island floras results from their isolation. Their vegetation, developed in the absence of herbivores, often has little resistance to grazing, as exemplified by St. Helena. The ecosystems are often rather fragile, with few species occupying a wide range of ecological niches to which they are not fully adapted. Continental plants, many of which are adapted to continual disturbance of their habitats, can thus invade and profoundly alter the character of oceanic island vegetation once they are deliberately or accidentally introduced. Thus, 66% of the species on Tristan da Cunha come from other areas, and New Zealand, with 1,200 native and 1,700 alien species, shows the same pattern, which can be repeated on very many islands throughout the world. Finally, most islands are of small area and the populations of their often unique species numbered in tens or hundreds of individual plants. Consequently, each new development catering for tourists seeking the tranquility of islands, or improving the economic lot of the inhabitants, or utilizing their location to improve transoceanic travel, is a direct threat to plants found nowhere else in the world. Because island floras constitute a series of natural experiments which have much to tell us about plant evolution and migration, and because of the role of plants in determining the aesthetic quality and habitability of islands, these problems are of great concern to conservationists. Although extinction, as a necessary part of evolution, has played its part in the development of the unique plant life of islands it would be tragic if we missed any opportunity to prevent unnecessary acceleration of the process by our own activity.

D.M.M.

CHAPTER SEVEN

Man and the Green Planet

The Influence of Man on Plants and their Environment

Early Man, as a hunter and a gatherer of such plant products as fruits, leaves, roots and timber, for food and shelter, would, like any other animal, have had a relatively small effect on the plants he lived with. Human influence extended as land was cleared for cultivation and grazing by domesticated animals, following the Neolithic Revolution. More subtly, perhaps, Man's preference for, and cultivation of, particular food plants, such as the cereals, resulted in selection of types increasingly suited to his needs.

On a global scale, however, these activities initially had little impact, as we can still see today in the few cases where primitive tribes, with small populations, practice shifting agriculture in the tropical forests. During the last few centuries, however, the increasing human population has been making ever greater demands on the world's resources, to support the requirements of successive civilizations. In the Old World, the inhabitants gradually began to look beyond Eurasia and northern Africa to replace their diminishing sources of supply. Today, the interaction between Man and his green planet is so extensive and varied that this chapter can only outline some of the more obvious trends.

Despite numerous newspaper headlines, prophecies of ecological catastrophe and the everyday evidence of our eyes, numerous plant groups and the vegetation of some parts of the world remain relatively unaffected by Man. Indeed, if this were not so we should be dependent upon historical information for much that has been covered in this book. The reasons are basically twofold. Firstly, economic considerations mean that certain areas and groups of plants are not worth exploiting, by whatever methods, either because they are too far from consumer-areas or because the natural resources they contain are of insufficient interest or value for modern needs. Secondly, over the past century, conscious decisions have been taken by Man to conserve particular areas or groups of plants in order that they may be maintained for future study and enjoyment.

The plant cover of extensive areas of the world has, however, been modified to a greater or lesser extent by Man. The development of human living areas, towns and cities, and the communication networks linking them, stifles plant life under a cover of concrete and tarmacadam. Such centers of human activity also affect plant (and animal) life for great, often considerable, areas around them, as a result of industrial processes altering the atmosphere, water and soil that make up part of the plants' environment. Such pollution is also an increasingly recognized target for conservation groups. Despite this, the natural plant cover of the world has probably been most affected by the development of the normally accepted agricultural activities of Man. Obviously, clearing vegetation to make way for crop plants and intensive culture of animals has had a profound effect, as have such practices as irrigation, but the selective grazing of domestic animals in much less intensively managed vegetation has been no less devastating in altering the plant cover. A corollary of these activities has been the spread of species native to one part of the world to others, as an incidental result of Man's activities, and such weeds and aliens have profoundly affected the constitution and composition of the plant cover of many areas in which they now occur.

The development of crop varieties for particular regions may be considered among the more active forms of human involvement with the world's plant cover. The use of agents such as herbicides is another aspect of the same process. The introduction of crop plants to different parts of the world is also relevant, as are afforestation programs, which have often caused public concern because of the long-term effects on numerous communities, plant and animal, dependent upon the tree cover. On a lesser scale the development of gardens may be included here. They can be used to demonstrate and to study a much wider selection of the world's flora than is usual in any one area, and they can be used to cultivate species, and perhaps communities, extinct or almost so in the "natural" world. Gardens, then, are yet another means by which Man can study and preserve parts of the green planet he has been privileged to inhabit.

Opposite It is inevitable that Man's utilization of the Green Planet has required the clearance of forests to provide space for growing food crops. However, there are more ecologically and economically acceptable ways of doing it than the clear-felling, seen here in Indonesia, which leads to widescale erosion and impoverishment of the bare soil.

Domestication of Plants

All Man's staple food crops were domesticated in prehistoric times, and the unraveling of their origins provides a fascinating mixture of botanical and archaeological research. It seems unlikely that any higher plants growing wild today have the potential to displace the staple food crops even if plant breeders were to invest the time and effort necessary to domesticate them. Increase in the world's food supply is more likely to come from improvement of crops which have already been domesticated. In this respect, the domestication of plants was essentially completed before the historic period. However, the future may well see domestication of some nonfood crops: plants with oils or waxes of unusual composition, hence specialized industrial applications, or plants with unusual chemical constituents useful in medicine.

In addition, crop physiologists are actively investigating crop structure in relation to yield, so that plant breeders can breed for specified components of yield. Dwarf cereals, on which the "green revolution" (see p. 243) is founded, produce less straw and therefore a higher percentage of the total dry matter of the plant is contained in the grain. Okra-leaf cotton has deeply divided leaves with narrow leaf-lobes so that self-shading is reduced and leaves on lower parts of the plant can photosynthesize more efficiently, hence increasing the yield. Leafless peas carry out sufficient photosynthesis in stems, tendrils

This Egyptian fresco from one of the tombs at Thebes is a vivid illustration of the ancient roots of modern agriculture. It shows threshing, winnowing and porterage of grain by gangs of laborers some 3,500 years ago.

and pods to fill the seeds, so dry matter need no longer be expended in producing economically unnecessary leaves and yield may again thereby be increased. Insofar as domestication of plants involves establishment of genetic changes differentiating domesticated and wild populations, domestication of plants is evidently very much a continuing process.

The Origins of Modern Crops

Although the number of plant species domesticated by Man, and the families to which they belong, are small in comparison with the total number of species and families of angiosperms, the diversity of products which they yield is impressive. Domesticated plants supply carbohydrate, protein, edible fats and oils, vitamins and minerals, so Man can obtain a nutritionally complete diet from his domesticated plants alone. Plants also supply most other necessities of daily life (shelter, containers, clothes, medicines), as well as luxuries (spices, perfumes, stimulants, narcotics and hallucinogens). Other plants have been domesticated for purely aesthetic reasons; for example, most garden ornamentals. Domestication of plants is a process which began some 10,000 years ago with domestication of wheat and barley in the Middle East, extended in recent times to the development of industrial crops such as plantation rubber and the oil palm, and will continue in the future as new uses are found for wild plants. For example, attempts are now being made to domesticate those Mexican wild species of *Dioscorea* which produce steroidal saponins (used as starting materials for synthesis of compounds such as cortisone and the hormones used in oral contraceptives).

It is useful to distinguish between cultivation and domestication of plants. Cultivation involves deliberate sowing or other management of the plants by Man, but the plants cultivated do not necessarily differ genetically from wild populations of the same species. The first plantations of *Hevea* rubber and quinine in the Far East were established from seed collected in the wild in South America. At this stage in their history these crops were therefore cultivated but not yet domesticated. A domesticated plant is one which has been changed genetically, usually through conscious or unconscious human selection, so that it differs morphologically from its wild relatives and frequently is dependent on Man for its survival. Seed crops which have lost their dispersal mechanisms, eg legumes with indehiscent pods or cereals with non-brittle rachises, are examples of domesticated plants. Domesticated plants are thus necessarily cultivated plants, but cultivated plants may or may not be domesticated.

Some of the most striking changes which occur under domestication involve changes in qualitative characters such as loss of seed-dispersal mechanisms, development of new color forms of fruits, seeds or flowers, and loss of protective mechanisms such as thorns. These characters are mostly under fairly simple genetic control and could thus become established rapidly in cultivated populations, although the actual rate of establishment would depend also on the breeding system of the crop and the intensity of selection for the new variant character. Quantitative changes, notably in size and other components of yield, also occur under domestication. These characters are usually affected by numerous genes with individually small effects (polygenes), hence show a gradual response to selection under domestication, but a response which continues for a long time.

Ecological and Cultural Background to Early Plant Domestication

Recent studies of hunting and gathering societies, such as the Bushmen, suggest that they spend no more of their time procuring food than do farming peoples, and that hunters and gatherers may actually have more leisure-time than agriculturalists. It is therefore necessary to explain not only where and when but also why Man started to domesticate plants. No final answers to these questions can yet be given, but various theories have been proposed, and some evidence is now becoming available against which to test them.

One of the most stimulating (and, according to his critics, least testable) of these theories was put forward by the American geographer Carl Sauer. He suggested that agriculture began among people who already lived a settled existence (thus being able to give their crops year-round attention) and who already had an assured food supply; since if food were short, people would be unlikely to risk experiments with new methods of subsistence. Sauer further considered that the concept of vegetative propagation, ie planting roots, tubers or pieces of plant, was easier for primitive people to grasp than the idea of sowing seed to obtain a crop. Agriculture based on vegetatively propagated root crops should have preceded seed agriculture. Root crops in general contain abundant carbohydrate but little protein or fat, which suggested to Sauer that root-crop cultivation began among people who were getting their protein and fat from other sources, probably fishing. Sauer also considered that forests would be easier to clear than grasslands, since many trees can be killed by ring-barking and burned off once they are dead, whereas the fibrous mat of roots developed beneath savanna grassland can be broken up only with specialized tools. According to Sauer, therefore, plant domestication began with cultivation of root crops by sedentary fishermen in the forested tropics.

Subsequent workers have argued against or added to various aspects of Sauer's theories. The fact that a seed will germinate to produce a plant was probably known to most hunters and gatherers and does not demonstrate that vegetatively propagated crops were domesticated earlier than seed-propagated crops. If agriculture did begin in the ecological environment suggested by Sauer, then the first domesticated plants may not have been food plants at all but "technological" plants such as cotton (for manufacture of fishing nets and lines) and gourds (as fish-net floats and containers), both of which are propagated by seed.

Man's activities, particularly once he had ceased to be nomadic and had adopted a sedentary mode of life, resulted in a disturbance of his immediate environment. Certain plants, notably those with weedy tendencies, are particularly well adapted to living in the disturbed conditions found along trackways, around living sites and, above all, on rubbish heaps. "Camp-following weeds" of this sort probably included squashes and pumpkins, amaranths, hemp and sunflowers. The "rubbish heap theory" of the origin of plant cultivation suggests that useful and exploited weeds, continuously available around human habitations, would have been natural subjects for early experiments in cultivation and would subsequently develop into the first domesticates. Prehistoric rubbish heaps would also provide an environment where discarded plant material brought back to the settlement might continue to grow. Any form of selection practiced while gathering the wild material, such as selection for large seeds, early maturity or retention of seed on the plant after maturity, would be reflected in the characteristics of the population developing around the settlement. This would initiate the process of genetic divergence between the natural wild populations and the populations associated with Man, which would ultimately result in domestication.

A strikingly different point of view from Sauer's on the ecological environment of early domestication was put forward by the British prehistorian V. Gordon Childe. He considered that plant cultivation began in semi-arid upland areas, under conditions of increasing aridity following the end of the last Ice Age. He thought that this progressive desiccation would bring Man, the larger animals and the less drought-resistant plants into increasingly intimate contact in the shrinking moist areas, which would prompt experiments by Man on ways of managing and increasing the numbers of useful plants and animals in his immediate environment (the "oasis theory" of the origin of agriculture). Childe's theories were based on his detailed knowledge of the prehistoric Near East. Subsequent work has produced little evidence for any marked post-Pleistocene desiccation in the Near East, and it has been

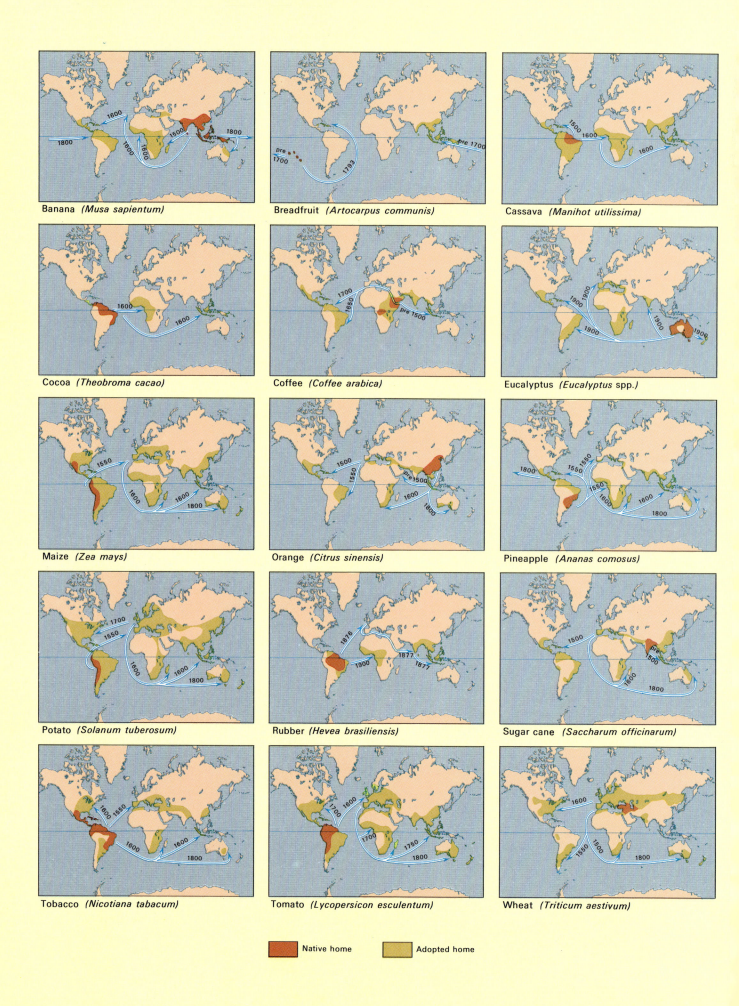

Banana *(Musa sapientum)*

Breadfruit *(Artocarpus communis)*

Cassava *(Manihot utilissima)*

Cocoa *(Theobroma cacao)*

Coffee *(Coffee arabica)*

Eucalyptus *(Eucalyptus spp.)*

Maize *(Zea mays)*

Orange *(Citrus sinensis)*

Pineapple *(Ananas comosus)*

Potato *(Solanum tuberosum)*

Rubber *(Hevea brasiliensis)*

Sugar cane *(Saccharum officinarum)*

Tobacco *(Nicotiana tabacum)*

Tomato *(Lycopersicon esculentum)*

Wheat *(Triticum aestivum)*

Native home   Adopted home

Series of maps showing the countries of origin (red) and routes of distribution (arrows) to new areas of cultivation (brown) of some of the major crop plants. In most cases the dates given are approximate; some are also conjectural, and the subject of speculation and controversy (notably sugar cane).

suggested that population pressure, not desiccation, may have been the main stimulant to experiments in cultivating plants. In present-day hunting and gathering societies, population growth stops below a level at which the food supply is depleted, so that human numbers and the carrying capacity of the environment remain in equilibrium. However, this equilibrium is upset by migration of one group into the area of another. Post-Pleistocene changes in sea level, following melting of the continental ice sheets, could have displaced coastal fishing populations and created conditions inland in which food supplies had to be increased. According to these theories, plant domestication began in semi-arid environments with an open plant cover which would need little clearing for cultivation, and was based on locally available and previously exploited plants, mainly seed-propagated annual grasses and legumes.

To assess the merits and demerits of these rival theories on the origins of plant domestication, evidence from both botany and archaeology must be used.

Botanical Criteria for Determining the Place of Domestication of a Crop

The first and most obvious guide to where a crop was domesticated comes from its present distribution. However, the area of many crop plants has been greatly extended by Man and provides only a very rough guide to their place of domestication. Wild plants are often more restricted in area, so if the wild ancestors of a particular crop can be determined, their distribution may pinpoint more precisely just where the crop was domesticated. *Nicotiana tabacum*, source of most smoking tobacco, is not known in the wild but is cultivated throughout the tropics and subtropics. Distribution of the crop itself thus merely suggests that it could not have been domesticated in temperate latitudes. Cytogenetic studies have shown that *N. tabacum* is a tetraploid derived from hybridization between wild *N. sylvestris* and one of four wild species in the section Tomentosae, followed by chromosome doubling (see Speciation, p. 105). These wild species are all confined to South America, and their present distributions suggest that *N. tabacum* originated, and was domesticated, in the eastern foothills of the South American Andes.

This argument assumes, however, that the distribution of the wild relatives of a crop has remained essentially unchanged since that crop was domesticated, and this assumption is not always justified. The closest wild relative, and immediate wild ancestor, of the domesticated tomato (*Lycopersicon esculentum*) is the cherry tomato (var. *cerasiforme*). This has a pantropical distribution as a weed of disturbed ground. Much of this wide range results from spread of the cherry tomato, often by Man, in historic times, but it has proved impossible to distinguish these areas of recent introduction from the original, pre-agricultural range of the cherry tomato. All other wild species in the genus are confined to western South America but, for various reasons, it is unlikely that the tomato was domesticated there. It seems more probable that tomatoes were domesticated in Mexico, following a northward spread of the weedy cherry tomato.

One reason for concluding that the tomato was domesticated in Mexico, not western South America, is that Mexican tomatoes vary much more in size, shape and color than those of South America. In the early decades of this century, the Russian botanist N. I. Vavilov observed, while engaged on his extensive collecting and classification of cultivated plants, that they were not uniformly variable throughout their range. Instead, they usually showed greatest variability in one particular part of that range, which Vavilov called the center of diversity. He suggested that, in many crops, the center of diversity corresponded to the center of origin, ie the area in which that crop had been domesticated (see Centers of Origin, Diversity and Variability, p. 220). In recent years, Vavilov's generalizations have been criticized, particularly on the grounds that various factors, of which the period under domestication is only one, may produce morphological diversity. Hybridization, ec-

ological diversity and, in recent times, the activities of plant breeders also affect variability. Nevertheless, if modern cultivars are ignored and appropriate allowances made for the effects of other factors, then the distribution of variability within a particular crop may provide a useful guide to the area in which it was domesticated.

Vavilov found that the centers of diversity for different crops often coincided, and this enabled him to propose eight primary centers of domestication. He considered that agriculture had developed independently in each of these. Among Vavilov's primary centers are lowland areas of tropical forest, suggested by Sauer as early centers of agricultural development, and semi-arid upland areas, as favored by Childe and his followers.

Subsequent work has tended to confirm that centers of domestication for different crops often coincide, ie that crops were domesticated as agricultural complexes rather than as isolated curiosities. However, there is much less agreement on the extent to which domestication in Vavilov's eight centers was truly independent. Even though different crops were domesticated in the different centers, the idea of cultivating plants could have spread from one area to another. Determining the relative age of plant domestication in different parts of the world, and the influences and interactions between different regions, is the task of archaeology.

Primitive and modern varieties of wheat. The primitive domesticates are: (a) emmer (*Triticum dicoccum*), once grown in the Near East, Africa and Europe; (b) einkorn (*T. monococcum*), once grown in Turkey and Europe; and (c) *T. timopheevi*. Principal modern wheats are (d) bread wheat (*T. aestivum*) and (e) macaroni or durum wheat (*T. durum*).

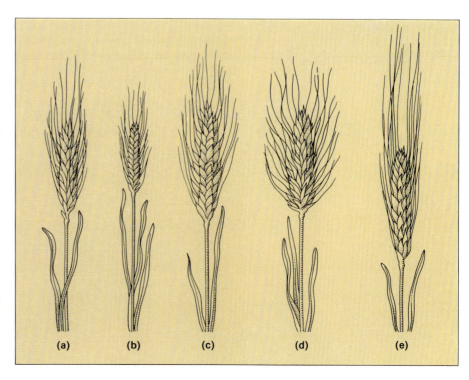

(a)   (b)   (c)   (d)   (e)

Archaeological Evidence on Early Plant Domestication

In an archaeological site, different occupation levels show up as bands in the soil profile, with the oldest occupation at the bottom. If plant and animal remains associated with the various occupation levels are recovered, then it may be possible to determine whether the people living at that site at a particular time were hunters and gatherers or farmers. Archaeological interest in, and techniques for, recovering prehistoric plant and animal remains have advanced considerably in recent years. Soil and other debris removed during the course of the excavation is now routinely screened to recover large plant and bone fragments. Material passing through the finest mesh may then be subjected to flotation in liquids, which enables the smaller organic materials, such as fish bones, small seeds, etc, to be separated from inorganic soil particles by their different specific gravities. In addition, samples may be taken for analysis for microscopic plant remains such as pollen grains or the silica bodies left after decay of grass leaves.

The chances of organic material being preserved in a particular site depend on local environmental conditions. In semi-arid regions and in cave sites out of reach of percolating groundwater, plant material may be preserved by desiccation. Another common form of preservation is by carbonization. Overheating near a hearth, in an oven, or as a result of a house fire may convert organic material to carbon, which is resistant to decay, while preserving many details of shape and surface sculpturing of the carbonized material. Anaerobic conditions also inhibit decay, and some amazingly well-preserved material has been recovered from

sites that are waterlogged.

Conditions of preservation thus vary from site to site. Plant material is most likely to be recovered from sites in dry areas, and least likely to be recovered from sites in hot humid areas such as the lowland tropical forests. In addition, some types of plant material are more likely to be preserved than others. Pollen grains have a tough outer covering which is particularly resistant to decay. Hard-coated seeds and wood fragments survive longer than soft plant parts such as leaves and flowers. Vegetative storage organs such as potato and manioc tubers decay rapidly and seedless fruits like bananas and breadfruit also leave little trace in the

The early Iron Age tomb caves in Mali, Africa, once inhabited by the "Teller" people, provide evidence of the early domestication of plants. The structures visible are shelters and millet granaries.

archaeological record. Despite the utmost care in the recovery of archaeological plant materials, the record will, therefore, inevitably be biased, and this must be borne in mind when assessing archaeological evidence on plant domestication.

Macroscopic remains are the most useful in showing whether a particular group of people, living at a particular time, possessed domesticated plants or not. When such specimens are not available, evidence on plant domestication is more indirect. If the post-Pleistocene vegetational history is well known (see The Quaternary Period, p. 80), the pollen record may suggest when agriculture reached an area. In northwest Europe many different sites have shown a similar pattern of changes around 3000 BC. At about this time the total tree pollen decreases, there is a corresponding temporary increase in pollen of grasses, heaths and ribwort plantain (*Plantago lanceolata*), and there is a good deal of charcoal. This suggests that the native climax forest was cut and burned, cereals were cultivated for a short time, producing a local increase in grass and weed pollen, then the fields were abandoned. Cereals produce abundant pollen, so cereal cultivation is usually detectable in the pollen record, but many vegetatively propagated root crops seldom flower, and their flowers produce limited quantities of sticky pollen

This map shows the areas where the earliest domestication of plants and cattle is thought to have taken place.

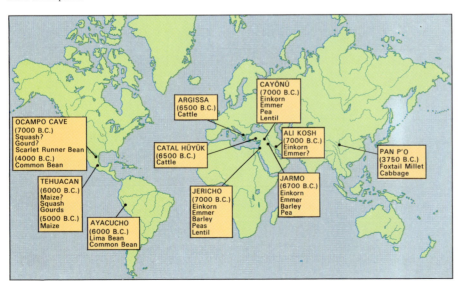

OCAMPO CAVE
(7000 B.C.)
Squash?
Gourd?
Scarlet Runner Bean
(4000 B.C.)
Common Bean

TEHUACAN
(6000 B.C.)
Maize?
Squash
Gourds
(5000 B.C.)
Maize

AYACUCHO
(6000 B.C.)
Lima Bean
Common Bean

ARGISSA
(6500 B.C.)
Cattle

CATAL HÜYÜK
(6500 B.C.)
Cattle

JERICHO
(7000 B.C.)
Einkorn
Emmer
Barley
Peas
Lentil

CAYÖNÜ
(7000 B.C.)
Einkorn
Emmer
Pea
Lentil

ALI KOSH
(7000 B.C.)
Einkorn
Emmer?

JARMO
(6700 B.C.)
Einkorn
Emmer
Barley
Pea

PAN P'O
(3750 B.C.)
Foxtail Millet
Cabbage

which is not wind-borne. The pollen record, like the archaeobotanical record in general, is thus biased towards particular types of crops.

Other indirect evidence relevant to plant domestication may come from prehistoric art. Plants may be represented in paintings on cave walls, house murals or pottery, in sculpture, or in carvings or models. Sometimes the representations are extremely realistic, as in some of the pots of the Mochica culture of coastal Peru, which not only provide some of the best evidence for prehistoric domesticated potatoes, but even suggest that the elaborate freeze-drying process employed today by the Indians at high altitudes in the Andes was already invented by AD 700. Other representations are so stylized that there is considerable argument about what plant was actually intended, for example the so-called "pineapples" identified in murals from Pompeii, in Egyptian pottery and in carvings from the Assyrian city of Nineveh.

Specialized tools may sometimes indicate plant domestication. Archaeological evidence of ploughs or ploughing implies cultivation (often cereal cultivation) even if no plant remains can be recovered. Stone hoes (which may imply cultivation) are sometimes distinguishable from stone axe blades (which do not imply cultivation) by their different wear marks. Flint blades used in sickles for reaping cereals acquire a characteristic sheen from silica in the grass stems, though it is not possible to determine whether the cereals reaped were domesticated or wild.

Long archaeological sequences covering the transition from hunting and gathering to agriculture are available for two of Vavilov's primary centers of domestication. In the Middle East, sites yielding early evidence of plant domestication occur in the foothills of the mountains of southeast Turky and Iran. In parts of this area, stands of wild cereals are so dense that it has been calculated that one family working over a three-week harvest period could reap more grain than they would consume in a year, and thus would have no incentive to turn to plant cultivation. In support of this, the sites producing the earliest domesticated plants are in marginal areas where there are no dense stands of wild cereals. The crop seeds were apparently sown in areas of locally high water table, in one instance so close to a swamp that seeds of sedges (*Scirpus* spp.) were mixed with the harvested grain. The first unequivocal remains of domesticated plants date to about 7000 BC, but for many centuries after that domesticated plants constituted less than half the total seeds recovered, the remainder being wild grasses and legumes. The shift to plant cultivation was probably so slow because rainfall was marginal for successful farming. High-yielding cultivated cereals would probably be lost to drought three years in every five, whereas gathered wild plants would produce low but reliable yields even in unfavorable

years. Only after the development of irrigation freed plant cultivation from dependence on rainfall did the proportion of domesticated plants increase dramatically, but irrigation in turn led to problems of soil salinity, resulting in much agricultural land being abandoned.

The second long sequence comes from the highland valleys of south-central Mexico. The earliest inhabitants of this area were apparently hunters of small game and collectors of wild plant products such as avocado and guava fruits and seeds of the grass *Setaria*. They were semi-nomadic, spending the rainy season in large bands in the mountains and in the dry season splitting up into small hunting groups. Some time around 7000–5000 BC they apparently started cultivating some of the plants they had hitherto gathered (avocados, chili peppers) and also acquired domesticated maize and squash. These crops were planted on the valley floor, but the people remained semi-nomadic, returning to their fields only at harvest time. As in the Middle East, the start of plant cultivation did not lead to a rapid switch in subsistence habits. In Mexico it took 2,000 years for agricultural products to form even 20% of the diet. The phrase "Neolithic (or agricultural) revolution" should therefore be replaced by "Neolithic evolution".

Comparable data are not yet available from the wet tropics and (because of the inherently biased nature of the archaeobotanical record, discussed above) they may never be available. However, recent excavations in the Far East, particularly in Thailand and Indonesia, have produced specimens of rice which may date back to 5000 BC. Rice is another seed-propagated crop, but the vegetatively propagated crops presumed to have been domesticated in this part of the world, such as bananas, yams (*Dioscorea*), coco-yams (*Colocasia*) and sugar cane, would leave little trace archaeologically.

The archaeological data therefore support those who consider that domestication of plants began in semi-arid areas with seed-propagated crops. They do not rule out the possibility that plant domestication, including both seed and vegetatively propagated crops, began at least as early in the lowland tropical forests, but there is no convincing evidence so far that developments in the tropical forests influenced plant domestication in the dry highlands.

The question of how many truly independent beginnings there were to the domestication of plants remains vexed. In widely separated areas, the first domesticated plants appear at more or less the same time: 7000 BC in the Middle East, 5000 BC in Mexico, around 6000 BC in the Andes, perhaps 5000 BC in the Far East. Those who argue that Man is not an inventive animal see this as indicating contact between these areas, such that developments in one triggered off similar experiments with local plant materials in the others. However, if plant domestication were proved to begin at

different times in different parts of the world, it could be argued that that also indicates diffusion from a single early center. Since ideas seldom leave identifiable archaeological traces, the hypothesis of a single origin of the concept of plant domestication can never be disproved. However, botanical evidence of domestication of different species in different centers, and archaeological evidence on differences in artifacts, makes it seem much more likely that Man started experimenting with cultivation and domestication of plants not once but many times, and in several different parts of the world.

B.P.

Crop Improvement and the Green Revolution

Although crop improvement is as old as agriculture itself, the organized, purposeful activity of plant breeding is no more than about 200 years old, and scientific plant breeding started only at the beginning of this century. The evolution of wild plants, primitive crop improvement and scientific plant breeding have much in common: the essential features are generation of genetic variability, genetic recombination, differential reproduction among the new progeny (ie selection for improved adaptation) and isolation of the products. So wild plants evolve new varieties and, eventually, new species (see Speciation, p. 105); similarly, the plant breeder generates new cultivars and, sometimes, new crop species. The main difference lies in the nature of the selection applied. Natural selection for improved reproductive performance is always present, both in the wild and in the plant-breeder's plots; but it is supplemented by the selective pressures of the agricultural environment (very different from those of nature) and by the conscious decisions of the improver to retain one feature and discard another.

Crop improvement, therefore, whether conscious or unconscious, has, by imposing new norms of adaptation, had two very important effects. It has greatly modified the morphology, physiology, chemistry and cytogenetical structure of the plants and, in doing so, has often wrought changes so profound that the products are quite unfitted for survival in the wild state; they have become wholly dependent upon the agricultural environment. Thus the cereals have mostly lost essential dormancy and seed-dispersal mechanisms, many crops have lost essential protective poisons and others (for example, the bananas) have lost their seeds.

Scientific plant breeding, therefore, started from a baseline of crops already much altered from their wild ancestors, but has had the advantages of some scientific understanding of crop evolution in the past and of access to a great variety of scientific techniques by which the current phase of crop evolution may be accelerated. The plant breeder, in fact, is an applied evolutionist.

Crop improvement is as old as cultivation, and the primitive wheat growing shown (**above left**) is already the product of a long process of somewhat casual selection by Man. The ears of wheat, however, vary greatly in size, height and other characters. Modern plant breeders have exploited their understanding of genetics to utilize this variability present in such so-called land-races to produce a diversity of strains with different desirable qualities, but each being kept as uniform as possible for ease of cultivation (**above right**).

There are, maybe, 200 crops (depending upon definition), of which perhaps half have received at least some attention from plant breeders. They cover a great range of taxonomy, morphology, biology and end-use. Despite their diversity, certain common principles emerge as to the breeding of them. These principles cover: objectives, breeding plans (strategy), breeding methods (tactics), and techniques.

Objectives

The two main components of the plant breeder's objectives are yield and quality. There are many routes to high yield, including: improved partition of nutrients, so that the desired seeds, fruits, tubers, fibers, etc, are increased at the expense of such unwanted elements as waste straw, stem, leaf and rind; improved adaptation to the normal seasonal cycle of temperature, day-length, moisture and radiation; improved adaptation to environmental stress, such as cold, drought and wind, so that more plants survive to give better yields in due season; reduced harvest losses, such as grain shedding in cereals; and reduced losses due to pathogens and pests, including viruses, fungi, bacteria, insects, nematodes and birds. The relative importance or potential of these routes varies between crops, but it may be asserted that none can ever safely be ignored.

Quality is best defined simply as adaptation to end-use. It covers an enormous array of characters such as nutritive value of foodstuffs, aptitude for freezing and processing in vegetables, bread-making quality in wheat, malting quality in barley, appearance, smell, taste and storage ability of fruits, fiber length, strength and fineness in cotton, sugar-content of cane and beet, and so on.

Assessment is sometimes subjective (the smell and taste of a strawberry), sometimes objective (chemical or physical measurements related to quality).

Many plant-breeding programs are biased toward either yield or quality, but none can ever wholly ignore one or the other because there is always an economic "trade-off" between the two.

Breeding Plans

Breeding plans constitute the strategy of a program and they depend primarily upon the breeding and reproductive system of the crop. Outbreeders, adapted genetically to cross- (as opposed to self-) fertilization, typically have variable populations in which individuals are highly heterozygous (see Speciation, p. 105) and, if inbred, there is severe inbreeding depression; they all have crossing mechanisms—sex in animals and in a few plants, and a great array of morphological, physiological and genetic techniques to promote crossing in the remainder. By contrast, many plants, including some crops, are tolerant of inbreeding, are often naturally self-pollinated and tend to exist as inbred lines.

These facts dominate breeding strategy: outbreeders *must* (ultimately) be crossed whereas inbreeders *can be* (and often are) selfed. On top of this, life cycle is important. Annuals must obviously be seed-propagated whereas perennials can sometimes (but not always) be made into clones. So the following plans (somewhat simplified) emerge:

Inbred annuals: Cross different pure-line varieties, inbreed and isolate new, pure lines, with new combinations of characters. Examples are wheat, barley, cotton, tomato, tobacco and soya bean.

Outbred annuals: either recombine variable outbred populations in various ways to produce new outbred populations; or inbreed and cross chosen inbred lines to produce "hybrid" varieties with restored heterozygosity. Examples here are sugar beet (H), maize (H), onions (H), rye, many brassicas (H) and mustards, sunflowers (H), broad beans and many ornamentals (H). (H means that "hybrid" varieties are

used and are sometimes important.)

Seed-propagated outbred perennials: as with outbred annuals except that fully developed "hybrid" varieties are lacking. Examples are coconut, oil palm, nutmeg, cacao, ryegrass, some clovers and lucerne (alfalfa).

Clonal perennials: cross heterozygous clones (plants of the same genetic constitution derived by vegetative reproduction from the parents); then select new clones from the variable progeny. Potatoes, cassava, sweet potato, rubber, sugar cane and many fruit trees provide examples.

This list deserves a comment. "Hybrid" varieties and clones, inasmuch as they replace heterogeneous field populations by uniform ones, are inherently at risk to new diseases or other hazards; the great epidemic of Southern leaf blight in the United States was a direct (and predicted) consequence of the "hybrid" maize-breeding strategy. However, despite the risks, the economic attractions of clones and "hybrid" varieties are often great. The agricultural gains are sometimes real, especially in relation to uniformity, but so are the hazards.

Fashion and commercial pressures from those who want to sell expensive seed have not been without influence in the matter, and this influence is now extending even to the inbred annuals. Here, there are substantial efforts to replace pure lines by "hybrids" in, for example, several cereals and (with some success) in tomatoes; however, it may be doubted on genetic grounds whether they offer sufficient agricultural advantage over pure lines to justify the inevitably more expensive seed.

Breeding Methods

Breeding methods are the tactics of a program and are determined once the strategy has been defined. The questions that must always be answered are: what parents are to be used? In what combinations are they to be crossed? How many progeny are to be raised? and what selection patterns are to be followed?

Biometrical genetic principles, based on statistical analysis of breeding data, offer general guidance to the breeder, but most practical decisions have to be made on the basis of plausible assumption or general experience. Fortunately, most characters the breeder has to deal with are inherited either in a (highly predictable) Mendelian manner or show a substantial "additive" component—which means, roughly, that "like breeds like". So it nearly always makes sense to concentrate on already good varieties as the principal parents (but with a few intuitively selected "fancies" included), to bring in exotic parents for special characters (eg disease resistance), to raise as many progeny as resources permit—recalling that selection is often, inevitably, rather inefficient—and to put one's faith on trials results rather than on appearance.

Within this framework, the practice is to generate several hundred different inbred lines, populations or clones from any one cross (depending on the breeding system) and reduce these by selection over the years to the few promising survivors which enter commercial trials. Since the breeder will be generating a continuous stream of new crosses, he will have several thousands or even tens of thousands of different entities on hand in any one season. For annuals it normally takes six to eight years from making the cross to identifying a promising line, but official/commercial testing and multiplication may double this time-scale. Long-lived perennials take even longer. Plant breeding is not a quick process.

Disease resistance, mentioned above, deserves some elaboration because it constitutes a substantial part of many breeding

programs and poses special problems. Many resistances are caused by Mendelian major genes. They are often specific to certain strains of the pathogen and are ineffective against others. They come from a variety of sources (established cultivars, primitive cultivars and wild species) and are readily handled by back-crossing. Against some diseases, notably viruses and some soil-borne pathogens, they have been very successful; against others, notably the air-borne fungi such as mildews, rusts and potato blight, they have, in varying degree, failed. The reason for failure is that host and parasite have evolved complementary genetic systems, so that a race-specific resistance evokes virulence in the pathogen by selection of mutants, recombinants or latent strains. The result is what have been aptly called "boom and bust" cycles. The alternative is to look for (or construct by breeding) race-nonspecific

resistance, sometimes referred to as "field" or "durable" resistance. This seems generally to be controlled by many genes acting together and the evidence so far suggests that it should, indeed, be durable. The principle has only fairly recently become widely accepted but it seems to be both important and generally applicable.

Techniques

A vast amount of effort and ingenuity has been expended upon what might be called the nuts and bolts of plant breeding, the technology of crossing, growing, harvesting, storing and testing. A few examples of major elements in this technology are: growing parents and making crosses reliably in and out of season (glasshouse lighting and pollen storage may be necessary); sowing or planting large numbers of small plots in a short time (this usually demands special equipment, the design of which has attracted much ingenuity); the application of special environmental stresses to permit selection for resistance to them (which may necessitate the use of environment chambers, wind tunnels or chemically treated plots); refined glasshouse or laboratory techniques of infection and/or artificial epidemics (since selection for disease resistance can rarely rely wholly on natural infection); equipment especially adapted to plot work along with portable, sometimes automatic, weighing devices at harvest time; chemical sophistication including, for example, various kinds of chromatography and automatic analyses for quality assay; and computer data storage and analysis (for example, of pedigrees, descriptions and trials results). Such techniques developed very rapidly in the 1960s and 1970s, and the contemporary breeder has access to much more and better information than his predecessor. However, it may sometimes be easier to get the information than to make good use of it.

So far we have been concerned with what might be called "mainstream" plant breeding. Some crops present special problems or opportunities. Thus, many clonal crops raised for produce other than seed often display seed or pollen sterility, which may be merely troublesome (as in potatoes and sugar cane), may present acute difficulties (as in some tropical root crops), or may even make breeding very nearly impossible (as in bananas, in which seedless plants have to be bred by seed to yield a seedless product). As to opportunities, cytogenetic understanding sometimes permits chromosome manipulations of great elegance. Thus, means have been found for reconstructing bread wheats with whole or parts of alien chromosomes in them, and for recombining the genetic

On a plant-breeding station many thousands of plants are grown in trials, and pure inbred strains with desired qualities are developed from selected plants. Here individual plants are isolated in pollen-proof cages or bags to prevent cross-pollination from other plants. **Above left** Wheat growing in outdoor plots. **Left** Rice growing indoors under controlled environmental conditions.

material of chromosomes which would never otherwise pair. Again, new allopolyploids (fertile hybrids with a doubled chromosome number derived from a sterile hybrid) can sometimes be made and one such has recently entered agriculture as a new crop species: this is triticale (*Triticosecale*), a wheat-rye derivative having some of the merits of both parents. Radish-brassica hybrids (*Raphanobrassica*) are moving the same way and others will no doubt follow.

Within crop species, induced autopolyploids (plants with doubled chromosome numbers produced from diploid parents) have sometimes been useful, although not as often as was once hoped; examples are red clovers, some ornamentals and (perhaps) rye. Gene mutations induced by radiation or chemical methods have their uses, though perhaps not so many as was once hoped. However, "mutation breeding" has contributed some valuable dwarfing genes in a number of crops (eg barley and several fruits and ornamentals), disease resistance and biochemical changes. The technique is now an established part of the plant-breeder's repertoire, useful in certain well defined, though rather limited, circumstances. The foregoing are established techniques of proven value. In the future, tissue- and cell-culture methods, allied with good cyto-genetic control, may have a contribution to make.

The basic material, of natural evolution and of plant breeding alike, is genetic variability. This is a time of "genetic erosion"—of loss of variability occasioned by destruction of natural ecosystems by Man in his search for more land, grazing, timber and firewood; and by displacement of primitive agricultures, with their great diversity of crop material, by more "advanced" Euro-American-style systems dependent upon few cultivars. It is for this reason that efforts are being made to conserve such genetic resources (see p. 268) for future breeding.

Economic Factors

Agricultural advance depends upon the genotype-environment-phenotype triangle; the agronomist seeks to improve the environment by improved weed, pest and disease control, by improved fertilizing and so forth; the breeder improves the genotype and the joint result is a better phenotype.

However, agronomy and breeding tend to be interactive rather than simply additive in effect: genotype and environment tend to be mutually adapted. Thus, plant breeding objectives must change in response to the state of the agronomic art, and vice versa. The most striking example is the joint emergence of short-strawed cereals and heavy fertilizing. So it is sometimes impossible to apportion an improvement except to agronomy and breeding jointly.

Agricultural markets are generally in a state of near-perfect competition, so that the innovating farmer who first uses an excellent new variety will do very well out of it but the market will quickly ensure that, as the innovation spreads, either prices fall or supply is restricted. Cost-benefit analysis shows that society as a whole reaps the benefit, the farmers' share being restricted to a fair profit and no more.

The Green Revolution

The green revolution can best be discussed in the general socioeconomic context of plant breeding as summarized above. It grew out of the Rockefeller Foundation's Mexican agricultural program which started shortly after World War II. One of the achievements of this program was the breeding of dwarf wheats using, as a genetic source of dwarfness, the Japanese (Norin) stocks previously identified as valuable in the United States and since very widely used elsewhere. The Mexican wheats were designed for high fertility and

did outstandingly well in Mexico because agriculture in that country was already moving into a transition from peasant to high-fertility farming. Subsequently, they were found to be widely adaptable and did well also in many places in southern and western Asia—so long as they were given the right growing conditions.

An exactly parallel situation developed in rice a little later. The key component was the semidwarf "Dee-geo-woo-gen" from Taiwan; and its descendants, once again, performed outstandingly well, given high fertility.

By the later 1960s the way seemed clear for very substantial yield increases in these, the two most important crops in the world. Unfortunately, the achievement is unlikely to equal expectation because the new varieties are part of a "package deal" of which the other part is high fertility, and because inbred varieties newly grown on a huge scale must tend to generate new disease problems (this is, in fact, happening). The outcome is likely to be substantial and continued success in those agricultures which are highly mechanized, but non-success—perhaps even outright failure—elsewhere.

In practice, the achievement was very uneven. In some areas, farmers were unable or unwilling to adopt the new technology. Elsewhere, farmers successfully adopted the "package" of new variety and enhanced husbandry, yields rose and lower grain prices benefited the mass of consumers. The benefits to some communities were indeed great but they were accompanied by adverse social effects whereby the rich tended to become richer and the poor poorer. So the "green revolution", hailed at first as a great "break-through", partly failed and attracted much adverse socio-economic criticism. In addition, there were (and still are) risks of generating disease epidemics which could, of course, be peculiarly damaging to poor rural communities. In the past decade, however, the emphasis has shifted away from revolutions to a more cautious, pragmatic approach. A chain of internationally supported agricultural research insititutes in the tropics is now breeding diverse varieties of food crops (not only cereals) adapted to varying levels of husbandry. The socio-economic circumstances of the farmers are fully taken into account. We are seeing here the beginning of a fundamental attack on the problem of tropical food supplies, an attack which includes plant breeding as a substantial component of broadly based agricultural research and development.

N.W.S.

Wheat is usually self-fertilizing, so the production of a "hybrid" between two strains has to be achieved artificially. The unripe stamens are removed from the florets in one ear (**far left**), creating what is effectively a female flower. The anthers are then removed from the other ear (**left**) and the pollen is used to pollinate the female flowers in the first ear, thus achieving a cross between the two ears.

Plant Resources of the Future

Consideration of the plant resources of the future may be divided into three parts: the harvesting and use of plants that grow wild; the preservation and improvement of established crop plants; and the development of new crop plants.

The Use of Wild Plants

The harvesting of wild plants seems to have a relatively limited potential for the future, chiefly because so many plants that were once wild are now cultivated (The Origins of Modern Crops, p. 239). The main exception is timber, although even this is currently becoming a cultivated crop in temperate regions, especially in Europe. In temperate North America, timber-cropping still depends to a considerable extent on natural regeneration after cutting, but ever-larger areas are being reseeded or replanted and carefully tended while the trees are growing.

Exploitation of tropical forests, in both the Old World and the New, has been delayed by the fact that so many different kinds of trees commonly grow intermingled. It is usually uneconomical to sort out the different woods, and only the more valuable kinds can profitably be transported the long distances to where they might be used. Even in the production of pulp for paper, mixed woods cannot be used economically, since the different kinds must be treated differently.

In tropical America, forests are now being clear-cut for charcoal and to prepare the land for agriculture. Clear-cutting is also under way in Pacific tropical regions such as Borneo, although only the most valuable species (notably members of the Dipterocarpaceae) are used. What had seemed to be limitless resources are proving to be distinctly finite, and at present rates nearly all the remaining tropical forest will be destroyed within a few decades.

Because of increased population pressure, the traditional cyclic slash-and-burn agriculture of forested tropical regions is breaking down. Larger areas are being cleared, the period of use is being prolonged, and the recovery time curtailed. Much of the land now being cleared will not return to forest after its short-term agricultural potential has been exploited. Any traveler in the tropics can see former forest land in the process of being eroded away. It will not produce either forest or agricultural crops again in the foreseeable future.

Established Crop Plants

The common crops in temperate regions have nearly all been greatly improved during the past century by breeding and selection for increased yield, resistance to disease and tolerance of environmental adversities such

as low temperature and drought. One might at first suppose that little remains to be done to maintain and improve them, but this is not the case.

Breeders must be constantly at work to keep one step ahead of disaster. By planting large fields of disease-resistant crops, we impose a powerful selection on their fungal parasites: it is now becoming an anticipated pattern that cultivars initially disease-resistant will within a few years be attacked by new strains of old kinds of fungi. Furthermore, newly produced cultivars may prove to be unexpectedly susceptible to some of the currently less common pathogens.

Recent events in the culture of maize provide a case in point. For several decades most of the maize grown in the United States has belonged to a limited number of commercial hybrids. Most of these cultivars had one parent in common, a strain called "Texas male sterile", popular in breeding work because it produces virtually no pollen and thus does not have to be detasseled (emasculated) when used as the female parent of a hybrid. The susceptibility of "Texas male sterile" to the usually minor disease, southern leaf blight (caused by *Helminthosporium maydis*), was not recognized until 1970. Unusually moist growing conditions in that year favored explosive growth of the fungus, threatening the entire crop. It was only the timely onset of drier weather that restricted the crop loss to as little as 15%. Other strains have now been substituted for "Texas male sterile" in the production of hybrid corn in the United States, but these may have their own unsuspected weaknesses.

Maize is one of the world's most important crops, providing food for Man and animals and raw materials for industry, but the protein of maize grains is notably deficient in the amino acids lysine and tryptophan. A few strains of maize produce much more than the usual amount, but there are problems. High-

In many tropical countries covered with rain forest the local populations clear small areas of forest for cultivation of staple crops, moving on to new areas within a few years as the soil becomes exhausted of nutrients. Shown here are cultivated and fallow sweet potato gardens surrounded by forests near Mount Hagen in New Guinea. Sweet potatoes (*Ipomoea batatas*) provide a low-protein staple food in the tropics and their more extensive use in the future depends upon ecologically acceptable developments of the traditional methods of shifting agriculture, which generally tend to preserve the ecosystem from long-term or permanent damage.

lysine maize has a relatively soft endosperm, so that the grain does not store well, and its yield is about 10% less than comparable low-lysine strains. Intensive efforts are now under way to produce high-lysine maize without the attendant disadvantages.

Only within the past two or three decades have scientific plant breeders turned their attention to the tropics. As noted previously, the resulting "green revolution" has not been an unqualified success. One of the most serious problems associated with the widespread use of a few high-yielding cultivars of any crop is the depletion of the gene pool for future breeding programs, and increased efforts must be made in future to conserve these vital genetic resources (see p. 268).

New Crop Plants

It is, of course, impossible to predict exactly the species of plants that will, in the future, become important crops—unless one has a reliable crystal ball. However, by considering the trends of current research and by extrapolation from the requirements of today's world, we can make some reasonably informed guesses as to the crop plants that will figure largely in the decades to come.

One of the essential raw materials of civilization is pulp for paper. Forests in temperate and cool regions are being cut for

A plantation of fast-growing *Gmelina arborea* trees established in Brazil. The trees are only $2\frac{1}{2}$ years old and much research is at present being undertaken with such fast-growing trees in the hope of reversing the continued decline of natural forests due to exploitation by Man. These trees, which are native to India and Southeast Asia, will be felled at an age of 7–10 years and the wood used to produce pulp for turning into paper. The plantations are actually coppiced, new vigorous shoots growing from the stumps, which will grow into new trees.

In other areas, experiments are being performed in an effort to conserve at least some of the natural habitat of tropical forests, yet still be able to crop the trees. In Uganda, for instance, fast-growing *Maesopsis eminii* trees are being planted within natural forests that are only partly felled.

pulp much faster than they are reproducing, and the great mixture of species in tropical forests is not suitable for pulp operations; the supply is beginning to run out.

One possible new source of pulp is fast-growing trees that can be grown in tropical countries, on land otherwise not well suited to agriculture. *Gmelina arborea* (family Verbenaceae), from tropical Asia, is one of the most promising. A single very large plantation of *Gmelina* has recently been established in Brazil and should soon come into production. Another potential pulp plant is kenaf (*Hibiscus cannabinus*), an African annual herb of the same family (Malvaceae) as cotton. It can be grown in closely packed stands in many of the same places and in much the same way as cotton. In the more northern part of its potential range, such as in the southeastern United States, it does well vegetatively but does not set seed. After the leaves have been killed by frost, the stems can be harvested by mechanical means.

Ramie (*Boehmeria nivea*, family Urticaceae) is an East Asian species with very long, tough bast fibers. It has been cultivated for thousands of years in China and Japan, so that it can scarcely be considered a new crop; but if the technical problems of extraction and preparation of the fibers can be met it could become much more important than it has been in the past. The fibers are said to have eight times the tensile strength of cotton and three times the durability in fabrics, and they have other desirable qualities as well.

Unfortunately, the fiber bundles are not easily extracted from the stem, and a tenacious coating of gum must be removed before they can be used. Efforts are being made to find more effective ways to extract the fibers and especially to remove the gum without weakening them.

Several kinds of plants that are not now extensively cultivated store long-chain fatty acids in their seeds. These fatty acids, forming a sort of liquid wax, are resistant to degradation at high temperatures, and can be treated chemically to produce something comparable to sperm-whale oil. Among the plants that have recently attracted attention for their production of such long-chain fatty acids are species of *Simmondsia* (family Simmondsiaceae), *Crambe* (family Cruciferae) and *Limnanthes* (family Limnanthaceae).

Simmondsia, popularly called jojoba, is a desert shrub of the southwestern United States. It has the advantage of growing well in an otherwise non-agricultural habitat— and the reciprocal disadvantage of low yield per hectare. It is now being tested in experimental plantations in Israel. Erucic acid is produced by many members of the Cruciferae, or mustard family. With appropriate chemical treatment it can be used for the same purposes as jojoba oil. In *Crambe abyssinica* as much as 50% of the seed oil is erucic acid. The crop potentiality of erucic acid plants is being explored.

Some species of *Limnanthes* native to the western United States have seed oil similar to that of *Simmondsia* and may have greater potential as crop plants. They are annual and suitable for mass cultivation. It is believed that a *Limnanthes* crop that will yield as much as 1,000kg per hectare (900lb per acre) can be fairly quickly developed.

Finally, the latex of several species of *Euphorbia* from various parts of the world contains large amounts of hydrocarbons similar to those of petroleum. Since the supply of petroleum is finite and not replaceable, some of these species may have a bright agricultural future.

A.C.

Man's Impact on Particular Vegetation-types

As we saw earlier in Chapter Five some zonal and azonal vegetation-types, such as scrub and heath respectively, have been greatly extended as a result of Man's activities, while there is a strong suspicion that many of the world's grasslands may have originated in this manner. It is not possible in an account of this sort to document all the ways in which the world's natural vegetation has been affected by Man. Consequently, we shall consider briefly here only three vegetation-types which have experienced human influence in different ways and to rather different degrees: *tundra*, which has most recently been affected by Man as a rather casual consequence of his other activities; *forest*, which has been cleared for grazing since prehistoric times and which now, in its tropical strongholds, is facing final decimation as a result of logging and agricultural programs; *grassland*, much of which has provided the agricultural wealth of mankind as a result of its grazing value for his domestic animals, and for which there is probably the greatest amount of information on active management practices and their effect.

D.M.M.

Tundra

Apart from the domestication and herding of reindeer in the tundra of Eurasia, Man has had only limited, local effects upon the vegetation and plants of the tundra. In the Antarctic, Man's relatively recent presence has led to the accidental introduction of many alien species across natural barriers to migration, especially to the oceanic islands, probably ever since his first explorations. For example, the native flora of the subantarctic island of South Georgia comprises 26 species but an additional 54 alien species have been introduced, mostly from the Northern Hemisphere.

By far the greatest changes in tundra regions are arising as a result of the extraction of minerals and fossil fuels, such as oil from under the North Slope of Alaska. Wherever the vegetation cover is broken by tracked vehicles to expose the peat beneath, the reflective properties of the ground change completely. More solar radiation is absorbed, leading to the thawing of a much thicker active layer above the permafrost. Persistent vehicle passage over the same route furthers this process and erosion spreads. Roads are now commonly constructed of thick layers of gravel, deposited on top of the vegetation, using local materials. A problem for the future may well arise if smelters are built to process minerals, since mosses and lichens which are highly sensitive to atmospheric pollution (see p. 253), make up a far higher proportion of the vegetation than in temperate regions.

Trampling by tourists, which is especially common in alpine tundra, introduces a dimension of instability into habitats that are otherwise relatively stable, so that in popular mountain areas, such as parts of the Cairngorms in Scotland, the vegetation has been eroded away to reveal the bedrock beneath. Tourism is increasing in all tundra areas, both polar and alpine, and even in the Antarctic, where there has been a great increase in the number of visits from cruise ships in recent years. Certain antarctic moss communities are vulnerable, especially the moss peat banks (see Antarctic Dominion, p. 219), which grow at rates of only 1–3mm (0.04–0.1in) per year. It is only as a detailed understanding of the life cycles and rates of growth of tundra plants is developed, together with an understanding of rates of change and recovery from disturbance, that management plans can be evolved to ensure the preservation of tundra ecosystems alongside ever increasing human activity.

N.J.C.

Destruction of evergreen tropical forest in Brazil for the construction of Road BR158, one of the principal highways being driven through the Amazon Basin. Since the photograph was taken in 1969, all the forest in the vicinity has been largely destroyed for ranch-type agriculture and squatter farming.

Forests

Tropical rain forest, once believed to be inexhaustible, is the main source of hardwood timber in the world. Since the 1950s it has been felled faster than it grows, so although forests are potentially a renewable natural resource, they are at present being eroded. In 1981 the rate of destruction was estimated to be 50 hectares (124 acres) per minute. Southeast Asia is the main present-day resource, and will at present rates be largely logged out before the end of the

The effect of tracked vehicles upon tundra. Marks persist upon the tundra even when vehicles passed only in winter, on top of hard-packed snow, in this instance about 10 years before.

century. Large-scale exploitation has commenced in Latin America. The main resource in Africa has gone. The quantity of wood (roundwood, fuelwood and charcoal) felled in 1973 was 2,500 million cubic metres (87,500 million cubic feet), having increased 17% since 1962. Roundwood was 69% of the total (45% of this was coniferous). The

production in 1973 of hardwood in South America, Africa and Asia was 204 million cubic metres (7,200 million cubic feet), 300 million cubic metres (10,600 million cubic feet), 552 million cubic metres (19,500 million cubic feet), respectively. In Southeast Asia, Indonesia produced 134,000 cubic metres (4,630,000 cubic feet) and Malaysia 32,000 cubic metres (1,130,000 cubic feet), representing 58% and 180% increases over 1962, respectively. These figures demonstrate how, in the world as a whole, wood for fuel is still a major usage (31% of the total), and how there is a huge and increasing demand for tropical hardwoods. For the developing countries where tropical rain forest occurs, they are the source of a major export commodity and there is also pressure to clear land for cash crops (oil palm, rubber, etc) as well as increasing pressure for agricultural land to feed the growing populations.

The destruction of tropical rain forests will soon begin to cause extinction of species of animals and then of plants. In the longer view, Man is depleting the diversity of life of the planet and the reservoir of species his descendants may need to develop and cultivate as a source of complex molecules when fossil fuels have become exhausted. Other forests are also under pressure. The Boreal coniferous forests supply most of the North Temperate Zone with construction timbers, and the most accessible stands are nearing exhaustion. These are also the world's main source of paper pulp, for which the long fibers of conifers are superior. Virgin conifer forests still exist, but restrictions on cutting are now applied in Scandinavia because, at the high latitudes where these forests occur, growth is slow and the forest could disappear unless present trends are curtailed.

As noted earlier (see Summer Deciduous Forests, p. 164), the agriculturally productive grasslands of western Europe and other regions result, to a considerable extent, from the destruction of the deciduous forest which they formerly supported. The value of the disappearing forests, both for their timber and their effect on the general environment, has been well known for a number of years, especially in regions with the longest history of severe exploitation, so that programs to reintroduce trees (afforestation) have frequently been undertaken and are worthy of some consideration here.

T.C.W.

Afforestation

Afforestation is the creation of new areas of forest or woodland on sites that have been without tree cover for a long period, often hundreds of years or even since prehistoric times. Such sites may have been used for agriculture and allowed to deteriorate or they may have changed because of climatic factors, as in the case of sand-dune invasion, or as a result of industrial dereliction such as slag heaps in a mining area. In nearly all cases, afforestation is in fact restoration of tree cover after a long interval and in conditions totally different from those when the forest was there before.

The purpose of afforestation is often to satisfy the growing demand for industrial wood but other objectives, such as flood control, shelter, prevention of erosion or the amelioration of the climate (particularly in arid countries) are also important. It is estimated that there are about 100 million hectares (250 million acres) of newly created forest in the world and that this figure will be doubled by the end of the 20th century. Large programs of afforestation have been carried out in many countries; the Tennessee Valley scheme restored the "dust-bowl" to fertility in the 1930s and in the EEC countries it is proposed to afforest 5 million hectares (12 million acres) of "surplus" agricultural land. But it is in the USSR and China that really massive programs of afforestation for timber production and shelter are now taking place.

The techniques required for afforestation often differ considerably from those used in the regeneration of existing forests. The conditions are nearly always harsh, the soil impoverished, exposure to wind or sun severe, and the risk of damage by animals and pests much greater than in the more stable forest ecosystem. In wet uplands the land must be drained and impacted soils must be broken up by some form of cultivation to enable the small tree roots to penetrate. On the arid slopes of hills in hot climates, terracing may be necessary to conserve any rain that may fall and in rocky terrain it is sometimes essential to import soil to enable the newly planted trees to make a start. Application of fertilizers, particularly phosphates, is nearly always necessary.

The species of tree selected for afforestation are not necessarily the same as those of the indigenous forests of the region. Fast-growing species are needed to satisfy industrial needs and they must also be hardy to withstand the rigorous conditions as pioneers. This normally means that conifers will form the bulk of the crop that is planted and these are often exotic species not present in the natural forests of the country. Thus Sitka spruce (*Picea sitchensis*), from North America, is the principal species used in afforesting the wet uplands in Britain. The Monterey pine (*Pinus radiata*) from California is the main species used in New Zealand and Australia and the Cuban pine (*Pinus caribaea*), from the southeastern United States, is extensively used in South Africa.

In some countries where conditions permit, broad-leaved trees such as oak, poplar, eucalyptus, alder, and birch are included in the initial planting; the more rigorous the conditions, however, the less chance there is for the successful establishment of a diverse crop in the original planting. Normally, afforestation must be regarded as the pioneer stage in the restoration of a forest ecosystem which may take 50 or 100 years to re-establish.

Nevertheless, even a monoculture of a pioneer coniferous species brings a great environmental gain to the site. Protection from grazing and disturbance, from wind and erosion, and the conservation of moisture, start the chain of events in the ecological development of the site from a plantation to a forest with its diversity of composition and structure and flora and fauna.

P.F.G.

Grasslands

Although in many parts of the world, Man has cleared the forests by felling and burning in order to provide grazing land for his domestic animals, it is the animals themselves that have maintained the non-forested conditions by their grazing of tree and shrub seedlings which persist in trying to recolonize their natural habitats; the development of scrub and woods on former chalk grasslands in Britain, as a result of the decimation of rabbits by myxomatosis in the late 1950s, provides an obvious indication of this. Furthermore, grasslands, whether natural or the result of human activities such as those just described, have formed the basis of Man's pastoral wealth and consequently have received a great deal of attention during the period of applied scientific studies of vegetation.

Indeed, 1860 saw the initiation of experimentation at Rothamsted in England into long-term productivity and botanical composition of pastures and meadows under various fertilizer treatments, because of the need to obtain maximum productivity from a grassland sward. Since this time, research has been carried out into virtually all aspects of the nature and methods of increasing pasture productivity through detailed studies of the effect of grazing, palatability of different species and varieties, nutrient requirements and deficiency in soils, and general management practice. The basic effect of grazing is to reduce the leaf area of the plants and, in so doing, to alter leaf- and root-growth and levels of stored carbohydrate. But this is by no means the complete effect and a whole series of additional features is altered: the microenvironment of the grassland is changed; trampling and the return of dung and urine to the soil are introduced; the disposal of seeds is modified; and, since the component species vary in terms of their growth rhythms and palatabilities, their interrelationships may be greatly altered.

It is, therefore, impossible to simulate the total effects of grazing by mere defoliation in cutting experiments, but it is important to know what general features are affected. The immediate growth potential is lowered, but,

A scene in the foothills of the Tararua range, New Zealand, showing how the application of fertilizer can increase the productivity of this low grade pasture. The pastures on the right have had fertilizer applied, producing a much lusher and greener pasture that will support more sheep per hectare than the unfertilized pasture on the left.

Overgrazing and poor cultivation of grasslands can aggravate the effects of erosion, as has occurred here on a grand scale in South Dakota, United States, to produce the "Big Badlands".

up to a certain grazing level, the formation of tillers is encouraged. Heavy grazing or severe defoliation reduces the growth of roots and the amount of stored carbohydrate in the perennating organs. Continuous grazing will often change the growth-form from erect tussock grass to a decumbent form. In the alteration of the microenvironment by grazing, several generalizations are possible. Regular grazing increases the penetration of light and, while the total area for photosynthesis is reduced, there is some evidence that this is partly compensated by increased photosynthesis in better light intensities. The depth to which moisture penetrates the soil, the amount of litter, temperature of the soil surface and the degree of erosion of the soil are also affected by grazing and associated trampling.

Selective grazing, where the animals choose the more palatable grasses and herbs, often has a pronounced effect upon grassland composition. Most grasses are palatable when eaten in the early stages of their growth, but several develop a tougher leaf anatomy and hairs or spines along leaf margins and veins as they grow older and

thus become less attractive to the grazing animal. For example, Yorkshire fog (*Holcus lanatus*) and tall fescue (*Festuca arundinacea*) become less palatable as they develop, while mat grass (*Nardus stricta*) is wiry and unattractive from the earliest stages of growth. In cultivated pastures, a concentration of grazing on the tasty species produces a sward which is dominated by coarse grasses and rosette herbs. Tor grass (*Brachypodium pinnatum*) and stemless thistle (*Cirsium acaulon*) are frequently dominant in the chalk grasslands of northwest Europe, while the grasslands in the temperate and subtropical woodland zone of Australia were fomerly dominated by the kangaroo grass (*Themeda australis*) before sheep grazing encouraged the dominance of the bunch spear grass (*Heteropogon contortus*).

Similarly, overgrazing causes three main types of compositional change. Firstly, an increase in unpalatable herbs such as wormwood (*Artemisia absinthium*) is a characteristic sign of overgrazing in steppe and prairie grasslands, and sometimes a change to scrub is caused, eg the formation of creosote bush (*Larrea divaricata*) communities from *Bouteloua* grasslands in the southern prairies of Texas. Secondly, increased levels of nitrogen and phosphorus from dung and urine encourage weeds, especially thistles (*Cirsium* spp., *Carduus* spp.). Thirdly, localized sward

erosion permits the growth of unproductive annual species. Sudden cessation of grazing on heavily grazed pastures, similar to the situation which occurred in Britain after the ravages of myxomatosis on the rabbit population in the late 1950s, also produces marked sward compositional changes. In the chalk downlands, for example, the coarse grasses grew rapidly at the expense of the finer, more palatable grasses, overall specific diversity was reduced and gradual scrub development, in the form of hawthorn (*Crataegus monogyna*) began.

In the high-grade ryegrass pastures and leys, the periodicity of grazing has been studied in detail and changes in the relative percentages of the sward components are well documented. If the ley is grazed continuously throughout the year, white clover (*Trifolium repens*) increases at the expense of ryegrass and other useful grasses. Those pastures that are rested from mid-March to mid-April show an increase in ryegrass, cocksfoot and weed grasses, while white clover declines; and a rest period from mid-April until mid-May encourages the dominance of ryegrass. The maintenance of the best balance of clover, ryegrass, cocksfoot, other useful grasses, and weed species is achieved when the ley is grazed "on and off" from early spring to early autumn. This means that once the grass has grown to a height of 100mm (4in), the stock are turned into the pasture and the concentration of animals is monitored to enable the herbage to be consumed in the minimum time period. The stock are then removed, the grass allowed to grow to 100mm (4in) and the cycle begun again. In some intensive grass producing regions, the "on and off" grazing is alternated with mowing, this management practice forming an intricate four-yearly rotational system which benefits not only the stock, but also the pasture.

A knowledge of the intensity of grazing, the relative palatability of species and detailed methods of pasture management are thus important tools for the increase of primary productivity of grasslands. In a similar manner, an understanding of the nutritional requirements of pastures is also of importance. Of all the nutrients which affect botanical composition, nitrogen and phosphorus are the most important because they are the two most common essential elements that are deficient in virgin soils. In some situations, this deficiency is partly remedied by burning, which leads to the rapid replenishment of the soil by nutrients locked away in dead leaves and litter, and in others, the introduction of legumes, with their symbiotic nitrogen-fixing bacteria, increases nitrogen levels. By far the easiest method of correcting nutrient imbalance and increasing grass productivity, however, is via the addition of organic or inorganic phosphate and nitrate fertilizers. Other nutrient deficiencies are much more subtle and difficult to detect, and a detailed soil analysis is desirable before corrective treatment can be prescribed. The discovery, in 1938, that the reason why many Australian soils were incapable of supporting good pasture was due to copper deficiency, prompted a period of research into the role of minor nutrients in grassland development. Deficiencies in molybdenum, zinc and sulfur in several geographical regions of Australia were later demonstrated and, by the early 1950s, nutrient imbalances were being corrected and low-productive native grasses being replaced by high-productive introduced grasses and legumes. Parallel examples are to be found in grasslands in many of the developed countries and an extensive knowledge of the nature of deficiency and fertilizer treatments is now available to the grassland farmer.

With the vast knowledge of the various factors responsible for affecting grassland productivity, the agriculturalist is thus able to define detailed grazing programs, including precise data on carrying capacity, ie the numbers of stock per unit area that a grassland will support for a given period of time without having adverse effects upon the grass sward. Such details have been available in many regions and for many types of cultivated grasslands for the past 20 years, and secondary production of meat and milk has become a sound, scientifically-based enterprise. More recently, an International Biological Program scheme for the study of primary productivity of semi-natural, un-cultivated grasslands throughout the world has started to publish basic data on all aspects of the subject, with the future aim of increasing both primary and secondary production levels to optimal conditions.

D.W.S.

Pollution

As we have seen in the preceding sections, Man has affected, subtly or otherwise, the vegetation of our green planet for millennia. Frequently these changes take place so gradually, or are accepted as so much a part of Man's environment, that they can pass unnoticed by many people. However, the current large numbers of Man, his tendency to live at an increasing density, especially in large conurbations, and his continuing expansion of industrial activity have forced on to the attention of many the ways in which the natural environment has been interfered with, polluted, by the wastes and other by-products of modern society.

Pollution, the unnatural and usually unfavorable change in natural biological cycles as a direct or indirect result of Man's activities, takes place by altering the chemical and physical composition and the radiation levels of the environment, resulting in changes in the number of organisms and their energy-transfer relationships. Green plants have a key role in pollution by virtue of their importance as the primary fixers of solar energy for utilization by animals and by decomposer organisms. Disruption of the fundamental biochemical processes carried out by plants is reflected elsewhere in nature.

Pollution can be classified according to either the nature of the pollutants (solids, liquids or gases) or the types of habitats affected by pollution—terrestrial or aquatic (freshwater or marine). Plants can be involved in pollution in a number of different ways.

Direct Effects
The presence of substances in toxic concentrations may cause death or decreased vigor of plants growing near emissions from industrial facilities such as iron-sintering plants, aluminum smelters and nuclear reactors. These occurrences are usually restricted to the immediate locality, especially downwind of the source.

Indirect Effects
As a result of atmospheric wind movements, many potentially phytotoxic substances (ie plant poisons) are distributed over great distances from their point of release, leading to a general increase in their concentration on a worldwide scale. This is a problem particularly with gaseous pollutants and radioactive isotopes.

Increase in Population Numbers
Increase in the concentration of nitrates and phosphates, elements essential for plant growth but usually present in limited amounts in nature, can lead to a rapid increase in the numbers of aquatic organisms, which, in turn, put an excessive demand on oxygen supply as they decay.

Food Chains
Certain plants have the ability to accumulate elements and compounds which they themselves do not require for healthy growth and development. These are passed on to animals (including humans) which feed upon them and in turn may concentrate the substances to toxic levels, sometimes with disastrous results. For example, DDT can be concentrated in food chains which start with plankton and end with the death of predatory birds.

Quality of the Environment
Plants produce oxygen in photosynthesis and, although they respire some of this, there is a net increase of oxygen produced which maintains the level in the atmosphere. It has been estimated that at least 70% of atmospheric oxygen is produced by phytoplankton in the oceans and the remainder by terrestrial and freshwater plants. Any serious depletion in the numbers of plants in these habitats will have serious consequences for all other living organisms.

Atmospheric Pollution

Atmospheric pollutants consist of two main components. One is particulate material, solid particles of dust and smoke, the other is gaseous.

Particulate Material

The chief components of particulate material are carbonaceous and tarry deposits, road dust, pollen, unburned materials from fires, and numerous unidentifiable substances from manufacturing industries. There is also release of smoke, dust or spray from chemical and metallurgical processes, quarrying operations, and agricultural practices.

In areas of high dust and smoke, deposition of particles on plant leaves reduces the effective rate of photosynthesis, decreases the heat-exchange capacity of the leaves, and interferes with the normal action of stomatal mechanisms. The effect of smoke and particles on leaves varies from species to species. Short-day plants with photoperiods of less than 12 hours actually benefit from the lowered light intensity (see Developmental Responses to Light, p. 124). Rhubarb (*Rheum* sp.), for example, grows well in industrial areas. For most plants, however, the ability to photosynthesize efficiently is doubly impaired: by the decrease in light intensity and by the coating of particles on their leaves. Examples are the bay laurel (*Laurus nobilis*) and many conifers, including Scots pine (*Pinus sylvestris*), Norway spruce (*Picea abies*) and firs (*Abies* spp.).

Gaseous Pollutants

The major phytotoxic gaseous pollutants are sulfur dioxide, fluorides, oxides of nitrogen, ozone and peroxyacetyl nitrate (PAN). Other serious but localized pollutants are sulfur trioxide, nitric acid, hydrocarbons and hydrochloric acid. Most of these gases are produced by manufacturing industries, exhausts from internal combustion engines and the burning of fossil fuels. In high concentrations all of these compounds kill plants, but toxic concentrations are found mostly in specific localities. Gaseous pollution has varying effects both on individual plants and on plant populations. Lower plants in particular exhibit varying degrees of susceptibility, some plants being so susceptible that they are used to indicate the degree of aerial pollution. The species diversity and distribution of epiphytic lichens, for

Atmospheric pollution from a large soda ash and chemical works in northwestern England. Smoke acts as a pollutant in several ways: it deposits particles of matter such as soot on plant leaves; it contains poisonous chemicals such as sulfur dioxide; and it reduces the light intensity from the sun.

example, are related to varying levels of sulfur dioxide in the atmosphere and a number-scale has been produced to relate the concentration of sulfur dioxide to the incidence of the lichens *Lecanora* and *Parmelia* species in England and Wales.

Cell damage in leaves has been reported with sulfur dioxide concentrations of 0.5ppm (parts per million), but this level is extremely high and localized. Concentration in industrial areas is generally from 0.01 to 0.22ppm, and at these levels the effects of sulfur dioxide on higher plants may include leaf scorch, premature ageing, the yellowing and blanching which are symptomatic of chlorosis, leaf lesions, localized death of tissue (necrosis) and reduced growth. Trees, cereals, fruits and vegetables have all suffered widespread damage from sulfur dioxide pollution. Fumes from a copper-smelting plant at Redding, California, damaged 50 square kilometres (20 square miles) of natural vegetation and orchards. Emissions from the Washoe smelter at Anaconda in Montana destroyed a large area of timber plantation up to 30km (20mi) away. The trees mainly affected were Douglas fir (*Pseudotsuga menziesii* var. *glauca*), lodgepole pine (*Pinus contorta* var. *latifolia*), subalpine fir (*Abies lasiocarpa*) and the timber pine

(*Pinus flexilis*). In Wawa, Ontario, there was an increase in the sulfur content of air and of sulfate in soil and water near an iron-sintering plant, and a marked reduction was recorded in both the number and abundance of epiphytic lichens and bryophytes (liverworts and mosses). The effect of pollution was noted up to 56km (35mi) from the source. Trees and other vegetation downwind of an aluminum smelter on Holy Island, off Anglesey, in North Wales, have suffered severe scorch damage with subsequent death of leaves and branches.

Fluorides are thought to cause the collapse of leaf tissue by acting as a cumulative toxin. Several hundred hectares of western yellow pine (*Pinus ponderosa*) forest near Washington in the United States were destroyed by fluoride pollution.

Petrol and diesel engine exhausts produce a number of phytotoxic gases. Lead is added to fuel as an antiknock agent and about 75% is exhausted into the atmosphere mixed with chlorinated and brominated hydrocarbons. The effect of lead on plants is not clearly understood but it is known to reduce plant size, root growth and leaf pigment content. Experiments applying varying lead concentrations in *Lemna minor*, a small aquatic floating plant, show that low lead concen-

trations induce enlargement of the mito-chondria (bodies within the cell responsible for respiration), and also chloroplast abnormalities while at higher levels crystalline inclusions are observed in the cells. Vehicle exhausts and combustion of fossil fuels also produce ethylene gas, which affects the abscission area of leaves and causes early death and leaf fall. It also influences the action of auxins, the compounds associated with plant growth and development. Ethylene produces petal curl in carnations (*Dianthus* spp.) and dries and discolors orchid sepals.

Photochemical action by ultraviolet light on vehicle exhaust fumes produces the highly toxic peroxyacetyl nitrate (PAN) and the secondary pollutants formaldehyde, ozone, nitric acid and hydrocarbons. This is a particular problem in the large conurbations of the United States. For example, there is a high density of motor vehicles in Los Angeles, where geographical and meteorological factors prevent dispersal of exhausted gases, as a result of which bright sunlight frequently produces this photochemical effect. Lettuce (*Lactuca sativa*), oats (*Avena sativa*), and pinto bean (*Phaseolus vulgaris*) are sensitive to PAN concentrations as low as 20ppb (parts per billion). Tobacco (*Nicotiana tabacum*) and wheat (*Triticum sativum*) are intermediate in reaction to PAN, while cucumber (*Cucumis sativum*) and cotton (*Gossypium hirsutum*) suffered no adverse effects after several hours exposure to 100ppb PAN and are regarded as insensitive. The symptoms of PAN injury are usually glazing, bronzing or bleaching of leaf surfaces. Internal injury produces collapse of the mesophyll cells around the stomatal areas.

Dead trees resulting from sulfur dioxide poisoning near a fertilizer factory in Ireland.

Lichens are particularly sensitive to atmospheric pollution, some species hating it, some loving it. Shown here is the distribution in the British Isles of two species of lichen with these extremes of tolerance. **Below left** *Lecanora conizaeoides* can tolerate quite high levels of pollution so that between 1953 (red and gray cross-hatching) and 1973 (the original area plus red and white cross-hatched area) its distribution increased. **Below right** *Lobaria pulmonaria* is pollution-sensitive so that between 1950 (gray and cross-hatched areas) and 1973 (cross-hatched area only) its distribution has decreased.

Inland Fresh Waters

Pollution of rivers, lakes and ponds is invariably a consequence of human activity, resulting from the disposal or dispersal of effluents by the easiest and cheapest method, or from ignorance or malpractice. Freshwater pollutants can be placed broadly in a number of categories.

Airborne Solubles

These substances are dissolved in rain and ultimately find their way into underground

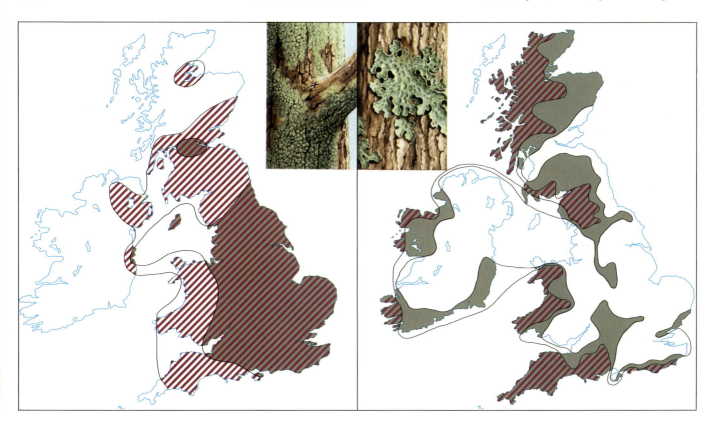

Absence of reeds allows boats to approach banks, destroying reeds and banks

Bird life killed off by bacteria from algae

Treated sewage effluent rich in phosphates, encourages growth of algae and plankton

Dense algae and plankton cut out light to underwater plants

Mud loosened by lack of weeds: sediment washed downstream causing silting-up

Reduction of weeds: fish lose spawning sites

Man is making ever increasing demands on his environment for leisure purposes. Recent researches have indicated that such natural areas as the Norfolk Broads, England, used for many decades for boating, angling and ornithology, cannot escape Man's pollution. Phosphate rich sewage entering the water stimulates algal growth, which blocks out light and reduces oxygen concentration, causing other aquatic plants to die and reducing the food and spawning grounds for fish. The dead and decaying algae are breeding grounds for bacteria that kill birds. Turbulence from the increasing traffic of boats kills underwater plants near the banks, allowing the boats to reach the banks. This breaks the banks down, further destroying plant life.

watercourses. Sulfur dioxide in the atmosphere is highly soluble and dissolves in rainwater and open water to produce dilute sulfuric acid. This increased acidity of the water can bring about changes in the flora and fauna.

Inert Suspended Solids

In the main, these result from flooding and erosion following heavy rain. In addition, effluents of finely divided inert matter are produced as washings from root-crop preparation, china-clay mining, coalmining and other forms of mining and quarrying. The concentration of inert pollutants usually declines by settlement as they proceed downstream from their source. The effects on the aquatic vegetation are twofold. Firstly, the reduction of light penetration by turbidity decreases or eliminates photosynthesis by aquatic plants. Secondly, sedimentation of solids can cover aquatic rooted plants and prevent their growth. The aquatic rooted plants *Elodea canadensis*, *Ranunculus aquatilis* and *Myriophyllum* spp., and the moss *Eurhynchium riparioides* have been eliminated by burial. They are generally replaced by floating-leaved species such as *Potamogeton natans* and *P. pectinatus*.

Poisons

These may be organic or inorganic compounds. The organic poisons include phenols, tar-acids, dyes and detergents. Inorganic poisons include cyanides, heavy metal salts, acids and sulfides. The majority of poisons are the products of industrial processes and waste by-products and their biological effects depend on the initial concentration, dilution and temperature. Generally, poisons have a greater effect on aquatic organisms during summer. The acidity of the water (pH) affects the toxicity of poisons. For example, cyanides and sulfides are more poisonous in acid than in alkaline waters. In acid waters, cyanides release the highly toxic hydrocyanic acid gas, and sulfides form the poisonous hydrogen sulfide, both of which are soluble in water. In alkaline waters, heavy metals have a tendency to separate

out from solution, thus reducing the chance of incorporation into the tissues of aquatic plants. Some plants are natural accumulators of particular nonessenetial elements and have been used in mineralogical surveys. The effects of poisons on plants are dependent upon concentration as well as length and frequency of exposure. In most rivers where poisons are discharged there is a dead zone, but aquatic life, though not necessarily the original flora and fauna, reappears as the poisons become diluted downstream.

Appreciable quantities of the heavy metals copper, zinc, lead, nickel, chromium, cadmium, silver and mercury may kill aquatic vegetation and prevent self-purification of the water body. An instance of this occurred when old lead mines were reopened in North Wales causing pollution of the rivers Afon Rheidol and Afon Ystwyth. The vegetation of these rivers was reduced to two species of green algae, *Batrachospermum* sp. and *Vaucheria* sp. During the early 1920s, after mining had ceased, starwort (*Callitriche verna*), water crowfoot (*Ranunculus aquatilis*) and numerous algae and bryophytes were found to have reappeared.

Synthetic detergents have adversely affected the growth of aquatic rooted plants. *Ranunculus aquatilis*, *Potamogeton pectinatus* and *P. densus* are seriously damaged by concentrations of 2.5ppm. On the other hand, 5ppm appeared to have no effect on the alga *Cladophora* sp. or the moss *Eurhynchium* sp.

Organic Wastes

The major source of organic pollution is discharge of crude or partially treated domestic and industrial sewage. The food-processing industries produce large quantities of organic residues which are commonly discharged into rivers or along the coasts. The immediate effect of organic pollution is to place a demand on the dissolved oxygen in the water leading to anaerobic conditions which produce phytotoxic sulfides and ammonia. A high concentration of suspended solids also results, reducing light penetration. If the organic pollution is severe, even algal species disappear, owing to total deoxygenation of the water. Further downstream an algal succession occurs as the pollutants and their effects decrease and the water becomes reoxygenated. The blanket weed (*Cladophora glomerata*), a filamentous green alga, thrives in water mildly polluted with organic matter, often developing very large growths.

Bryophytes and higher aquatic plants are completely eliminated by severe organic pollution, for a number of reasons. In the immediate area where anaerobic conditions prevail, deoxygenation and the presence of toxic gases are sufficient to kill the plants. High turbidity prevents photosynthesis and suspended solids can smother the plants during settlement. Sewage fungus (*Sphaerotilus natans*) also smothers rooted aquatic plants. Even further downstream where the pollutants become diluted and the normal water plants can re-establish themselves (eg *Potamogeton natans*) they are still in danger of being smothered by the banket weed. Return of the normal flora of rivers occurs only where recovery is completed, which may be a long distance downstream.

Eutrophication (Excessive Enrichment)

Plants require certain essential chemical elements for healthy growth. These include the major elements carbon, hydrogen, oxygen, nitrogen, sulfur, phosphorus, magnesium, potassium and calcium, together with several minor or trace elements. The normal low concentration of nitrogen and phosphorus in water limits the growth of aquatic organisms, especially phytoplankton and algae. If there is an excess of these two elements they become pollutants, either by promoting *algal blooms* (the rapid growth of algae visibly identified by floating mats of scum on lakes and reservoirs), or by making the water unfit for human consumption. The numerous sources of excessive nitrate and phosphate additions to waters include farm wastes, fertilizer application and leaching, food-processing factories or purified sewage effluent.

The River Mersey after its passage through Manchester and associated industrial areas, England. Such rivers are typically heavily polluted with inert suspended solids that exclude light and cover rooted plants, poisons and detergents that kill large and small forms of plant life, and organic wastes that stimulate algal and bacterial growth, which reduces the oxygen concentration to levels that will not support other plant life.

Algal blooms have more serious effects in lakes and ponds since these are normally isolated water bodies. They also occur in rivers but the effects are lessened because the bloom is not confined but is swept away constantly by the current. An algal bloom causes the water to become supersaturated with oxygen during the day and almost depleted of oxygen during the night, low enough to kill fish. Since algae need light for photosynthesis they remain on or near the surface and during blooms they can blanket the surface, eliminating rooted aquatic plants by preventing light from reaching them. Algal growth continues until essential nitrogen or phosphorus once again becomes limiting to plant growth. Death of the algae releases a large amount of organic material into the water and aerobic decay depletes the dissolved oxygen available, producing stagnant conditions. Partly decayed plant material sinks to the bottom as semidecomposed ooze, or may be washed onshore where it becomes a nuisance owing to its smell and interference with leisure pursuits.

There are numerous examples of eutrophication problems in European and North American Lakes. In the 1880s Lake Zürich was polluted by the introduction of sewage and still receives large amounts of domestic effluent and salts from industrial wastes. Algal blooms have been observed there since 1896. Fure Lake, near Copenhagen, receives 24 tonnes (23.6 tons) of nitrogen and 4 tonnes (3.9 tons) of phosphorus annually and the resultant algal blooms producing increased turbidity have confined higher plants to shallower water. Excessive sewage pollution of Oslo Fjord in 1955 deoxygenated the deeper water, producing large quantities of sulfides. Lakes Erie and Michigan have almost become "dead" lakes through sewage and industrial pollution.

Marine Pollution

The problems produced by pollutants in rivers are by no means all resolved there, and many substances not oxidized, settled or absorbed in rivers ultimately reach the sea. Coastal industries and towns discharge untreated industrial effluents and raw sewage directly into the sea. Although dilution of pollutants does occur, the littoral zone around coasts and estuaries is affected. At present, occasional blooms of dinoflagellates, which are poisonous, are eaten by shellfish, and some people have died from the resulting food poisoning. The sea is also subject to oil pollution from collisons, from oil rigs and from leakage and washing out of storage tanks of tankers at sea.

A number of factors influence the effect of oil pollution on marine plants: the type of oil, whether light fuel oil, crude diesel oil or emulsified oil; the emulsifiers used to disperse it; and which stage of the plant life cycle is present when the oil strikes. The majority of marine plants can suffer considerable damage and still recover. Emulsified oil clings more readily and adheres to plants. Susceptibility is greatest in plants with shallow roots and no stored food reserves such as seablight (*Suaeda maritima*) and glasswort (*Salicornia* spp.), which are killed more quickly than the more resistant perennials with large food reserves. Plants which are known to have survived numerous successive oilings are mud rush (*Juncus gerardii*), sea milkwort (*Glaux maritima*) and species of spurrey (*Spergularia* spp.). Emulsified oil from the tanker Torrey Canyon adhered to laver weed (*Porphyra umbilicalis*) off Cornwall, England, and the more delicate red algae suffered enormous damage. Off Puerto Rico, heavy crude oil eliminated large quantities of algae; however, marine grass is even more susceptible and *Thassalaria* beds showed continued degeneration months after the oil spill. Off the Californian coast crude diesel oil seriously affected the large brown alga *Macrocystis* and killed areas of cord grass (*Spartina alternifolia*) and glasswort (*Salicornia bigelowii*). The oil refinery at Milford Haven in South Wales has been involved in various incidents concerning leaks and oil spillages. These affected a salt marsh at Pembroke causing extensive damage to plants.

Terrestrial Pollution
Addition of Sewage Sludge to Soils

Sewage sludge, extensively used as an organic fertilizer to improve soil structure, has been shown to contain high concentrations of the heavy metals copper, nickel, zinc and cadmium which may have originated from electroplating industries. In 1968 sewage sludge applied to horticultural fields in Nottinghamshire in England affected the growth of roses to such an extent that

rose production on the fields was stopped. Other instances of heavy metal contamination of agricultural land by applying sewage sludge have produced poor crop performance and effectively prevented normal cropping for several years.

Farm Effluents

Intensive farming methods have given rise to large quantities of animal waste products. In addition, the production of silage for winter feeding of livestock produces a toxic liquid effluent (a complex mixture of carbohydrates, nitrogenous substances, fatty acids, phenols and salts) which is difficult and expensive to dispose of. If it is allowed to enter water courses, aquatic organisms are seriously affected because of the toxic compounds and the very high demand for oxygen which they impose. A common disposal practice is to spray silage effluent onto fields; breakdown of the organic components eventually occurs in the soil but crops may be seriously affected.

Mining and Quarrying

The extraction of ores, minerals and fossil fuels is becoming more widespread as demand for declining resources increases. Even famous beauty spots are threatened. Industries associated with mining and quarrying produce additional pollution in the form of spoil heaps, slag heaps and tips. The natural succession of plants is disrupted for

many years since most waste materials are inert, consisting chiefly of ash or crushed rock, and few higher plants can grow on them because of the low content or absence of organic matter. For plants attempting to recolonize, spoil heaps from heavy metal quarrying or mining produce additional hazards, and in areas where these industries still operate atmospheric pollution spreads the contaminants over a wide area. Land contaminated by aerial pollution from extraction industries is usually rendered sterile except for the establishment of a few tolerant species. Tips for colliery workings are topographically very high in relation to the surrounding land and have acute slopes producing an almost "alpine" microclimate unsuited to most unspecialized plant species.

Biocides

Until the 1930s, pesticides were generally inorganic compounds such as copper sulfate, fluorides, arsenates and natural compounds of plant origin, for example, pyrethrum from certain species of *Chrysanthemum*, nicotine from tobacco, and rotenone from tropical legumes. Apart from arsenic compounds, none of these is highly toxic in terms of accumulation in food chains. Progress in organic chemistry has led to the development of synthetic organic pesticides, the classic example being DDT. The latest insecticides are organochlorine compounds (eg dieldrin, endrin and aldrin), organophosphates such as parathion, and organomercury compounds, for example verdasan.

A rubbish tip on the shores of a lagoon at Point Barrow, Alaska. Visually this is the most obvious type of pollution, excluding any plant life from the immediate vicinity. However, the problem is more serious in the long term when the poisonous products of such tips are slowly released into the environment, affecting plant life over a far greater area.

Although insecticides are applied to control insect pest populations, some are nonbiodegradable and remain in soils for years, causing damage to wild and cultivated plants. Initial large applications of organochlorine compounds may retard plant growth and produce mutations. Persistent compounds such as endrin and dieldrin have contaminated carrots, sugar beet and soybeans after the soil had been previously treated a year or years before. Since the more persistent biocides are not readily degraded or excreted, they pass through food chains and can have serious effects on higher animals, including Man. The systemic insecticides demethion and endothion are taken up by plants and translocated to areas where the plant sap becomes toxic to aphids and other plant predators. They protect the plants against insect attack but are themselves poisonous to mammals.

Herbicides range from the inorganic sodium and potassium chlorates which effectively kill all plants, to the derivatives of phenoxyacetic acids (eg 2,4-D) and nitrogen compounds such as simazine, which preferentially eliminate dicotyledonous plants from mixtures with monocotyledons. Some compounds kill on contact with plants, others are transported within the plant and their effect is by hormonal action. Most herbicides are not persistent and have little effect on animal populations. However, as a result of continued use of herbicides on agricultural land, roadside verges, railway and motorway embankments and drainage dikes, plant associations in marginal areas are greatly affected and the species diversity has been reduced. The indiscriminate application of herbicides to water in order to restrict weed growth has had even more serious effects on plant populations than in terrestrial situations since there is a very close interdependency between aquatic plant and animal communities. In 1965 paraquat was applied to a series of small lakes in Nottinghamshire in England which supported well-established communities of aquatic plants including Canadian pondweed (*Elodea canadensis*). curled pondweed (*Potamogeton crispus*), water milfoil (*Myriophyllum spicatum*) and amphibious bistort (*Polygonum amphibium*). The paraquat eliminated most of the higher plants and algae. A year later *Elodea* and *Myriophyllum* were still not re-established, and two years later there were still no algal blooms observed. Most of the primary producers in the lake ecosystems were eliminated.

Radiation

All living organisms are constantly exposed to small amounts of naturally occurring radiation present in the biosphere as a result of the decay of naturally occurring radioactive isotopes. With the advent of atomic fission, Man has vastly increased the quantity of readily available radioactive materials on the planet and consequently has raised the level of natural radiation by unnatural means. Unlike organic and inorganic compounds, radiation cannot be treated and destroyed. Radioactive elements may have very long half-lives and can persist for many years in active concentrations. Any leakage from nuclear power stations, stores of waste materials, or nuclear fall-out has serious effects on biological food chains. Radiation is not only persistent but also cumulative.

Algae have high resistance to radiation damage but they accumulate radioactive ions. In the River Clinch, Tennessee, algae were found to have concentrations 10,000 times the radiation of the water in which they were growing. The use of algae to remove radioactive isotopes from water has been suggested and experiments carried out at Hanford on the Columbia River showed that algae concentrated the potassium isotope ^{32}P to 300,000 times that of the river water concentrations. Despite the enormous quantities of radioactive isotopes extracted by the algae no genetic mutants have been found. Marine algae examined after the atomic tests on Bikini Atoll revealed no serious deleterious genetic effects. Investigations of low-level radioactive wastes in the Columbia River, derived from plutonium-

producing works, revealed a number of radioactive elements with short half-lives in the water. These included isotopes of copper (^{14}Cu), manganese (^{86}Mn), sodium (^{24}Na) and silicon (^{31}Si), together with the longer half-life isotope of phosphorus (^{32}P) at the discharge point. Further downstream ^{32}P was found to predominate in the water, presumably owing to the more rapid decay of the other elements. However, it was found that stream organisms had concentrated the short half-life elements to a considerable degree. The greatest concentration appeared in plankton and filamentous algae, organisms which derive their nutrients directly from the water. Detritus-feeders of the next trophic level contained the next highest concentration, followed by small and newly hatched fish and culminating in the predatory fish. Fish which fed entirely on plant matter had a much higher concentration than other species.

Mosses and lichens are known to become more radioactive than higher plants, possibly because their mode of nutrition is to derive chemical elements dissolved in rainwater directly through their leaves or thallus, but little work has been done on these lower plants to determine subcellular effects. Higher plants appear to be more sensitive to radiation than bacteria or algae, with gymnosperms the most sensitive plants of all. An experiment was carried out in the United States to study the effect of directly releasing radioactivity from an unshielded reactor on a forest ecosystem. At a distance of 300m (1,000ft) from the reactor, all pine trees were killed within one year. Premature leaf senescence occurred in shrubs and hardwoods in a zone several hundred feet beyond the dead pines. In the following spring an area of leafless trees extended for 1,000m (3,300ft), surrounded by a forest in full leaf. Although most of the trees beyond 210m (700ft) from the reactor began to

produce leaves later on, the growing period was shortened. Within 210–240m (700–800ft), the growing tips of the woody vegetation were killed, and trees and shrubs were dependent upon adventitious buds for further growth. The growing tips of these buds were also killed in time, and oak and hickory were also affected and died.

J.O.R.

Weeds and Aliens

Together with crop plants, which owe their status to deliberate choices made by people seeking food and other useful products, the plants considered in this section comprise those most closely linked to Man. Aliens owe their existence in areas beyond their natural range to deliberate or accidental transport by Man, while weeds are, by definition, plants which occur in places where they are inconvenient or harmful to human activities, usually because they interfere with the cultivation of plants considered desirable; however, a sense of almost moral disapproval extends the category to species, often the same as those that invade cultivated ground, which occupy other sites grossly modified by Man. Weeds and aliens belong to a great diversity of unrelated plant families but very many of them share characteristics in their growth and reproduction with species important in the initial colonization of natural habitats (see Colonization and Establishment, p. 145). D.M.M.

Aerial crop spraying, much used for the application of fungicides and pesticides, is a technique of great value for increasing productivity but one which must be carefully controlled to prevent harm to adjacent crops and natural vegetation by wind-drift of the chemicals employed.

Weeds

A weed can probably best be defined as a plant which Man considers to be out of place or otherwise "undesirable"; this implies that a given species may or may not be a weed, depending on the circumstances. On uncropped land, for example, the only plants to be undesirable may be those that are thorny, sting or cause allergies; hence under these circumstances the great majority of plants would not be regarded as weeds. However, on agricultural land it is usual to regard any plant other than the crop as a weed. This latter view is an oversimplification; in grassland, for example, some self-established herbs appear to be desirable and under wet tropical conditions some ground cover, in addition to the primary crops, may be essential to prevent soil erosion. Furthermore, in the highly mechanized agriculture currently practiced in many parts of the world some of the most troublesome weeds are crop plants that have persisted from an earlier crop in the rotation, such as potatoes in a following cereal crop.

Species of *Ceratophyllum* are among the most widespread and troublesome submerged aquatic weeds, mainly in stagnant water, sometimes choking the intakes of hydroelectric power stations. Here, *Ceratophyllum submersum* grows as submerged masses in Lake Rawa Tombor, Java.

Classification and Characteristics of Weeds

For the purposes of control techniques, weeds are frequently grouped by habit into annuals, biennials and perennials; however, this classification is not hard and fast, since certain species may vary their habit, depending on the circumstances. It is of greater practical importance to know whether propagation is mainly by seed, eg charlock (*Sinapis arvensis*), by underground organs, eg ground elder (*Aegopodium podagraria*) or by both, eg couch grass (*Agropyron repens*). An alternative method is to classify species as herbaceous or woody—a difference that is important in chemical weed control. Further alternative groupings are terrestrial and aquatic species and parasitic and nonparasitic species. Aquatic and parasitic species tend to be extremely serious weeds in the tropics—for example, water hyacinth (*Eichhornia crassipes*) has posed serious problems by preventing navigation and by blocking drainage and irrigation channels, while the equally attractively flowered parasitic witchweed (*Striga* spp.) may decimate the yields of many crops (eg sorghum, maize and millet). Parasitic weeds may also be important in temperate areas, eg dodders (*Cuscuta* spp.) on clover and lucerne, the broomrapes

(*Orobanche* spp.) on beans, tobacco etc and mistletoe (*Viscum* spp.) on numerous tree species. Part of the success of *Striga* and *Orobanche* results from their producing vast numbers of seeds that generally do not start to germinate until activated by an exudation from a potential host.

The characteristics that make a species successful as a weed are essentially those that make any plant or animal successful, ie the ability to exploit its environment. On agricultural land this usually means a species that spreads rapidly and easily and is not easily killed either by the operations carried out in growing the crop or by natural, biological or environmental factors. Hence important factors in weediness are the production of numerous seeds or other propagules (especially those that germinate over a long period), effective dispersal and rapid establishment followed by vigorous growth under a wide range of conditions. A weed of agricultural land must be able to complete its life cycle in at least one of the crops in a rotation; thus we find that on market garden land, which is frequently cultivated and on which the crops are of short duration, the most successful weeds are annuals that germinate, grow, flower and set seed in a comparatively short time, whereas in orchards and grassland the successful weeds tend to be perennials that do not need constantly to re-establish them-

selves from seed. The growth-form of a plant may also be of importance in its success as a weed; for example, on well-mown lawns and hard-grazed pastures, plants that grow close to the surface, eg daisy (*Bellis perennis*) and dandelion (*Taraxacum officinale*) have a competitive advantage. Species introduced into a new area may be particularly successful, at least initially, as their natural enemies may not be present.

Studies of a number of weedy plants and their nonweedy relatives have resulted in attempts to define a "weediness syndrome" of characters. For example, the perforate St. John's wort (*Hypericum perforatum*) is a relatively "normal" member of grassland and wood-margin communities in Europe and western Asia. Toward the end of the last century it was accidentally introduced, probably in hay or other packing materials, into Australia, New Zealand and North America, where it seemed to show a complete change of character, rapidly occupying considerable areas—100,000 hectares (247,000 acres) in Victoria, Australia; 700,000 hectares (1.7 million acres) in the western United States—and often proving injurious or fatal to stock eating it. Comparison of European material with the aggressive migrants showed that the latter were much more vigorous and taller growing, producing more stolons in a given period and more flowers and seed per plant than the nonweedy "stay-at-homes". Similarly, weedy populations of *Eupatorium microstemon* in southern North America showed a combination of phenotypic plasticity, annual habit, rapid flowering, lack of strict photo-periodic requirements, self-pollination and low pollen productivity, all distinguishing them from nonweedy representatives of the species; clearly, the weediness depends not only on the opportunities provided by new habitats but also on the presence of sufficient genetic variation to allow some populations to exploit the opportunities offered to them by Man.

Factors governing the seriousness of a particular weed in a given area are (in addition to those listed above): types of crops grown, methods and standards of husbandry, type of soil and amount and distribution of precipitation. Indeed, weeds may be so dependent on their environment that they can be useful indicators of the nutrient status, moisture and acidity of the soil—mouse-ear chickweed (*Cerastium arvense*) is characteristic of chalk soils, spurrey (*Spergula arvensis*) of sandy soils and rushes (*Juncus* sp.) of poorly drained land.

Harm Caused by Weeds

The damage done by weeds is not fully understood but it may exceed that more obviously attributable to pests and diseases. Weeds reduce the yields and affect the quality of crops, spread crop pests and diseases, make harvesting more difficult, poison humans and animals and cause allergic reactions.

REDUCTION OF CROP YIELDS. Weeds reduce the yields of crops mainly by competing with them for water, nutrients and light. Competition is a complex interaction and depends, among other things, on the type of crop (including its variety), the spacing, density and time of sowing of the crop, the types and amounts of nutrients present (eg soils deficient in nitrogen favor nitrogen-fixing crops such as clover, while under similar conditions a crop unable to fix nitrogen would find itself at a competitive disadvantage against most weeds), the species of weed (including probably its ecotype) and its ability to excrete or secrete substances toxic to other plants (see Allelopathy, p. 147). In general, weed competition is most serious during the first quarter to third of a crop's life; competition during this period is likely to have an irreparable effect; later weed growth is more likely to interfere with harvesting than to affect yields adversely.

CROP QUALITY. A number of weeds can affect crop quality by tainting the produce; for example, the wild onion (*Allium vineale*) causes off-flavors in milk. Weed contamination of crops grown for seed may render the produce unsaleable.

VECTORS OF DISEASES AND PESTS. There are numerous instances of weeds acting as hosts of crop pests and diseases; thus, for example, *Barbarea vulgaris* is the alternate host of the fungus causing stem rust of wheat, while fat hen (*Chenopodium album*) is a host of the virus causing sugar beet yellows. Many weeds are alternative hosts of insect pests. In addition, weed growth in and around crops may favor larger pests such as slugs and rats.

HARVESTING. This may be made more difficult if weeds delay ripening of the crop, cause lodging (flattening) or simply get in the way.

POISONING. In Man this is now less common than formerly, when weeds such as darnel (*Lolium temulentum*) could be a serious contaminant of flour. Livestock poisoning is more of a problem, for while animals generally avoid naturally growing poisonous weed plants, such plants become more attractive when they wilt and this makes ragwort (*Senecio jacobaea*), for example, particularly dangerous in hay. Numerous plants, both crops and weeds, cause hay fever, while others, such as poison ivy (*Rhus radicans*) in the United States, cause intense skin irritation.

Beneficial Effects of Weeds

It is customary to destroy non-crop plants indiscriminately, partly because of the difficulty of selectively controlling harmful species while leaving both the crop and harmless species undamaged, and partly because it is not altogether clear which non-crop plants are harmless or even beneficial. The beneficial effects of extraneous non-crop plants are not well documented and there seems to be a need for a special term for plants of this sort. A number of plants such as *Tagetes* and *Polygonum* species will suppress nematodes (eelworms) that adversely affect crops. Weeds add to the organic matter in the soil and thus help to improve its chemical and physical properties. Decaying shoots, and especially roots, have an important effect on the drainage and aeration of the soil both directly and by encouraging the spread and activity of earthworms. Plants vary widely in their ability to concentrate nutrients; for example sun spurges (*Euphorbia* spp.) accumulate boron, and such plants may be of considerable importance in recycling nutrients. The importance of providing soil cover to prevent soil erosion has already been mentioned. Finally, there is reason to think that, just as weeds may act as hosts of crop pests and diseases, so may they also act as hosts for the predators of crop pests and diseases.

Weed Control

Weed control may be achieved indirectly by shifting the dynamic ecological balance existing between a crop and its weeds, eg the encouragement of optimum crop growth by good seedbed preparation, good water management, the correct use of manures and fertilizers and by efficient pest and disease control. Direct control may be effected by: preventive methods, fire, and cultural, ecological, biological and chemical means.

PREVENTIVE METHODS. These comprise the use of clean seed to avoid the introduction of weeds on to the farm, the rotation of crops to hinder the buildup of individual weed species, the prevention of seeding of weeds in the field and on neighboring land, the cleanliness of equipment and clothing to prevent seed dispersal, the proper composting of manure and the prevention of the spread of weed seeds by runoff of rain and in irrigation water.

FIRE. Almost certainly the oldest method of weed control, fire is now generally most useful for removing cereal stubble after harvest, for clearing forest land and for disposing of bulky debris, though it may be used for sterilizing seedbeds. However, burning heaps of vegetation can cause problems as the ash of certain species is toxic. In general, it is better to compost plant residues than to burn them, so that the organic matter can be returned to the soil—proper composting generates enough heat to kill most weed seeds and other propagules.

CULTURAL METHODS. Although still paramount on a worldwide scale, the traditional fallowing, in which the land is left idle for a season or more so that weed control cultivations can be carried out, is now uncommon in Europe and North America. The simplest method, removing weeds by hand, is not only extremely laborious but may be ineffective, as some plants break when pulled up and then grow again, while the shoots of others are capable of producing viable seed even when uprooted before the flowers open. Hoes and cultivators cut weeds, bury them or drag them to the surface to be killed by the sun and wind. Their repeated, properly timed use will exhaust the food reserves in the roots of perennials, weakening and eventually killing them, while such cultivation will also stimulate seeds and other propagules to grow so that they can be killed before the crop is planted.

ECOLOGICAL CONTROL. This is the modification of the environment to favor the crop at the expense of weeds. Obvious examples include draining pastures to eradicate rushes and, at the other extreme, growing rice under swamp conditions to drown out weeds. Weed growth may also be affected by modifying the light regime of the crop (by shading or increasing light depending on the crop and weeds involved). Liming the soil will inhibit calcifuge weeds, while applications of fertilizers may be used to enhance selectively the competitive ability of desirable species. Mulching may be considered a variety of ecological control as it affects the physical condition of the soil as well as smothering weeds—a drawback of this method is that mulches frequently encourage pests and diseases. In recent years black plastic sheeting has been used as a mulch to control weeds and has been very successful in high-value crops, especially those grown in arid areas, eg pineapple, as the mulch conserves moisture as well as controlling weeds.

BIOLOGICAL CONTROL. This is usually thought of as the control of weeds by insects specially introduced for the purpose. However, much

Sheets of black polythene placed between rows of pineapples in Hawaii to suppress weed growth and conserve moisture.

biological control goes on naturally by insects, higher animals and plants. The best-known and most successful use of insects for weed control is the use of the cinnabar moth (*Cactoblastis cactorum*) in clearing huge areas of the prickly pear (*Opuntia* spp.) in Australia. There are, however, limitations in using insects for weed control. Among higher animals being used for weed control are geese, for the selective removal of weeds (especially grasses) in a number of crops, sheep in young coconut plantations and cattle in older coconut plantations, and the grass carp (*Ctenopharyngodon idella*) for the control of aquatic weeds. There seems great scope for making much more use of larger animals, especially in the tropics where labor to tend them is abundant and where there is great need for the protein they could provide.

Finally, plants may also be used to choke out noxious weeds and prevent the ingress of undesirable species—in arable land smother crops may be planted between the main crops in a rotation, while in tropical plantation crops, such as rubber and oil palm, legumes are frequently cultivated between the trees.

CHEMICAL CONTROL. While chemicals have long been used for total vegetation control on railways, storage yards etc, it is only since the development of selective herbicides during the last two or three decades that chemical weed control has become really widespread. There are now more than 200 chemically different herbicides available singly or as mixtures in thousands of commercial formulations. In recent years the value of herbicide sales in Britain and the United States has exceeded the sales values of insecticides and fungicides combined.

One of the main advantages of chemical weed control is that it is possible to control weeds selectively in the crop row without hand labor. The other main advantage of chemical weed control is that it is now possible to control a number of highly injurious weeds that were formerly virtually uncontrollable in crops; before the advent of the grass-killing herbicide dalapon, East African couch grass (*Digitaria scalarum*) could be so damaging in coffee plantations that it has even been known for trees to be dug up and replanted in clean land.

Many herbicides have a considerable residual effect (ie they continue to kill weeds for a period after they have been applied). As most weed seeds germinate in the surface layers of the soil, the absence of soil disturbance when weeds are controlled chemically results in far fewer weeds being induced to germinate, so that weed infestations become progressively less severe. This has been most evident in citrus orchards in the United States, where oil and other sprays have been the sole method of controlling weeds for many years—during the first year of a program of chemical weed control, six or more sprayings of a contact herbicide may be necessary, while by the third year only two or three applications may be required.

The use of herbicides is not without its

drawbacks, but these have been less apparent than with pesticides. This is partly because modern herbicides are not very poisonous to animals and partly because they usually decompose or are rendered inactive in the soil. Another factor is that plants reproduce more slowly than insects and disease-causing organisms, so that the development of resistance within a species has taken much longer to show up, and until recently it appeared that selection by herbicides was solely between species. It is thus important to use the range of herbicides available, since the repeated application of the same herbicide may produce a more intractable problem than that originally present.

Changing cropping patterns resulting from the use of herbicides have had a marked effect on weed problems—on the one hand the herbicides available for use in cereals mean that they now need no longer to be used in a rotation, while on the other hand the monoculture of cereals made possible by the use of herbicides has resulted in wild oats (*Avena* spp.) becoming a much more serious problem.   B.N.B.

Aliens

The natural dispersal of plants (see Dispersal and Migration, p. 142) has been one of the factors important in enabling species to extend their range over the period since they first evolved. Nevertheless, direct information on the efficacy of dispersal in any group of plants is notoriously difficult to obtain. However, plants thought to be growing outside their natural range are usually considered to have achieved this as a result of Man's activities, and such plants are termed aliens. In consequence, the mode of dispersal of aliens is assumed, by definition, to be by human mediation and therefore they have constituted an increasingly important phytogeographical component of the Earth's plant cover as men have expanded the distances and frequencies of their travels over the world. Since their mode of dispersal is known, studies of aliens have been valuable in providing information on the rates and patterns of such factors as establishment (see p. 145) and competition (see p. 146) important in ecology and phytogeography.

In many cases, Man has deliberately transported plants from one region to another, either for their economic value, for example crops, grasses for pasture-improvement and vegetables, or for ornament, as in the case of garden plants. It is also evident, however, that many aliens have been transported unwittingly by travelers.

Until the relatively recent improvements in seed cleanliness, many aliens were introduced as contaminants of seed-samples deliberately carried from one place to another; for example, the creeping buttercup (*Ranunculus repens*) was brought to Tierra del Fuego in grass seed from England, while the foxglove (*Digitalis purpurea*) arrived there in the same manner from Chile, where it had been introduced from Europe. Hay and straw packing for machinery, ballast in ships and similar means certainly resulted in the movement of aliens, one unusual example being the introduction of certain northern lichens to the subantarctic island of South Georgia on wooden packing cases used by whaling expeditions from Scotland and Norway. Aliens in several floras, such as that of Britain, have been carried in raw wool from South America and Australasia. These species, a high proportion of which have spiny and other fruits similarly adapted to animal dispersal, have been found in the vicinity of wool mills or elsewhere, since wool waste (shoddy) is often used as a mulch or manure in agriculture. Man unwittingly carries aliens himself, especially on footwear, while vehicle tires are undoubtedly an important dispersal agent; indeed surveys of car-washes in Australia have demonstrated the important role of the whole vehicle in dispersal.

Most aliens possess the group of characters associated with "weediness". The most important of these is self-compatibility, which enables the plant to reproduce by self-pollination, but the ability to exploit open and disturbed ground by, for example, prolific seed production or vegetative reproduction, tolerance of a range of environmental conditions and vigorous or rapid growth are features of many common aliens. Clearly, weeds, aliens and "natural" colonizing plants share many features such as those just noted, since the problems they face are essentially the same.

There is a good deal of evidence that aliens are dependent upon Man not only for their dispersal but also for the provision of suitable habitats in their new homes. There are essentially no definite examples of aliens occurring in vegetation not affected by human activities in the areas to which they have been carried; naturally open habitats, such as coastal beaches and river-shores, seem to be the only areas where aliens can establish themselves beyond human influence. Wherever aliens have been found in apparently "natural" vegetation, detailed studies have shown that subtle effects of human activity, such as selective grazing by domestic animals, have been in operation. However, aliens have been divided into two groups, transient or established, depending upon their behavior after introduction.

Transient aliens are those which survive in their new areas for only a short time unless there are repeated introductions, and most of these are associated with places strongly influenced by Man, such as gardens and similarly heavily cultivated ground. Established aliens persist for a significant period after their introduction and it is these which have colonized habitats least modified by Man. Interestingly, however, aliens conform to the principal climatic zones with which, as observed earlier (see p. 117), the world's major vegetation-types are associated. Thus, for example, the majority of the aliens in

Australia and California originate in the Mediterranean region, which has a similar climatic regime, while cool-temperate areas such as much of New Zealand and southernmost South America have derived most of their aliens from northwest Europe.

Some of the clearest ideas about the rates of dispersal of plants have been provided by aliens. Some have spread extraordinarily rapidly, like the mayweed or pineapple weed (*Chamomilla suaveolens*), which spread throughout Britain within 25 years of its introduction from northeast Asia towards the end of the 19th Century. A particularly well-documented example is the Oxford ragwort (*Senecio squalidus*). A native of light volcanic soils in Sicily, it was brought to the Oxford University Botanic Garden prior to 1690; by the mid-18th century it had

The yellow-flowered Californian poppy (*Eschscholzia californica*) is an alien in Chile where it grows in areas disturbed by Man, such as the roadside and railway embankment seen here south of Santiago. In its native California it occurs in natural habitats similar to the shrub-covered hills in the background of this picture. Interestingly, the part of Chile in which it has become naturalized has the same Mediterranean type of climate as its country of origin.

escaped from the garden and become naturalized on the city walls, where it remained until 1879 when the railway system reached the city. The light ballast of the rail-track beds clearly simulated its native volcanic soils and provided a route by which it rapidly spread to other areas, beginning with the neighboring county of Berkshire in 1883. By

1940 it had spread throughout much of England, but in a discontinuous manner, initially leap-frogging many areas and demonstrating that dispersal is not necessarily the orderly progression that one might imagine.

The occurrence of aliens in semi-natural and artificial vegetation obviously brings them to the notice of ecologists attempting to describe and interpret the composition and development of plant communities. The phytogeographer, trying to understand why plants, either as species or as components of vegetation, occur in those parts of the world that they do, must take an interest in weeds. In addition to the light they throw on the processes of plant migration etc, mentioned above, aliens need to be identified when determining the origins and affinities of the floras of different regions. Clearly, aliens are of no value in unraveling the natural migrations of plants and, since this often depends upon comparisons of whole floras (see p. 210), it is important to be able to differentiate aliens, dispersed by Man, from non-aliens. This is particularly important where aliens constitute a substantial part of the flora of a region. In truth, it is often difficult to determine which are aliens in many areas, but some regions, especially islands (see p 231), give an indication of the possible extent of the problem; for example, of the total floras of New Zealand, the Falkland Islands and Tierra del Fuego, respectively 59, 36 and 23% are aliens. Since most of these aliens came from Europe, any analysis which failed to recognize their true status would give a completely false impression of the origins and development of the native floras of such regions. Aliens, like weeds, with which they share many features, provide a graphic illustration of the ways in which Man, wittingly or unwittingly, has modified the occurrence of plants on this planet in such a way that they can both help and hinder his understanding of the processes involved; therein lie the reasons for their importance to the subject of this book.

D.M.M.

Conservation

Conservation has meant different things to different people: the preservation of wild plants and animals in natural ecosystems or even detached from their natural environment in gardens or zoos; the wise use of natural resources like coal, oil, water and minerals; and—more recently—the maintenance of our environment in such a state that the quality of life is maintained or enhanced. All these goals stem from the (often belated) realization that, whereas plant and animal populations are self-regulating, the human population is not, and that the human race has the power to

change its environment and all it contains. Consequently, people have a duty to ensure that any necessary modifications will not destroy the environment, because such destruction (apart from the havoc it will wreak on other organisms) is to the detriment of humanity. Man's survival ultimately depends on how wisely he chooses to exercise his crucial role in the environment.

In this account we shall concentrate on the relatively restricted, although crucially important, area of natural and semi-natural ecosystems, with their essential communities of plants and animals. It is worth first looking briefly at the history of conservation.

History of Conservation

As soon as primitive Man became more than a simple food-gatherer, some of his practices began to have deleterious effects upon natural ecosystems. This is attested by the increasingly well-documented correlation between primitive Man's development of hunting skills and the extinction of many large animals and birds, the apparently indiscriminate use of fire, and the considerable effects of Neolithic cultures on western European vegetation (see p. 164). At the same time, some practices were conservation-oriented, either deliberately or inadvertently. The use of terracing in the Middle East, for example, was a conscious attempt to conserve soil and water. "Slash and burn" agriculture, used in many areas of tropical rain forest, was also (but unintentionally) conservationist in its effects, as long as populations were small, since a newly cleared area was only utilized for a few years before abandonment and so had time to recover between clearances.

Although a few writers (Pliny among them) commented on the destructive effects of forest clearance and of goat grazing in the Mediterranean region, there was no serious concern about conservation in the early and classical civilizations. This may have been because people viewed wilderness as an enemy, or because they felt that natural resources had been created especially for their use, or because the resources seemed vast in relation to human populations.

The expansion of available resources that resulted from the spread of Europeans over the Earth's surface did nothing to alert people to their potential for destroying their environment. However, some voices were raised, as a result of the obvious effects of over-exploitation in a few places. In Europe, the depletion of timber resources with the advent of the Industrial Revolution led to some sound ideas on the replenishment of forests. In Africa and North America, timber extraction, overgrazing and the introduction of exotic species stimulated some people to write about the problem of depletion of natural resources, and some positive results ensued, such as the setting up of a National Park system in the United States.

The Importance of Conservation

Even today the importance of conservation is not fully realized, nor are the reasons for it: aesthetic, scientific and economic, among others. Aesthetic, since more and more people in developed countries are turning back to the out-of-doors for their recreation, and without active conservation the areas they use will not exist in their present state for long. Scientific, because, despite the tremendous spate of ecological work in recent decades, we still know very little about natural ecosystems and their use; once a species is destroyed its potential is lost forever. Economic, because, as we have belatedly realized, effective conservation can mean long-term economic advantages that more than balance the short-term economic gains of rapid and uncontrolled exploitation.

Even with the more general acceptance of the need for conservation, there is still an educational problem. To many people, conservation of an area means leaving it alone, without any interference. Such a view, even in existing true wilderness, is totally unrealistic. Since the spread of agriculture and pastoralism and, more recently, that of industry, their effects have become so all-pervading that nowhere in the world can one say that there is no effect of human activity. Further, Man has drastically altered the prospects for many species of plants and animals. Some have been totally extinguished, and it has not always been rare creatures that have suffered; the fantastically abundant passenger pigeon was eliminated in a few decades. Others have been made rare as their habitats have been destroyed. Yet others, formerly of rather restricted range, have been offered immense scope for spread. For example, weeds have frequently originated as species of rather restricted open habitats and spread into arable land. Further, the genetic potential of many species has been widened: species that were formerly isolated have been brought into contact and allowed to hybridize; types that are useful have been consciously or unconsciously selected; and new types of plant and animal have been created by breeding techniques and selection. Many plant communities, such as those of western Europe (grasslands and heath, for example), were initially the product of agricultural and other human management, and have been artificially perpetuated by such means as fire and general clearance. Indeed, a "laissez-faire" attitude to the European countryside would eventually result in a return of woodland. The demand for the preservation of such man-made landscape as grassland is mainly a response to the belief that the scenic and biological variety which it introduces into the landscape is in itself desirable.

Diagram illustrating the main environmental resources available to mankind.

POTENTIALLY INEXHAUSTABLE
RESOURCES

solar
energy

the atmosphere

water

POTENTIALLY EXHAUSTABLE
BUT RENEWABLE RESOURCES

human
populations

common
plants

unpolluted
air

common
animals

usable
water

essential
soil nutrients

some
ecosystems

soil

POTENTIALLY EXHAUSTABLE
AND IRREPLACEABLE RESOURCES

some sources
of timber

ground
water

some
minerals

rare plants
and animals

some
ecosystems

oil

coal

marine
fisheries

Conservation Management

The scope of management must differ with the location of the community, its size, its past history and its "naturalness" or otherwise. Some areas will respond well to a policy of something approaching "zero management"; others will demand intensive management if they are to remain in the desired state. Ideally, conservation policies should be based on detailed scientific investigation of the area together with a complete understanding of the biology of all animals and plants involved.

In practice, our understanding is nowhere near complete and most conservation has to be based on subjective assessment, often carried out in a short time, with minimum finance and assistance, and based on a generalized knowledge of similar ecosystems. Because, in such cases, mistakes will inevitably be made, the options should be kept as open as possible and where potential alternative management requirements exist, experiments should be set up to compare their efficiency. A decision to set aside an area for its conservation interest involves a long-term commitment that is not immediately obvious to the casual observer.

Size of Reserves

The question of the size of nature reserves has been insufficiently investigated. In many cases, the acquisition of a nature reserve, instead of being based on established scientific criteria and a proper knowledge of the area, is the result of a threat to the area, or a local pressure group, or a generous donor, or some other factor that often results in a less than ideal choice. Size has rarely entered into the considerations. This means that many potential reserves are of limited size because all the surrounding country is in other use. Yet size may be crucial in ensuring that the populations of many species are large enough to maintain adequate genetic variability and to reduce, as far as possible, the risk of extinction. Again, lack of knowledge hampers rational decisions. Besides large areas in reserves, a case could be made for a number of smaller reserves near enough together for some species-migration to occur between them. A small area may be sufficient to conserve a specialized community of limited extent, but woodland, for example, requires much greater areas for effective conservation. In all this the conservationist is often faced with the pressures of having to act quickly. Wherever possible, of course, carefully considered decisions should be made, based on the maximum possible information.

Conservation in Developing Countries.

The kind of conservation just discussed could be, and is by some, regarded as a luxury, involving essentially a very small proportion of the population of a sophisticated human society, freed from the necessity of wondering where the next meal is coming from. Its justification lies in the retention of our heritage for future generations who, without our action, would be denied the choice. It also lies in the provision of a reservoir of natural and semi-natural communities for scientific investigation and, possibly in the long term, for exploitation and even economic return if the areas are used for recreation.

Many people regard conservation in purely altruistic terms, as an expression of our duty to see that we do not destroy other forms of life than ourselves. It is not easy to muster support for this view. People question its relevance to countries where there is no tradition of admiring nature for its own sake, where population increase has put intolerable pressures on land, where food (especially protein) is in short supply and where political considerations are likely to take precedence over a concern for wildlife. Ecologists from developed countries faced with these considerations may suppose that the next few decades will inevitably result in the virtual destruction of some of the richest and most diverse ecosystems in the world. What should be their attitude? Can the governments of countries under these pressures be persuaded that their collections of plants and animals are irreplaceable and should be saved for all mankind? Can they be convinced that to do so is in their own long-term interest? Can enough information be gathered and presented to make a supportable case? Scientists have a responsibility to see that the answers to these questions are in the affirmative, and their governments have a responsibility to see that enough money is available to enable them to carry out programs of rapid research to back up their arguments and perhaps to provide compensation to the countries concerned for holding land free from agricultural or industrial development.

One of the most constructive ways in which progress can be made is by exploring multipurpose use of land. It has, however, the serious danger that in attempting to extract a variety of uses from land, none is effective and the conservation aim is not achieved; great care has to be taken to see that multipurpose use does not become multipurpose abuse. One only has to consider briefly the possible uses to which a single area can be put to realize how complex the problem is. The question is whether the different demands can be resolved in such a way as to maintain sufficient habitat diversity, minimum disturbance for wildlife, and at the same time allow some kind of exploitation, either by removing a crop, such as timber, or by grazing with stock, or by allowing recreation. The constraints imposed by these different uses mean that effective multiple use is only really appropriate for large areas where the different uses can be kept as distinct as possible.

Left *Tecophilaea cyanocrocus* is a most beautiful plant believed to have been exterminated in its wild habitat in the Chilean Andes by collectors. It is known in cultivation and much prized by growers of small bulbs for its superbly colored flowers. For such a plant future survival may only be in gardens.

Far left A great many plants, especially in Mediterranean-type floras, are very rare, only occurring in a few localities, but because they are predominantly montane, or growing on cliffs, their populations have remained relatively steady and they are not under any immediate threat. Much of the endemic flora of the countries of southern Europe falls into this category. Because of their small population size, the sites of these plants need constant monitoring to ensure they remain safe. Such species are particularly vulnerable to collectors since one collector could easily eliminate a whole population. Illustrated here is *Aeonium simsii* from Gran Canaria in the Canary Islands, which is confined to a few steep barrancos towards the center of the island.

In North America large areas of outstanding natural beauty have been designated National Parks, where both plants and animals are protected. Pictured is mixed coniferous forest in Banff National Park in the Rocky Mountains, Canada.

Although this kind of multiple use has been shown to work reasonably well in some developed countries, notably in a restricted way in the National Parks of the United States, only developing countries now have sufficiently large areas. Here it is essential that something of material benefit is extracted from the area and that some of the benefits are immediate and obvious; people who are most likely to put pressure on the area will probably not appreciate the benefits of tourism, for instance, since these are hardly apparent to the local population.

Often, protein is the item in shortest supply, and therefore a possible immediate benefit is the controlled exploitation of grazing animals. This involves education; introduction of domestic animals will modify the ecosystem, but the use of wild animal meat may not be popular. Ranching of wild game has not met with great success except when the herbivore concerned has become semidomesticated. There are possibilities for the future in this, and also great potential for controlled cropping of wild animals; this is especially valuable where the agricultural use of surrounding land has concentrated populations of herbivores into small areas where they overgraze and damage the system. This kind of management could be very valuable in the future, and is perhaps the best way to ensure that threatened ecosystems are maintained in something approaching their original state.

The Future of Conservation

The contrasting examples discussed may indicate that conservation is not a simple concept which can be clearly defined or easily applied. Even in western countries, despite the massive publicity and educational campaigns of the last few decades, it is essentially a middle-class preoccupation. Obviously 50 years have made a difference. The early conservationists of the 1930s such as Paul Sears, Aldo Leopold, William Vogt, G. V. Jacks and R. O. Whyte, were largely unheeded; now few can be unaware of the problems that face our environment and what conservation is all about. Yet, when short-term economic gain or political expediency are involved, conservation still has to take a back seat. A few tens of thousands of hectares of not very productive land may be set aside in countries like those of western Europe which annually cover far more with concrete, but the total area properly conserved is miniscule. How much more difficult it is, therefore, to make a case for conservation in countries where starvation is just around the corner, and where the only people who can enjoy national parks are foreign tourists. But our environment will survive only so long as we do not extract from it more than it can give. For millennia each individual has taken from his environment what he needs and has come to regard as his rightful share, and so long as populations remained small this caused no problems. Population increase means that there are more individuals than there are shares.

In the face of this what can the conservationist do? One of the primary problems that has to be faced is the accelerating rate of change. Western Europe, with its slow initial spread of population and relatively benign climate, has escaped the worst excesses of ecosystem destruction and thus, although relatively little natural vegetation exists, people have learned to modify nature in such a way that they can coexist with it. The ability of plants and animals to survive in changed circumstances is remarkable. Even here, however, the speed and extent of change is increasing and more and more ecosystems and with them species, are being threatened. Other areas of the world are undergoing much more rapid and dramatic change. In 200 years, North America has gone through changes which took 4,000 years in Europe. In the tropics the time-scale is even shorter: removal of vegetation here means instantaneous removal of most of the minerals in the ecosystem, instant soil degradation and severe erosion. It is against such rapid and imminent disaster that action has to be planned.

Ultimately, the use to which any land is put is likely to be determined by economic, social and political criteria. What the conservationist can do is to attempt to make a clear case for the potential benefits that may come from conservation in the long term, as against the immediate benefits that come from exploitation. Although to some the importance of conservation is very clear, to many it appears a luxury. What the conservationist has to do is to show that this is not so. Until decision-makers are convinced of this the conservationist can fight a rearguard action by delaying, if only briefly,

some irreversible action, so that time is allowed for its consequences and possible alternatives to be considered. Ultimately, it must be shown that to neglect conservation is to endanger Man himself. Although in the face of an increasing world population and increasing materialism, political, economic and social pressures demand immediate returns from the land and arguments for long-term planning carry little weight, they must be advanced. Natural environments will continue to diminish and ultimately the results may well backfire on the human race, and even decide its very survival.

s.w.

Genetic Resources in Plants

An important aspect of conservation concerns those plants, such as crops, upon which Man most obviously depends for his survival. This has been the focus of much international concern and action in recent years, and the conservation of the world's genetic resources is therefore dealt with in this section.

The plant breeder's task is truly without end, because of the continuing changes in the many facets of the environment to which crops must be made to respond. Success depends on the genetic resources from which the building blocks for constructive breeding can be selected.

The term "genetic resources" denotes the plant materials available to plant breeders for the genetic adaptation of plants of economic importance. Its introduction in the 1960s was due to the widespread realization that the traditional genetic resources of crops, the ancient land races from which modern crop varieties had derived, were threatened and needed husbanding and preservation like any other valuable resource liable to be used up.

The greatest manageable diversity of resources is an obvious need for the world as a whole, not only for the present but for the foreseeable future. Thus the emphasis is on *use* rather than on assembling collections for their own sake. But since not even current utilization—let alone future needs and opportunities—can be assessed, it is sensible to assemble and preserve a significant representation of the available genetic diversity, especially if this can be achieved

The primitive races of cultivated plants that are the genetic resources from which new varieties are created (and without which they cannot be created) must be located and collected wherever they still exist. Notes on the type of crop, its condition and locality are also vital for further researches. Above all, discussions are held with the local farmers who have spent a lifetime raising these primitive crops and know their characteristics.

without imposing an inordinate burden on the present generation.

The natural genetic diversity, or *gene pool* of economic species—as distinct from variation induced by physical or chemical mutagens—has three distinct categories. These are the *primitive cultivars* or land races of traditional agriculture, the *advanced cultivars* produced by plant breeders in the last 100 years, and the wild or weedy species related to domesticates. Of these groups, the land races have received particular emphasis in recent years because they are rapidly vanishing, giving way to advanced cultivars, even in the remote and until recently little-developed areas where they remained relatively undisturbed.

Land Races

Archaeological evidence shows that many of the species which were domesticated from about 8000 BC onward spread rapidly throughout the then known world (see Domestication of Plants, p. 238). Migration played a major and continuing role in the diversification of crops, for wherever they went they were modified by the environment and by the cultural practices of different peoples.

Evidently the genetic diversity of land races has two dimensions: *between* and *within* sites and populations; both are maintained by natural selection in balance with the environment, and, thanks to the inherent genetic diversity, are responsive to environmental changes. In the absence of strong selection pressures leading to a loss of heterogeneity, genetic diversity was preserved over long periods, and to this day the remaining land races are rich storehouses of such diversity. This is available for selection,

Phoenix theophrasti, the Cretan date palm, is virtually confined to a small river valley leading to the sea in Crete. It is very closely related to the cultivated date palm, differing in its hard fruits and the stems suckering from the base. To some extent it is threatened by the great numbers of tourists it attracts, in particular campers using the stems for firewood and visitors preventing regeneration by driving their vehicles into the center of the grove. Conserving near relatives of crop plants is particularly important since they can be used for breeding such factors as disease resistance into the cultivated species.

either for direct use—usually in the country of origin or in others with similar conditions—or, more commonly, to provide parental material for constructive breeding anywhere in the world.

As parental material, land races can make two distinct kinds of contribution. Firstly, they may contain one or more genes which code for a specific character of economic importance for which effective selection procedures can be set up in hybrid progenies from crosses with locally adapted and productive varieties. Resistance to diseases or pests are the most common examples. Over many years, the International Rice Research Institute, Los Baños, Philippines, made a systematic search for resistance to diseases and pests of rice in the Institute's germ-plasm collection (numbering many thousands of varieties), and succeeded in finding resistance to most of the parasites. Many of the resistance genes discovered were incorporated in varieties of rice now grown in a number of the rice areas of southern Asia.

An American plant explorer, collecting in the home of the pea plant in Near Eastern countries and in Ethiopia, found, in only a few collections made in a remote part of Ethiopia, resistance to a damaging root rot of peas, *Fusarium solani* forma *pisi*. An even more significant discovery was made in 1963 when the world collection of barley of the United States Department of Agriculture was searched for resistance to the barley yellow dwarf virus, which had been identified as the cause of severe damage to cereal crops in many parts of the world. The only source of resistance was found in some of the many hundreds of samples collected in Ethiopia. As far as is known this is due to one gene for resistance, and this is still the only known source.

Apart from such resistance factors, there are many other characteristics that can, and have been, obtained from land races, and those of nutritional or of industrial significance are likely to grow in importance in years to come.

The second major contribution of land races is the great diversity of adaptations they have acquired in a multitude of environments, including stress conditions of various kinds. Crops are grown over a wide range of temperature, water-supply and soil conditions and, as we know from experiments with plants grown in environments of known and defined differences, plant populations tend to respond adaptively to such environmental selection pressures. Over hundreds of generations, adaptive complexes must have been built up in many land races of crops, and, since there must also have been natural selection for productivity, such adaptations hold a good deal of promise for enterprising plant breeders. Evidence for such adaptive complexes for productivity comes from the fact that all the early high yielding "advanced" cultivars were selected from indigenous land races. So far, such ideas have been applied by few plant breeders, but they have included the most successful

ones—for example, breeders in the USSR applying the ideas of N. I. Vavilov and, in Mexico, workers under the leadership of Norman Borlaug.

It has already been emphasized that land races are now a vanishing asset in regions where they prevailed as recently as 30 years ago. In most cases, this is as it should be, since under present-day conditions of high-input production methods and of ever growing demands for higher productivity, they should be replaced by high-yielding varieties wherever possible—even by high-yielding selections from the land races themselves. At the same time, the displaced land-race populations must not be lost, but should be preserved as breeding materials, with as high a level of diversity and integrity as can be contrived.

There are in existence extensive collections of crops, and some of these, in the USSR, the United States and elsewhere, are very large indeed. Most have two important defects: firstly, they contain for the most part "advanced" varieties, with only a small proportion of primitive land races, and these are often kept as individual selections rather than as populations representing the original diversity; and, secondly, many collections have been maintained under conditions necessitating frequent regeneration, which has caused biological and mechanical contamination and loss or change through natural selection. It is therefore essential that whatever land-race material is still available in the field, and especially material which is likely to be replaced by improved varieties, should be collected and preserved as rapidly and efficiently as possible.

Where is such material still to be found? In the 1920s and 1930s the Russian geneticist and plant geographer N. I. Vavilov discovered areas in which crops had great genetic diversity (see Centers of Origin, Diversity and Variability, p. 220), and which had not as yet been reached by modern methods of cultivation and plant improvement. We have come to call them centers, or regions, of genetic diversity. They are situated in parts of Asia, Africa and Latin America, mostly in mountainous and less accessible regions that have remained relatively isolated and undeveloped.

Political independence, the drive for economic development, and population expansion triggered off, or accelerated, agricultural development and the introduction of modern production methods, including the replacement of traditional varieties. The introduction of the new high-yielding varieties, as part of a "package" of high fertilizer application, plant protection, intensive cultivation and, where possible irrigation—the so-called green revolution—was the last, and most effective, step towards the displacement of the traditional varieties which had started in most of the countries in the 1920s or earlier. In some regions, such as most of the Near East, land races of the cereals have all but disappeared except in the more isolated or disadvantaged areas. But

Collected seeds of "primitive" races of crop plants are sent to one of the world's conservation centers or "gene banks" where they are stored under strictly controlled conditions for use in the future. At intervals samples are examined and tested for germination (as here) and, if necessary, they are planted out and the seed harvested for future storage.

collections of many other crops, in the Near East and elsewhere, can still be made, although time is fast running out.

The most hopeful sign is the rapidly growing interest and concern shown by the countries themselves in attempting to preserve what is their genetic heritage, and most of them welcome collaboration with other countries. The location of the remaining resources, and the threat to their continued existence, have become fairly well known in recent years, and the world has been alerted to the need for rapid and comprehensive action.

Scientific and technological problems of *genetic conservation*—the preservation of valuable genetic stocks—have been worked out, including the ecological and population-biological principles of sampling in the field, the classification and evaluation for use in plant breeding, and the organization and distribution of computer-based records, an indispensable aid in making preserved materials available and useful.

Advanced Cultivars

The advanced cultivars, the products of scientific plant breeding in the last 100 years, are the most widely used genetic materials of present-day plant breeders, except for the infusion of specific characters that can be obtained only from primitive cultivars or from wild species. The advanced cultivars have evolved through gradual improvements, and many of the earlier ones are now "museum pieces". Yet their significance should not be underrated. The advanced cultivars may incorporate most of the mutually adapted gene complexes which were assembled in their predecessors by earlier scientific breeders. However, it is likely that the early varieties—selected prior to the widespread use of fertilizers, intensive culti-

vation or irrigation—possessed adaptive characteristics that later became redundant, but which may still be needed for less favored environments and low-input agricultural systems. Similarly, resistance to currently rare biotypes of parasites may be required again in the future. The most convincing reason for retaining representative collections of obsolete cultivars is that they have been—or should have been—described and documented for many characteristics associated with performance. The question arises, in view of their enormous number, which to preserve: judicious culling is inevitable. In fact, such obsolete varieties tend to be fairly well represented in existing germplasm collections of the major crops.

Wild Species

Wild species related to domesticated crops are important genetic resources which no doubt will be extensively used in the years to come. It is well known that many wild species are highly resistant to diseases which attack their domesticated relatives—in some instances, wild relatives are the only effective sources of resistance. The difficulties arising from sterility barriers between species are being overcome by cytogenetical methods which facilitate the transfer of the chromosome segment upon which the resistance gene is located from a chromosome of a wild species to one of a related crop species. This has been effected by introducing the resistance-bearing chromosome of the wild donor into the domesticated receptor species, and by radiation-induced chromosome breakage of both the alien and some receptor chromosome, resulting in the translocation of the relevant chromosome segment. Such transfers have been carried out for resistance to leaf rust (*Puccinia recondita*) from a wild grass, *Aegilops umbellulata*, to common wheat, and, subsequently, similar transfers from several other wild relatives to wheat have been made.

Little use has so far been made of wild progenitors as sources of ecological adaptations, especially for the less favored environments that could be made productive if suitable varieties could be bred—in fact, the opposite of the "green revolution" varieties, which require the best of everything to be entirely successful. The fact that the gene pools of wild ancestors are largely unexplored is no doubt due in part to the difficulty and novelty of such projects and in part to the poor representation of wild relatives in existing collections.

There are many wild species that have not been domesticated but are extensively used by Man, and some of these are acutely threatened in their native habitats. Foremost among these are the forestry species which

Traditional races of crop plants, like this wheat being winnowed by hand in Turkey, may not produce the bountiful yields given by modern varieties, but are becoming increasingly important as sources of genetic diversity.

have their genetic reservoirs in natural forests. In many parts of the world these are rapidly shrinking, due to exploitation or land clearing. This pressure has in recent years greatly intensified, especially in the tropics. Indeed, it has been predicted that by the end of the 1980s some of the major timber-producing countries, such as Malaysia and the Philippines, will have no remaining accessible virgin forests.

This trend is worldwide and threatens not only forestry species but others which have their genetic sources in natural forests—coffee in Ethiopia and many fruit species in Southeast Asia, for example. It must not be forgotten that the tropical forests are the sole remaining reservoir for species which may become economically useful in the future.

Many of the forage species have become domesticated, but others, including most of the pasture and rangeland species, have retained their character of wild species, with "provenances" being recognized—as in forestry—but with little if any breeding for superior performance. Few such species are seriously threatened in their native habitats, yet collections representing diverse genetic resources have proved immensely useful in pasture establishment and improvement in many parts of the world. The production of pastures in Australia, New Zealand and the United States is largely based on grasses and legumes selectively introduced from the Mediterranean region, Africa, Latin America and Southern Asia. Substantial collections have been established, and these should be maintained for use elsewhere—and, of course, by future generations.

Conservation, Evaluation and Utilization

It is clear that the genetic resources described, and especially the land races of traditional agriculture, are not only essential materials for plant improvement but valuable and irreplaceable resources for the future. Two conclusions follow: material should be preserved with the highest practicable integrity, and description and evaluation should be systematically recorded for the benefit of potential users. There is therefore only limited benefit to be derived from collecting activities which are not linked with measures for the long-term preservation of the collected material. Of equal importance is the safekeeping of existing collections, many of which are inadequately housed and hence require frequent regeneration with all the attendant risks already described. Although there are areas requiring further research, for the majority of agricultural crops the scientific and technological problems of conservation are essentially solved.

Conservation is both safest and cheapest if life processes are reduced to a minimum. This is the case in seeds preserved under conditions appropriate to each species. Live plants are subject to genetic change and to losses through parasites, climatic extremes or human error, but in some plants seeds are either not produced or cannot be stored for extended periods. The great majority of seed-reproduced economic species can be safely stored for long periods—decades, and probably centuries—without undue genetic change or loss of viability.

The Panel of Experts on Plant Exploration and Introduction of the Food and Agriculture Organization (FAO) of the United Nations specified long-term storage conditions for "conventional" seeds—ie those which can be stored at low temperatures and low moisture content, at the level of $-18°C$ ($-0.4°F$) or less, at a moisture content of $5\% \pm 1\%$, in airtight containers. This has been adopted by the International Board of Plant Genetic Resources (see below) as standard conditions for long-term seed storage. A number of storage laboratories are now at one or other of these levels, and more are being built or reconstructed.

The viability of stored seed must be checked periodically and, when it drops, seed must be regenerated under conditions which enable the entire population to survive.

Many domesticated species, including fruit and forest trees and many tropical species, cannot be stored at low moisture contents. For these, research is in progress to discover suitable storage conditions. Until the methodology of maintaining tissue cultures over long periods and of regenerating plants from them is more fully developed, such species (and some, such as bananas, which produce no seeds) have to be maintained vegetatively in live conditions. This is expensive and presents major difficulties—for example, how to keep them safe from virus attack.

Genetic conservation of economic plants is not an end in itself but is directed towards its utilization in plant improvement, although not solely in the immediate future. Indeed, concentration on current problems has tended to obscure the long-term problem. But the placing of plants in safekeeping is only a part of genetic conservation, the others being arrangements for multiplication, distribution to users, evaluation of the potential uses and an effective form of documentation and information. Multiplication and distribution must be closely linked, but not necessarily by the institutions equipped for, and carrying out, long-term conservation. This is the case now and is being developed on a worldwide scale. Evaluation calls for the widest possible participation—by specialist institutions, by agronomists and by plant breeders—and their results should be incorporated in the information system.

The scientific and technological problems have been, or are about to be solved, largely through initiatives and activities by the FAO and its panel of experts, and by the International Biological Program. These were taken over and greatly expanded by the International Board for Plant Genetic Resources. The Board was established in 1974 at FAO Headquarters in Rome, with a sizeable staff partly supplied by FAO, and the capacity to raise funds. It supports national and regional efforts in exploration and conservation of genetic resources and aims at establishing a worldwide network dedicated to collecting preserving and using the world's plant genetic resources.   O.H.F.

Gardens

Wittingly or otherwise, the majority of people obtain some idea of the world's plants in the living state from observing them in gardens. Prosperous civilizations have always found the resources to develop gardens which, in addition to the cultivation of culinary and medicinal herbs, provided aesthetically pleasing settings by the judicious planting of a diversity of species. The extent of the floristic diversity available to garden-makers naturally increased as the world became increasingly explored. Indeed, many expeditions, especially those sent from Europe after the Middle Ages, had as a prime or principal purpose the collection of material for cultivation in gardens (see Plant Exploration, p. 12), which emphasizes the fascination that "exotics" (plants from other regions), have always had for gardeners.

The first reliably documented gardens were those of ancient Egypt, Persia and Syria. Water, shade, resting places and, above all, scented retreats from the dusty world, were essential in such hot climates and the early gardeners sought to provide these in a pleasant form. As the influences of Persian gardening spread westward they were adapted by Greece and Rome and, further extended with the Roman Empire, which inevitably carried to its outposts favorite decorative species, such as the horse chestnut (*Aesculus hippocastanum*), as well as developing the concept of the garden as an outside extension of the house.

As Roman influence waned in the West, the Persian gardens were taken eastward to the Indian subcontinent. Their practices, and plants, were carried throughout the Islamic world, ultimately returning to Europe with the Moorish conquest of Spain in the 8th century. While the East and its empires flourished, the West lived through the Dark Ages and gardening was only practiced in those few outposts of Christian civilization, the monasteries, which mostly concentrated on growing vegetables and medicinal herbs. From the 14th century onward the Renaissance gradually saw the redevelopment of creative gardening in Europe. The Italians, in particular, influenced most of the continent with their blend of horticulture and art to give many of the formal gardens which still persist, a trend later developed by the Dutch, masters of bulb cultivation, and extended to North America.

The 18th century saw the development of landscaped, more "natural" gardens, which reached their finest development in the work

Above Bird of paradise flowers (*Strelitzia reginae*) in the National Botanic Gardens of South Africa, Kirstenbosch.

Left The controlled solar environment of the climatron in the Missouri Botanical Garden, St. Louis, United States.

of the Englishman "Capability" Brown (1716–1783), who left the most notable and longlasting impact of Man upon nature. Such gardens sought, through careful plantings, to produce apparently natural parklands in which there was no obvious sign of human formalization. This period saw the flowering of the great era of plant introductions, initially to Europe, by collectors searching the world for plants of horticultural interest—an era which still continues in various forms. Although the major effort to bring exotic plants into cultivation was stimulated, and often financed, by owners of large gardens, whether private or state institutions, the results have eventually percolated throughout the whole of society, so that a wide range of people can now enjoy and learn from the fruits of these endeavors by using material generally available from the businesses selling seeds and plants, as well as by visiting the major gardens themselves.

Botanic Gardens

As we have seen above, gardens started as places where people found shelter and pleasure. Throughout much of their history this has remained essentially true, the ancient gardens of Persia and China, and those of the Italian Renaissance, contained useful and beautiful plants, but the plants themselves were secondary to the concept of the garden as a whole. Although this view of gardens continued, and still persists of course, the medieval herb gardens gradually gave rise to the development of the physic garden, for example London's Chelsea Physic Garden (established 1673), devoted to great collections of medicinal plants. This is not to say that the early physic gardens lacked design; the pleasure garden tradition remained important, as can still be seen in the botanical garden at Padua (established 1542).

As the European countries, notably Holland, Spain, Portugal and Britain, entered their colonial phase in the 16th century, what later became botanic gardens were used to test the economically important plants coming to light through their explorations. The Oratava Botanic Garden on Tenerife in the Canary Islands, for example, was founded in 1788 to acclimatize the new and useful plants sent back by the Spanish conquerors of the New World. The tropical botanic gardens at Singapore, Peradeniya in Sri Lanka and Bogor in Java are examples of what were, in effect, outposts of colonial governments through which new plants were sought and tested for their food, fiber and other economic potential. Continued exploration and enlarged intellectual horizons led to the botanic garden as we know it today.

What started in the physic garden—the idea of plants for a particular use—soon developed into the gathering of plants for ornament, curiosity and show. The orangery, introduced to overwinter citrus trees, soon became the stovehouse and eventually the greenhouse and the great conservatory, housing exotics from all over the world. While rudimentary systems for organizing plants in a taxonomic sense had existed since ancient times, the Linnaean system, introduced in 1753, provided the method by which order could be established and reliable nomenclature achieved at a time when it was desperately needed (see Nomenclature, p. 29).

Most modern botanic gardens serve several purposes. All provide the opportunity to see large collections of living plants labeled as to name and provenance, the hardy ones out of doors, the tender sorts under glass, usually within a specific arrangement or plan. The arboretum, whether a separate entity or included within the bounds of a wider botanic garden, is devoted to woody plants (trees and shrub). It also has its origins in the pleasure gardens of old and the medieval orchard (a grove of trees planted to provide shade and shelter), and these functions are still important—great groves of woody plants providing pleasure, as well as much botanical information, whether they are planted with exotics, as in many parts of Europe, or are protected areas of natural forest, with few exotic plantings, as in some arboreta of the United States of America.

The botanic garden, therefore, acquaints the visitor with a larger measure of botanical diversity than is likely to be encountered elsewhere. Interpretation of this diversity is often pursued through explanatory signs, guides, associated literature or indoor museum displays.

Many botanic gardens offer courses of study in various aspects of the botanical sciences and arts, thereby carrying the interpretative function a step further. Some award diplomas or certificates or are affiliated with institutions of higher learning through which university degrees are conferred.

The larger botanic gardens, and nearly all that are associated with universities, are centers of botanical research and professional training. Although systematic botany remains the principal research focus, a number of gardens also conduct research in such fields as biochemistry, ecology, and paleobotany, and participate in a growing number of cross-disciplinary studies in environmental science. Many of these gardens also maintain great herbaria (collections of preserved specimens) and libraries. International commerce in specimen loans and in supplying bibliographic data or actual books on loan has facilitated botanical research all over the world.

Increasingly, botanic gardens have accepted their role in the conservation of species in danger of extinction. Clearly, the answer wherever possible must be to conserve the habitat where the threatened species occur. Only then can the species maintain themselves in the long term and continue to evolve. Cultivating threatened species in gardens or maintaining them in seed-banks (see Conservation, Evaluation and Utilization, p. 270) is no substitute, being far more expensive and, in addition, very uncertain over an extended period.

Even the oldest extant botanic garden is only about 500 years old and few species have been maintained in cultivation for more than two or three centuries without renewed collecting from the wild. In large botanic gardens the continued loss every year of some taxa means that it is very unlikely that more than a small number of species extinct or threatened in the wild can be maintained in the really long-term. Even more crucial is that in most cases only a fraction of the genetic diversity of species can be held in cultivation and, when grown like this, genetic change in annuals and biennials is a constant factor eroding and altering the original gene-pool, so that the species evolves into a garden plant and may lose its power to maintain itself against competition if reintroduced into the wild.

Nevertheless, botanic gardens have an important role to play in conservation, and some of the problems can be mitigated by ensuring that a species is grown in more than one institution. When a species is lost in the wild or field conservation is a practical impossibility, botanic gardens necessarily become the holding grounds for the future, but here the emphasis should be on reintroducing the species into a protected area within its original range and habitat as soon as possible. By growing a selection of species, rare and declining in the wild, and bulking up their numbers using modern horticultural techniques, botanic gardens are important in providing propagating material to remove the pressure by collectors on remaining wild populations, and in educating the public about threatened plants and the vital need for conservation.

Conservation Again

This book amply shows that the planet we are privileged to inhabit has developed its present green cover over millions of years of plant evolution. Since an inevitable concomitant of evolution is extinction, it may be wondered why so much concern has been expressed, often violently, in recent years about the extinction of plants (and animals). Quite simply, the reason is that Man has demonstrated to himself his ability to exert a major, possibly *the* major, influence on the planet Earth. As an intelligent being, he has come, in part, to appreciate the moral consequences of actions which lead to that terrible finality—the extinction of species (20,000–30,000 species of flowering plants, about 10% of the total, are now threatened or dangerously rare) and plant communities (the once vast tropical rain forests, for example, may effectively follow the dinosaurs within 50 years) as a result of his actions over a relatively short period. Further, he begins to perceive that these actions may quickly destroy his livelihood and very existence. Since much of mankind, however, is mortally concerned with getting a meal tomorrow, whatever the cost to the survival of the world's plant cover, and therefore the animals (including Man) that ultimately depend on it, there is no simple solution to the dilemma.

Nevertheless, Man has been shown in this chapter to be capable not only of destroying the world's plant cover, but also of developing methods by which it can be maintained and utilized for his physical and mental sustenance. If our children, and their children, can recognize in the world they will inherit the *Green Planet* described in this book, then Man has a future.

D.M.M.

Biographies

The following are some of the major figures in the development of plant ecology and plant geography. Cross-references to other entries within this section are given by asterisks. For further information, see the Index.

Braun-Blanquet, Josias (1884–1980)

Born in Chur, Switzerland, and educated in Zürich, where he also spent his early professional life as Chef de l'Herbier at the Rübel Institute for Geobotany, Braun-Blanquet spent most of his life in France. He taught at the University of Montpellier and, in 1929, founded the Station Internationale de Géobotanique Méditerranéene et Alpine (SIGMA). His book *Pflanzensoziologie*, first published in 1928, was of fundamental importance to the development of the Zurich-Montpellier school of phytosociology, of which Braun-Blanquet became the leader, and the students of which have been prominent in describing and mapping the vegetation of so many parts of the world.

Brongniart, Alexandre (1770–1847)

A French geologist, Brongniart's studies of the layers of Cretaceous sedimentary rocks in the Paris basin, with their different series of fossil plants, led him to support, though hesitatingly, the Catastrophic Theory of his contemporary and compatriot Georges Cuvier. On this theory it was proposed that a series of floods had extinguished life at different times in the past. Brongniart's botanist son, **Adolphe-Théodore Brongniart (1801–1876)**, developed a classification system for botanical fossils, and is considered to be one of the founders of paleobotany.

Brown, Lancelot ("Capability") (1716–1783)

An English pioneer of landscape gardening, Brown trained as a gardener and developed his theories on design at Stowe in Buckinghamshire, under the influence of William Kent. Brown's ideal landscape was a complete reversal of the formality of earlier gardens. He sought to imitate nature, exploiting the natural features of the terrain, or even creating them when necessary. He worked on the grand scale, raising hills, damming rivers to form lakes, moving estate villages *en bloc*.

The typical Brown park had a lawn near the house, planted with apparently random groups of trees, leading down to a serpentine stretch of water and, beyond that, more parkland, the whole landscape enclosed by a "fringe" of thickly planted trees to obscure the boundaries. The ha-ha, or sunken fence, allowed cattle to graze within the park, seemingly in an Arcadian setting. Brown gained his nickname from the way in which he referred to a park's "capabilities", or possibilities for improvement.

Brown, Robert (1773–1858)

At heart a taxonomist, this British botanist was responsible for reintroducing the Natural System of classification through his dissatisfaction with *Linnaeus' Sexual System and through his close links with the French school of Augustin *de Candolle and de Jussieu.

From an early date, Brown's taxonomic work relied heavily on the microscope. He described for the first time the cell nucleus in plant tissues, and

streaming protoplasm, as well as "Brownian movement"—the universal phenomenon of "random" molecular collisions which he observed in his research on pollen grains. Also from his microscopic studies on Proteaceae began the science of palynology (the study of pollen and spores).

Clements, Frederic Edward (1874–1945)

An American botanist who developed the vigorous school of ecology at the University of Minnesota, Clements' *Research Methods in Ecology* (1905) was the first comprehensive treatment of the application of experimental methods in the field. He insisted on the dynamic nature of vegetation: that each plant community is part of a succession which changes continuously towards an optimum, which he called the climax. This concept—partly shared by the Englishman Arthur *Tansley—proved important in Anglo-American ecology for a long period. Clements also believed that a plant community was comparable to a superorganism, with a life history and a correlation of its components paralleling those of an individual organism. He also carried out genecological studies, using the behavior of plants under controlled conditions to evaluate environmental factors in the field.

Darwin, Charles Robert (1809–1882)

This English naturalist studied medicine at Edinburgh and later theology at Cambridge, where his desire for travel and his interest in natural sciences developed through his friendship with Adam Sedgwick, Professor of Geology, and John Henslow, Professor of Botany, and through his reading of *Humboldt and Herschel. As a result of reading Lyell's *Principles of Geology* (1830) his attention was focused on the fossil evidence for the very long terrestrial time-scale and the changes that had occurred in animal and plant structure in that period.

Darwin became naturalist on *H.M.S. Beagle*, which circumnavigated the world from 1831–36. He called this "... by far the most important event in my life and has determined my whole career ...". At the time of Darwin's voyage the evolution-

Darwin, Lyell and Joseph Hooker debating the proposed joint publication of the findings of Darwin and Alfred Wallace on the origin of species.

ary scene was largely Biblical. The French zoologist *Lamarck, in his *Philosophie zoologique* (1813), was the first to draw together the evidence for evolution as an explanation of life. Darwin's grandfather, Erasmus Darwin (1731–1802), English poet and naturalist, proposed in his book *Zoonamia* (1794–96) a common ancestry for all life and this, together with the views put forward in 1838 by Thomas Malthus in his *Essay on the Principle of Population*, helped Darwin to crystallize his views on natural selection and the survival of the fittest.

Darwin's first notes were written in 1837, and in 1856 he began to write his book on natural selection. When, in June 1858, he received a set of notes from the naturalist Alfred Russell *Wallace, whose research had led him to strikingly similar conclusions on the origin of species through natural selection, Lyell and Joseph *Hooker recommended joint publication before the Linnean Society, which occurred on July 1, 1858. The painstakingly documented and brilliantly deduced *Origin of Species* created the anticipated furore of disagreement when it was published. Subsequent editions carried modifications, incorporating the theories for use and disuse of parts and the effects of external conditions.

From 1836 to his death, Darwin's work was influential and prolific, including the popular account of his voyage, *Coral Reefs* (1842), *Volcanic Islands* (1847), *Geology of South America* (1846), *Fertilization of Orchids* (1862), *Effects of Cross and Self Fertilization* (1876) and *Action of Worms* (1881). His *Descent of Man* (1871) first emphasized Man's descent from animals and his ability to inherit culture. The later theories of mutation, the discovery of chromosomes and Mendel's laws of heredity enhanced rather than detracted from Darwin's work.

De Candolle, Alphonse Louis Pierre Pyramus (1806–1893)

Professor of Botany at Geneva from 1842, Alphonse De Candolle succeeded his illustrious father Augustin, the last 10 volumes of whose *Prodromus* he completed. His *Géographie botanique raisonée* (1855) has been rightly considered to have brought the subject of plant geography to age. In this he posed the question of why plant species occurred in some areas and not in others, and then provided surprisingly modern criteria for

finding the answers. Of particular importance, he was impressed by the Darwinian theory of evolution and by the data relating to the Earth's history, all pointing to environmental and biological change.

De Candolle, Augustin Pyramus (1777–1841)

Born in Geneva, where he studied under the eminent Swiss botanist J. P. Vaucher, De Candolle later lived in Paris (1796–1816). There he worked with the French zoologists Jean *Lamarck and, later, Georges Cuvier, whose assistant he became at the Collège de France, gaining a sound introduction to classification, zoology, geology and fossil history. He published numerous monographs on families as diverse as the Cruciferae, Leguminosae, Compositae and Liliaceae, as well as revising Lamarck's *Flore Française* (1805–1815). His *Théorie élémentaire de la botanique* (1813) developed the theory that natural classification is based on discovering the symmetry of form in plants, especially of floral organs, and that physiology is not relevant. This book also helped to establish standard Latin botanical nomenclature.

In 1816 De Candolle took the chair of Natural History at Geneva, where he re-established the Botanic Garden, established a Natural History Museum and founded the Conservatoire Botanique. Here he also began his most famous publication, *Prodromus Systematis Naturalis Regni Vegetabilis* (1824–1873), completing 7 of the 17 volumes before his death. This work dealt not only with classification but included information on such subjects as ecology, phytogeography and agronomy—a trend continued by his son Alphonse.

Douglas, David (1798–1834)

A Scottish plant explorer, Douglas's many introductions were of immense significance to European horticulture. In 1820 he was taken onto the staff of the Glasgow Botanic Garden and two years later he was sent by the Royal Horticultural Society of London on the first of his several expeditions to North America and Hawaii. Over 200 new species were collected by him, including conifers such as *Pinus lambertiana*, *Pseudotsuga taxifolia*, *Abies grandis*, *A. nobilis* and *Picea sitchensis*, and shrubs such as *Garrya elliptica*, *Mahonia aquifolium* and *Ribes sanguineum*.

Douglas met an untimely and horrific death in Hawaii when he fell into a pit for trapping wild animals and was savaged by an ensnared bull.

Drude, Carl Georg Oscar (1852–1933)

Professor of Botany and Director of the Botanic Garden at Dresden, Drude was a student and follower of his compatriot *Grisebach. Amongst his numerous publications, probably the most important are *Handbuch der Pflanzengeographie* (1890) and *Oekologie der pflanzen* (1913). In these he presented an outline of the world's vegetation on the basis of formations, which he conceived of as areas, of any size, carrying vegetation of more or less uniform physiognomy, but also taking into account floristic features, environmental factors and biological relationships. Against this background, in 1910 he began with *Engler *Die Vegetation der Erde*, an unfinished series of monographs on various parts of the world.

Du Rietz, Gustaf Einar (1895–1967)

Born in Stockholm, Sweden, Du Rietz attended Uppsala University where, after 13 years as a lecturer, he succeeded Rutger *Sernander as professor of plant ecology (1934). Du Rietz's doctoral thesis on the methodological basis of modern phytosociology reflected Sernander's in-

Augustin Pyramus de Candolle, a botanist who helped develop the natural system of classification.

fluence and he was a leading figure in the Uppsala School of phytosociology, contributing also to many aspects of plant geography. Du Rietz was also a notable taxonomist, with studies of several groups to his credit, including Southern Hemisphere *Euphrasia*, while his rather precise definitions of the infraspecific categories, subspecies, variety and forma in 1930 have greatly assisted later taxonomists.

Engler, Heinrich Gustav Adolf (1844–1930)

A distinguished German taxonomist and phytogeographer, with a particular interest in the plants of Africa and Asia, Engler was professor of the Museum and Botanical Garden at Berlin-Dahlem. Amongst his noteworthy taxonomic publications is the 20-volume *Die Natürlichen Pflanzenfamilien* (1887–1899), in which, with K. Prantl, he provided a systematic account of the families and genera of the world according to a classificatory system which has been much followed by later workers. Together with *Drude he edited the unfinished monographic series *Die Vegetation der Erde*, begun in 1910, and made important contributions to the study of endemism.

Farrer, Reginald (1880–1920)

An English naturalist, gardener, plant explorer and alpine enthusiast, Farrer collected specimens from the European Alps and traveled with William Purdom on the Chinese-Tibetan border (1914–15) and with E. H. M. Cox in Upper Burma (1919–20). His best-known introductions include *Viburnum fragrans*, *Gentiana farreri*, *Meconopsis quintuplinervia*, *Clematis macropetala*, *Rosa farreri* (the threepenny bit rose), *Rhododendron calostrotum*, *R. sperabile* and *R. caloxanthum*.

He published several books, including *The Gardens of Asia* (1904), *Among the Hills* (1911), *On the Eaves of the World* (1917) and *The Rainbow Bridge* (1921). *The English Rock Garden* (1919) did a great deal to popularize the then undeveloped art of rock gardening.

Forbes, Edward A. (1815–1854)

An English biologist who recognized the importance of glaciations in determining the distributions of plants and animals at higher latitudes, Forbes considered that different geographical components in the British flora (and fauna)

resulted from successive migrations across land bridges exposed as seawater was locked up in the glaciers. Forbes' use of the term Pleistocene as equivalent to the Glacial Epoch is still followed.

Forrest, George (1873–1932)

One of an illustrious line of Scottish plant explorers, Forrest was one of the most prolific collectors of the Chinese flora. After a short period of working in a chemist's shop in his home of Falkirk, he visited Australia and South Africa, coming back to work for a year or so in the Herbarium at the Edinburgh Botanic Gardens. From there he went to China, probably as the first collector to work for a syndicate rather than a horticultural society or a botanic garden.

In all, he made seven expeditions to Yunnan from 1904 to 1932, using native collectors whom he had trained. His collections were large and important, including 300 species new to science, 31,000 impeccably prepared herbarium specimens and vast quantities of seed. Among the many beautiful plants he introduced were *Gentiana sinoornata*, several primulas (including *Primula malacoides*), *Pieris forrestii*, *Abies forrestii*, *Hypericum patulum* and various rhododendrons such as *Rhododendron repens* and *R. giganteum*.

Fortune, Robert (1813–1880)

After training at Edinburgh Botanic Gardens, Fortune became a highly successful plant explorer. He traveled and collected plants in China and Japan (1843–1862), initially for the Royal Horticultural Society and later for the East India Company. He toured Anhwei Province in China (dressed in native disguise) to procure tea plants for the East India Company for their Indian tea plantations. He used the newly devised Wardian cases to transport thousands of specimens. In China he faced many dangers and wrote vividly about his experiences in several books, including *Three Years Wandering in the Northern Provinces of China* (1847) and *A Journey to the Tea Countries of China* (1852).

Fortune's plant introductions were instrumental in creating greater interest in the chrysanthemum and also included such species as *Anemone japonica*, *Weigela rosea*, *Prunus triloba*, *Mahonia bealei*, *Primula japonica*, *Rhododendron fortunei*, *Jasminum nudiflorum*, *Dicentra spectabilis* and many hostas.

Gray, Asa (1810–1888)

A leading American botanist and taxonomist, Gray was a strong advocate of the natural system of classification and of Darwinism. His monograph on the relationship of the plants of Japan to those of North America made him a pioneer in the study of discontinuous plant distributions. In 1835 he became curator/librarian of the New York Lyceum of Natural History, issuing *Elements of Botany* (1836). With Professor Torrey he published *Flora of North America* (2 vols, 1838–43) and in 1842 became natural history professor at Harvard, creating and building up the herbarium and library.

Despite his Presbyterian background Gray led pro-Darwinism in America, being a confidant of *Darwin and a member of the innermost circle of his correspondents over many years. Gray's other publications included *Botany of the Northern United States* (1848), *Genera Florae Americae Boreali-Illustrata* (1848–49) and *Darwiniana* (1876).

Grisebach, August Heinrich Rudolph (1814–1879)

An eminent German taxonomist, Grisebach's major contribution to plant geography is set out in his *Vegetation der Erde* (1872). In this he recognized 60 major vegetation units in the world, introducing for the first time the concept of floristic provinces; these are areas of more or less homogeneous vegetation, both floristically and physiognomically, usually delimited by natural barriers. Grisebach pointed out the importance of climate, geological history and topography in determining plant distributions.

Hooker, Sir Joseph Dalton (1817–1911)

After graduating from Glasgow University in 1839 with a degree in medicine and an enthusiasm for botany, Hooker was appointed Assistant-Surgeon and Naturalist aboard the exploration ship *Erebus* and took part in the scientific explorations of Australasia, Antarctica and the Falkland Islands. This work culminated in the *Flora Antarctica* (1844–47), *Flora Novae Zelandiae* (1855) and the *Flora Tasmaniae* (1860). His introduction to the latter laid the foundations for all later work on the phytogeography of the temperate Southern Hemisphere.

Hooker sailed to India in 1847 and his work there was published as the *Flora of British India* (1872–96). He returned to England and became Assistant Director to his father, Sir William Jackson Hooker, at Kew in 1855, taking over as Director in 1865 and continuing the editorship of *Icones Plantarum* and the *Botanical Magazine*. He was knighted in 1877.

Hooker, Sir William Jackson (1785–1865)

An English botanist who was the first director of the Royal Botanic Gardens, Kew, and an authority on cryptogamic botany, Hooker became Professor of Botany at Glasgow in 1820 and editor of Curtis's *Botanical Magazine* in 1827. In 1841 Kew Gardens ceased to be a royal possession and Hooker was appointed Director. Inspired by Joseph Banks's vision of Kew as a great clearing house for the diffusion of botanical knowledge and the distribution of economic and ornamental plants, Hooker devoted himself to this task for the remainder of his life. He corresponded with all the principal botanists of the day and amassed an enormous herbarium which was later purchased for Kew. He also wrote voluminously, his *Icones Plantarum* appearing between 1827–54.

Humboldt, Baron Alexander Von (1769–1859)

A German scientist and plant explorer, Humboldt was originally trained as an engineer and subsequently became a geologist and a mining expert. His yearning for scientific exploration and adventure led him to join Aimé Bonpland, a French surgeon with a great interest in botany, on an expedition to South America. They left Europe in June 1799, eventually landing at Cumaná in Venezuela. For the next five years they traveled extensively across the continent, visiting Colombia, Ecuador, Peru, Mexico and Cuba, documenting their progress and collecting plant specimens to send or take back to Europe. In 1804 they started on their homeward journey bringing with them some 32 cases of specimens.

For the next 23 years Humboldt lived in Paris, writing up the various aspects—geological, climatological, botanical and ethnographic—of his expedition. Owing to the great range of his knowledge, few expeditions have been so exhaustively described; in particular, there was an enormous number of new botanical descriptions. Of the many plants that were brought back, the large number of different forms of *Dahlia variabilis* aroused an interest in this plant that has persisted ever since. Humboldt spent the last 25 years of his life writing his magnum opus *Kosmos* (published

Frank Kingdon-Ward, a plant explorer noted for his work in East and Southeast Asia.

1845–62), which might be described as a synoptic view of the Universe.

Kingdon-Ward, Frank (1885–1958)

One of the most eminent English plant explorers of this century, Kingdon-Ward's major expeditions were to western China, Tibet, Sikkim, Assam and Burma. He wrote several books describing his travels and the many species of plants he encountered. He was most successful in procuring new types of rhododendrons and primulas, the latter being his favorite plants. His name is marked by the yellow *Rhododendron wardii* and *Lilium wardii*.

Lamarck, Jean Baptiste Pierre Antoine de Monet, Chevalier de Lamarck (1744–1829)

A French naturalist and biologist, Lamarck put forward various arguments for a serious theory of evolution, 50 years before Darwin.

His early interest in botany is reflected in the *Flore française* (1778), published when he was in charge of the herbarium at the *Jardin du Roi* (later the *Jardin des Plantes*) in Paris.

In his *Natural History of Invertebrates* (1815–22) Lamarck developed a more detailed classification of the invertebrates, which had been left in an incomplete form by *Linnaeus. Lamarck's most notable work, however, is probably his *Zoological Philosophy* (1809), in which he sets out his theory of "the inheritance of acquired characteristics". His ideas, although largely displaced by those of Darwin, have been upheld by some scientists in recent years, notably the Soviet biologist Trofim Denisovich Lysenko.

Linnaeus, Carolus (1707–1778)

Born Carl von Linné, in Råshult, Sweden, Linnaeus attended Lund University (1727) and about a year later he moved to Uppsala University, where he began to lecture and conduct field courses in 1730. He undertook an exploratory journey through Lapland in 1732 under the auspices of the Uppsala Academy of Sciences; he travelled 7,400km (4,600mi) and collected hundreds of new plant species. On his return he published *Flora Lapponica* (1737). Later he traveled extensively in Germany, Holland, England and France before returning to Uppsala in 1740 as Professor of Natural History.

During this period he continued to develop his ideas on the classification of plants, and his publications included *Genera Plantarum* (1737) and his greatest work, *Systema Naturae* (1735–58), which contained his system for classifying plants according to the number of

stamens in their flowers. Although we no longer use exactly this system, Linnaeus was the first to recognize that flower structure was of far greater value than other characters for classifying plants. The credit for developing the system of ranked taxonomic grouping, with species grouped into genera, genera into families etc, attributed to Linnaeus, was built on the earlier work of others such as Ray, Tournefort and Artedi.

Linnaeus' other great achievement was his binomial system for naming plant and animal species, using one Latin word to signify the genus and another to distinguish the species within that genus. Each species was thus given one, and only one, proper binomial name that could be recognized throughout the scientific world, irrespective of local language usages.

Meyen, Franz Julius Ferdinand (1804–1840)

Professor at the Friedrich Wilhelms-Universität of Berlin, Meyen was an early successor to *Humboldt and had also travelled rather widely. In his *Grundriss der Pflanzengeographie* (1836) he delimited vegetation regions on the basis of 20 physiognomic growth-forms and explicitly rejected the use of floristic data for recognizing such regions. While he admitted that mutability of species would help in understanding plant distributions he was unable to believe that such evolutionary change was possible.

Mutis, José Celestino (1732–1808)

A Spanish botanist and plant collector, Mutis spent much of his life in Colombia, South America. He arrived in Bogotá in 1760 where he was employed as physician to the new Viceroy but his correspondence with *Linnaeus over the next 17 years increasingly turned him to botany. He was first professor of natural history at the University of Bogotá and, in 1783, organized a large "Expedición Botánica". His students and associates collected from all regions of the country although he never traveled far himself; he remained in Bogotá to supervise the botanic garden and a team of artists executing a large number of plant drawings, eventually totalling 7,000 drawings of 2,700 species, now largely held in Madrid and many of which are still unpublished.

Raunkiaer, Christen (1860–1938)

A Danish ecologist concerned with the ecological classification of plants in terms of survival mechanisms, Raunkiaer postulated that the earliest flowering plants originated under tropical conditions whence they spread to areas with unfavorable periods, where they evolved a variety of structural and functional survival mechanisms.

Using over 1,000 species, he characterized different environments according to the proportion of species with different bud positions in relation to the soil surface; in the tundra, for example, buds of most species would be buried below the surface in winter, whereas in the tropics they are freely exposed. Although elaborated by some, and partly superseded by others, Raunkiaer's classification is still being used. He published *Life Forms of Plants and Statistical Plant Geography* (1934).

Rübel, Eduard (1876–1960)

A noted Swiss phytogeographer to whom the Geobotanische Institut ander ETH Stiftung Rübel in Zürich is a lasting tribute, Rübel developed, together with M. Brockmann-Jeroch, an ecological-physiognomic classification of vegetation (*Lignosa*, *Herbosa* etc) which was based on the climatic formations of *Schimper. His extensive phytosociological studies in many parts of Europe and North Africa, some in conjunction with *Braun-Blanquet, contributed to the developing Zürich-Montpellier School of Phytosociology which is so important today.

Schimper, Andreas Franz Wilhelm (1856–1901)
Born at Strasbourg, where he attended university, Schimper's early interest in crystallography led him into histological studies of crystalline inclusions in plant cells, including starch-containing organelles. From this developed his physiological viewpoint of phytogeography, when he entered that field, and this is reflected in his major work *Pflanzengeographie auf physiologischer Grundlage* (1893), translated into English 2 years after his death. This richly illustrated description of world vegetation types was based on his studies of plants of special sites, such as deserts, mangrove swamps etc, which impressed him with the importance of physiological adaptations.

Sernander, J. Rutger (1866–1944)
Professor of plant ecology at Uppsala University until 1934, Sernander was a pioneer in developing the Uppsala School of phytosociology, in which the stratification of vegetation was emphasized. From his studies on Swedish peat deposits he was also instrumental in the widespread acceptance of the Boreal, Atlantic and other post-glacial climatic stages which were later related to Scandinavian pollen zonation. This "Blytt-Sernander" scheme developed out of earlier studies by Axel Blytt on the elements of the Norwegian flora which he considered had immigrated at different times during alternating wet and dry periods.

Skottsberg, Carl Johan Fredrik (1880–1963)
Born in Karlshamn, Sweden, Skottsberg was appointed lecturer (1907) and then Keeper of the Herbarium in Uppsala University until 1915, when he began to plan and build Gothenburg's botanical garden, remaining its Director until 1948. His taxonomic and phytogeographical

Linnaeus studying plants by night.

knowledge was firmly based on his extensive field studies. He was botanist on the Swedish Antarctic Expedition (1901–1904) and leader of the Swedish Magellan Expedition (1907–1909) and the Swedish Pacific Expedition (1916–1917), but also made many other journeys of shorter duration.

Although Skottsberg's more than 200 publications range widely over his fields of interest, his major contributions undoubtedly reflected his love of the southern temperate zone and his works on the floras and vegetation of the Falkland Islands, Juan Fernández and Easter Islands, Tierra del Fuego and the southern Andes are fundamental to other studies in these regions.

Tansley, Sir Arthur George (1871–1955)
A prominent figure in the development of plant ecology and nature conservation in the British Isles, Tansley considered that ecology is primarily concerned with studying plant associations, the species and individuals composing them and their relations to each other and the physical environment, a view fundamental to much British and American ecology. He was largely responsible for the formation in 1913 of the British Ecological Society and he wrote many books, including *Types of British Vegetation* (1911), *The British Islands and their Vegetation* (1939) and *Britain's Green Mantle* (1949).

Turesson, Göte Wilhelm (1892–1970)
A Swedish botanist, Turesson founded the science of genecology, concerned with the genetical adaptation of plants to their environment. His ideas were developed from extensive field experiments in southern Sweden, growing a range of species which, he found, consisted of a number of genetically different ecotypes, each adapted to, and resulting from selection by, local environmental conditions.

Turesson studied in the United States, returning to Sweden to teach at Lund University when he was 23. In later life he worked at Uppsala.

Vavilov, Nikolai Ivanovich (1887–1942)
A Russian geneticist and agricultural botanist, Vavilov is noted for his work on the origins of cultivated plants, particularly cereals and legumes. With the aid of numerous expeditions to Central and South America, Southwest Asia and China, he firmly established the concept of the Centers of Origin of crop-plants, now more usually known as Centers of Diversity. This concept is based upon the fact that most variation in native ancestors of crop-plants is exhibited in certain very restricted areas of the world. His law of Homologous Series was based on his observations on the parallel evolution of many minor characteristics of related species and genera. Vavilov, whose theories were opposed to those of *Lysenko (who believed in the inheritance of acquired characteristics) is reported to have died in a Soviet concentration camp.

Wallace, Alfred Russell (1823–1913)
An English naturalist, Wallace propounded the theory of the origin of species through natural selection, independently of and at the same time as *Darwin. As a result of his work in the Malay Archipelago (1854–62) and a chance reading of the *Essay on the Principle of Population* by Thomas Malthus he published in 1855 a significant pre-Darwin contribution to the evolutionary theory. His notes, which arrived on Darwin's desk on June 18, 1858, persuaded Darwin to communicate a joint paper to the Linnean Society on July 1, 1858. Thus, "Darwinism" might well have been "Wallacism". Wallace's magnanimity, however, made

him a lifelong correspondent and supporter of Darwin.

Wallace's *Contributions to the Theory of Natural Selection* (1870) allowed, in contrast to Darwin, a spiritual influence to Man's evolution. His other works include *Geographical Distribution of Animals* (1876), which pioneered modern zoogeographical science, *The Malay Archipelago* (1867), *Island Life* (1880) and *Tropical Nature* (1878).

Warming, Johannes Eugen Bülow (1841–1924)
Professor of botany at the University of Copenhagen, this Dane effectively set out the scope and general principles of ecology as it is recognized today. His *Plantesamfund* (1896), rewritten for translation as *Oecology of Plants* (1907), provided an ecological basis for classifying communities utilizing the habit, density, phenology etc of the plants and their adaptation to physically or physiologically available soil moisture. Warming's contention that communities are not theoretical, but natural entities requiring recognition and study in relation to the environment, is still the basis of much ecological study.

Willdenow, Carl Ludwig von (1765–1812)
Professor of botany at the University of Berlin, Willdenow's principal interests were taxonomic and, among other publications, he edited the 4th Edition of *Species Plantarum* and wrote a flora of the Berlin region. In his *Grundrisse der Kräuterkunde* (1792), however, he devotes a chapter to phytogeography and this constitutes the first attempt to explain the world distribution of species in a scientific manner. He recognized natural communities of plants, most obviously those characterized by one prominent species, and suggested former connections between landmasses now widely separated, drawing attention to the limitations imposed by the available means of dispersal.

Willis, John Christopher (1868–1958)
Born at Birkenhead, England, Willis was appointed Director of the Royal Botanic Garden, Ceylon, in 1896. Here he wrote the first edition of his *Dictionary of Flowering Plants and Ferns* (1897), of which the seventh edition (edited by Airy Shaw) was published posthumously in 1966. His statistical studies of the endemic flora of Sri Lanka led to a belief that the data belied natural selection and that evolution was merely a process of differentiation and a function of time. His books on *Age and Area* (1922) and *The Birth and Spread of Plants* (1949) reflected these views but caused other workers to focus attention on the problems posed by endemism. He was Director of the Botanic Garden at Rio de Janeiro (1912–13) and wrote a monograph on the Podostemaceae.

Wilson, Ernest Henry (1876–1930)
An English plant explorer, Wilson earned the nickname "Chinese Wilson" for the wealth of plants that he introduced from China—1,000 new species and 16,000 herbarium specimens. First a nurseryman gardener, then Curator at Birmingham Botanic Gardens, later a student at Kew, Wilson visited China twice between 1899 and 1905 for Veitch's nursery, initially to obtain specimens of the handkerchief tree (*Davidia involucrata*). From 1907 to 1918 he collected plants in China and Japan for Professor Sargent of the Arnold Arboretum, Massachusetts, United States. After being appointed Assistant Director of the Arnold Arboretum (1919) he visited Australia, New Zealand, Central and South America and India. In 1927 he was appointed Keeper at the Arboretum on Sargent's death.

Glossary

Abiotic   Devoid of life.
Abundance   Concerning the contribution (eg in numbers, biomass of a species) made by a species, relative to others, in a community.
Acid   A substance that releases hydrogen (H^+) ions in solution and has a pH of less than 7.
Acuminate (of leaves)   Narrowing gradually by somewhat concave edges to a point.
Adaptability   Ability to adjust to environmental conditions.
Adaptation   The process, genetical or physiological, by which an organism becomes adjusted to its environment.
Adsorption   The process of holding molecules of gases or liquids onto a solid surface.
Adventive   A plant introduced by Man into an area where it is not native.
Aeration   Provision of air.
Aerenchyma   A tissue composed of loosely packed, thin-walled cells with large air spaces between. It is found in stems, leaves and roots of plants that grow in oxygen-starved conditions, particularly in water or waterlogged soils.
Aerial root   A root that originates above ground level.
Aerobic   Requiring oxygen for life.
Afforestation   The planting of trees to produce a forest.
Agamospermy   Reproduction by seed formed without sexual fusion of gametes.
Albedo   That amount of incident radiation which is reflected by a surface.
Algal bloom   The turbid appearance of water due to the presence of large numbers of unicellular and filamentous algae.
Alkaline   Pertaining to substances that release hydroxyl (OH^-) ions in solution and have a pH of more than 7.
Allele   One of the alternative forms of a gene.
Allopatry   (of populations, species and other taxa) With distributions that do not overlap.
Allopolyploid   A polyploid organism in which one or more sets of chromosomes come from dissimilar populations, usually of different species.
Amphidiploid   An allopolyploid.
Anaerobic   Able to live without oxygen.
Angiosperm   A plant producing seeds enclosed in an ovary. A flowering plant.
Anion   An ion with a negative charge.
Annual   A plant that completes its life cycle from germination to death within one year.
Anther (of flowers)   The terminal part of the male organs (stamen), usually borne on a stalk (filament) and developing to contain pollen.
Anthropogenic   Caused by Man.
Apoplast   See Casparian band.
Appressed   Pressed closely to a surface and pointed to the apex of the organ.
Arborescent   Tree-like in size and appearance.
Aspectation   The situation where species growing together have their main growth periods at different times of the year, thus reducing mutual interference.
Aufnahme   Sampling unit in phytosociological studies of vegetation—also called a relevé.
Autochory   Dispersal of diaspores (seeds, spores etc) by the plants' own mechanisms, as in explosive dehiscence of balsam (*Impatiens*) fruits.
Autopolyploid   An organism that has more than the normal two sets (diploid) of chromosomes, all derived from the same parent species. (cf Allopolyploid, Amphidiploid.)
Autotoxicity   Poisoning by substances produced by changes within the organism.
Autotrophic   Able to synthesize food substances from inorganic materials. (cf Heterotrophic.)
Auxin   A substance that controls plant growth.
Awn   A stiff, bristle-like extension to an organ, usually at the tip.

Basal   Borne at or near the base.
Benthic   Pertaining to the bottom of the ocean or a lake.
Berry   A fleshy fruit without a stony layer, and containing one or more seeds.
Biennial   A plant that completes its life cycle in more than one, but less than two, years and which usually flowers in the second year.
Biocide   A substance that kills living organisms.
Biosystematics   The study of the classification and relationships of organisms using the techniques and concepts of cytogenetics.
Biotic   Pertaining to life or living organisms.
Bisexual (of flowers)   Containing both male and female reproductive organs.
Bract   A leaf, often modified or reduced, which subtends a flower or an inflorescence in its axil.
Breccia   A type of rock composed of angular fragments cemented together by some material such as lime.
Bryoid   Pertaining to bryophytes (mosses, liverworts) and plants such as filmy ferns (eg *Hymenophyllum*) resembling them.
Bryophyte   A moss or liverwort.
Bulb   An underground organ comprising a short, disk-like stem, bearing fleshy scale-leaves and buds, and surrounded by protective scale-leaves; it acts as a perennating organ and is a means of vegetative reproduction.
Bulbil   A small bulb or bulb-like organ often produced on aboveground organs.

Calcareous   Containing lime.
Calcinomorphic (of soils)   Developed over calcareous rocks which largely control their characteristics.
Calcrete   The lime-crust found over limestones in arid areas when any water with its calcium carbonate moves upward by evaporation.
Calyx   Collective term for all the sepals of a flower.
Capsule   A dry fruit which normally splits open to release its seeds.
Capitulum   An inflorescence consisting of a head of closely packed stalkless flowers.
Carnivorous plant   A plant that is capable of catching and digesting small animals such as insects.
Carotenoids   Fat-soluble pigments including carotenes (yellow and orange) and xanthophylls (yellow); they are found in chromoplasts and chloroplasts of plants.
Carpel   One of the flower's female reproductive organs, comprising an ovary and a stigma, and containing one or more ovules.
Casparian band   A layer of thickening, mainly suberin (qv), in the cell walls of a sheath of cells known as the endodermis, which is located between the outer cortical layer and inner stele (water- and nutrient-conducting layer) in all roots and many stems of higher plants. The Casparian band prevents the direct passage of water and solutes through the apoplast (ie between the cell walls and through the intercellular spaces), which must in consequence enter the cells to allow passage from the cortex to the stele.
Catastrophic selection   The elimination of ecologically marginal populations, except for one or more exceptionally adapted individuals, by an environmental extreme.
Cation   An ion with a positive charge.
Cerrado   Brazilian term for savanna-woodland.
Chert   A form of quartz resembling flint.
Chlorenchyma   Parenchymatous cells that contain chloroplasts.
Chloridoid   Resembling or pertaining to the grass genus *Chloris*.
Chlorophyll   The green pigment of plant cells necessary for photosynthesis.
Chloroplast   A plastid containing chlorophyll; found in algal and green plant cells.
Chlorosis (of green parts of a plant)   Turning yellowish or whitish.
Chromoprotein   A light-absorbing substance comprising an absorbing chromophore attached to a protein which modifies the wavelength.
Chromosomes   Thread-like strands of DNA and proteins which occur in the nucleus of living cells and carry the units of heredity, the genes.
Chronistics   A discipline concerned with attempts to determine relationships between taxa with reference to an assumed evolutionary time-scale.
Cladogram   A branching diagram representing the phylogenetic relationships between taxa (the branches) in which the vertical scale is considered to indicate time or evolutionary advancement.
Cline   A gradual change in any measurable character with populations, species etc.
Clone   A group of plants which have arisen by vegetative reproduction from a single parent and which are, therefore, genetically identical.
Co-dominant   One of two or more species that together dominate a vegetation type.
Columnar   Column-shaped or with column-shaped constituents.
Compound   Consisting of several parts; eg a leaf with several leaflets or an inflorescence with more than one group of flowers.
Conifer   A cone-bearing tree, eg pine, larch, fir etc.
Consociation   A part of an association dominated by one species.
Constancy   The occurrence of essentially the same association of species in different localities.
Convergent evolution   The process by which similar structures are developed independently in unrelated organisms; often shown by organisms inhabiting similar environments.
Corolla   All the petals of a flower; it is normally colored.
Cross-fertilization   See Fertilization.
Cross-pollination   See Pollination.
Cryopedogenesis   Development of soils under cold conditions.
Cryptogam   A general term applied to plants that do not produce seeds, eg mosses and ferns.
Cryptophyte   Plants with their dormant buds and other surviving organs situated below the soil or water surface.
Culm   The stem of a grass or sedge.
Cuticle   The waxy or fatty layer on the outer surface of epidermal cells.
Cycadophyte   Member of the phylum Cycadophyta, the cycads; they belong to the general group knows as gymnosperms.
Cytology   The study of cells and their internal structure.

Deciduous (of plants)   Shedding their leaves seasonally.
Dehiscent   Splitting open to discharge the contents. (cf Indehiscent.)
Desmids   A group of freshwater green algae in which the cell-wall is in two halves, united by a narrow connection.
Diagenesis   A general term for the processes by which sedimentary rocks are formed.
Diaspore   A dispersal propagule, eg a seed, fruit, spore etc.
Diatoms   A class of unicellular marine or freshwater algae with the brown pigment isofucoxanthin, as well as chlorophyll, in the chloroplasts. The cell-wall is in two halves, one overlapping the other, is made of pectic materials impregnated with silica and is finely sculptured.
Dicotyledon   A member of one of two subclasses of angiosperms; a plant whose embryo has two cotyledons. (cf Monocotyledon.)
Dinoflagellate   A member of the algal class Dinophyta—single-celled aquatic algae which move by the action of two flagella.
Dinophyta   See Dinoflagellate.
Dioecious   Having male and female flowers borne on separate plants.
Diploid   Having two sets of chromosomes in the nucleus of its cells, one set from each parent.
Dormancy   The condition of being inactive, usually during unfavorable conditions.
Dorsal   Upper.

Edaphic   Pertaining to the soil.
Eluviation   Removal of fractions of the clay-humus complex in soils by podsolization or laterization.
Endemic   A population, species, genus or family

which is of a more restricted distribution than is expected for that rank.

Endozoochory (of fruits, seeds etc) Dispersal within animals, eg in the gizzard or gut.

Enzyme A protein which mediates (catalyzes) a specific biochemical reaction.

Epharmony Concerned with unrelated species.

Epibionty The phenomenon of being an "old" endemic with a restricted distribution resulting from contraction of a former wider area.

Epidermis The outer protective, usually single-celled, layer of many plant organs, particularly leaves and herbaceous stems.

Epiphyte A plant that grows on the surface of another, without being dependent upon it for nutrition.

Epizoochory (of fruits, seeds etc) Dispersal outside animals, eg by hooks attaching them to fur or wool.

Ericoid Like *Erica*—a dwarf shrub with small, inrolled leaves.

Etiolated (of plants) Pale in color and straggling in appearance; caused by exclusion of light.

Eukaryote Any organism that has a membrane-bound nucleus, which includes all fungi, green plants and algae (except the blue-green algae).

Eulittoral Referring to the zone and its inhabitants lying between upper and lower tide limits.

Eustatic Concerning worldwide changes in sea-level.

Eutrophic (of lakes etc) Rich in nutrients. (cf Oligotrophic.)

Evapotranspiration The loss of water from an area, both via the vegetation and the ground surface.

Evergreen With leaves all the year round.

Exine The outer layer of the wall of a pollen grain.

Facies Appearance or aspect.

Fell-field See Fjeldmark.

Fertilization The fusion of male and female reproductive cells (gametes). *Cross-fertilization* occurs between gametes from separate plants; *self-fertilization* occurs between gametes from the same plant.

Festucoid Resembling fescue-grass (*Festuca*).

Fjeldmark Areas of rock-debris and open, mineral soil with sparse plant-cover, usually at high elevations with cold conditions.

Flower The structure concerned with sexual reproduction in angiosperms. Essentially it consists of the male organs (androecium), comprising the stamens, and female organs (gynoecium), comprising the ovary, style(s) and stigma(s), usually surrounded by a whorl of petals (the corolla) and a whorl of sepals (the calyx).

Flurapatite A rock-forming silicate in which fluorine is an important constituent.

Foliose Bearing leaves.

Foraminifera Minute marine organisms with perforated shells.

Forb A non-grasslike herb.

Forma (form) A taxonomic division ranking below variety and used to distinguish plants with trivial differences.

Fruit Strictly the ripened ovary of a seed plant and its contents. Loosely, the whole structure containing ripe seeds.

Fruticose Resembling a shrub.

Gamete The haploid reproductive cell (male or female) that is produced during sexual reproduction. At fertilization the nucleus of the gamete fuses with that of another gamete of the opposite sex to form a diploid zygote that develops into a new individual.

Genotype The genetic constitution of an organism.

Genus A taxonomic rank grouping together more or less closely related plants. The genus title is the first word of the species binomial. A genus may be divided into subgenera, sections and series, in descending order of hierarchy.

Geomorphology The study of the physical features of the Earth's surface in relation to the underlying geological structure.

Geotropic Pertaining to plant-growth in relation to gravity.

Ginkgophyte Member of the phylum Ginkgophyta, now including only the maidenhair tree (*Ginkgo biloba*), but with many fossil representatives; included in the general group called gymnosperms.

Glabrous Without hairs or projections.

Globose Spherical, rounded.

Graminoid Resembling a grass.

Graptolite An extinct marine animal found as a fossil in rocks of Paleozoic age.

Gymnosperm A seed plant in which the seeds are not enclosed in an ovary; conifers are the most familiar example.

Gyttja Lake sediments, usually black, containing much organic matter.

Habit The characteristic mode of growth or occurrence of plants; the form and shape of a plant.

Haematite Iron ore in the form of ferric oxide.

Halophyte A plant adapted to growth under highly saline conditions.

Haploid Having within the nucleus of its cell(s) one complete set of chromosomes, as in the gametes or sex cells.

Hardy Able to withstand extreme conditions, usually of cold.

Haustoria Organs of various parasitic or symbiotic plants that are used to draw nutrients from the host plant.

Herb (adj. herbaceous) A plant that does not develop persistent woody tissue above ground and either dies at the end of the growing season or overwinters by means of underground organs, eg bulbs, corms, rhizomes.

Heteromorphy Occurrence in different forms.

Heterospory The condition of having spores of two kinds, usually termed microspores and megaspores.

Heterotrophic Unable to synthesize food from simple inorganic substances and thus requiring a source of organic food, as in the fungi.

Heterozygous Having different forms (alleles) of a gene at the same locus on homologous chromosomes. (cf Homozygous.)

Homozygous Having identical alleles at the same locus on homologous chromosomes. (cf Heterozygous.)

Hornblende A blackish to dark green mineral composed principally of silicates of calcium, iron and magnesium; a constituent of granite and other rocks.

Hybrid The offspring of two plants of different taxa, most often species.

Hydromorphic (of plants) With features related to and developed under conditions of abundant water.

Hydrophyte An aquatic plant.

Hygrochastic (of organs) Opening in moist air and closing in dry air, eg capsules.

Hygromorphic = Hydromorphic.

Hygrophyte = Hydrophyte.

Illite A clay mineral, similar to mica; the dominant mineral in shales and mudstones.

Illuviation Redeposition of leached colloidal materials in different soil horizons.

Imbricate (of sepals, petals, leaves) Overlapping, as in a tiled roof.

Incompatible Of plants between which hybrids cannot be formed.

Indehiscent Not opening to discharge the contents. (cf Dehiscent.)

Inflorescence Any arrangement of more than one flower, eg capitulum, corymb, cyme, panicle, raceme, spadix, spike and umbel.

Insolation Exposure to the sun.

Interglacial Period between glaciations in which temperatures rise sufficiently to support warmth-loving plants and closed forest. (cf Interstadial.)

Interspecific Occurring between members of different species. (cf Intraspecific.)

Interstadial Period between glaciations in which temperatures do not rise sufficiently to support warmth-loving plants and closed forest. (cf Interglacial.)

Intraspecific Occurring between members of the same species. (cf Interspecific.)

Ion An atom or group of atoms that has become electrically charged by gain or loss of negatively charged electrons.

Irradiance Density of light radiation falling on a given surface.

Isopleth Lines on a map connecting places with the same values for rainfall, temperature, height and other such features.

Isostasy The rise and fall of land relative to sea-level as a consequence of the flexibility of the Earth's crust.

Isotope One of two or more forms of an element differing in atomic weight.

Kaolinite Clay with a low exchange capacity—the adsorption of calcium ions is not much greater than that of potassium ions.

Krummholz A term applied to trees at the timberline on mountains, which have their trunks almost horizontal and branches vertical to form a dense scrub.

Lamina The flat blade of a leaf.

Laterite Red or yellowish iron-containing tropical clay soil.

Leaching (of soils) Removal of soluble salts by percolation of water.

Leaf An aerial and lateral outgrowth from a stem which makes up the foliage of a plant. Its prime function is the manufacture of food by photosynthesis. It typically consists of a stalk (petiole) and a flattened blade (lamina).

Lenticel A pore through the bark which allows gaseous exchange.

Lepidendroid Resembling *Lepidodendron*, a giant, fossil "horsetail".

Leptophyll (of leaves) Up to $25mm^2$ in area.

Liana A woody climbing vine.

Lignin One of the most important constituents of the cell-wall in secondary tissues.

Lignotuber See Xylopodium.

Lithosol A young soil developed over solid rock. (cf Regosol.)

Littoral On or pertaining to the shore.

Loess Wind-deposited silt forming deposits up to 50m (160ft) thick; very porous and forming a highly fertile topsoil.

Macromolecule A molecule containing large numbers of atoms.

Macrophyll (of leaves) Between 18,225 and 164,025mm^2 in area.

Malacophyllous With soft leaves.

Maquis A term used in some Mediterranean areas (eg Corsica) for sclerophyllous scrub.

Materia medica Substances used as remedies in medicine, and their study.

Matorral Scrub.

Megaphanerophyte Plants with at least some dormant buds borne on twigs up to more than 30m (98ft) above the ground.

Megaphyll (of leaves) Over 164,025mm^2 in area.

Meiosis Two successive nuclear divisions of a diploid cell in the course of which the chromosome number is reduced from diploid to haploid; the haploid cells often subsequently develop into gametes. (cf Mitosis.)

Meristem A group of cells capable of dividing indefinitely.

Mesic With adequate or high moisture availability or needs.

Mesophanerophyte Plants with at least some dormant buds borne on twigs 8–30m (26–98ft) above the ground.

Mesomorphic (of plants) With features related to adequate moisture availability.

Mesophyll The internal tissues of a leaf which are concerned with photosynthesis. Also, a term used to describe leaves of size between 2,025 and 18,225mm^2—the size range for most leaves from broad-leaved trees of temperate climates.

Mesophyte A plant having moderate moisture requirements.

Mesotrophic Adapted to living in mesic (non-extreme) conditions.

Metazoa A subkingdom containing multicellular animals with differentiated tissues.

Mica A shiny mineral composed principally of aluminum silicate.

Microclimate Local climatic conditions.

Microhabitat The localized environmental conditions of a population or even an individual.

Micronutrients Inorganic elements required in minute amounts for plant growth, eg boron, chlorine, copper, iron, manganese and zinc.

Microphanerophyte Plants with at least some dormant buds borne on twigs up to 2–8m (6.5–26ft) above the ground.

Microphyll (of leaves) Between 225 and 2,025mm² in area.

Microtopography The detailed structure of the surface of a body including a very small part of the Earth's surface.

Mitochondrion An organelle, surrounded by a double membrane, found in eukaryotic cells; contains enzymes vital to cellular metabolism.

Mitosis A form of nuclear division in which the chromosome complement is duplicated exactly and the daughter nuclei remain haploid, diploid or polyploid. The form of division that occurs during normal vegetative growth. (cf Meiosis.)

Monocarpic Fruiting only once and then dying.

Monoclimax A climax vegetation in which only one species dominates.

Monocotyledon One of two subclasses of angiosperms; a plant whose embryo has one cotyledon. (cf Dicotyledon.)

Monoculture Cultivation of a single crop.

Monoecious Having separate male and female flowers on the same plant. (cf Dioecious.)

Monopodial With a single main axis from which branches or appendages arise. (cf Sympodial.)

Montmorillonite A form of clay with a high exchange-capacity, ie it adsorbs calcium more than potassium ions.

Mor Humus formed under acid conditions. (cf Mull.)

Morph Form.

Mull Humus formed under alkaline conditions. (cf Mor.)

Mutation An inherited change in the structure of a gene from one allelic form to another.

Nanophanerophyte Plants with at least some dormant buds borne on twigs up to 25cm–2m (10in–6.5ft) above the ground.

Nanophyll (of leaves) With an area not exceeding 225mm².

Nothophyll (of leaves) Between 2,025–4,500mm² in area.

Nunatak Isolated areas of rock projecting through land-ice or snow.

Nut A dry, single-seeded and non-opening (indehiscent) fruit with a woody pericarp.

Oligotrophic (of lakes etc) Poor in nutrients. (cf Eutrophic.)

Orthoclase A form of felspar in which the crystals have two lines of cleavage at right angles.

Orthogenesis Gradual evolution of related groups of organisms in the same direction due to inherited "tendencies".

Ovary The hollow basal region of a carpel, containing one or more ovules and surmounted by the style(s) and stigma(s). It is made up of one or more carpels which may fuse together in different ways to form one or more chambers (locules).

Ovoid Egg-shaped.

Ovule The structure in the chamber (locule) of an ovary containing the egg cell. The ovule develops into the seed after fertilization.

Paleobotany The study of fossil plants.

Paleomagnetism Magnetism remaining in rocks from the period when they were formed.

Paleontology The study of fossil organisms.

Palmate (of leaves) With more than three segments or leaflets arising from a single point, as in the fingers of a hand.

Palynology The study of pollen.

Pampas Argentinian grasslands (prairies).

Pan A hard layer in soil.

Pangenesis A theory of inheritance in which the cells of an organism were thought to contain particles that circulate freely within it and are transmitted from parent to offspring.

Papilla A small blunt hair or projection.

Pappus A ring of fine hairs (often feathery, sometimes scale-like) developed from the calyx, and crowning fruits of members of the Compositae and some other flowering-plant families.

Páramo Upland treeless plateau in tropical South America.

Parasite A plant that obtains its food from another living plant to which it is attached.

Parenchyma A tissue made up of thin-walled, living, photosynthetic or storage cells, which is capable of division even when mature.

Pathogen An agent that causes disease.

Peat An accumulation of plant material incompletely decomposed due to lack of oxygen, usually as a result of waterlogged conditions.

Perennate To live over from season to season.

Perennial A plant that persists for more than two years and normally flowers annually.

Periglacial Of conditions in areas under the strong influence of, but beyond the margins of, a glacier.

Petiole The stalk of a leaf.

pH A value of hydrogen-ion concentration which gives a measure of the acidity or alkalinity of a solution. (See Acid, Alkaline.)

Phanerophyte Plants with dormant buds exposed on twigs high above the ground. (cf Mega-, Meso-, Micro- and Nanophanerophytes.)

Phenecotype A population adapted to its local environment by a physiological modification not involving permanent, heritable change.

Phenetic Relating to all the features of an organism.

Phenology The study of periodical phenomena of plants, eg the flowering and fruiting times in response to climate.

Phenotype The characteristics of an individual or a population as determined by the genotype interacting with the environment.

Phloem That part of the tissue of a plant which is concerned with conducting food material. In woody stems it is the innermost layer of the bark. (cf Xylem.)

Photoperiodism A mechanism whereby organisms respond to duration and timing of light and dark periods.

Photosynthesis The process by which green plants manufacture sugars from water and carbon dioxide by converting the energy from light into chemical energy with the aid of the green pigment chlorophyll.

Phototropism Movement and orientation of plants in response to light.

Phyllosphere (phyllophane) The environment produced for microorganisms on the external surfaces of living leaves. It is the result of an interaction of nutrients and surface structures of the leaf with gases and water, especially dew, from the atmosphere.

Phylogeny The study of the evolutionary history of groups of organisms.

Phylum A major division of the plant (or animal) kingdom eg Bryophyta, Pterophyta, Anthophyta (flowering plants).

Physiognomy The overall appearance of a community.

Phytochorion A floristic division of the world.

Phytochrome A pigment in green plants that is associated with the absorption of light—a photoreceptor.

Phytocoenology The description and delimitation of plant communities.

Phytocoenon A phytosociological unit of vegetation derived by comparing samples of vegetation—an abstract community. (See Phytocoenose.)

Phytocoenose A plant assemblage occupying a particular piece of vegetation—a "real" community. (See Phytocoenon.)

Phytotoxin A substance poisonous to plants.

Phytotron A building in which plants can be grown under a series of carefully controlled environmental conditions.

Pinnate (of leaves) Compound, with leaflets in pairs on opposite sides of the midrib.

Planosol Continental soil formed on a flat surface and subject to periodic heavy waterlogging so that a hard pan forms by compaction of the clay washed down from the upper, A, horizon.

Plate tectonics Movement of the plates comprising the Earth's crust by the formation of new material at one edge and subduction into the earth's central magma at other margins—the process has resulted in continental drift.

Pneumatophores Extensions of the roots of plants of swampy habitats, such as mangroves, which grow up out of the water to ensure adequate aeration.

Pollen Collective name for the pollen grains, ie the minute spores (microspores) produced in the anthers.

Pollen sac The chamber (locule) in an anther where the pollen is formed.

Pollination The transfer of pollen grains from stamen to stigma. *Cross-pollination* occurs between flowers of different plants of the same species; *self-pollination* occurs between flowers of the same plant, or within one flower.

Pollinium A mass of pollen grains produced by one anther-lobe, cohering together and transported as a single unit during pollination, as in the orchids.

Polyclimax The occurrence of several apparently stable vegetation types in an area, due to edaphic and such factors other than climate.

Polygene One of a number of genes determining the same character.

Polyploid An organism that has more than the normal two sets of chromosomes in its somatic cells.

Polytopy Term applied to distributions consisting of several disconnected areas.

Porphyrin A rock in which large isolated crystals are set in a fine matrix.

Prisere A succession of plant communities developed on an originally bare, non-vegetated surface.

Progymnosperm An early or presumed ancestor of the gymnosperms.

Prokaryote Any organism lacking a membrane around its nuclei, plastids and other cell organelles, as in bacteria and blue-green algae. (cf Eukaryote.)

Prothallus The more or less independent sexual phase (gametophyte) of certain primitive plants such as ferns and club mosses.

Proximal Near to.

Psammosere A succession of plant communities developed during the colonization of sandy areas.

Pseudocopulation The attempted copulation by male insect visitors with a part of a flower which resembles the female of the insect species, as in the orchids.

Psilophyte Any member of the phylum Psilophyta; notably the whisk ferns (*Psilotum*, *Tmesipteris*), which are the only vascular plants to lack roots and true leaves.

Pteridophyte A collective term that encompasses the lower vascular plants, ie ferns, club mosses and their allies, and horsetails.

Pteridosperm Any member of the "seed-ferns" (Pteridospermales), gymnospermous seed plants dominant in the Carboniferous period, but now extinct.

Pyritic sand A form of sand rich in iron sulfide.

Recombination The production of combinations of allelic forms of genes in gametes different from those in the plant producing them, as a result of the independent segregation of chromosomes and the exchange of chromosome segments during meiosis.

Refugia Habitats that have escaped drastic changes in climate, enabling species and populations to survive, often in isolation.

Regosol A young soil developed over unconsolidated material, eg rock fragments etc. (See lithosol.)

Rhachis The major axis of an inflorescence or of a fern-frond.

Rhizome A horizontally creeping underground stem which lives over from season to season (perennates) and which bears roots and leafy shoots.

Root The lower, usually underground, part of a plant. It anchors the plant in the soil and absorbs water and mineral nutrients by means of the root hairs.

Rosette A group of leaves arising closely together from a short stem, forming a radiating cluster on or near the ground.

Ruderal (of plants) Occurring on disturbed ground and, particularly, rubbish dumps.

Salverform Trumpet-shaped.

Samara A dry fruit that does not split open and has part of the fruit wall extended to form a flattened membrane or wing.

Saprophyte A plant that cannot live on its own, but which needs decaying organic material as a source of nutrition.

Scale A small, often membranous, reduced leaf frequently found covering buds and bulbs.

Scape A leafless flower-stalk.

Schist Metamorphic rock with layers of different minerals and splitting into irregular thin plates.

Sclerenchyma A tissue composed of cells with thickened cell-walls, often woody (lignified), and which give mechanical strength and support.

Sclerophyllous With leathery leaves.

Scree Mountain slope covered with loose stones.

Seed A unit of sexual reproduction developed from a fertilized ovule, which either lies naked on the ovuliferous-scale as in gymnosperms (including conifers) or is enclosed in the fruit as in angiosperms (flowering plants).

Self-fertilization See Fertilization.

Self-incompatible Incapable of self-fertilization, usually because the pollen-tube cannot germinate or grows very slowly.

Self-pollination See Pollination.

Sepal A floral leaf or individual segment of the calyx of a flower; usually green.

Sere A type of plant succession. (See Prisere, Psammosere.)

Sessile Without a stalk, eg leaves without petioles or a stigma without a style.

Shrub A perennial woody plant with well developed side-branches that appear near the base, so that there is no trunk. They are less than 10m (30ft) high.

Sierozem Gray soils of arid deserts which have a high calcium content and feebly differentiated horizons.

Solum A collective term for the A and B horizons of a soil profile, ie those between the litter on the surface and the partly weathered base rock or substrate beneath.

Solifluction Movement of wet soil down a slope.

Sp. Abbreviation for species (singular).

Species The basic unit of classification. Species are grouped into genera and variations may be categorized into subspecies, variety and forma (form) in descending order of hierarchy. A species name consists of two units (binomial): the genus title and specific epithet, both italicized and only the initial letter of the genus part capitalized, for example *Betula pendula*.

Sporangium A hollow structure in which spores are formed.

Sporophyll A fertile leaf or leaf-like organ.

Spp. Abbreviation for species (plural).

Stand A uniform group of plants growing in a continuous area.

Steppe Strictly, the natural grasslands of Eurasia.

Sterile Not involved in reproduction; not bearing sex organs.

Stigma The receptive part of the female reproductive organs of flowering plants on which the pollen grains germinate; the apex of a carpel.

Stipule A leafy appendage, often paired, and usually at the base of the leaf-stalk.

Stolon A shoot that roots at the tip and produces new plants.

Stomata Pores that occur in large numbers in the epidermis of plants, particularly leaves, and through which gaseous exchange takes place.

Stratigraphy Concerned with the relative positions and sequence of geological strata.

Stromatolite Precambrian fossils formed in dolomite or limestone and interpreted as blue-green algae formed at the advent of life on Earth.

Style The elongated part of a carpel or ovary bearing the stigma at its apex.

Sub-Antarctic Strictly, refers to the area between about 50°S and 60°S latitude, but may be extended further northwards in discussions of phytogeographical affinities.

Suberin A complex fatty substance that occurs in the walls of certain types of cells, eg cork and those in the Casparian band (qv), rendering them impervious to water.

Subspecies A taxonomic division ranking between species and variety. It is often used to denote a geographical variation of a species. Abbreviation ssp.

Succession The sequence of changes in the vegetation of an area from the initial colonization to the development of the climax communities.

Succulent With fleshy or juicy organs containing reserves of water—hence stem-succulents, leaf-succulents.

Sudd Water plants with floating stems from which tall leafy shoots grow.

Sympatric Of populations and species that occur together in the same area.

Sympodial (of stems or rhizomes) With the apparent main stem consisting of a series of usually short axillary branches. (cf Monopodial.)

Syntaxon Any level of the hierarchical classification of vegetation determined by phytosociological methods.

Synusia A group of plants occupying a specific habitat within an ecosystem and having a similar role, though taxonomically unrelated.

Synzoochory The deliberate carrying of diaspores (seeds etc) in the mouths or on the limbs of animals, eg ants and birds.

Taiga Northernmost coniferous forest, with open boggy and rocky areas between.

Taxon Any taxonomic group, such as a species, genus, family etc.

Terpene A large group of hydrocarbons found in essential oils.

Terra rossa Red, often acid and clay, soils developed over hard Paleozoic limestone in the Mediterranean basin.

Thalloid, thallose Flattened; not distinctly div-

ided into root, stem and leaves.

Therophyte An annual plant.

Tiller A shoot produced from the base of the original stalk(s).

Timberline The upper limit of tree growth on mountains.

Topocline A gradual change in the characteristics of plant populations in relation to a topographical gradient.

Transpiration Movement of water vapor out from plant leaves or stems.

Tree A large perennial plant with a single branched and woody trunk and with few or no branches arising from the base. (cf Shrub.)

Trilobites Paleozoic marine arthropods.

Trophic Pertaining to nutrition.

Tuber An underground stem or root that lives over from season to season and which is swollen with food reserves.

Turgor The rigidity of cells resulting from their uptake of water.

Turion A short, scaly branch produced from a rhizome.

Ultraviolet radiation Radiation of wavelengths below 300nm, just beyond the visible spectrum.

Undershrub A perennial plant with lower woody parts, but herbaceous upper parts that die back seasonally.

Understory Lower layer of arborescent vegetation in a community, such as shrubs in a forest.

Vacuole The cavity in the protoplasm of a cell, containing gases or fluid.

Var. Abbreviation for variety. (See Species.)

Varve Layers of sand and silt deposited in a lake each year, used in constructing a chronology of such sediments.

Vascular Possessing vessels; able to conduct water and nutrients.

Vegetative reproduction Production of offspring without sexual reproduction or use of sexual apparatus.

Vernalization The process of exposing seedlings to low temperatures which is necessary if subsequent flowering and fruiting is to be effective.

Vicariance A term applied to the occupation of discontinuous areas by closely related taxa.

Viviparous (of seeds) Germinating before becoming detached from the parent.

Weed A plant growing in cultivated ground where it is not wanted.

Xeromorphic Possessing characteristics such as reduced leaves, succulence, dense hairiness or a thick cuticle, which are adaptations to conserve water and so withstand extremely dry conditions.

Xerophyte A plant which is adapted to withstand extremely dry conditions.

Xylem The woody fluid-conveying (vascular) tissue concerned with the transport of water about a plant.

Xylopod(ium) A distended woody stem-base or rootstock used for water storage in some trees and shrubs of arid regions—also known as a lignotuber.

Illustration Credits

Unless otherwise stated, all the illustrations on a given page are credited to the same source. Reference numbers denote page number followed by position on page, where relevant (t = top, b = bottom, r = right, l = left, c = center)
Photographs:
Abbey Garden Press: 32. Ashmolean Museum: 15. P. Barnard: 64t. G. Bateman: 76c, 77, 79, 81, 82, 203. Biofotos: 45, 95t, 102l, 149, 200, 229l, 231. S. Bisseror: 236. G. S. Boulton: 48. B. Bracegirdle: 126t. J. K. Burras: 54tl. T. V. Callaghan: 174t, 176br, 177. Cambridge University Press: 36l. N. J. Collins: 138b, 173, 174b, 175b, 176bl, 225, 249, 258. P. R. Crane: 76, 78. Crown Copyright 1982: Image made by Space Dept., RAE Farnborough. Courtesy Space Frontiers: 68. Down House, Royal College of Surgeons of England: 274. Equinox Archive: Jacket (inset), 8, 12b, 13, 14, 16b, 17tl, tr, 18, 19bl, 19br, 22b, 27, 30, 33, 34, 35, 51, 52, 53l, 92, 98cb, tr, 120, 140br, 143t, 171. Mary Evans Picture Library: 12t. Food and Agricultural Organisation: 244l (E. Bennett), 244r (F. Botts), 268r, 269. Werner Forman Archive: 242. D. J. Fox: 164, 166, 186. S. W. Fox: 86. G. Frame: 6. J. B.

Free: 168. Geological Museum, London: 62. R. B. Gibbons: 55tl, 58, 147, 162, 184, 189t, 226, 254, 257. C. H. Gimingham: 190. C. Grey-Wilson: 54tr, 55tr, 57tr, 98br, 156, 183, 185b, 187t. Sonia Halliday Photographs: 270. Robert Harding Associates: 215. R. M. Harley: 249. D. Harris: 42, 53c, 54b, 55b, 121b, 127t, 137bc, 144t, 230t, 247. F. N. Hepper: 26 (by permission of Kew Gardens), 139b, 98bl. V. H. Heywood: 98lc, c, 179. Michael Holford Photographs: 238. Alan Hutchison Library: 126b. Imitor: 66b. International Rice Research Unit: 245b. Irish Tourist Board: 196. L. Kasasian: 262. D. M. Keith-Lucas: 47, 56tr, 66t, 117, 163, 187b, 204. R. H. Kemp: 248. J. Kennedy: 63. M. Khan: 178. P. Kramer: 23. W. S. Lacey: 100, 206. K. Lewis/B. John: 93. D. Mabberley: 185t, 212, 213, 230b. Mansell Collection: 19tr, 275, 277. G. A. Matthews: 56tl, c, 83, 94, 96b, 138t, 197. D. M. Moore: 37, 60, 105, 116b, 128, 129, 167, 182, 189, 218, 222, 224, 229r, 234, 263, 272b. P. B. Moore: 50b. C. Muller: 148b. Natural History Photographic Agency: 143b (A. Bannister). Natural Science Photos: 2–3 (C. Walker), 24 (M. Freeman), 44t (C. Walker), 44b (P. Burton), 79 (J. K. Burton), 112 (I. Bennett), 161 (I. Bennett), 201 (P. Burton), 267 (C. Walker). New York Botanic Garden: 272t. P. H. Nye: 130t, 131br, 132, 133, 136t, 137tl, tc, tr, cr, bl, br, 139b, 252. Oxford Scientific Films: 116t, 165. C. Parker: 260. Popperfoto: 276. M. C. F. Proctor: 192. D. Prue: 122.

125t. Ann Ronan Picture Library: 16t, 20. Royal Botanic Gardens, Edinburgh: 17b. R. Schmid: 146. B. Seddon: 53r, 54tc, 140bl. Shell: 259. D. W. Shimwell: 50t, 181, 251. N. W. Simmonds: 121t. Space Frontiers Ltd.: 88. Spectrum Colour Library: 198. H. E. Street: 41tr. A. M. Synge: 266, 268l. U.S. Department of Agriculture: 246. D. W. H. Walton: 119. G. J. Waughman: 150, 151, 193, 194b, 195, 199, 207, 210. Weed Research Organisation: 56b, 57tl. T. C. E. Wells: 22t. B. D. Wheeler: 175t. B. A. Whitton: 141t, 255t. S. R. J. Woodell: 101, 142, 211. T. W. K. Young/D. N. Pegler: 115. Zefa Picture Library: Jacket (background), 87, 180, 208, 232. Zentralinstitut für Kulturpflanzenforachung: 245t.

All artwork © Equinox (Oxford) Limited

The Publishers have attempted to observe the legal requirements with respect to the rights of the suppliers of photographic materials. Nevertheless, persons who have claims are invited to apply to the Publishers.

Subject
Index

The following is an index to subjects and terms to be found in the text. A **bold** number indicates that following a heading in the text a substantial amount of information is to be found on the subject concerned. An *italic* number indicates an illustration and an *italic* number in parentheses indicates that the caption itself contains information on the subject concerned.

Index of Plant Names

The following is an index to common plant names and scientific names (species, genera, families etc) cited in the text. An *italic* number indicates an illustration.

Bibliography

The following list of titles indicates those books published in the English language during the last 20 years, which deal with aspects of the ecology and geography of plants pertinent to themes developed in *Green Planet*. Titles marked with an asterisk indicate books concerning both animals and plants.

ANDERSON, J. B. (1981). *Ecology for Environmental Sciences*. Arnold, London.

BANNISTER, P. (1976). *Introduction to Physiological Plant Ecology*. Blackwell, Oxford.

BARBOUR, M. G., BIRK, J. H. & PITTS, W. D. (1980). *Terrestrial Plant Ecology*. Addison-Wesley, London.

BIRKS, H. J. B. & BIRKS H. H. (1980). *Quaternary Palaeoecology*. Arnold, London.

CAIN, S. A. (1944). *Foundations of Plant Geography*. Harper and Row, New York.

COLLINSON, A. S. (1977). *Introduction to World Vegetation*. Allen and Unwin, London.

*COX, C. B., HEALEY, I. N. & MOORE, P. D. (1973). *Biogeography, an ecological and evolutionary approach*. Blackwell, London.

*DAJOZ, R. (1977). *Introduction to Ecology*. Hodder and Stoughton, London.

DAUBENMIRE, R. (1968). *Plant Communities. A textbook of plant synecology*. Harper and Row, New York.

DAUBENMIRE, R. (1978). *Plant Geography, with special reference to North America*. Academic Press, London.

ETHERINGTON, J. R. (1975). *Environment and Plant Ecology*. Wiley, New York.

EYRE, S. R. (1968). *Vegetation and Soils, a world picture*. Edition 2. Arnold, London.

GOOD, R. (1974). *The Geography of the Flowering Plants*. Edition 4. Longman, London.

GREIG-SMITH, P. (1982). *Quantitative Plant Ecology*. Edition 3. Blackwell, London.

JONES, G. (1979). *Vegetation Productivity*. Longman, London.

KELLMAN, M. C. (1975). *Plant Geography*. Methuen, London.

*MAY, R. M. (1976). *Theoretical Ecology, Principles and Applications*. Blackwell, London.

McCLEAN, R. C. & IVIMEY-COOK, W. R. (1973). *Textbook of Theoretical Botany*. Volume 4. Longman, London.

MILES, J. (1979). *Vegetation Dynamics*. Wiley, New York.

MUELLER-DOMBOIS, D. & ELLENBERG, H. (1974). *Aims and Methods of Vegetation Ecology*. Wiley, New York.

*ODUM, E. P. (1971). *Fundamental Ecology*. Edition 3. Saunders, Philadelphia.

*PIELOU, E. C. (1979). *Biogeography*. Wiley, New York.

*POOLE, R. W. (1974). *An Introduction to Quantitative Ecology*. McGraw-Hill, New York.

*RICKLEFS, R. E. (1976). *The Economy of Nature*. Chiron, Portland.

*RICKLEFS, R. E. (1979). *Ecology*. Edition 2. Chiron, Portland.

*SEDDON, B. (1971). *Introduction to Biogeography*. Duckworth, London.

STOTT, P. (1981). *Historical Plant Geography*. Allen and Unwin, London.

WALTER, H. (1973). *Vegetation of the Earth, in relation to climate and the eco-physiological conditions*. Springer-Verlag, New York.

*WATTS, D. (1971). *Principles of Biogeography. An Introduction to the functional mechanisms of ecosystems*. McGraw-Hill, London.

WHITTAKER, R. H. (1975). *Communities and Ecosystems*. Edition 2. MacMillan, New York.